# physica status solidi

VOLUME 16 · NUMBER 1 · 1966

# physica status solidi



Volume 16

1966

AKADEMIE-VERLAG · BERLIN

Subscriptions and orders for single copies should be addressed to

AKADEMIE-VERLAG GmbH, 108 Berlin, Leipziger Straße 3 – 4 or to Buchhandlung KUNST UND WISSEN, Erich Bieber, 7 Stuttgart 1, Wilhelmstr. 4 – 6 or to Deutsche Buch-Export und -Import GmbH, 701 Leipzig, Postschließfach 160

Schriftleiter und verantwortlich für den Inhalt: Professor Dr. Dr. h. c. P. Görlich, 102 Berlin, Neue Schönhauser Str. 20 bzw. 69 Jena, Humboldtstr. 26. Redaktionskollegium: Dr. S. Oberländer, Dr. E. Gutsche, Dr. W. Borchardt. Anschrift der Schriftleitung: 102 Berlin, Neue Schönhauser Str. 20. Fernruf: 426788. Verlag: Akademie-Verlag GmbH, 108 Berlin, Leipziger Str. 3–4. Fernruf: 220441, Telex-Nr. 011773. Postscheckkonto: Berlin 35021. — Die Zeitschrift „physica status solidi" erscheint jeweils am 1. des Monats. Bezugspreis eines Bandes MDN 60,—. Bestellnummer dieses Bandes 1068/16. Jeder Band enthält 2 Hefte. Gesamtherstellung: VEB Druckerei „Thomas Müntzer", Bad Langensalza. — Veröffentlicht unter der Lizenznummer 1310 des Presseamtes beim Vorsitzenden des Ministerrates der Deutschen Demokratischen Republik.

# Contents

**Review Articles**



**Original Papers**

**Short Notes**

# Author Index

# physica status solidi



Volume 16 · Number 1 · Pages 1 to 348 and K 1 to K 104

July 1, 1966

AKADEMIE-VERLAG · BERLIN

Subscriptions and orders for single copies should be addressed to

AKADEMIE-VERLAG GmbH, 108 Berlin, Leipziger Straße 3 – 4 or to Buchhandlung KUNST UND WISSEN, Erich Bieber, 7 Stuttgart 1, Wilhelmstr. 4 – 6 or to Deutsche Buch-Export und -Import GmbH, 701 Leipzig, Postschließfach 160

**Editorial Note:**

"physica status solidi" undertakes that an original paper *accepted for publication* before the 8th of any month will be published within 50 days of this date unless the author requests a postponement. In special cases there may be some delay between *receipt* and *acceptance* of a paper due to the review and, if necessary, revision of the paper.

Schriftleiter und verantwortlich für den Inhalt: Professor Dr. Dr. h. c. P. Görlich, 102 Berlin, Neue Schönhauser Str. 20 bzw. 69 Jena, Humboldtstr. 26. Redaktionskollegium: Dr. S. Oberländer, Dr. E. Gutsche, Dr. W. Borchardt. Anschrift der Schriftleitung: 102 Berlin, Neue Schönhauser Str. 20. Fernruf: 426788. Verlag: Akademie-Verlag GmbH, 108 Berlin, Leipziger Str. 3–4, Fernruf: 220441, Telex-Nr. 011773, Postscheckkonto: Berlin 35021. — Die Zeitschrift „physica status solidi" erscheint jeweils am 1. des Monats. Bezugspreis eines Bandes MDN 60,—. Bestellnummer dieses Bandes 1068/16. Jeder Band enthält zwei Hefte. Gesamtherstellung: VEB Druckerei „Thomas Müntzer" Bad Langensalza. — Veröffentlicht unter der Lizenznummer 1310 des Presseamtes beim Vorsitzenden des Ministerrates der Deutschen Demokratischen Republik.

# *Review Article*

phys. stat. sol. **16**, 3 (1966)

*Physical Research Department,*
*The National Cash Register Company, Dayton, Ohio*

# Magnetization Curling

By

A. AHARONI[1])

## Contents



## 1. Introduction

The theory of micromagnetics [1] should, in principle, give a self-consistent description of all magnetization processes. This theory assumes continuous matter and localized magnetic moments, which should be a good approximation as long as any characteristic length that comes out of the problem is very large compared to the lattice spacing of the crystal. But as far as these assumptions

---

[1]) On Sabbatical leave from the Weizmann Institute of Science, Rehovoth, Israel.

are sufficient approximations, no further simplifying assumptions need be made in order to formulate the problem in terms of a set of nonlinear partial differential equations. However, the solution of these equations, even for rather simple cases, is difficult, and has never been satisfactorily approached. Actually, the full hysteresis curve has been calculated by micromagnetics theory, only for one highly unrealistic case [2][2]), and some aspects of it have been studied for very few, almost equally unrealistic cases ([1], Chap. 6, [4, 5]). A much more extensive work has been carried out for cases in which the differential equations are linearized, for which cases one can use standard techniques involving known analytical solutions. The first of these cases has been the study of the approach to saturation in a material containing point, line, or plane imperfections [6] represented by a local concentration of forces, later extended in various respects [7 to 10]. In this case, the linearization is achieved by the assumption that a large magnetizing field is applied, so that deviations of the magnetization vector from the direction of the field are small, and second order terms in the direction cosines along axes perpendicular to the field can be neglected. The main idea is that complete saturation cannot be obtained in any finite field.

Mathematically, the same set of linearized partial differential equations (with different boundary conditions) applies to the physically completely different problem of evaluating the nucleation field. In contradistinction to the previous case, it is assumed here that a large enough, but finite, magnetic field applied in the $z$-direction (say) saturates the sample, namely, it makes the magnetization vector parallel to $z$ everywhere in the material. The validity of this assumption has been discussed elsewhere [11] and will not concern us here. Suppose now that once the sample is saturated, the external field is slowly reduced. Instability of the magnetization to saturation state should ultimately occur, at a certain critical value for the field known as the "nucleation field". In this case the linearization is due to the smallness of the deviations from the saturated state, when instability just starts. The equations are obtained [12] by considering all possible deviations from saturation, and minimization of the energies involved, taking into account the exchange, magnetostatic, and anisotropy energies. The nucleation field is then the least negative eigenvalue, $H$, of the following set of equations ([1], Chap. 5, see also [13]):

$$- C \nabla^2 \alpha + (g_{11} + M_s H) \alpha + g_{12} \beta + M_s \frac{\partial U}{\partial x} - M_s \frac{\partial U_0}{\partial z} \alpha = 0 , \qquad (1\text{a})$$

$$- C \nabla^2 \beta + g_{12} \alpha + (g_{22} + M_s H) \beta + M_s \frac{\partial U}{\partial y} - M_s \frac{\partial U_0}{\partial z} \beta = 0 \qquad (1\text{b})$$

in the material, while on its boundary

$$\frac{\partial \alpha}{\partial n} = \frac{\partial \beta}{\partial n} = 0 . \qquad (2)$$

Here $M_s$ is the magnitude of the spontaneous magnetization, directed almost along the $z$-axis with small direction cosines $\alpha$ and $\beta$ along the $x$ and $y$ axes respectively; $C$ is the exchange constant, $n$ is the outward normal, and $g_{ij}$ are the coefficients of the quadratic terms in $\alpha$ and $\beta$ in the expansion of the magnetocrystalline anisotropy energy density, assumed independent of position. The potential $U$ is that due to the transverse magnetization $M_s (\alpha \boldsymbol{i} + \beta \boldsymbol{j})$ and

[2]) It is interesting to note, though, that the slab calculation has been recently found to fit experiment, see [3].

fulfills ([1], Chap. 5, see also [13]) Poisson's equation

$$\nabla^2 U_{\text{in}} = 4\,\pi\, M_s \left(\frac{\partial \alpha}{\partial x} + \frac{\partial \beta}{\partial y}\right) \tag{3}$$

inside the ferromagnetic material, and the Laplace equation

$$\nabla^2 U_{\text{out}} = 0 \tag{4}$$

outside, with the boundary conditions

$$U_{\text{in}} = U_{\text{out}}\,, \quad \frac{\partial U_{\text{in}}}{\partial n} - \frac{\partial U_{\text{out}}}{\partial n} = 4\,\pi\, M_s\, \alpha_n\,. \tag{5}$$

Here $M_s \alpha_n$ is the component of the magnetization $\boldsymbol{M}$ perpendicular to the surface, namely $\alpha_n = l\,\alpha + m\,\beta$, where $l$, $m$ are the direction cosines of the outward normal, $n$. The other potential, $U_0$, in (1) is that due to surface charges in the reference state of uniform magnetization along $z$. In the case of an ellipsoid, $(1/M_s)\,\partial U_0/\partial z = \text{const} = N$, the demagnetizing factor along $z$.

In the special case of cylindrical symmetry, Brown's linearized equations (1) through (5) have a particular solution with cylindrical symmetry, for which $U \equiv 0$. This solution is known as the *magnetization curling* mode. It has been first found by Brown [12, 14] for a sphere and for an infinite circular cylinder, with some speculations about a general prolate spheroid. Almost at the same time, this very mode for the same cases was studied by Frei, Shtrikman, and Treves [15], who gave it the name "magnetization curling". They did not approach the problem by looking for a solution of (1) through (5) as Brown did, but rather *assumed* this particular distribution of the magnetization vector in space, in the same way as certain domain configurations are chosen intuitively in the usual domain theory studies, when the energy is minimized under constraints. It turned out, however, that this particular choice led to an eigenmode of Brown's equations, namely, a true energy minimum, which by a strange coincidence was identical with the mode studied by Brown [12, 14] — a short while before. It was later shown that under certain conditions, this magnetization curling mode is the lowest eigenmode of Brown's equations, in an infinite cylinder [16], and sphere [17] with strong indication that the same holds true in prolate or oblate spheroids [18]. These cases will be studied in the next section, which deals with cylindrically symmetric configurations. It contains a complete treatment of an oblate spheroid, a first approximation for a chain of spheres, and certain aspects of the more complicated problem of a scratch on the surface of a cylinder. In Section 3 cylindrically symmetric particles will be discussed, in which the external magnetizing field is applied at an angle to the axis of symmetry. The applied field thus destroys the complete symmetry, yet the essentials of the curling modes are conserved. Section 4 will be a study of the non-cylindrically symmetric case of an infinite rectangular bar, in which the magnetization curling mode becomes complicated, but can be approached by upper and lower bounds. Finally, in Section 5, the theoretical results will be compared to the relevant experimental studies of ferromagnetic materials.

## 2. Cylindrical Symmetry

### *2.1 General*

In the special case when the ferromagnetic material's shape has a cylindrical symmetry about the $z$-axis, the surface charges in the state of saturation along $z$ will be independent of the coordinate $\varphi$. Therefore, in cylindrical coordinates,

$r$, $\varphi$, $z$, the potential $U_0$ that these charges give rise to, will also be a function of $r$ and $z$ only. For this case we shall set $g_{11} = g_{22}$, $g_{12} = 0$ in (1), and expand $\alpha$, $\beta$, and $U_{\text{in}}$ in Fourier series of the form

$$\alpha = \frac{1}{2}\sum_{m=1}^{\infty} A_m(r,z)\cos[(m+1)\varphi - \varphi_m] + \frac{1}{2}\sum_{m=1}^{\infty} B_m(r,z)\cos[(m-1)\varphi - \varphi_m] + \\ + A_0(r,z)\cos\varphi + B_0(r,z)\sin\varphi\,, \tag{6a}$$

$$\beta = \frac{1}{2}\sum_{m=1}^{\infty} A_m(r,z)\sin[(m+1)\varphi - \varphi_m] - \frac{1}{2}\sum_{m=1}^{\infty} B_m(r,z)\sin[(m-1)\varphi - \varphi_m] + \\ + A_0(r,z)\sin\varphi - B_0(r,z)\cos\varphi\,, \tag{6b}$$

$$U_{\text{in}} = \sum_{m=1}^{\infty} V_m(r,z)\cos(m\varphi - \varphi_m) + V_0(r,z)\,, \tag{6c}$$

where $m$ are integers, $\varphi_m$ are constants. Note that the terms with $m = 0$ are taken separately, in a somewhat different combination, because this is the way the separation of the $\varphi$ variable turns out to be in its simplest form.

Substituting (6) in (1), and taking the appropriate linear combinations [16] of the two equations, one obtains an equation of the form $\sum a_m(r,z) \times \times \cos(m\varphi - \varphi_m) = 0$, which implies $a_m \equiv 0$ for every $m$, and a similar equation for $\sin(m\varphi - \varphi_m)$. This yields the following equations for $m \geqq 1$:

$$\left[\frac{\partial^2}{\partial r^2} + \frac{1}{r}\frac{\partial}{\partial r} - \frac{(m+1)^2}{r^2} + \frac{\partial^2}{\partial z^2} - \frac{g_{11}}{C} - \frac{M_s H}{C} + \frac{M_s}{C}\frac{\partial U_0}{\partial z}\right] A_m(r,z) + \\ + \frac{M_s}{C}\left[\frac{m}{r} - \frac{\partial}{\partial r}\right] V_m(r,z) = 0\,, \tag{7a}$$

$$\left[\frac{\partial^2}{\partial r^2} + \frac{1}{r}\frac{\partial}{\partial r} - \frac{(m-1)^2}{r^2} + \frac{\partial^2}{\partial z^2} - \frac{g_{11}}{C} - \frac{M_s H}{C} + \frac{M_s}{C}\frac{\partial U_0}{\partial z}\right] B_m(r,z) - \\ - \frac{M_s}{C}\left[\frac{m}{r} + \frac{\partial}{\partial r}\right] V_m(r,z) = 0\,, \tag{7b}$$

and for $m = 0$:

$$\left[\frac{\partial^2}{\partial r^2} + \frac{1}{r}\frac{\partial}{\partial r} - \frac{1}{r^2} + \frac{\partial^2}{\partial z^2} - \frac{g_{11}}{C} - \frac{M_s H}{C} + \frac{M_s}{C}\frac{\partial U_0}{\partial z}\right] B_0(r,z) = 0\,, \tag{8a}$$

$$\left[\frac{\partial^2}{\partial r^2} + \frac{1}{r}\frac{\partial}{\partial r} - \frac{1}{r^2} + \frac{\partial^2}{\partial z^2} - \frac{g_{11}}{C} - \frac{M_s H}{C} + \frac{M_s}{C}\frac{\partial U_0}{\partial z}\right] A_0(r,z) - \\ - \frac{M_s}{C}\frac{\partial V_0(r,z)}{\partial r} = 0\,. \tag{8b}$$

Similarly, by substituting (6) in (3) one obtains for $m \geqq 1$

$$\left[\frac{\partial^2}{\partial r^2} + \frac{1}{r}\frac{\partial}{\partial r} - \frac{m^2}{r^2} + \frac{\partial^2}{\partial z^2}\right] V_m(r,z) = 2\pi M_s\left[\frac{\partial}{\partial r}(A_m + B_m) + \frac{m+1}{r}A_m(r,z) - \\ - \frac{m-1}{r}B_m(r,z)\right], \tag{9a}$$

and for $m = 0$

$$\left[\frac{\partial^2}{\partial r^2} + \frac{1}{r}\frac{\partial}{\partial r} + \frac{\partial^2}{\partial z^2}\right] V_0(r,z) = 4\pi M_s\left(\frac{1}{r} + \frac{\partial}{\partial r}\right) A_0(r,z)\,. \tag{9b}$$

This set of equations replaces (1) and (3) and should, therefore, be solved together with (4) and the boundary conditions (2) and (5). In a true cylindrically symmetric case, $U_{\text{out}}$ can be expanded similarly to (6c), and the boundary conditions are such that each integral value of $m$ can be treated separately.

It is seen from the derivation of the basic equations (1) through (5), that dropping the positive term of transverse magnetostatic self energy is equivalent to writing $U_{\text{in}} = 0$ in (7) and (8) and ignoring (9) and (5). This makes equations (7) and (8b) similar to (8a) with at most a difference in the term with $(m \pm 1)^2 r^{-2}$, with the same boundary conditions (2). One can therefore, apply to these equations the general theorem of Titchmarsh [19] as used by Aharoni [18] for the special case of prolate and oblate spheroids, namely when $\partial U_0/\partial z = \text{const}$. For the use of Titchmarsh's theorem, the assumption $\partial U_0/\partial z = \text{const}$ is not necessary and it applies equally well to the more general case discussed here, and implies that for $m \neq 1$ the eigenvalues $H$ of (7) with $V_m = 0$ are never more positive than the least negative eigenvalue of (8a). On the other hand, dropping a positive term in the energy obviously makes nucleation easier, i.e., makes $H$ less negative. It can thus be concluded that all eigenvalues of (7) for $m \neq 1$ and of (8b) are more negative than some eigenvalue of (8a).

Now, higher modes might be of interest for studying the approach to saturation in certain cases when the boundary conditions require using them [20], although in the cases studied so far [6 to 10] their need has not been encountered. They might be usefull for dynamic problems ([1], Chap. 7), or in the study of nucleation in non-cylindrically symmetric cases by expanding in the complete set of eigenfunctions for a cylindrically symmetric case (a possibility which has never been tried), when the boundary conditions cause mixing of modes. However, for study of nucleation problems, which is our concern here, the only eigenvalue $H$ of physical significance is the least negative one. Since in a true cylindrically symmetric case there is no mixing of the modes for different values of $m$, all those which have been shown to yield more negative eigenvalues than (8a) can, therefore, be left out. One is thus left with $m = 1$ in (7) and (9a) and with (8a). Carrying out the necessary expansion of (4) and keeping just the term that can fulfill the boundary condition (5) and using all the foregoing expansions in (2) and (5), one is thus left with the following two sets of differential equations with boundary conditions for determining the nucleation field as the least negative eigenvalue $H$ of either:

$$\left[\frac{\partial^2}{\partial r^2} + \frac{1}{r}\frac{\partial}{\partial r} - \frac{1}{r^2} + \frac{\partial^2}{\partial z^2} - \frac{g_{11} + M_s H}{C} + \frac{M_s}{C}\frac{\partial U_0}{\partial z}\right] B_0(r, z) = 0 \tag{10a}$$

with the boundary condition

$$\frac{\partial B_0}{\partial n} = 0\,, \tag{10b}$$

or:

$$\left[\frac{\partial^2}{\partial r^2} + \frac{1}{r}\frac{\partial}{\partial r} - \frac{4}{r^2} + \frac{\partial^2}{\partial z^2} - \frac{g_{11} + M_s H}{C} + \frac{M_s}{C}\frac{\partial U_0}{\partial z}\right] A_1(r, z) + \frac{M_s}{C}\left[\frac{1}{r} - \frac{\partial}{\partial r}\right] V_{\text{in}}(r, z) = 0\,, \tag{11a}$$

$$\left[\frac{\partial^2}{\partial r^2} + \frac{1}{r}\frac{\partial}{\partial r} + \frac{\partial^2}{\partial z^2} - \frac{g_{11} + M_s H}{C} + \frac{M_s}{C}\frac{\partial U_0}{\partial z}\right] B_1(r, z) - \frac{M_s}{C}\left[\frac{1}{r} + \frac{\partial}{\partial r}\right] V_{\text{in}}(r, z) = 0\,, \tag{11b}$$

$$\left[\frac{\partial^2}{\partial r^2} + \frac{1}{r}\frac{\partial}{\partial r} - \frac{1}{r^2} + \frac{\partial^2}{\partial z^2}\right] V_{\text{in}}(r, z) = 2\pi M_s\left[\frac{\partial}{\partial r}(A_1 + B_1) + \frac{2}{r} A_1(r, z)\right], \tag{11c}$$

that apply in the ferromagnetic material, with

$$\left[\frac{\partial^2}{\partial r^2} + \frac{1}{r}\frac{\partial}{\partial r} - \frac{1}{r^2} + \frac{\partial^2}{\partial z^2}\right] V_{\text{out}}(r, z) = 0 \tag{12}$$

and the boundary conditions

$$\frac{\partial A_1}{\partial n} = \frac{\partial B_1}{\partial n} = 0\,, \tag{13a}$$

$$V_{\text{in}} = V_{\text{out}}\,, \quad \cos(\varphi - \varphi_1)\left(\frac{\partial V_{\text{in}}}{\partial n} - \frac{\partial V_{\text{out}}}{\partial n}\right) = 4\,\pi\, M_s\, \alpha_n\,. \tag{13b}$$

The last condition apparently contains $\varphi$, both explicitly on the left hand side, and implicity in $\alpha_n$, the component of magnetization normal to the boundary, which depends on $\varphi$ according to (6a) and (6b). However, this apparent $\varphi$-dependence is due only to the inadequacy of the cartesian components $\alpha$ and $\beta$ to describe the magnetic vector in the case of cylindrical symmetry. Actually, if one considers the components in cylindrical coordinates, $\alpha_\varphi$ and $\alpha_r$, a cylindrical symmetry of the outer surface implies that $\alpha_\varphi$ does not contribute to $\alpha_n$, which is solely composed of the normal component of $\alpha_r$, in the $r\,z$ plane. And since it is seen from (6) that $\alpha_r$ has the same $\varphi$-dependence as $U$, this dependence should cancel out of (13b) when the appropriate $\alpha_n$ is introduced.

The three-dimensional problem has thus been reduced to two two-dimensional ones (10), which is known as the magnetization curling mode, and (11) through (13). The latter is much more complicated and has been rigorously studied only in the simplest case of an infinite circular cylinder [16], or slight modifications of it [2, 11, 21]. Some estimation for its eigenvalues has been given [18] for a prolate spheroid, in which case it yields the rotation in unison and another mode called buckling. For a sphere it has been shown [17] that this buckling mode gives higher eigenvalues than the curling. This result will be generalized in the following for oblate spheroids. Results for other cylindrically symmetric particles will be then briefly discussed.

### 2.2 Oblate spheroid

For an ellipsoid, $\partial U_0/\partial z$ = const in (10) and (11). In this case (10a) is of the form

$$(\nabla^2 + \text{const})\, B_0 = 0\,,$$

and its separation in oblate spheroidal coordinates, $\xi$ and $\eta$, leads to [22]

$$B_0 \propto S_{1n}(-\,i\,c, \eta)\, R_{1n}^{(1)}(-\,i\,c, i\,\xi) \tag{14}$$

with

$$c^2 = \frac{R^2}{C}\,\frac{-g_{11} - M_s\,(H - N\,M_s)}{1 + \xi_0^2} = -\,\pi\, S^2 \frac{h_n}{1 + \xi_0^2}\,. \tag{15a}$$

Here $n$ is a positive integer, $R$ is the radius of the spheroid in a direction perpendicular to the $z$-axis, $S$ and $h_n$ are the reduced radius and reduced nucleation field respectively [18],

$$N = M_s^{-1}\frac{\partial U_0}{\partial z} \tag{15b}$$

is the demagnetization factor along $z$, and $\xi_0$ is the value of $\xi$ on the spheroidal surface of the particle. The latter can be expressed [22] in terms of the ratio of the spheroid axes, $q$, according to

$$q = \xi_0 (\xi_0^2 + 1)^{-1/2} . \tag{15c}$$

Equation (10b) is then fulfilled by

$$\left[\frac{\mathrm{d}R_{1n}^{(1)}(-i\,c,\,i\,\xi)}{\mathrm{d}\xi}\right]_{\xi\,=\,\xi_0} = 0 \tag{16}$$

which, for a given $\xi_0$, determines $c$, from which the eigenvalue $H$ is obtained according to (15).

For numerical computations it is simpler to transform (16) to the angle function. Using standard transformations (see equations (4.2.3) and (3.1.3b) of [22]) the equation becomes

$$\sum_{r=0,\,1}^{\infty}{}' \, d_r^{1n}(-i\,c)\,\frac{\mathrm{d}P_{r+1}^1(i\,\xi_0)}{\mathrm{d}\xi_0} = 0 \,, \tag{17}$$

where the $P_{r+1}^1$ are the associated Legendre functions, as defined [23] in the complex plane,

$$P_n^m(i\,\xi) = (\xi^2 + 1)^{m/2}\,\frac{\mathrm{d}^m P_n(i\,\xi)}{\mathrm{d}\xi^m} \,, \tag{18}$$

and the prime on the summation indicates that $r$ takes only even or odd values, according to $n$ being even or odd.

Using tabulated values [22] for the coefficients $d_r^{1n}(-i\,c)$, and computing the Legendre functions from the algebraic equation implied by the definition (18), equation (17) has been solved numerically to yield $\xi_0$ for various values of $c$. From these values, and (15), the nucleation field parameter, $-S^2\,h$, has been computed as a function of the spheroid elongation, $q$. It turns out that the smallest $|h|$ for a given elongation is obtained for $n = 1$, and that $S^2|h_\mathrm{n}|$ is a monotonic decreasing, but slowly varying function of $q$, between the limits $-S^2\,h_\mathrm{n} = 1.424$ for $q \to 0$ (an infinite plane) and $-S^2\,h_\mathrm{n} = 1.379$ for $q \to 1$ (a sphere). For any oblate spheroid the value is thus

$$-S^2\,h_\mathrm{n} = 1.4$$

to within 2%, for nucleation by the curling mode.

For the other equations that have to be considered, (11) through (13), one solution is

$$A_1 = 0,\; B_1 = \mathrm{const}\,, \quad V_\mathrm{in} = \frac{1}{2}\,r\,B_1\left[-\frac{1}{M_\mathrm{s}}\,g_{11} - H + N\,M_\mathrm{s}\right]. \tag{19a}$$

This is the coherent rotation. After using it in (13b) with the appropriate solution of (12) in oblate spheroidal coordinates, it yields for the nucleation field

$$-h_\mathrm{n} = \xi_0\left[(1 + \xi_0^2)\left(\frac{\pi}{2} - \tan^{-1}\xi_0\right) - \xi_0\right]. \tag{19b}$$

Being an eigenmode, this solution should be orthogonal to all the other eigenmodes of equations (11) through (13), and can thus be treated independently of them. In particular, if $A_1$ and $B_1$ are expanded in series of any complete orthogonal set of functions, the term $B_1 = \mathrm{const}$ can be dropped out of the series. One can thus write (using the notation of reference [22]) for the most general solution of (11) through (13), which excludes the coherent rotation mode,

$$A_1 = \sum_{n=2}^{\infty} \sum_{k=1}^{\infty} a_{kn} S_{2n} (-i\, c_{kn}^{(2)}, \eta)\, R_{2n}^{(1)} (-i\, c_{kn}^{(2)}, i\,\xi), \tag{20a}$$

$$B_1 = \sum_{n=1}^{\infty} \sum_{k=1}^{\infty} b_{kn} S_{0n} (-i\, c_{kn}^{(0)}, \eta)\, R_{0n}^{(1)} (-i\, c_{kn}^{(0)}, i\,\xi) + \sum_{k=2}^{\infty} b_{k0} S_{00} (-i\, c_{k0}^{(0)}, \eta)\, R_{00}^{(1)} (-i\, c_{k0}^{(0)}, i\,\xi), \tag{20b}$$

namely without the $k = 1$ term in the second sum of (20b), which is a constant independent of $\xi$ and $\eta$. Here $c_{kn}^{(m)}$ are the zeros, at $\xi = \xi_0$, of the derivative of $R_{mn}^{(1)}$, in increasing order, so that the expansion (20) fulfills (13a) for every term.

To solve the problem rigorously one should use (20) to calculate the potential, and obtain a secular equation for the coefficients $a_{kn}$ and $b_{kn}$ from the boundary conditions (13b). Since this is complicated, one might start by looking for a lower bound, which is easily obtained by dropping out the transverse self magnetostatic energy. This is equivalent to writing $V_{\text{in}} = 0$ in (11) and dropping (11c), (12), and (13b). In this case each term in (20) is an eigenmode of the differential equations, the determinant of $a_{kn}$ and $b_{kn}$ is diagonalized, and the eigenvalue $c$, defined in (15a), is simply one of the $c_{kn}^{(m)}$ of (20). Now, according to the above mentioned theorem of Titchmarsh, all $c_{kn}^{(2)}$ are larger than the curling eigenvalue, $c_{11}^{(1)}$, and are thus of no interest. One should, therefore, check just $c_{1n}^{(0)}$, for $n \geqq 1$, and $c_{20}^{(0)}$. Note that $c_{10}^{(0)} = 0$ has been excluded from (20), before neglecting the transverse self magnetostatic energy.

These eigenvalues have been computed using a similar procedure to that mentioned above for the curling eigenvalue. It should be noted that the available tabulation [22] of the coefficients $d_r^{mn}$ does not suffice for any reasonable accuracy in the computation of $c_{20}^{(0)}$, but is definitely sufficient to show that it is larger (for any given elongation, $q$) than $c_{11}^{(0}$. The latter can be computed with a much better accuracy, only there is no point in doing so, since it turns out to be always *larger* than the eigenvalue for curling.

It has thus been shown that a lower bound to all the eigenvalues of (11) through (13), except for the coherent rotation, is larger than the curling eigenvalue, for any oblate spheroid. Therefore, for an ideal oblate spheroid it is only the curling or coherent rotation that can take place. Using the above mentionedvalues of $S^2 h_n$, computed for the curling mode, and comparing to the coherent rotation nucleation, (19b), one can compute the critical reduced radius, $S$, for the turnover from curling to coherent rotation. The results are plotted in Fig. 1 as a function of the spheroid

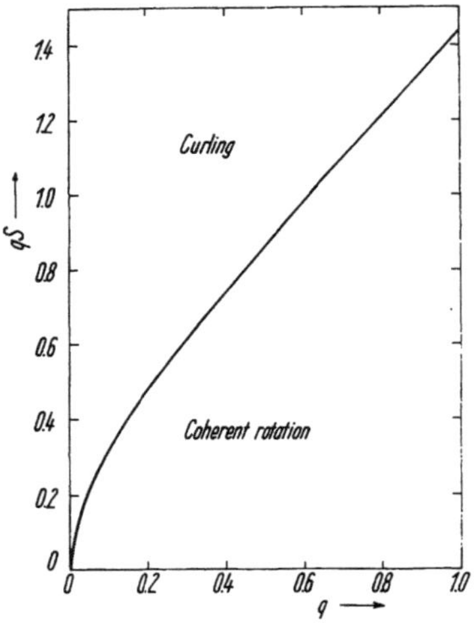


Fig. 1. The critical reduced small radius,
$$q\,S = R_1 M_s \left(\frac{1}{2}\, C\right)^{-1/2},$$
for the turnover from coherent rotation to curling in an oblate spheroid, as function of its elongation, $q$

elongation, $q$. Since no other mode is possible, deviation from saturation should be by the curling mode above the line for critical radius, and by coherent rotation below that line, as marked on the figure.

### 2.3 *Prolate spheroid*

A treatment similar to the one presented here for the oblate spheroid, has been published [18] for a prolate spheroid. In this case, however, the lower bound which is obtained by dropping the transverse self magnetostatic energy, turns out to be *larger* than the curling eigenvalue, in contradistinction to the case of an oblate spheroid. Therefore, the eigenvalues of (11) through (13) cannot be disregarded for the prolate spheroid, even aside from the coherent rotation mode which is a particular solution of these equations for any ellipsoid. For a prolate spheroid, one has therefore to consider three modes for magnetization reversal, namely the magnetization curling, the coherent rotation, and this other mode, called magnetization buckling. This latter mode is complicated and its eigenvalue has never been rigorously computed, except for the limiting case [16] of an infinite circular cylinder. However, in that limiting case, the buckling eigenvalue was very close to the coherent rotation one (for radii for which it is smaller than the curling one), so that one can just assume the coherent rotation is a good approximation to the buckling in acicular particles, and even a better approximation for non-acicular particles.

From the above mentioned computations of a lower bound to the buckling eigenvalue, it is possible to get a lower bound to the critical size for coherent rotation. An upper bound for this size is obtained by comparing the coherent rotation eigenvalue to that of curling. These results are reproduced in Fig. 2, where below the lower curve coherent rotation is definitely the easiest of all possible modes. Above the upper curve, at least the curling is easier than coherent rotation, so that it would no longer take place. It is seen that the curves are close enough together to give an idea of what the critical size is, below which the extensively studied coherent rotation mode takes place. It should be particularly noted that the two extreme points on the *lower* curve (i.e., the sphere, $q = 1$, and the infinite circular cylinder, $q = \infty$) are exact, which seems to indicate that the exact curve is probably much closer to the lower than to the upper curve in this figure.

The turnover from buckling to curling is probably just above the upper curve, as this is the case for the infinite cylinder. The buckling can thus be considered just as an insignificant transition region between the coherent rotation and the magnetization curling modes.

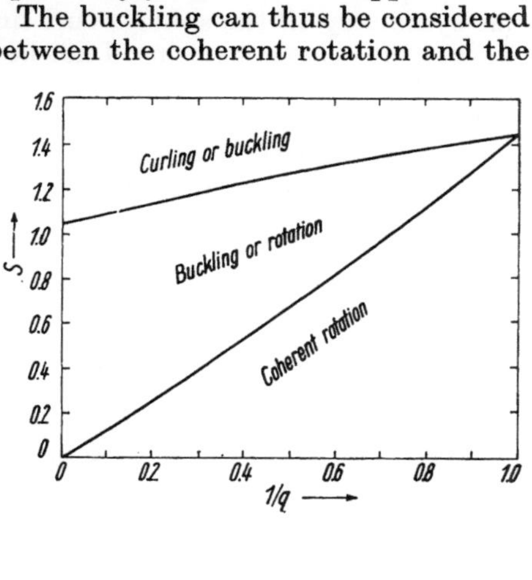


Fig. 2. Upper and lower bounds for the critical reduced small radius $S = R\,M_s \left(\frac{1}{2}\,C\right)^{-1/2}$ for coherent rotation in a prolate spheroid, as a function of its elongation, $q$. The modes that are theoretically possible in each of the regions between curves are marked on the figure (after Aharoni [18])

### *2.4 Chain of spheres*

The model of a chain of identical ferromagnetic spheres, with a negligible exchange coupling at the contact, has been introduced in order to account for certain properties of the so-called ESD particles [24] which are made [25] by electrodeposition of iron into mercury. Two modes have been studied [24] originally for this model, the coherent rotation and a mode called a non-symmetric fanning, approximated for most purposes by the so-called symmetric fanning. The latter is most probably a good approximation to the buckling mode (as can be seen for example by comparing to the buckling mode in an infinite cylinder), or at least as good an approximation as the whole model of a chain of spheres is for the actual "peanut" shape [25] of the particles studied. According to the foregoing Section 2.1, there remained just one other possible mode, namely the curling (equation (10)). A first approximation to the curling mode eigenvalue in a chain of spheres has been recently calculated [26], and will be briefly reviewed here. An extension of this calculation to a field applied at an angle to the chain axis will be given in the following Section 3.

The potential $U_0$ to be substituted in (10), is that due to the surface charges on all the spheres in the state of magnetization to saturation along the $z$-axis, which is the chain axis. It is also assumed here to coincide with the direction at which the magnetic field, $H$, is applied. One should thus use the well known form of a potential due to a magnetized sphere, in its form as an inside potential for the self energy of each sphere, and in its form as an outside potential for the interaction between spheres. If $R$ is the radius of each sphere, and if the chain is taken to be infinite in length, the potential inside the sphere whose center is the origin of coordinates, is

$$U_0 = \frac{4\pi M_s z}{3} + \sum_{n=-\infty}^{\infty}{}' U_{0,n}(r, z)\,, \tag{21a}$$

where the prime indicates that $n = 0$ is excluded from the summation, and where

$$U_{0,n} = \frac{4\pi}{3} M_s (z - 2nR)\,[(z - 2nR)^2 + r^2]^{-3/2}\,. \tag{21b}$$

Substituting (21) in (10), one has a differential equation for the magnetization in the sphere $r^2 + z^2 \leqq R^2$, with its boundary condition. However, because of the symmetry all the spheres curl simultaneously in the same nucleation field so that the other spheres need not be considered once the interaction is included as in (21).

In contradistinction to all cases considered before, in this case $\partial U_0/\partial z$ is not a constant, and therefore no analytical solution of (10) is known. However, if the $\partial U_0/\partial z$ term is considered as a perturbation, first-order perturbation theory leads to the following eigenvalue [26]:

$$h_n = \frac{H_n + g_{11} M_s^{-1}}{2\pi M_s} = -\frac{\mu^2}{\pi S^2} + \frac{2}{3} - \frac{\zeta(3)}{3} + \frac{\zeta(5)}{10(\mu^2 - 2)}\left[\frac{\mu^2}{3} + \frac{1}{2} - \frac{5}{3\mu^2}\right] =$$
$$= 0.335 - 1.379\, S^{-2}\,. \tag{22a}$$

Here

$$S = R M_s \left(\frac{C}{2}\right)^{-1/2}, \tag{22b}$$

$\zeta$ is the zeta function, and $\mu = 2.0816$ is the smallest root of

$$(\mu^2 - 2) \tan \mu + 2 \mu = 0 . \tag{23}$$

The symmetric fanning [24] gives $|h_n| = 0.250$, independent of $S$, so that if these were the only possible modes, nucleation should take place by curling for $S > 1.53$ and by fanning for $S < 1.53$. However, Jacobs and Bean [24] have also calculated what they call a "nonsymmetric fanning", which is evidently a better approximation to the rigorous buckling mode. The details of this calculation are not clear from their paper, but they give a result of $H_c = 1750$ for an infinite chain, which should correspond to $|h_n| = 0.162$, if the 2700 Oe corresponds to 0.250. This pushes the critical size up to $S = 1.67$, and it might be further pushed if the buckling and curling are calculated more accurately. If one also takes into account that the whole model of a chain of spheres is just an approximation to the actual shape of the particles, that the curling nucleation field does not necessarily coincide with the coercive force (in particular for an aggregate of particles with a size distribution), and that calculating the reduced radius, $S$, from the measured radius, $R$, contains the exchange constant, $C$, which is not known with any reasonable accuracy, the reported lack of size-dependence [25] in the coercive force of ESD particles for $1.08 \leqq S \leqq 2.54$ does not seem significant. This will be further discussed in Section 5.2.

### *2.5 Other cases*

In order to take into account, at least approximately, the effect of surface roughness on the nucleation field, a model of a semi-toroidal scratch on the surface of an infinite circular cylinder has been proposed [11]. This model conserves the cylindrical symmetry, so that the curling mode eigenvalue can be calculated from the relatively simple equation (10), with $U_0$ a function of space that can be expressed [11] as a double integral. Unfortunately, it turned out [27] that even this problem is too complicated to be handled rigorously, the convergence being rather slow. As a preliminary crude approximation, we shall, therefore, try here to replace the inhomogeneous demagnetizing field $\partial U_0/\partial z$ of equation (10) by a constant, proportional to its value in the center of the scratch, $r = z = 0$. If one further assumes (10b) is fulfilled on the whole (rather than the scratched) cylinder, the effect of the scratch (to this approximation) is to make the nucleation field, $H_n$, more positive by the amount $\delta H$ which is proportional to $-\partial U_0/\partial z$ at $r = z = 0$. It should be noted, though, that to the same approximation one gets the same change, $\delta H$, for the buckling or any other mode, and not only for the curling mode.

Using the general expression [11] for $U_0$, one obtains thus

$$\delta H \propto \frac{4\pi}{3} N_{\text{eff}} M_s , \tag{24a}$$

where

$$N_{\text{eff}} = \frac{3}{2} a^2 \int\limits_{-\pi/2}^{\pi/2} \frac{\sin^2 \theta \, (1 - a \cos \theta) \, d\theta}{(1 + a^2 - 2 a \cos \theta)^{3/2}} \tag{24b}$$

and $a$ is the ratio of the radius of the scratch to the radius of the cylinder (see Fig. 3, insert). It should be particularly noted that the same expression holds, if the semi-toroid is assumed to project out of the surface of the cylinder, as in the dotted lines in Fig. 3, insert.

The integral in (24) is obviously elliptic. Using the substitution $\theta = \pi - 2\psi$, it is transformed to the standard form of elliptic integrals of the first and second kind, yielding

$$N_{\text{eff}} = (1 + 2a^2)(1 - a)\left[K(k) - F\left(\frac{\pi}{4}, k\right)\right] - \\ - (1 - 2a^2)(1 + a)\left[E(k) - E\left(\frac{\pi}{4}, k\right)\right] - a\,(1 - 2\,a^2)\,(1 + a^2)^{-1/2} \quad (25\text{a})$$

with

$$k^2 = 4\,a\,(1 + a)^{-2}\,. \quad (25\text{b})$$

This $N_{\text{eff}}$ has been computed from (25) for different values of $a$. In order to compare with experiment, it is plotted in Fig. 3 not as a function of $a$, but as a function of the volume, $V$, of the semi-toroid, divided by the cube of cylinder diameter. Also plotted on this graph are experimental results [28] for the reduction in the absolute value of the nucleation field in iron whiskers, as a function of the surface-defect volume.

It should be noted that for this particular model the curve cannot in principle be extended beyond the plotted value $V/(2\,R)^3 = 0.71$, since this is obtained for $a = 1$, namely the toroid radius equals the cylinder radius. Larger defect volumes can thus be obtained only for scratches which are more elongated in the $z$-direction. Also, we have assumed the boundary conditions hold on the whole rather than scratched cylinder. This is the kind of approximation used by Néel [29] to study surface roughness on thin films, but can hardly be used for $a \to 1$, and can be justified only for $a \ll 1$. Bearing this in mind, the $N_{\text{eff}}$ curve shape seems similar enough to the experimental trend, so that one can expect a more rigorous calculation to yield essentially the experimental curve. As a further justification for this hope, Fig. 4 plots the same demagnetizing factor, $N_{\text{eff}} = (3/4\pi\,M_s)\,(\partial U_0/\partial z)_{r=z=0}$ for a different model of a straight-cut cylindrically symmetric scratch. This model is used as an illustration, in spite of the infinite demagnetizing field at the corner (it is only logarithmic [11] and anyway does not show at the center $r = z = 0$), mainly because its elongation in the $z$-direction can be varied as an additional parameter. If ones bears in mind that toward the end of the curves shown in Fig. 4 the approximation in the boundary conditions is obviously wrong, these curves show a remarkable similarity to the experimental results and one can expect the rigorous curve to depend almost solely on $V$, and be relatively insensitive to the exact shape of the scratch. It thus seems that it will be possible to calculate

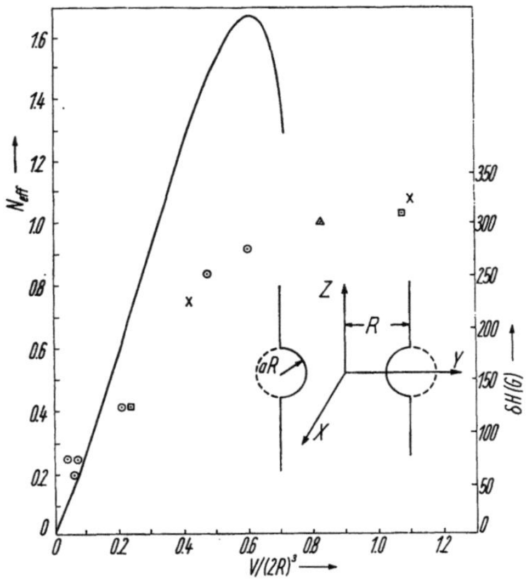


Fig. 3. Theoretical demagnetization coefficient at the center of a semitoroidal scratch on an infinite circular ferromagnetic cylinder, as a function of scratch volume, $V$. For not too large values of $a$, one can assume this demagnetization is essentially proportional to the reduction in nucleation field, $\delta H$, experimental values of which are reproduced from ref. [28]

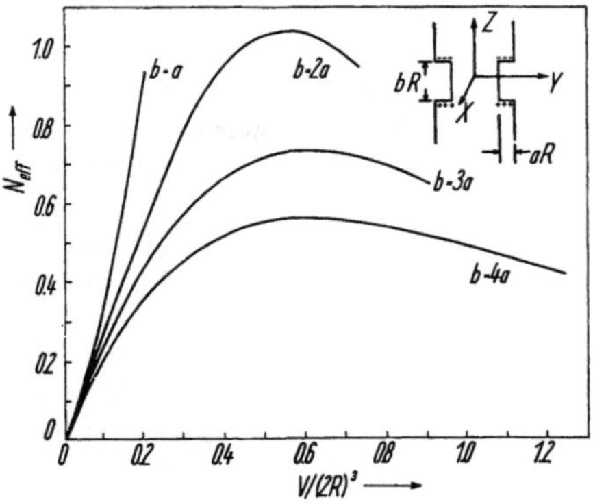


Fig. 4. Same as Fig. 3, for the scratch shown in the insert

the nucleation fields at least in whiskers, if a way is found to overcome the mathematical difficulties of dealing with a highly non-homogeneous demagnetizing field.

To conclude this section, it seems adequate to mention that curling eigenvalues have also been calculated in cylindrically symmetric cases in which $g_{11}$ of (10a) varies in space in one way or another [21], and in cases when (10b) is modified [2] due to surface anisotropy.

## 3. Quasi Cylindrical Symmetry

When the external magnetic field is applied at an angle to the axis of cylindrical symmetry, the complete symmetry is destroyed. However, there still remains an eigenmode which involves essentially curling of the magnetization vector, namely it involves no change in the self magnetostatic energy, to a first approximation in small deviations. This has been first illustrated for the case of an infinite cylinder [30] but can be easily modified for any prolate or oblate spheroid. The curling in this case is around the direction of the magnetization vector just before nucleation, which does not necessarily coincide with either the direction of the applied field or the axis of symmetry, since before nucleation can start the magnetization rotates in unison towards the applied field, according to the model of Stoner and Wohlfarth [31]. As an illustration of the procedure involved, we shall develop here the differential equation for the case of a chain of spheres, already mentioned in Section 2.4, and study an approximate solution of it. In this case it is known [24] that besides the coherent rotation there can exist the fanning mode. However, the fanning yields just one jump of the magnetization at a certain field, and therefore curling cannot nucleate from any intermediate stage of it. Assuming there is no other mode, curling might nucleate only at some stage of the coherent rotation, provided this nucleation is possible *before* nucleation by fanning takes place.

Consider a chain of $n$ identical ferromagnetic spheres of radius $R$, which lies in the $x\,z$ plane. Let the $z$-axis be chosen in the direction of the applied field, which makes an angle $\psi$ with the chain axis. Before nucleation, the magnetization is at an angle $\theta_0$ to the field, $H$, and this angle is the solution of Stoner and Wohlfarth's equation

$$2\,H \sin\theta_0 + \pi\,M_s \sin 2\,(\theta_0 + \psi) \sum_{i=1}^{n} \left(1 - \frac{i}{n}\right) i^{-3} = 0\,. \tag{26}$$

We consider the deviation from this state of saturation at angle $\theta_0$ to the $z$-axis to be such that the direction cosines of the magnetization are

$$\alpha = \sin\theta_0 + \varepsilon(r)\cos\theta_0 \sin\theta \sin\varphi\,, \tag{27a}$$

$$\beta = \varepsilon(r)\,(\sin\theta_0\cos\theta - \cos\theta_0 \sin\theta\cos\varphi)\,, \tag{27b}$$

$$\gamma = \cos\theta_0 - \varepsilon(r)\sin\theta_0\sin\theta\sin\varphi + \frac{\varepsilon^2}{2\cos\theta_0}\left[(\cos\theta_0\cos\theta + \sin\theta_0\sin\theta\cos\varphi)^2 - 1\right] \tag{27c}$$

in the sphere whose center is in the origin, and similar expressions with the appropriate translation of origin are assumed to hold in the other spheres. Here $r$, $\theta$, $\varphi$ are the spherical coordinates and $\varepsilon$, which is a measure of the amplitude of curling deviation, is assumed small and a function of $r$ only. The last term in (27 c) should be included in order to make $\alpha^2 + \beta^2 + \gamma^2$ constant to a second order in $\varepsilon$. It does not appear *explicitly* in Shtrikman and Treves' study [30] of the infinite cylinder, only because of the different choice of coordinate systems, but would appear on rotation of their system. The mode reduces to the usual curling when $\psi \to 0$.

To a first approximation in $\varepsilon$, the deviation is perpendicular to the radius vector and the divergence of $\boldsymbol{M}$ is zero. This mode thus bears the characteristic features of the curling mode, namely that it produces no first-order change in the magnetostatic self-energy. The second order term in the potential needs not to be calculated, since according to the reciprocity theorem of Brown [32], the interaction of the second order potential and the zero order magnetization can be expressed as an interaction of the zero order potential and the second order magnetization. Using the general expression thus obtained ([32], p. 128) for the particular case of equation (27), and using for the zero order potential (namely the potential of the magnetization (27) for $\varepsilon = 0$) essentially the expression in (21), with the appropriate changes implied by the rotation of the coordinate system, one obtains that the change in the self magnetostatic energy from the state $\varepsilon = 0$, to a second order in $\varepsilon$, is given by

$$\delta W_M = -\left(\frac{4}{3}\pi M_s\right)^2 n \int_0^R [\varepsilon(r)]^2 r^2\,\mathrm{d}r + \\ + \frac{2\pi M_s^2 R^3}{3\cos\theta_0} \sum_{i=-n}^{n}{}' (n - |i|) \int_0^R [\varepsilon(r)]^2 F_i(r, \theta_0, \psi)\, r^2\,\mathrm{d}r\,, \tag{28}$$

where the prime indicates that $i = 0$ is excluded from the sum, and

$$F_i(r, \theta_0, \psi) = \\ = \int_0^{2\pi}\int_0^{\pi} [(\cos\theta_0\cos\theta + \sin\theta_0\sin\theta\cos\varphi)^2 - 1]\left\{\frac{\cos\theta_0}{r_i^3} - \frac{3}{r_i^5}\times\right. \\ \times [(r\sin\theta\cos\varphi + 2iR\sin\psi)\sin\theta_0 + (r\cos\theta - 2iR\cos\psi)\cos\theta_0]\times \\ \left.\times (r\cos\theta - 2iR\cos\psi)\right\}\sin\theta\,\mathrm{d}\theta\,\mathrm{d}\varphi\,, \tag{29a}$$

$$r_i^2 = r^2 + 4i^2R^2 + 4iRr(\sin\theta\cos\varphi\sin\psi - \cos\theta\cos\psi)\,. \tag{29b}$$

The change in the exchange and interaction with the external field energies is

$$\delta(W_{ex} + W_H) = \frac{1}{2}C\int[(\nabla\alpha)^2 + (\nabla\beta)^2 + (\nabla\gamma)^2]\,\mathrm{d}\tau - HM_s\int\delta\gamma\,\mathrm{d}\tau\,,$$

where the integration is over the whole chain, but can be obiously reduced to $n$ times the integration on one sphere. Using (27) and carrying out the integration over the angles, one obtains that to a second approximation in $\varepsilon$

$$\delta(W_{ex} + W_H) = \frac{4\pi}{3}Cn\int_0^R\left[r^2\left(\frac{\mathrm{d}\varepsilon}{\mathrm{d}r}\right)^2 + 2\varepsilon^2\right]\mathrm{d}r + \frac{4\pi}{3\cos\theta_0}nHM_s\int_0^R[\varepsilon(r)]^2 r^2\,\mathrm{d}r\,. \tag{30}$$

Neglecting the magnetocrystalline anisotropy energy, the function $\varepsilon(r)$ should be such as to minimize the sum of (28) and (30). This leads to the following differential equation:

$$\frac{d^2\varepsilon}{dr^2} + \frac{2}{r}\frac{d\varepsilon}{dr} - \frac{2\,\varepsilon}{r^2} - \frac{\varepsilon\, M_s}{C}\left(\frac{H}{\cos\theta_0} - \frac{4\pi M_s}{3}\right) - \frac{M_s{}^2 R^3 \varepsilon}{2\,C\cos\theta_0}\sum_{i=-n}^{n}{}' \frac{n-|i|}{n} F_i(r,\,\theta_0,\,\psi) = 0\,, \tag{31a}$$

where $F_i$ is defined in (29), with the boundary condition

$$\left(\frac{d\varepsilon}{dr}\right)_{r=R} = 0\,. \tag{31b}$$

To first order in perturbation theory, equations (31) yield the following eigenvalue:

$$-\frac{H_n}{\pi M_s\cos\theta_0} + \frac{4}{3} - \frac{2\,\mu^2}{\pi S^2} =$$

$$= \frac{R^2}{2\pi\cos\theta_0}\left[\frac{1}{2} - \frac{\sin^2\mu}{\mu^2} + \frac{\sin 2\mu}{4\mu}\right]^{-1}\int_0^R\left[\frac{\sin\left(\frac{\mu r}{R}\right)}{\frac{\mu r}{R}} - \cos\left(\frac{\mu r}{R}\right)\right]^2\cdot\sum_{i=-n}^{n}{}'\frac{n-|i|}{n}F_i\,dr\,. \tag{32}$$

Here $\mu$ is the smallest root of (23) and the reduced radius, $S$, is defined in (22b).

The integration involved in (29) is rather complicated, although it can be carried out analytically, as will be seen in the following. As a starting point we shall therefore study a rough approximation in which $F_i(r,\,\theta_0,\,\psi)$ is replaced by a constant, which equals its value at the center of the sphere, $r = 0$, namely by

$$F_i(0,\,\theta_0,\,\psi) = -\left(\frac{\pi}{i^3 R^3}\right)\left[\frac{1}{3}\cos\theta_0 - \cos\psi\cos(\psi+\theta_0)\right]. \tag{33}$$

It should be noted that to this approximation (32) is exact, as is always the case for a first order perturbation involving a constant. Substituting (33) in (32) one obtains

$$-\frac{H_n}{\pi M_s\cos\theta_0} + \frac{4}{3} - \frac{2\,\mu^2}{\pi S^2} = \sum_{i=1}^{n}\frac{n-i}{n\,i^3}\left[\frac{\cos\psi\cos(\psi+\theta_0)}{\cos\theta_0} - \frac{1}{3}\right]. \tag{34}$$

For $\psi = 0$ the coherent rotation (26) yields one jump, so that the curling nucleation takes place at $\theta_0 = 0$. For this case, and for an infinite chain, $n = \infty$, one obtains from (34)

$$\frac{H_n}{2\pi M_s} + \frac{\mu^2}{\pi S^2} = \frac{2}{3} - \frac{\zeta(3)}{3} = 0.266\,,$$

whereas in (22a) the value is 0.335. The present zero order approximation is thus considerably below the first order approximation of Section 2.4, and will be improved in the following. However, it is worthwhile to follow first this simpler calculation, which gives some idea of the shape of the curves, and will also aid in the more complicated calculation of the first order approximation.

For $\psi \neq 0$, the curling mode nucleates after rotation of the magnetization an to angle $\theta_0$, given by (26). For nucleation, the field $H$ in (26) should be identical with $H_n$ of (34). Eliminating $\theta_0$ between these two formulas yields the nucleation field as a function of the angle $\psi$, the reduced radius $S$, and the chain length, $n$. For $n = \infty$ the result is

$$\frac{3\,H_n}{\pi M_s\zeta(3)} = (1.328 - 6.884\,S^{-2})\,(1.443 - 2.295\,S^{-2})\,[(1.443 - 2.295\,S^{-2})^2 - \\ - \sin^2\psi\,(1.885 - 4.590\,S^{-2})]^{-1/2}\,. \tag{35}$$

This formula is rather similar to the curling nucleation field in an infinite cylinder [30] but with different coefficients. The main difference, however, is that for an infinite cylinder the nucleation field is always negative and approaches zero only for $S \to \infty$, while here the nucleation becomes positive for rather small values of $S$.

For $S = 1.560$, the coefficient of $\sin^2 \psi$ vanishes and $H_n$ does not depend on $\psi$. For $S$ larger than this value, $|H_n|$ increases for increasing $\psi$, while for smaller $S$ the field $|H_n|$ decreases with increasing $\psi$. The values of $H_n$ computed from (35) are plotted in Fig. 5. Also plotted is the nucleation by fanning, which is identical with the curling curve for $S = 1.634$. All the curves are discontinued at the intersection with the coherent rotation curve, because for larger angles than shown the magnetization reverses by coherent rotation before curling (or fanning) can take place. Curling and coherent rotation which yield $|H_n|$ larger than that of the symmetric fanning mode are shown dashed. They have no physical significance, because the magnetization reverses by fanning before they have a chance to take place. To this approximation, if there is no other mode, magnetization in an ideal chain of spheres should reverse by fanning for $S < 1.634$, and by curling for $S > 1.634$. This is somewhat larger than the critical radius mentioned in Section 2.4, and results from the poorer approximation used here in replacing $F_i$ of (29) by a constant.

To get a better approximation, one should carry out the integrations in (29). It turns out that after certain transformations and long algebraic manipulations, this can be done analytically. When substituted in (32), the result is

$$- \frac{H_n}{\pi M_s \cos\theta_0} + \frac{4}{3} - \frac{2\mu^2}{\pi S^2} =$$

$$= \left[\frac{\cos\psi \cos(\psi+\theta_0)}{\cos\theta_0} - \frac{1}{3}\right] \sum_{i=1}^{n} \frac{n-i}{n\, i^3} + \frac{\frac{1}{3}\mu^2 + \frac{1}{2} - \frac{5}{3\mu^2}}{8(\mu^2-2)} \times$$

$$\times \left[\frac{3\cos\psi\cos(\psi+\theta_0)}{\cos\theta_0} + 3\cos^2(\psi+\theta_0) - \frac{7\cos\psi\cos^3(\psi+\theta_0)}{\cos\theta_0} - \frac{3}{5}\right] \sum_{i=1}^{n} \frac{n-i}{n\, i^5}\,. \tag{36}$$

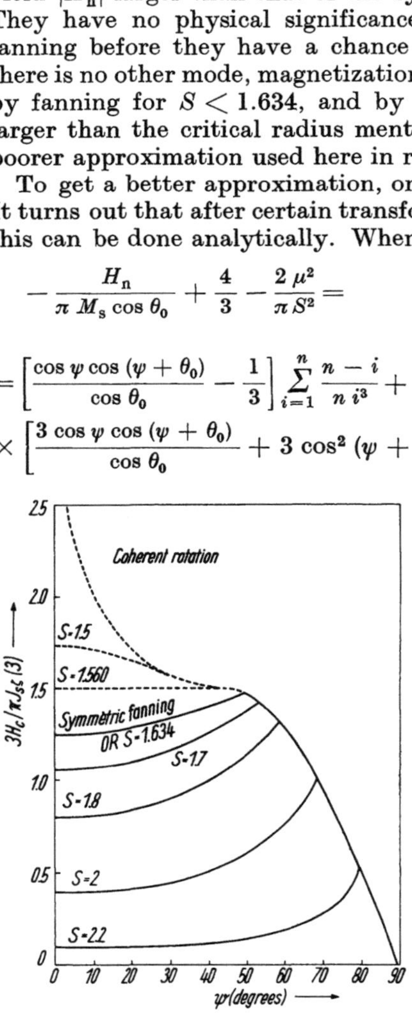


Fig. 5. The zeroth approximation for the nucleation field, $H_n$, of an infinite chain of spheres, plotte das a function of the reduced radius $S = R M_s \left(\frac{1}{2} C\right)^{-1/2}$ and the angle $\psi$ between the applied field and the chain axis

For $\psi = 0$, one should take $\theta_0 = 0$, and then (36) reduces to the first order relation (22), if $n = \infty$. For other values of $\psi$, one should use in (36) the angle $\theta_0$ given by (26) for the same value of the field, $H = H_n$. In this case, elimination of $\theta_0$ involves the roots of a fourth order polynomial in $\tan\theta_0$. Therefore, if one is interested in improving the zero order approximation (35), the best procedure seems to be successive numerical solutions of (26) and (36) to get $H_n$ and $\theta_0$ for every value of $\psi$ and $S$ (or also $n$), starting with (35). A rough check shows that the shape of the curves thus obtained cannot differ much from those in Fig. 5. The second order perturbation is also not too difficult to calculate since $F_i$ is known analytically, and so are all the eigenfunctions of the unperturbed equation in (31).

The critical size for changeover from curling to fanning also depends somewhat on $n$, but very slightly so. Actually by equating $-H_n$ of (36), with $\psi = 0$, to $H_c$ calculated by Jacobs and Bean [24] for the symmetric fanning, the critical size for $n = 4$ turns out to be some 1.5% lower than for $n = \infty$. Even for $n = 2$ one obtains about 1.50 for the critical size.

## 4. Infinite Rectangular Cylinder

### *4.1 General*

When the ferromagnetic crystal does not have a cylindrical symmetry, the $\varphi$ variable is no longer separable, and one has generally a three-dimensional problem instead of the two-dimensional one treated in the foregoing. However, there is another case in which the number of dimensions can be readily reduced to two, namely a cylinder of any cross section, infinite in one dimension (which will be taken as the $z$-axis). In this case, if the field is applied along the $z$-axis, one can eliminate the $z$-dependence from the nucleation differential equations (1) through (5) by writing $\alpha$, $\beta$, and $U$ as sinusoidal functions of $z$, with amplitudes that depend on $x$ and $y$ only. In contradistinction to the cylindrically symmetric case in which the periodicity in $\varphi$ implied that $\cos\,(m\,\varphi - \varphi_0)$ involve $m$, an integer, here one can write $\cos\,(k\,z - z_0)$ for any real value of $k$.

This general procedure can be found in the literature only for two cases: an infinite circular cylinder [2, 16, 21] and a plate [33] infinite in *two* directions. Both these cases are further simplified by the possibility of eliminating still another dimension by the techniques discussed above, so that the differential equations are actually one-dimensional and can be readily solved analytically, leaving only the algebraic problem of fitting the boundary conditions. In the more general case of a cylinder, the sinusoidal variation in the $z$-direction complicates an already complicated problem. We shall therefore add the simplifying assumption that $k = 0$, namely that there is no $z$-dependence. This assumption can be justified by considering the circular cylinder [16], in which the inclusion of $k$ did not affect the curling mode and changed only slightly the coherent rotation eigenvalue to that of the buckling. Hopefully, in the general cylinder the $z$-dependence will not modify the eigenvalues much either. With this assumption, the differential equations which yield the nucleation field in any infinite cylinder are (1) through (5), with $U_0 = 0$ and with $\nabla^2 = \partial^2/\partial x^2 + \partial^2/\partial y^2$. Or, going back to the variational problem from which these equations are derived, one is interested in minimizing the variable part of the free energy per unit length:

$$W = \frac{1}{2}\,C \int [(\nabla\,\alpha)^2 + (\nabla\,\beta)^2]\,\mathrm{d}S + \frac{1}{2}\int (g_{11}\,\alpha^2 + 2\,g_{12}\,\alpha\,\beta + g_{22}\,\beta^2)\,\mathrm{d}S + \\ + \frac{1}{2}\,H\,M_s \int (\alpha^2 + \beta^2)\,\mathrm{d}S + W_M\,. \tag{37}$$

Here $\mathrm{d}S = \mathrm{d}x\,\mathrm{d}y$, the integration is over the cylinder cross section, and the magnetostatic self energy can be written in terms of the potential, $U$, defined by equations (3), (4), and (5), as

$$W_M = (8\,\pi)^{-1} \int (\nabla\,U)^2\,\mathrm{d}S\,, \tag{38}$$

where this last integration is over the whole $x\,y$ plane. All other notations are as in Section 1.

In many cases, the $W_M$ term is the most difficult to deal with. In the cylindrically symmetric case, the curling mode is rather easy to handle mainly because $W_M = 0$, but this lucky situation cannot be extended to other cases. Other modes in previous sections have been estimated by a lower bound obtained when the magnetostatic term was dropped. Brown ([34], see also [1], Section 5.5) has generalized this kind of argument and noted that if the magnetostatic term $W_M$ is replaced by a smaller quantity, the resulting eigenvalue $H_n$ cannot be more negative than the correct eigenvalue of the problem. And if $W_M$ is replaced by a larger quantity, the resulting eigenvalue cannot be more positive than the correct one. This enables one to obtain upper and lower bounds to the nucleation field, and in case they turn out to be close enough to each other the rigorous value is not needed. Moreover, Brown [34] could give specific forms for such quantities to replace $W_M$, by proving that always

$$W_H \leqq W_M \leqq W_B + 2\,\pi \int \boldsymbol{M}^2\,\mathrm{d}S\,, \tag{39}$$

where $W_H$ is the interaction of the correct magnetization vector $\boldsymbol{M}$ with the field, due to an arbitrary potential, $\Phi$, which vanishes at infinity, is continuous in the whole $y\,x$ plane[3]) and has continuous derivatives *inside* the material:

$$W_H = -\,(8\,\pi)^{-1} \int (\nabla\,\Phi)^2\,\mathrm{d}S + \int \boldsymbol{M} \cdot \nabla\,\Phi\,\mathrm{d}S\,. \tag{40}$$

Here the first integration is over the whole $x\,y$ plane, while the second one is over the cylinder cross section[3]). Similarly, $W_B$ is the interaction of the correct $\boldsymbol{M}$ with an *arbitrary irrotational* field, $\boldsymbol{B}_1$:

$$W_B = (8\,\pi)^{-1} \int \boldsymbol{B}_1^2\,\mathrm{d}S - \int \boldsymbol{M} \cdot \boldsymbol{B}_1\,\mathrm{d}S\,. \tag{41}$$

It should be noted that the same argument about increasing or decreasing $W_M$ applies equally well to any other positive term in the energy. For example, in a specific case, Brown [33] has tried to drop the term $(\nabla\,\alpha)^2$, the effect of which is obviously to make $H_n$ equal or greater than the correct value. However, manipulating with $W_M$ rather than with any other term is more practical because it is generally more complicated to handle, and because of the specific useful forms (40) and (41).

Another useful way [33, 34] of obtaining upper bounds is to minimize the variable part of the free energy (37) under constraints, namely only for functions $\alpha$ and $\beta$ of a certain type rather than for all functions of space. It is obvious that if a certain eigenmode is found under constraint, this is still a permissible function for the general minimization problem, but the general problem considers also other functions, not allowed by the constraint, which might yield a still lower minimum. The nucleation field, $H_n$, calculated under constraint is therefore never more positive than the correct eigenvalue.

These two methods of evaluating upper and lower bounds to the nucleation field will be used in the following to estimate this field for the particular case of a cylinder with a rectangular cross-section.

### *4.2 Possible modes in a rectangular cylinder*

Consider an infinite cylinder with a rectangular cross-section $-a \leqq x \leqq a$, $-b \leqq y \leqq b$, where $b \leqq a$. This should be a good approximation to the experi-

[3]) Brown's theorem is formulated here for the particular case of an infinite cylinder, but it actually holds for the more general three-dimensional case [34].

mentally studied whiskers. Most of these experiments have been carried out for an easy axis along the $z$-axis, and this will therefore be assumed the case in the following, although many of the calculations could be rather easily extended to anisotropy in other directions as well. In particular, a film with an easy axis perpendicular to it and a field applied in the plane of the film (note that if the field is applied along the easy axis, this is a limiting case of an oblate spheroid discussed in Section 2.2) as in the experiment of Huber and Smith [35], could be readily obtained as the limiting case $a \to \infty$. However, this problem has been studied ([33], see also [1], p. 80) more rigorously than in the present approach, namely without neglecting the variation in the $z$-direction, and will therefore be ignored here. It is thus assumed that $g_{12} = 0$, $g_{11} = g_{22} > 0$, which reduces equation (37) to

$$W = \frac{1}{2} C \int [(\nabla \alpha)^2 + (\nabla \beta)^2] \, dS - \frac{1}{2} \lambda' \int (\alpha^2 + \beta^2) \, dS + W_M, \tag{42}$$

using the notation

$$\lambda' = - g_{11} - H M_s . \tag{43}$$

It is readily seen that if $\lambda' < 0$, the energy (42) is larger, for any $\alpha$, $\beta$, than in the saturated state, $\alpha = \beta = U = 0$. The state of magnetization to saturation in the $z$-direction is therefore stable at least until $\lambda'$ reaches zero, and all the eigenvalues $\lambda'$ must be positive. As in all nucleation problems, one is interested only in the smallest possible $\lambda'$ (that is, the least negative nucleation field, $H$), since if deviation starts at a certain value of the field, the initial conditions for nucleation do not exist any more when a more negative field is applied, and the other modes do not get any chance to nucleate.

Using symmetry considerations in the particular case of a square cross-section, $a = b$, Brown [36] showed that the eigenfunctions can be divided into four classes of odd and even functions in $x$ and $y$, which are mutually perpendicular to each other, so that each one can be treated separately. However, his proof did not use the assumption $a = b$ and can be equally well used in the more general case of a rectangular cross-section [37]. The argument is essentially as follows. If $\alpha$ and $\beta$ are eigenfunctions corresponding to a certain eigenvalue, $\lambda'$, then reflections of these functions in the $y$- and/or $x$-plane, are also eigenfunctions of the problem, corresponding to the same $\lambda'$. By taking the appropriate linear combinations of these, one ends up with four separated modes, the symmetries of which are listed in Table 1 (where e stands for even, o for odd).

Table 1

Symmetries of the four possible modes for nucleation in an infinite rectangular cylinder (e = even, o = odd) and the names by which they are referred to in the text

| $\alpha$ | | $\beta$ | | $U$ | | Name |
|---|---|---|---|---|---|---|
| in $x$ | in $y$ | in $x$ | in $y$ | in $x$ | in $y$ | |
| o | o | e | e | e | o | rotation, $y$-direction |
| e | e | o | o | o | e | rotation, $x$-direction |
| e | o | o | e | o | o | curling |
| o | e | e | o | e | e | anticurling |

The table also contains names to facilitate identification. The first two were named by Brown [36] rotation in quasi-unison, since most of the flux rotates in the general direction of $x$ or $y$ respectively. They reduce to the rotation in unison mode in a circular cylinder. In the magnetization curling mode, the flux is essentially in a closed path, which means a low magnetostatic self energy, although complete closure (which resulted in $W_M = 0$ for curling in a circular cylinder) is not possible for this configuration. The anticurling mode looks essentially like a flow out of the center, and therefore involves a high magnetostatic energy. Since in magneto-crystalline and exchange energies the anticurling is rather similar to the curling, one would expect that the difference in magnetostatic energies will make the anticurling eigenvalue larger than that of the other modes, so that it will never take place. Previous estimations of the magnetostatic self energy [37] failed to reveal this point, but better estimations, to be reported in the following, do prove that the anticurling is higher than the other modes (except probably for extremely small values of $b/a$, where the proof fails) and can thus be ignored.

In the following, upper and lower bounds for the eigenvalue, $\lambda'$, will be calculated for each mode in turn, starting with rotation in quasi-unison. In this case it turns out (as one would expect anyway from magnetostatic considerations) that the rotation in the $x$-direction is easier than that in the $y$-direction, except near the square $b \approx a$, when these two modes should converge to the same eigenvalue.

### *4.3 Rotation in quasi-unison*

If the rotation is in the $x$-direction one should choose for the arbitrary potential, $\Phi$, in (40) a function that has the symmetry of the true potential. Accordto Table 1, this $\Phi$ should, therefore, be odd in $x$ and even in $y$. The simplest choice is thus

$$\Phi = A\,x\,, \text{ for } |x| \leqq a\,,\ |y| \leqq b\,, \tag{44}$$

where $A$ is a constant. This function should now be continuously extended to the rest of the $x\,y$ plane, so that it would vanish at infinity. One would also like to use a harmonic function, namely with $\nabla^2 \Phi = 0$, outside the rectangle, although this is not required for the validity of Brown's theorem (39), since one would like $W_H$ to be as large as possible in order to approach the true eigenvalue by the lower bound as closely as possible, and maximization of (40) leads to the Laplace equation for $\Phi$. The first approach of Brown [36] for a square and of Aharoni [37] for the general rectangle, was to extend $\Phi$ of (44) to an ellipse circumscribed around the rectangle, with semi-axes $a\sqrt{2}$ and $b\sqrt{2}$, and to solve the Laplace equation outside this ellipse. More recently, Brown [38] improved the lower bound thus obtained by taking $\Phi$ outside the rectangle as the solution of the Laplace equation $\nabla^2 \Phi = 0$ with the boundary conductions of zero at infinity, and continuity on the rectangle with $\Phi$ of (44).

One way of solving the Laplace equation is to transform conformally the outside of the rectangle under discussion in the $z = x + i\,y$ plane, to the inside of another rectangle $|u| \leqq K$, $0 \leqq v \leqq 2\,K'$ in the $w = u + i\,v$ plane. The transformation is [39]

$$z = a - \frac{i\,b}{E - k'^2\,K}\left[Z\,(w, k) + \left(\frac{E}{k} - k'^2\right)w\right], \tag{45a}$$

where $E$ and $K$ are the complete elliptic integrals, $Z$ is the Jacobian zeta-function (the notations of Byrd and Friedman [40] are used here throughout), and the modulus $k$ is determined by the length to breadth ratio of the original rectangle, $a/b$, according to the relation [39]

$$(E - k'^2 K)\, a = (E' - k^2 K')\, b\,. \tag{45b}$$

Using the Fourier expansion of the zeta-function [40], one can immediately write down the solution of the Laplace equation which fulfills the above mentioned boundary conditions:

$$\Phi = a\,A\left[1 - \frac{v}{K'} - \frac{\pi}{K'\,(E' - k^2 K')}\right] \sum_{n=1}^{\infty} \frac{(-1)^n \sin\left(\frac{n\,\pi\,v}{K'}\right) \cosh\left(\frac{n\,\pi\,u}{K'}\right)}{\sinh\left(\frac{n\,\pi\,K}{K'}\right) \cosh\left(\frac{n\,\pi\,K}{K'}\right)}\,. \tag{46}$$

Equations (44) and (46) now define completely the arbitrarily chosen potential, $\Phi$, in the whole $x\,y$ plane. Substituting in (40), and carrying out the integrations, one obtains

$$W_H = \left(\frac{M_s\,A}{a}\right) \int \alpha\,\mathrm{d}S - \frac{k'^2 A^2}{8}\,(E' - k^2 K')^{-2}\,.$$

Maximizing this expression with respect to $A$, to obtain the largest possible $W_H$, yields

$$W_H = 2\,M_s^2\,\frac{(E' - k^2 K')^2}{a^2\,k'^2}\,[\int \alpha\,\mathrm{d}S]^2\,. \tag{47}$$

Brown [38] obtained the same equation (47) by studying a conformal transformation of the region outside the rectangle in the $z$-plane, to the outside of a unit circle in the $w$-plane.

Before proceeding to get a lower bound for $\lambda'$ by substituting $W_H$ for $W_M$ in (42), we note that in the case $\int \alpha\,\mathrm{d}S = 0$, the magnetostatic self energy is neglected altogether, and one would expect too low an estimation. Actually, this case of $\int \alpha\,\mathrm{d}S = 0$ gives tise to a lower bound of the form of the curling lower bound [37] which interferes with the curling estimation for $b \leqq 3^{-1/2}\,a$. To avoid this, we modify (44) to read

$$\Phi = A\,x + B\,(x^3 - 3\,x\,y^2)\,, \quad \text{for} \quad |x| \leqq a,\ |y| \leqq b\,. \tag{48}$$

In principle one should proceed as before, calculating the $\Phi$ outside, integrating and maximizing over $A$ and $B$. However, for the following argument the details are not necessary. It is sufficient to note that the construction of (40) is such that after maximization with respect to $A$ and $B$, the result should be of the form

$$\begin{aligned} W_H = {} & c_1[\int \alpha\,\mathrm{d}S]^2 + c_2[\int \alpha\,\mathrm{d}S] \int [\alpha\,(x^2 - y^2) - 2\,\beta\,x\,y]\,\mathrm{d}S + \\ & + c_3\,\{\int [(x^2 - y^2)\,\alpha - 2\,\beta\,x\,y]\,\mathrm{d}S\}^2\,, \end{aligned} \tag{49}$$

where the $c_i$ are constants, which depend on $a$ and $b$, but not on the coordinates, or on $\alpha$ and $\beta$.

Consider first the case $b \neq a$ and $\int \alpha\,\mathrm{d}S = 0$, when (49) contains only the term in $c_3$. Substituting this $W_H$ for $W_M$ in (42), and minimizing $W$ with respect to $\alpha$ and $\beta$, leads to the differential equations

$$C\,\nabla^2 \alpha + \lambda'\,\alpha = 2\,c_3\,(x^2 - y^2) \int [(x^2 - y^2)\,\alpha - 2\,\beta\,x\,y]\,\mathrm{d}S\,, \tag{50a}$$

$$C\,\nabla^2 \beta + \lambda'\,\beta = -4\,c_3\,x\,y \int [(x^2 - y^2)\,\alpha - 2\,\beta\,x\,y]\,\mathrm{d}S \tag{50b}$$

with the boundary conditions

$$\frac{\partial \alpha}{\partial n} = \frac{\partial \beta}{\partial n} = 0\,. \tag{50c}$$

Integration of (50a) over the cross-section, using the boundary conditions and the assumptions $b \neq a$ and $\int \alpha \, dS = 0$, yield

$$\int [(x^2 - y^2)\, \alpha - 2\, \beta\, x\, y]\, dS = 0\,, \tag{51}$$

i.e. the equations (50) are homogeneous. The general solution of (50b), which fulfills the boundary conditions and the symmetry requirements of Table 1, for this case of rotation in quasi-unison in the $x$-direction, is

$$\beta = B_1 \sin\left[\left(m + \frac{1}{2}\right)\frac{\pi x}{a}\right] \sin\left[\left(n + \frac{1}{2}\right)\frac{\pi y}{b}\right], \tag{52a}$$

where $m$ and $n$ are integers, yielding the eigenvalue

$$\lambda' = C\,\pi^2 \left[\left(m + \frac{1}{2}\right)^2 a^{-2} + \left(n + \frac{1}{2}\right)^2 b^{-2}\right]. \tag{52b}$$

The smallest eigenvalue is obtained for $m = n = 0$, and this equal the *upper* bound for the curling mode, as will be seen in the next section. This mode is thus of no interest, and one should therefore take $\beta = 0$ as the solution of (50b). The general solution of (50a), for the symmetry considered, is

$$\alpha = A_1 \cos\left(\frac{m\,\pi\,x}{a}\right) \cos\left(\frac{n\,\pi\,y}{b}\right) \tag{53a}$$

with $m$ and $n$ integers, yielding for the eigenvalue

$$\lambda' = C\,\pi^2\,(m^2\, a^{-2} + n^2\, b^{-2})\,. \tag{53b}$$

But this solution should also fulfill the conditions $\int \alpha \, dS = 0$ and (51), with $\beta = 0$. Substituting from (53a) and carrying out the integrations, these conditions imply that both $m \neq 0$ and $n \neq 0$. Therefore the lower bound eigenvalue of (53b) is larger than the upper bound for the curling mode, and as such has no physical significance.

One should thus consider only the case where $\int \alpha \, dS \neq 0$ for $b \neq a$, and any value for this integral, if $b = a$. For these cases we note that (49) involves maximizing over $B$ and therefore cannot be smaller than its value for any particular $B$, for example $B = 0$. Now decreasing $W_H$ might at most decrease the lower bound for $\lambda'$, but it is still a lower bound. One can therefore take $B = 0$, which reduces (48) to (44), and thus (49) is reduced to (47). Using this to replace $W_M$ in (42) and minimizing the energy with respect to $\alpha$ and $\beta$ leads to the homogeneous equation in $\beta$, the solution of which is (52). This has already been argued to be at least as large as the upper bound for curling, and as such can be ignored. The other equation is

$$C\,\nabla^2 \alpha + \lambda'\,\alpha = \frac{4\,M_s^2}{a^2\,k'^2}\,(E' - k^2\,K')^2 \int \alpha\, dS \tag{54a}$$

with the boundary condition

$$\frac{\partial \alpha}{\partial n} = 0\,. \tag{54b}$$

The possibility $\int \alpha \, dS = 0$ needs to be considered only for $b = a$. In this case [36] the general solution of (54) is (53) where the condition $\int \alpha \, dS = 0$ implied $m\,n \neq 0$, thereby making the smallest $\lambda'$ of (53b) larger than the upper bound for magnetization curling. One can thus assume $\int \alpha \, dS \neq 0$, and then integration of (54a) over the rectangle, and using (54b) and (45b) implies

$$\lambda' = \varrho\, M_s^2\,, \tag{55}$$

where

$$\varrho \geqq \frac{16}{k'^2}\,(E' - k^2\,K')\,(E - k'^2\,K)\,. \tag{56}$$

Fig. 6. Upper and lower bounds of the smallest eigenvalue $\varrho = -H_n M_S^{-1} - g_{11} M_S^{-2}$ for nucleation by rotation in quasi-unison ($x$-direction) in an infinite rectangular cylinder, as a function of breadth to length ratio in the rectangle, $b/a$

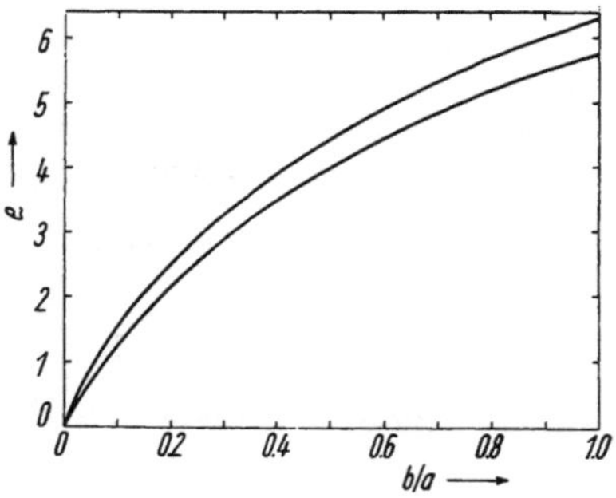


The inequality has been introduced to emphasize the fact that a lower bound has been calculated. Using (45b), this lower bound has been computed by Brown [38] as a function of $a/b$. The results are plotted in Fig. 6. As one would expect, this lower bound is considerably larger than the previous estimation [37] based on an ellipse circumscribed around the rectangle.

To get an upper bound it is sufficient to consider any pair of particular functions $\alpha$ and $\beta$, having the necessary symmetry. Choosing

$$\alpha = \text{const}, \quad \beta = 0$$

calculating $W_M$ and substituting in (42), one obtains [37] that

$$\varrho \leqq 2\, I\left(\frac{b}{a}\right), \tag{57}$$

where $\varrho$ is defined in (55), and

$$I(x) = 4 \tan^{-1} x + x \log(1 + x^{-2}) - x^{-1} \log(1 + x^2)\,. \tag{58}$$

This function is also plotted in Fig. 6. As is seen from the figure, the upper and lower bounds are so close to each other, that a rigorous solution is not required. An obvious way of improving the estimation is to actually calculate the coefficients $c_i$ in (49) and use it rather than (47), but the slight improvement hardly justifies carrying out the complicated integrations.

For rotation in quasi-unison in the $y$-direction, the same calculations can be carried out, interchanging $x$ and $y$, $\alpha$ and $\beta$, $a$ and $b$ consistently. The upper bound is then [37] the same as (57) with $a/b$ written for $b/a$, and the lower bound is [38] of the same form as (56), with $k$ and $k'$ interchanged. It is readily seen that these upper and lower bounds for rotation in the $y$-direction are respectively larger than the upper and lower bounds for the rotation in the $x$-direction, except for a square, $b = a$, when these bounds coincide. It stands to reason that the actual eigenvalue for rotation in the $y$-direction is larger than that of rotation in the $x$-direction, but formally this is proved only where the lower bound of the first mode is larger than the upper bound of the latter mode, i.e., for $b/a < \approx 0.9$. Even above this value for $b/a$, one can say that there should be some solution between the curves of Fig. 6, and that the lower bound for the rotation in the $y$-direction shows there is none below the lower curve. Therefore Fig. 6 describes upper and lower bounds for the lowest of *both* rotation in quasi-unison modes.

### *4.4 Magnetization curling*

To obtain an upper bound for this mode, we shall restrict the minimization of (42) only to functions $\alpha$, $\beta$ which fulfill $\alpha = \partial\psi/\partial y$, $\beta = -\partial\psi/\partial x$, and such that the normal component of the magnetization vector vanishes on the boundary of the rectangle [36, 37]. In this case $W_M$ vanishes, but one should introduce into (42) the restricting assumptions, via Lagrange multipliers. Minimizing $W$ with respect to $\psi$ then leads [36] to the differential equation

$$(C\,\nabla^2 + \lambda')\,\psi = 0 \tag{59a}$$

with the boundary conditions

$$\psi = 0, \quad \text{for} \quad x = \pm a \text{ or } y = \pm b\,. \tag{59b}$$

According to Table 1, for this case of curling, $\psi$ should be even in $x$ and in $y$. The most general solution for (59) is therfore proportional to

$$\cos\left[\left(m + \frac{1}{2}\right)\frac{\pi x}{a}\right]\cos\left[\left(n + \frac{1}{2}\right)\frac{\pi y}{b}\right]$$

with $m$ and $n$ integers, and with

$$\lambda' = C\,\pi^2\left[\left(m + \frac{1}{2}\right)^2 a^{-2} + \left(n + \frac{1}{2}\right)^2 b^{-2}\right]. \tag{60}$$

The smallest $\lambda'$ is obtained for $m = n = 0$. Using the notation

$$\lambda' = C\,\pi^2\,\sigma\,a^{-2} = C\,\pi^2\,\sigma^*\,b^{-2} \tag{61}$$

one can therefore write the upper bound as

$$\sigma \leqq \frac{1}{4}\,(1 + a^2\,b^{-2})\,, \quad \sigma^* \leqq \frac{1}{4}\,(1 + b^2\,a^{-2}). \tag{62}$$

This upper bound is plotted in Fig. 7.

To get a lower bound, we choose the arbitrary potential

$$\Phi = A\,x\,y, \quad \text{for} \quad |x| \leqq a,\ |y| \leqq b\,. \tag{63}$$

Outside the rectangle, $\Phi$ is required to be the solution of the Laplace equation, which vanishes at infinity and which has the same values as (63) on the rectangle. Using the transformation (45), this function can be readily expressed as a combination of two infinite series. After substituting in (40), carrying out the integrations, and maximizing over $A$, one obtains

$$W_H = 6\,M_s^2\,(E - k'^2 K)^2\,(E' - k^2\,K')^2\,\frac{[\int(\alpha\,y + \beta\,x)\,\mathrm{d}S]^2}{a^2\,b^2\,k^2\,k'^2} \tag{64}$$

which is the same result Brown [38] obtained from his transformation.

This expression contains $M_s^2$, which does not appear in the upper bound (62), and therefore the lower bound cannot be of the same form, and need be given as a function of two parameters. One would obviously prefer a somewhat worse estimation, if this can be expressed in terms of one parameter as the upper bound turns out to be, and this can be achieved using the following argument [37]. In

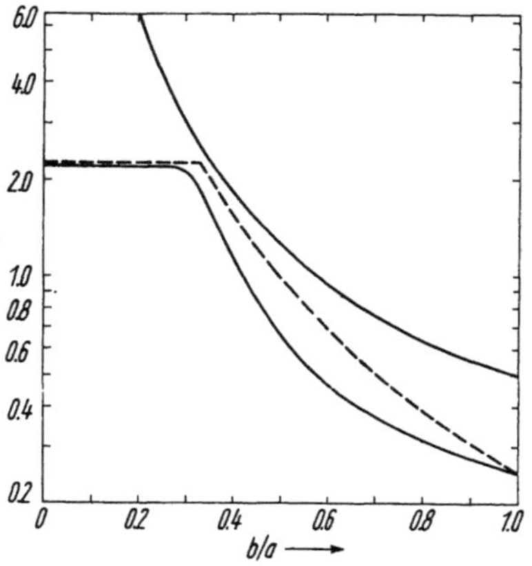


Fig. 7. Upper and lower bounds for the eigenvalue $\sigma = -(g_{11} + H_n\,M_s)\,(a^2/C\,\pi^2)$ for nucleation by magnetization curling in an infinite rectangular cylinder, as a function of breadth to length ratio of the rectangle, $b/a$. The dashed curve is the upper limit of all lower bounds that can be obtained with a single potential term, according to ref. [41]

plotting $\lambda'$ as a function of $b/a$,one is interested only in the smallest eigenvalue of all modes, and therefore the curling eigenvalue will certainly not be plotted in the region covered by the estimation for the rotation in quasi-unison (see Fig. 4 of [37]). In other words, one is not interested in solutions of the curling mode where they are above the lower bound (56) for rotation in quasi-unison. Rearranging (55) and (56), one is looking for $\lambda'$, only as long as

$$M_s^2 \geq \frac{\lambda' k'^2}{16\,(E' - k^2 K')\,(E - k'^2 K)} .$$

Therefore if the right hand side is substituted for $M_s^2$ in (64), the lower bound is not as good as when (64) is used, but it is still a lower bound. By substituting in (42) and minimizing with respect to $\alpha$ and $\beta$, one obtains the differential equations and boundary conditions

$$C \nabla^2 \alpha + \lambda' \alpha = F y ; \qquad \frac{\partial \alpha}{\partial n} = 0 , \tag{65a}$$

$$C \nabla^2 \beta + \lambda' \beta = F x ; \qquad \frac{\partial \beta}{\partial n} = 0 , \tag{65b}$$

where

$$F = \frac{3\,\lambda'}{4\,a^2\,b^2\,k^2} (E - k'^2 K)\,(E' - k^2 K') \int (\alpha\, y + \beta\, x)\, \mathrm{d}S . \tag{66}$$

If $F \neq 0$, it can be verified by substitution that the solution of (65) is

$$\alpha = \frac{F}{\lambda'} \left[ y - \frac{\sin(\mu y)}{\mu \cos(\mu b)} \right], \quad \beta = \frac{F}{\lambda'} \left[ x - \frac{\sin(\mu x)}{\mu \cos(\mu a)} \right], \tag{67a}$$

where

$$\mu = \left( \frac{\lambda'}{C} \right)^{1/2} . \tag{67b}$$

Substituting (67) in (66), carrying out the integrations and substituting for $\lambda'$ from (61), one obtains the following transcendental eigenvalue equation (for this case $F \neq 0$):

$$2 + \frac{\pi^2 \sigma}{3} \left( 1 + \frac{b^2}{a^2} \right) - \frac{\tan(\pi \sigma^{1/2})}{\pi \sigma^{1/2}} - \frac{\tan\left( \dfrac{b \pi \sigma^{1/2}}{a} \right)}{\dfrac{b \pi \sigma^{1/2}}{a}} = \frac{k^2 \pi^2 \sigma}{3\,(E' - k^2 K')^2} . \tag{68}$$

The lowest numerical solution of (68), with (45b), is the lower bound plotted in Fig. 7. It should be noted that the point $\sigma = \frac{1}{4}$, for $b/a = 1$, is not a solution of (68), even though the solution tends to $\frac{1}{4}$ for $b \to a$. It is a solution for the homogeneous equations (65), as will be seen in the following.

If $F = 0$, the general solution of (65) with the appropriate symmetry according to Table 1 is

$$\alpha = A_{mn} \cos\left( \frac{m \pi x}{a} \right) \sin\left[ \left( n + \frac{1}{2} \right) \frac{\pi y}{b} \right] ; \quad \sigma = m^2 + \left( n + \frac{1}{2} \right)^2 a^2 b^{-2} , \tag{69a}$$

$$\beta = B_{pq} \sin\left[ \left( p + \frac{1}{2} \right) \frac{\pi x}{a} \right] \cos\left( \frac{q \pi y}{b} \right) ; \quad \sigma = \left( p + \frac{1}{2} \right)^2 + q^2 a^2 b^{-2} , \tag{69b}$$

where $m$, $n$, $p$, $q$ are integers. It is readily seen that one can get eigenvalues that are not larger than the upper bound (62), only for

$$m = n = q = 0 . \tag{69c}$$

Substituting (69) in (66) and carrying out the integrations, it is seen that the condition $F = 0$ can be fulfilled only if *both* $A_{00}$ *and* $B_{p0}$ are different from zero. If they are both nonzero, the condition $F = 0$ determines only the ratio between them, which does not affect the eigenvalue, but none of them can vanish without the other vanishing too. This implies that the eigenvalues of (69a) and (69b) should be the same, namely

$$\sigma = \left(p + \frac{1}{2}\right)^2 = \frac{1}{4} a^2 b^{-2} \tag{70}$$

which can be fulfilled only for certain discrete values of $b/a$. For $b = a$, (70) implies $\sigma = \frac{1}{4}$, which is the value used in Fig. 7. All the other solutions of (70) are larger than the solution of (68), plotted in Fig. 7, which is therefore the plot of the smallest eigenvalue of (65).

The lower bound plotted in Fig. 7 is somewhat higher than the previous estimation [37], but not very much so. For $\frac{b}{a} \geqq \frac{1}{4}$, the upper bound is within a factor of 2 of the lower bound, but this ratio increases rapidly for smaller values of $\frac{b}{a}$, and tends to infinity for $\frac{b}{a} \to 0$. One can in principle improve the lower bound by replacing (63) by a more sophisticated choice of an arbitrary potential. It should be noted, however, that there are limitations inherent to the method used here, which will not allow [41] the lower bound to be larger than the line shown dashed in Fig. 7, for any choice of a potential, even if one happens to choose the correct potential of the rigorous problem. One cannot therefore gain much, in particular in the region $\frac{b}{a} \leqq \frac{1}{4}$, if *any* function is substituted for (63), provided the maximization of $W_H$ is done with respect to one amplitude only, like $A$ in (63). One can still, in principle, obtain larger lower bounds by maximizing over several amplitudes, but no simple expression could be guessed tor this purpose. It can be shown, however, that it is the lower bound and not the upper bound that needs improving, by considering the potential

$$\begin{aligned}
\Phi &= \sin\left(\frac{\pi y}{2b}\right) \sum_{m=1}^{n} A_m \sin\left(\frac{m \pi x}{a}\right), && \text{for} \quad |x| \leqq a,\ |y| \leqq b\,, \\
&= \sum_{m=1}^{n} A_m \sin\left(\frac{m \pi x}{a}\right) \exp\left[-\left(\frac{m\pi}{a}\right)(y-b)\right], && \text{for} \quad |x| \leqq a,\ y \geqq b\,, \\
&= -\sum_{m=1}^{n} A_m \sin\left(\frac{m \pi x}{a}\right) \exp\left[\frac{m\pi}{a}(y+b)\right], && \text{for} \quad |x| \leqq a,\ y \leqq -b\,, \\
&= 0, \quad \text{for} \quad |x| \geqq a\,,
\end{aligned}$$

which yields when substituted in (40), after maximizing with respect to all $A_m$,

$$W_H = 2\pi^2 M_s^2 \sum_{m=1}^{n} \frac{\left\{\int\left[\frac{m\alpha}{a}\sin\left(\frac{\pi y}{2b}\right)\cos\left(\frac{m\pi x}{a}\right) + \frac{\beta}{2b}\cos\left(\frac{\pi y}{2b}\right)\sin\left(\frac{m\pi x}{a}\right)\right] dS\right\}^2}{2m + \pi\left(m^2 b a^{-1} + \frac{1}{4} a b^{-1}\right)}. \tag{71}$$

The number of terms, $n$, is arbitrary, but since the sum in (71) consists of positive terms, $W_H$ will be larger (thereby making the lower bound closer to the rigorous eigenvalue) for larger $n$.

Proceeding as before, one obtains $\sigma^* = \frac{1}{4}$ as the lowest eigenvalue for the homogeneous equations. For the nonhomogeneous case, one obtains a determinant of order $n$ for the eigenvalue equation, which involves rather tedious computations even for $n = 2$. However, for the case $\frac{b}{a} \to 0$, one can evaluate the determinant analytically, even for $n \to \infty$. It turns out that for this case there is no solution below the homogeneous equations eigenvalue. Therefore, for $\frac{b}{a} \to 0$, the lower bound is $\sigma^* = \frac{1}{4}$, which is identical with the upper bound. For finite $\frac{b}{a}$, however, no way was found to overcome the complexity of the problem, and make the lower bound approach the upper bound.

### *4.5 The anticurling mode*

Consider the potential

$$\begin{aligned}
\Phi &= -\sum_{m=0}^{k} A_m \cos\left[\left(m + \frac{1}{2}\right)\frac{\pi x}{a}\right] - B\cos\left(\frac{\pi y}{2b}\right), && \text{for } |x| \leqq a,\ |y| \leqq b\,, \\
&= -B\cos\left(\frac{\pi y}{2b}\right)\exp\left[-\left(\frac{\pi}{2b}\right)(x-a)\right], && \text{for } x \geqq a,\ |y| \leqq b\,, \\
&= -B\cos\left(\frac{\pi y}{2b}\right)\exp\left[\frac{\pi}{2b}(x+a)\right], && \text{for } x \leqq -a,\ |y| \leqq b\,, \\
&= -\sum_{m=0}^{k} A_m \cos\left[\left(m+\frac{1}{2}\right)\frac{\pi x}{a}\right]\exp\left[-\left(m+\frac{1}{2}\right)\frac{\pi}{a}(y-b)\right], \\
& && \text{for } |x| \leqq a,\ y \geqq b\,, \\
&= -\sum_{m=0}^{k} A_m \cos\left[\left(m+\frac{1}{2}\right)\frac{\pi x}{a}\right]\exp\left[\left(m+\frac{1}{2}\right)\frac{\pi}{a}(y+b)\right], \\
& && \text{for } |x| \leqq a,\ y \leqq -b\,, \\
&= 0\,, && \text{for } |x| \geqq a,\ |y| \geqq b\,.
\end{aligned}$$

Substituting this potential in (40), carrying out the integrations, maximizing with respect to $B$ and all the $A_m$, substituting in (42) for $W_M$, and minimizing with respect to $\alpha$ and $\beta$, one obtains

$$(C\nabla^2 + \lambda')\,\alpha = \sum_{m=0}^{k} F_m \sin\left[\left(m+\frac{1}{2}\right)\frac{\pi x}{a}\right]; \qquad \frac{\partial\alpha}{\partial n} = 0\,, \tag{72a}$$

$$(C\nabla^2 + \lambda')\,\beta = G\sin\left(\frac{\pi y}{2b}\right); \ \frac{\partial\beta}{\partial n} = 0\,, \tag{72b}$$

where

$$F_m = M_s^2\,\pi^2 a^{-2}(2m+1)\left[1 + \left(m+\frac{1}{2}\right)\frac{\pi b}{a}\right]^{-1}\int \alpha \sin\left[\left(m+\frac{1}{2}\right)\frac{\pi x}{a}\right]\mathrm{d}S, \tag{73a}$$

$$G = M_s^2\,\pi^2 b^{-2}\left(1 + \frac{\pi a}{2b}\right)^{-1}\int \beta\sin\left(\frac{\pi y}{2b}\right)\mathrm{d}S\,. \tag{73b}$$

The solution of (72) is

$$\alpha = \sum_{m=0}^{k} F_m\left[\lambda' - C\pi^2 a^{-2}\left(m+\frac{1}{2}\right)^2\right]^{-1}\sin\left[\left(m+\frac{1}{2}\right)\frac{\pi x}{a}\right], \tag{74a}$$

$$\beta = G\left(\lambda' - \frac{C\pi^2}{4b^2}\right)^{-1}\sin\left(\frac{\pi y}{2b}\right). \tag{74b}$$

Substituting (74) in (73), and carrying out the integrations, the determinant of the coefficients of $F_m$ is diagonalized. Actually one obtains that for every $m$, either $F_{m'} = 0$, or

$$\left[\lambda' - C\pi^2 a^{-2}\left(m+\frac{1}{2}\right)^2\right]\left[1+\left(m+\frac{1}{2}\right)\frac{\pi b}{a}\right] = \frac{2b}{a}\pi^2 M_s^2(2m+1), \quad (75\text{a})$$

and similarly, either $G = 0$ or

$$\lambda' = \frac{C\pi^2}{4b^2} + \frac{2a}{b}\pi^2 M_s^2\left(1+\frac{\pi a}{2b}\right)^{-1}. \quad (75\text{b})$$

Besides the eigenvalues (75), one should therefore consider the homogeneous case, in which $G$ and all the $F_m$ vanish. The solution of the homogeneous equations (72), with the appropriate symmetry for anticurling, is then the same as (69), with $\alpha$ and $\beta$ interchanged. Substituting this solution in (73), the condition $G = 0$ obviously makes the eigenvalue for $\beta$ (see (69a)) larger than the *upper* bound for curling (see (62)). The eigenvalue for $\alpha$ (compare (69b)) is also larger than the upper bound for curling, if $q \neq 0$. For $q = 0$, the condition that all the $F_m$ vanish implies.

$$\int \sin\left[\left(p+\frac{1}{2}\right)\frac{\pi x}{a}\right]\sin\left[\left(m+\frac{1}{2}\right)\frac{\pi x}{a}\right]\mathrm{d}S = 0,$$

for every $0 \leqq m \leqq k$. Hence $p$ should be different from all the $m$, and its lowest value is thus $k+1$. The smallest eigenvalue from (69b) is then

$$\sigma = \left(k+\frac{3}{2}\right)^2. \quad (76)$$

For $k \to \infty$, this is again larger than the upper bound for the curling (see (62)), even when $b/a \to 0$, since for every value of $b/a$ one can choose $k$ that will make (76) larger than (62). One should therefore consider only the smallest of the nonhomogeneous eigenvalues, (75).

From (58) one obtains

$$I\left(\frac{b}{a}\right) \leqq \pi, \quad \text{for} \quad b \leqq a.$$

Using this relation in (75b),

$$\lambda' - \frac{C\pi^2}{4b^2} > 2\pi M_s^2 \geqq 2 M_s^2 I\left(\frac{b}{a}\right)$$

and therefore this lower bound is larger than the upper bound (57) for rotation in quasi-unison, and as such has no physical meaning. One is thus left with (75a) only, in which it is readily seen that the smallest possible value for $\lambda'$ is for $m = 0$, yielding

$$\lambda' = \frac{C\pi^2}{4a^2} + \frac{2b}{a}\frac{\pi^2 M_s^2}{1+\dfrac{\pi b}{2a}}. \quad (77)$$

Numerical computation shows that the coefficient of $2M_s^2$ in (77) is larger than $I(b/a)$, for $b/a \geqq 0.045$, and therefore for these values of $b/a$, the lower bound (77) is larger than the upper bound for rotation in quasi-unison.

It has thus been shown that at least for $b/a > 0.045$, the anticurling mode can never take place, since it yields eigenvalues that are larger than the upper bound of one of the other modes. It seems reasonable to assume that this is also the case for the smaller values of $b/a$, but this could not be proved with the particular choice of arbitrary potentials tried. In this region of $b/a < 0.045$, one has for a lower bound the larger[4]) of the two estimations, (77), and the previously discussed [37] $\sigma = 2.2$.

## 5. Comparison with Experiment

### *5.1 Fine particles*

In most experiments on fine particles, the data were taken on powders, composed of particles with distributions of shapes, sizes, and orientations, about which there is very little information. The magnetic interactions between the particles are very difficult to take into account, except by very rough and doubtful models [42]. Besides, the nucleation field is not measured for these particles, whereas the theory gives only nucleation. It has not advanced yet beyond this stage of nucleation (except when coherent rotation takes place), since this calls for the solution of non-linear differential equations. However, some of these difficulties can be relatively unimportant in certain cases, as will be argued in the following, so that one can expect the theory of fit certain experiments.

For the curling mode in cylindrically-symmetric ellipsoids, the magnetostatic potential vanishes during nucleation, according to the definition in Section 2. Therefore, there is no magnetic interaction between such parallel particles, at least during the first stages of magnetization reversal. For an infinite circular cylinder, with the field applied parallel to its axis, the non-linear curling equation was tackled, showing [16, 43] that the hysteresis curve is rectangular and the coercive force is identical with the nucleation field. One can thus assume that for rather elongated prolate spheroids, in the region of size and elongation where the curling is the easiest mode (Fig. 2), the calculated nucleation field should be a good approximation to the measured coercive force, which should be packing-independent, provided the particles are all parallel to the applied field and that the shape and size distribution are not too large.

In the experiment of Levi [44] on drawn iron wires in a copper matrix, the particles were reasonably parallel and extremely elongated, so that they can be taken as infinite cylinders for all practical purposes. However, their cross-section is far from being circular [25]. Since even nucleation, to say nothing of coercive force, has not been calculated for *elliptic* cylinders, comparison is difficult. The experimental coercive force shows a 1-p dependence on packing [44], but this might be due to lack of cylindrical symmetry. This ellipticity is probably also the main reason for the high scattering of data, when the coercive force (extrapolated to zero packing) is plotted as a function of radius, ignoring the different ellipticity of the cross-section in different samples. In this kind of plot [25], the experimental points all lie *below* [11] the theoretical nucleation field for an infinite *circular* cylinder, but one would anyway expect reversal in an elliptic cylinder to be easier than in a circular one, as is clearly indicated, for example, by the results on the rectangular cylinder. Nevertheless, the disagreement between theory and experiment is not much worse than between one experiment and another, and the tendency seems to follow a similar curve, which is all one could expect in view of the difference between a circle and an ellipse.

[4]) Note that for a particular choice of potential, one should take the smallest possible eigenvalue, but when two independent potentials are considered, one should take the larger result of the two. In a way, treatment of every potential can be regarded as a proof that there is no eigenvalue smaller than the lower bound thus obtained.

Data of Campbell et al. [45] on *unaligned* dilute iron particles (obtained by reduction of acicular oxide particles) also fit more or less the same curve as Levi's [25]. In view of the above-mentioned difficulties of comparing, it can thus be concluded that there is at least no obvious discrepancy between theory and these experiments.

Experimental data on other fine particle magnets have been recently reviewed by Wohlfarth ([42], p. 372). In some cases there are apparent indications of the possibility of magnetization reversal by curling like e.g. in fine $CrO_2$ particles which show a rather small [46][5]) packing dependence of the coercive force, or in the $\gamma$-$Fe_2O_3$ powders where the results definitely do not agree ([42], p. 372) with coherent rotation calculations. However, in these and in most of the other cases reviewed by Wohlfarth, the alignment is usually rather poor[6]), the distribution in size and shape rather large or unknown, and in general the published data are insufficient for any comparison with the theory. These objections are obviously removed in the case of measurement [49] of the coercive force for a *single* $\gamma$-$Fe_2O_3$ particle, with known radii. If the exchange constant was independently known, or if the measurements were carried out on several prolate spheroids with different radii, this could serve as a check of the theory. Unfortunately this measurement has been carried out before the theory was developed, and not repeated since, so that only one particle was measured. If one believes it is the curling that takes place, the result can be used to evaluate the exchange constant ([42], p. 372) (which at least turns out to be of the same order of magnitude as the exchange constant in other ferromagnetic materials), but there is no real check of the theory.

The most outstanding success of the theory is in correlating the different measurements of Rassmann and Henkel [50][7]) on wires of iron alloyed with 15% Ni and 12% Cr. With adequate heat-treatment and cold-drawing, ferromagnetic particles precipitate with various sizes and shapes. The hysteresis curve was found to be quite rectangular with the remanence 82 to 98% of the saturation magnetization, indicating the particles were reasonably well aligned parallel to the wire axis. It is not very clear from the data if the coercive force depended on packing, but apparently if there was such a dependence it was ignored, in accordance with curling in ellipsoids of revolution.

A calculation similar to the one reported here in Section 3, with the same assumptions, has already been carried out [30] for the curling nucleation field in an infinite circular cylinder, with the field applied at an angle to the cylinder axis. Assuming that at nucleation the magnetization jumps irreversibly to the energy minimum implied by the Stoner and Wohlfarth theory [31] of coherent rotations, these curves can be used to calculate the rotational hysteresis integral, as a function of the reduced radius, $S$, of the cylinder [30], and of the remanence as a function of $S$ and of the previously applied field [17]. Rassmann and Henkel [50] measured both the rotational hysteresis integral and the remanence curves, and by fitting to the theoretical values, calculated the reduced radius, $S$, for each sample. Using these values of $S$, and fitting the measured values of coercive

[5]) The same applies to certain cobalt nickel alloys, see [47]. In both cases they assume the lack of packing dependence indicates a high magnetocrystalline anisotropy.

[6]) One should always consider the possibility that ferromagnetic particles, even in dilute dacking, stick together to form aggregates, especially when one tries to align them in a magnetic field. See e.g. [48].

[7]) For an illuminating discussion of this experiment, see [42], p. 382.

force (with field parallel to the wire axis) to the theoretical nucleation field [17] for a prolate spheroid, they calculated the elongation $m_S$ of the spheroids. The values thus obtained for $m_S$ agreed reasonably well with the elongation, $m_N$, deduced from the anisotropy which was measured directly by a torquemeter. Both in the calculation of $m_N$ and $m_S$, the magnetocrystalline anisotropy was neglected (they [50] estimated it to be 1% of the shape anisotropy).

Also, using the same values of $S$, the plot of coercive force versus the angle between the wire axis and the applied field was in excellent quantitative agreement, at relatively small angles, with the theoretical curves [30] of the curling nucleation in an *infinite* circular cylinder, calculated under the above-mentioned assumptions. For larger angles, approaching the turnover from curling to coherent rotation, there are considerable deviations from the theoretical curves, the experimental maxima in coercive force being wider and lower than the theoretical ones. Evidently this is due to the finite elongation of the particles to which the coherent rotation (or actually the buckling) is more sensitive than the curling.

This remarkable agreement between theory and experiment would have been much more remarkable if electron micrographs were taken to see if the particles actually resemble prolate spheroids, and to measure their radii directly, which unfortunately was not done. Such a direct measurement could also yield [42], the exchange constant of the material, $C$, which appears in the relation (22b) between the radius, $R$, and the reduced radius, $S$. As it is, there is not any actual check of the theory, only of the self-consistency of certain parameters deduced from it, and this cannot always be considered as a sufficient proof. After all, one finds that certain measurements on thin films can be interpreted in terms of curling in infinite cylinders perpendicular to the plane of the film, [51] a model which seems rather difficult to swallow physically, although there are usually [52] dislocations perpendicular to the film. It should also be noted that the measurements of Rassmann and Henkel were carried out on a material with negligible magnetocrystalline anisotropy. Although the reason is not known, it seems that measurements of such materials fit the theory better than those of materials with an appreciable magnetocrystalline anisotropy, as will be discussed in Section 5.3.

### *5.2 ESD magnets*

In contradistinction to the fine particles discussed in the previous section, which are considered to have approximately an ellipsoidal shape, particles made by electrodeposition of Fe or Fe–Co into mercury, have a peanut shape [25]. This shape has led Jacobs and Bean [24] to propose the model of fanning in a chain of spheres. The predictions of this model are in reasonable agreement with all the available experimental data [25].

According to Fig. 5 here, fanning should take place for $S \leqq 1.6$, and curling should take place above this radius, in this zero order approximation. This critical radius is somewhat lower in the first order approximation, but if one compares to the "nonsymmetric fanning" (Section 2.4) it is pushed up to 1.67. Assuming that the largest radius on which experimental data are available [25], $2\,R = 305$ Å, corresponding to $S = 1.6$, one obtains from (22b) $R_0 = \left(\frac{C}{2}\right)^{1/2} \times M_s^{-1} = 95$ Å. Hence for iron $\frac{1}{2} C = 2.5 \times 10^{-6}$ erg/cm, which is not too far from the value $2.0 \times 10^{-6}$ erg/cm estimated by Kittel [53] for pure iron, and agrees with the value of $(2.5 \pm 0.5) \times 10^{-6}$ erg/cm deduced [54] from magnetic resonance experiments. Indeed, there are measurements that lead to lower value as well, which led Frei, Shtrikman, and Treves [15] to choose $R_0 = 60$ Å as a kind of averhage between known estimates of $C$, but this averaging cannot be reliable. Also, the 305 Å particles were not iron but 60% iron–40% cobalt alloy [25], and for this alloy the most recent estimation of $\frac{1}{2}$ $C$ is [55] $2.54 \times 10^{-6}$ erg/cm. Therefore its seems highly plausible that all the reported data are below the critical size for turnover from fanning to curling (even slightly above this size the difference between the two models is probably below experimental accuracy). To know this, one should either measure "chains" with still larger radii, or measure directly the exchange constant in iron and iron–cobalt alloys.

In considering the data for the peanut-shaped particles, one should also bear in mind that the chain of spheres model is not a very good approximation for the actual shape to begin with; in particular the cylindrical symmetry does not seem very good. There are a lot of irregularities and cross-links between particles, which have been demonstrated [56] to affect the coercive force and its angle-dependence. Also the alignment is rather poor [57], it being stated [56] that "95% of the particles lie with their axes within 20° of the alignment direction". Both this degree of misalignment, as well as the irregularities and cross-links present, certainly invalidate the estimation [26] of very small packing-dependence for curling in exactly parallel chains of ideal spheres, and one can easily conceive a considerable packing-dependence for curling under more realistic conditions. These also obviously affect the critical size, although their effect is difficult to estimate. Besides, even for ideal and parallel chains, the calculation of the nucleation field does not necessarily yield the coercive force, which might be considerably larger, especially as one approaches the size for which the nucleation becomes positive. Some approximation could probably be tried if the whole magnetization curve was known for one sphere, but even this is a formidable mathematical problem, and has never been tried.

It should also be noted that the particle axis does not coincide with a magnetocrystalline easy axis [58], a condition which has been assumed in all micromagnetics calculations, and is certainly an essential part for assuming a cylindrical symmetry. The effect of such an angle on the calculation is not clear, and actually since it takes an infinite field to saturate a crystal in any but an easy or a hard direction [59] the validity of the whole nucleation-field approach might be questioned (see Section 5.6). Experimentally it is found [60] that changing the direction and amplitude of the magnetocrystalline anisotropy considerably has very effect on the coercive force (except for changes that can be accounted for by the difference in saturation magnetization), a phenomenon that is difficult to understand with any model.

### 5.3 Whiskers

Theoretically, a perfect whisker should behave according to the results of Section 4, namely reverse its magnetization by rotation in quasi-unison for small cross-sections, and by curling for larger cross-sections. The turnover from rotation to curling is below the size for which any experimental data are available, so that one needs to consider the magnetization curling mode only, and one can write for the nucleation field from (43) and (61)

$$-H_n = g_{11}\, M_s^{-1} + C\, \pi^2\, \sigma\, a^{-2}\, M_s^{-1}\,, \tag{78}$$

with $\sigma$ between the upper and lower bounds of Fig. 7.

Experimentally it is found [55] that the coercive force of iron and iron-cobalt whiskers agrees fairly well with the nucleation field (78), for rather small values of $a$, namely in the region where the first term with $g_{11}$ ($= 2\,K_1$ in the usual notation of the magnetocrystalline anisotropy constant) is small compared to the second term. There is a rather large scattering of the experimental values, which might be attributed to finite length, inadequate averaging over the size distribution and clumping [55] of the particles. However, these approximations were not used for the larger whiskers which were measured individually, and in these whiskers relation (78) is not obeyed at all. The experimental results for the coercive force [55] continued to follow a curve rather similar to the second term in (78), even though in this size region this term is negligible with respect to the first one. Most intriguing is the fact that essentially one universal line is obtained for different compositions of iron-cobalt alloys (namely order of magnitude changes in the magnetocrystalline anisotropy) and includes some with $\langle 111 \rangle$ orientation. Thus, if one considers the 40—60 composition only, for which $g_{11}$ is very small, there seems to be a reasonable agreement between theory and experiment. However, such an approach is obviously misleading, since the coercive force is essentially the same for pure iron whiskers. More recently cobalt whiskers were also measured [61], yielding essentially the same results: For small sizes the coercive force fits fairly well the theoretical nucleation field for rotation in quasi-unison with a turnover to curling (if one takes [61] a negative $g_{11}$, to account approximately for the fact that the easy axis was perpendicular to the whisker long axis; strictly speaking the calculations of Section 4 should be modified before a comparison can be made). In larger whiskers, those with the easy axis parallel to the whisker long axis seem to have a smaller coercive force than those with the easy axis perpendicular to the long axis, or at an angle to this axis, which is very strange indeed.

The obvious difference between small and large particles is in the probability of having crystalline defects and surface irregularities, which the theory does not take into account, but which affect strongly the measured quantities. By comparing the energy involved, it has been shown [53] that in large enough particles the energy can be enormously reduced by subdivision into domains magnetized in different directions. The nucleation field calculation shows that there is a barrier which does not let them nucleate, in a perfect crystal with a smooth surface, until a certain field is reached, e.g. (78) in the case of smooth, rectangular infinite cylinder. However, there can be little doubt that once domain walls are somehow inserted into the material, at some imperfections where the barrier is for some reason lower, they can propagate very easily to reach lower energy states, and therefore the coercive force is no longer controlled by the nucleation field, unless it is the field required for domain nucleation at the imperfections. In soft materials like iron (and this is probably true to some extent for the cobalt whiskers too) it has been suggested [62] that domains nucleate at nonmagnetic inclusions and surface pits, where the demagnetizing field is large enough to overcome the $\frac{2\,K}{M_s}$ barrier. There were even some attempts to make this argument seem quantitative [63], but these calculations merely compare energies of certain configurations, with the wall assumed already there, without considering the problem of how the wall nucleates to begin with. However, the suggestion has been established by experiments to be discussed in the following. Therefore, to get an agreement between theory and experiment, one should either take the surface roughness into account in the theory, or do the experiments on smoother crystals. Otherwise one would expect the theory to fit experiment only for fine particles (in which the probability of finding sufficiently large irregularities is small), as is actually the case.

To obtain sufficiently smooth large crystals is practically impossible. Under rather general assumptions, Brown [64] studied the probability function and concluded that while it is practically a certainty to measure the theoretical $H_n$ in a cube or a sphere with a diameter of 1 μm, this probability becomes practically zero for a typical whisker, 1 cm long and 10 μm wide. There can be, however, another experimental approach, namely to hold domains nucleated in the "bad" parts of a whisker and not let them propagate into other regions. Then, in the smoother regions one might be able to measure the nucleation field of a perfect crystal.

Such an experiment has been performed by DeBlois und Bean [65]. In the size range where the overall coercive force (which is a measure of the domain nucleation at the *largest* imperfection in the crystal) is less than 1 Oe in the measurement of Luborsky and Morelock [55], DeBlois and Bean [65] found that in certain *parts* of the iron whisker the nucleation field can be 500 Oe or more. This is 90% of the theoretical value of 560 Oe, which is the first term of (78) for iron (the second term contributes [36] about 0.1 Oe for a 10 μm whisker) and $10^3$ to $10^4$ times the overall coercive force of the same [66] whiskers. In a later experiment [66] the local nucleation fields were found to depend drastically on electropolishing, showing that surface flaws play a dominant role in decreasing the magnitude of the nucleation field below its theoretical value. Extension of this experiment [28] showed a clear dependence of the local nucleation field on the volume of surface irregularities. The treatment of Section 2.5 leads one to believe that the theory will yield similar results. It thus seems plausible that the theory can be extended to account for the details of a DeBlois-type experiment quantitatively.

### *5.4 Bulk hard materials*

For hard materials, namely those for which $K \gg M_s^2$, surface imperfections cannot be very important, since their demagnetizing field is too small to overcome the barrier of $\frac{2K}{M_s}$. If the crystals have a prolate or oblate spheroidal shape, and if they are large enough so that the $S^{-2}$ term can be neglected, the results of Sections 2.2 and 2.3 imply that the nucleation field is

$$-H_n = \frac{2K}{M_s} - N M_s, \tag{79}$$

where $N$ is the demagnetizing factor (15b), along the field direction. The relation $K \gg M_s^2$ makes $H_n$ of (79) always negative since $N \leqq 4\pi$. The coercive force, although not calculated, should certainly be at least $|H_n|$, since at nucleation the reversal just starts, while at the coercive force half of the material is already reversed.

Experimentally, the coercive force is smaller than $|H_n|$ of (79). Moreover, it decreases with particle size (making the discrepancy larger for larger particles) much beyond the range where the theory predicts any size-dependence. For $BaFe_{12}O_{19}$ particles of the order of 1 μm, the experimental coercive force is more than half the theoretical value of $|H_n|$ at low temperatures, and their temperature dependences are rather similar, provided one takes into account the random distribution of orientation, and the "correction" of the second term in equation (79) [67]. However, for larger crystals, not only the coercive force is in disagreement with experiment, but domains were observed [68] to nucleate at *positive* fields of +500 to +2000 Oe, whereas (79) yields fo this material a nucleation field of −12000 Oe (using $N = 4\pi$, since the crystals were thin plate- lets) [68]. For MnBi, using the experimental values [69] for $K$ and $M_s$, equation (79) implies $-H_n = 29$ kOe, even with $N = 4\pi$, whereas the experimental values for coercive force range [53] from 12 kOe for 5-μm particles to 0.6 kOe for 100-μm particles.

Rathenau, Smit, and Stuyts [70] suggested that in hard materials domain walls nucleate in regions where the magnetocrystalline anisotropy constant $K$ is low, due to some structural defects. In particular, at dislocations the mechanical strains, via magnetostriction, modify $K$ locally. In these regions the barrier (79) is removed. Howeve·, there is another barrier [11] to be overcome before the region with low $K$ can be reversed, and that is the exchange interaction with spins in the other regions of the material, which are held by the full value of $K$. Several one-dimensional calculations, assuming zero anisotropy [4, 17, 71] or even negative local anisotropy [72] at the defective region, tried to make this argument somewhat more quantitative. Later, a better physical approach to a dislocation line was introduced, by assuming $K = 0$ in a *cylinder*, parallel to the applied field, and solving for the whole eigenvalue spectrum involved [11, 21]. The result is that the second term in (79) is unchanged to a first approximation, while the first term is reduced by an amount which depends on the radius of the cylinder, so that the nucleation field can be *positive* at not too large values for the radius of this cylinder. The value of +500 Oe observed [68] in certain single crystal platelets of $BaFe_{12}O_{19}$, is obtained theoretically [11, 21] for a defective region with a radius of about 350 Å, which is not unreasonable for a dislocation.

The importance of dislocations in hard materials can also be seen qualitatively from the observation [68] that domains always nucleate at the same place, or one of a few well defined points, in a given crystal. This does sound like the domains are nucleated at defects, although the nature of these defects has never been studied experimentally. Also one cannot be sure if the domains were actually nucleated at these particular points or if they were just not removed completely by the highest field applied to saturate the material, since saturation is particularly difficult to achieve in hard materials, as will be discussed in Section 5.6. In both cases it seems that the size-dependence of the coercive force should be somehow connected to the probability of having imperfections, which is larger in the larger particles.

Another qualitative support for this idea may be found in the now well-established experimental fact [73][8]) that by annealing $BaFe_{12}O_{19}$ particles at low enough temperatures so that particle growth is avoided, the coercive force is increased considerably, evidently due to removal of dislocations.

It thus seems fairly well established that in hard magnetic materials, the nucleation field, and therefore the coercive force, depends mainly on dislocations, and that a theory that takes them into account has a good chance of agreeing quantitatively with experiment. One important conclusion of the theory, that these dislocations will have relatively small effect in elongated particles [11], could not, unfortunately, be checked, for lack of any experimental data.

---

[8]) Apparently different results are reported in some cases, e.g. [74].

### 5.5 *Thermal fluctuations*

In comparing the theory with experiment, data on extremely small particles [75] were ignored. These particles are not ferromagnetic in the usual sense, since the direction of magnetization in an individual particle keeps changing in response to thermal agitation, and they behave essentially like huge paramagnetic atoms. Above a (rather small) critical size, the thermal agitation has a negligible effect and can be ignored for any reasonable measurement time [76]. Even particles to which one usually refers as "fine particles" — as discussed in previous sections —, are above the critical size for stable ferromagnetism. Therefore the thermal agitation was not taken into account in the theory, which thus applies only to particles above a certain size.

However, one might ask if a small region in a larger particle might not reverse its magnetization by thermal agitation, thus forming a "nucleus" for a reversed domain. To answer this question, it is necessary to know the energy barrier for such a reversal, namely all details of intermediate steps, and these are not known. For a certain model it was shown [77] that the probability for such a process is negligibly small at room temperature. The difference between a small particle and a small region in a large particle is that in the latter case there is an extra barrier due to the exchange interaction between this region and its surroundings. It is this extra barrier that is too larger to overcome (in this particular model), even if the anisotropy does not oppose reversal in that particular place, because of some structural defects. This calculation of probability used [77] a formula of Stacey [78] which Brown [76] later showed to be based on wrong assumptions. However, the correct probability function [76] does not differ much [79] from Stacey's and cannot change the result.

A more serious objection [80] to this estimation is that it is based on a particular mode of reversal. If there is an easier way than this to reverse the magnetization of part of the body, the thermal forces will find it. To improve this, Brown [80] studied the problem from a much more general point of view. However, he considered only magnetization reversal in the whole large particle, and ignored the main problem of reversal in small regions only, especially around defects. In this respect, the previous calculation [77] has not been generalized and is open to the same objections.

### 5.6 *Incomplete saturation*

The whole nucleation field approach is based on the assumption that the sample is completely saturated to begin with. It is not clear if this is possible in principle, in particular when defects are present. As has been mentioned in Section 1, the so-called approach to saturation has been studied from the standpoint that saturation is not possible in any finite field. If pursued consistently, such an approach calls for using only the non-linear equations in the calculation of the nucleation field [81]. Yet the nucleation-field approach is rather successfull in predicting the measured properties of fine particles and the discussion in previous sections makes it at least plausible that it can lead to agreement with large particle experiments in terms of the reduction in nucleation field due to imperfections. It therefore seems worthwhile to pursue this theory at least until some experiment will prove it definitely wrong.

However, even if one adopts the view that complete saturation is possible *in principle*, there is still the practical problem of whether saturation has been achieved in a particular experiment. If domains are observed in a positive field, it might be true nucleation at such a field, but it is also possible that the crystals "were never really saturated, but always contained vestigial domains" as already suggested by Brown [82] in 1945 as a possible explanation of the disagreement between theory and experiment.

In the somewhat analogous case of ferroelectricity, Stadler and Zachmanidis [83] take as a criterion for saturation a random nucleation of domains, assuming that nucleation at specific points means the domains were there at the "saturated" state, only were not visible. This can hardly be the criterion for ferromagnetism, where it has been argued in the foregoing that domains should nucleate only at crystalline defects or surface irregularities. And actually, in the relatively few cases where such an observation was tried (in most data on domains found in the literature, only part of the crystal was observed) domains seemed to form in specific places. It is thus difficult to give any criterion for saturation.

A rather sound experimental approach is that of Bates and Martin [62] who used for saturation 4 to 5 times the field in which the "spikes", at surface pits, "were too short to be distinguished from the clouds of colloid particles", although their field was most probably an order of magnitude too low in view of the later observation [84] that at the tip of an iron whisker domains still persist at a 6000 Oe field. It should be realized that the "disappearance" of domains at the assumed saturation depends on the sensitivity of the measurement. In describing their Hall-probe, Bate et al. [85] mention that it could detect a reversed region after it could no longer be seen by the Kerr and Bitter pattern techniques.

It has been estimated [11] that a field of at least $4\pi M_s$ is necessary to saturate the surface pits of a crystal. In cobalt, domains were seen [86] up to a field of $2\pi M_s$. Such fields were very seldom used in practice, and it is therefore difficult to compare the theory to any experimental data, unless it is for fine particles in which domain formation is not favorable anyway, or when the experimental arrangement is such that this problem is avoided, like in the experiment of DeBlois and Bean [65, 66] discussed in Section 5.3. The only experiment in which a really large field was applied yielded [87] a nucleation field of +6.4 kOe for a single crystal iron cube, which is remarkably close to the theoretical value for an iron sphere [11]. If such fields were used in other experiments, the result could yield much more valuable information. For example, in square platelets of Ni magnetized along a diagonal, domains generally nucleated at the corners of the other diagonal [88]. In most crystals this occurred in a positive field, but in others, domains nucleated only in negative fields. Moreover, this difference might be encountered in different runs on the same crystal [89]. This might be due [89] to the vertical component of the field being different in different runs, but it is also possible that the applied field was not sufficient to saturate the crystal at the sharper corners, and the reversed domains always remained there. A similar argument probably applies [11] to the experiment of Kooy and Enz [68], where the fact that domains did not always appear at the spot with largest nucleation field, leads one to suspect that these spots were not actually nucleation centers, but places that were saturated in some of the experiments and not in others. This is why the value of +500 Oe was used in Section 5.4 to estimate the equivalent size of the dislocation model, rather than +2000 Oe, and it stands to reason that this value for the nucleation field is also too high for the same reason, and lower values could have been measured with larger saturating fields. It should be noted, though, that in this case (as in essentially all hard materials) it is not the surface that is difficult to saturate, but the vicinity of crystalline imperfections.

The necessity of large fields to saturate hard magnetic materials has been demonstrated by Guillaud [90].[9] He showed that the coercive force of MnBi powder increased roughly from 500 to 5000 Oe when the maximum applied field was increased from 4000 to 20000 Oe, even though the saturation magnetization was the same. The latter fact only means that the reversed domains occupy a negligible portion of the volume of the material [86] at 4000 Oe applied field, but they are certainly there, and facilitate the reversal. Such an experiment certainly casts a lot of doubt on most published values of coercive force, since very few people seem to even try to measure coercive force as a function of the maximum applied field.[10] Even more remarkable in this respect is the experiment of Shur et al. [92] on MnBi crystals of 10 to 15 μm size. Such particles show a polydomain structure, and are completely saturated, in the sense that domains are not seen any more, at 5000 Oe. The domains were seen to "nucleate" at positive fields. However, once a field of 20000 Oe is applied to such a particle, the domains do not appear any more in any field, and the coercive force is increased by an order of magnitude. They could obtain the polydomain structure again only by applying a decaying ac field, or by cooling to liquid air temperature, where the magnetocrystalline anisotropy of this material vanishes. It should be noted, though, that using this larger field on larger (50 μm) particles did not affect [92] the multi-domain structure or the coercive force.

### *5.7 Conclusions*

The theory agrees fairly well with experimental results for fine particles, between the size region for superparamagnetism and that of multi-domain behavior. It seems to be possible to modify it to agree with experiments on larger particles, by taking into account surface roughness in soft materials and dislocations in hard materials, provided the experiments are done on previously saturated crystals.

## References


[1] W. F. Brown Jr., Micromagnetics, Interscience Publishers, New York 1963.
[2] A. Aharoni, E. H. Frei, and S. Shtrikman, J. appl. Phys. **30**, 1956 (1959).
[3] E. Goto, N. Hayashi, T. Miyashitu, and K. Nagakawa, Proc. Intermag. Conf., Washington, D.C., 1965, paper No. 5.2.
[4] C. Abraham and A. Aharoni, Phys. Rev. **120**, 1576 (1960).
[5] M. W. Muller, Phys. Rev. **122**, 1485 (1961).
M. W. Muller and A. Wehlau, J. appl. Phys. **32**, 2448 (1961).
[6] W. F. Brown Jr., Phys. Rev. **58**, 736 (1940).
[7] W. F. Brown Jr., Phys. Rev. **60**, 139 (1941).
[8] A. Seeger and H. Kronmüller, J. Phys. Chem. Solids **12**, 298 (1960); **18**, 93 (1961).
[9] A. Aharoni, Phys. Rev. **132**, 105 (1963).
[10] A. Aharoni, J. appl. Phys. **35**, 913 (1964).
[11] A. Aharoni, Rev. mod. Phys. **34**, 227 (1962).
[12] W. F. Brown Jr., Phys. Rev. **105**, 1479 (1957).


---

[9]) I am indebted to Dr. M. McCaig for drawing my attention to this work.

[10]) A strong dependence of the coercive force on the maximum applied field (and on the frequency of this field!) has also been recently reported by [91].

[13] S. Shtrikman and D. Treves, in Magnetism, Ed. G. T. Rado and H. Suhl, Vol. III, Academic Press, New York 1963 (p. 395).
[14] W. F. Brown Jr., Bull. Amer. Phys. Soc. (Ser. 2) **1**, 323 (1956).
[15] E. H. Frei, S. Shtrikman, and D. Treves, Phys. Rev. **106**, 446 (1957).
[16] A. Aharoni and S. Shtrikman, Phys. Rev. **109**, 1522 (1958).
[17] A. Aharoni, J. appl Phys. **30**, 70 S (1959).
[18] A. Aharoni, Phys. Rev. **131**, 1478 (1963).
[19] E. C. Titchmarsh, Eigenfunction Expansions Associated with Second Order Differential Equations, part II, Oxford University Press, 1958 (p. 88).
[20] A. Arrott, private communication.
[21] C. Abraham and A. Aharoni, Phys. Rev. **128**, 2496 (1962).
[22] C. Flammer, Spheroidal Wave Functions, Stanford University Press, Stanford (California) 1957.
[23] E. Jahnke and F. Emde, Tables of Functions, 4th ed., Dover Publication Inc., New York 1945 (p. 111).
[24] I. S. Jacobs and C. P. Bean, Phys. Rev. **100**, 1060 (1955).
[25] F. E. Luborsky, J. appl. Phys. **32**, 171 S (1961).
[26] A. Aharoni, J. appl. Phys. **35**, 347 (1964).
[27] C. Abraham, Ph. D. thesis, the Hebrew University, Jerusalem, 1964 (in Hebrew).
[28] A. Aharoni and E. Neeman, Phys. Letters (Netherlands) **6**, 241 (1963).
[29] L. Néel, C. R. Acad. Sci. (France) **255**, 1545 (1962).
[30] S. Shtrikman and D. Treves, J. Phys. Radium **20**, 286 (1959).
[31] E. C. Stoner and E. P. Wohlfarth, Phil. Trans. Roy. Soc. (London) **A 240**, 599 (1948).
[32] W. F. Brown Jr., Magnetostatic Principles in Ferromagnetism, North-Holland Publ. Comp., Amsterdam 1962 (p. 46).
[33] W. F. Brown Jr., Phys. Rev. **124**, 1348 (1961).
[34] W. F. Brown Jr., J. Phys. Soc. Japan **17**, Suppl. B-I, 540 (1962).
[35] E. E. Huber Jr. and D. O. Smith, J. appl. Phys. **30**, 267 S (1959).
[36] W. F. Brown Jr., J. appl. Phys. **33**, 3026 (1962).
[37] A. Aharoni, J. appl. Phys. **34**, 2434 (1963).
[38] W. F. Brown Jr., J. appl. Phys. **35**, 2102 (1964).
[39] H. Kober, Dictionary of Conformal Representations, 2nd ed., Dover Publication Inc., New York 1957 (p. 178).
[40] P. F. Byrd and M. D. Friedman, Handbook of Elliptic Integrals for Engineers and Physicists, Springer-Verlag, Berlin 1954.
[41] A. Aharoni, Proc. 1964 Conf. Magnetism, Nottingham (p. 686).
[42] E. P. Wohlfarth, in: Magnetism, Ed. G. T. Rado and H. Suhl, Vol. III, Chap. 7, Academic Press, New York 1963 (p. 363).
[43] W. F. Brown Jr., J. appl. Phys. **29**, 470 (1958).
[44] F. P. Levi, Nature **183**, 1251 (1959); J. appl. Phys. **31**, 1469 (1960).
[45] R. B. Campbell, H. Amar, A. Berkowitz, and P. J. Flanders, Proc. Conf. Magnetism and Magnetic Materials, Boston, 1956 (p. 118).
[46] F. J. Darnell, J. appl. Phys. **32**, 1269 (1961).
[47] R. B. Falk and G. D. Hooper, J. appl. Phys. **32**, 2536 (1961).
[48] F. D. Stacey, Austral. J. Phys. **13**, 196 (1959).
[49] A. H. Morrish and S. P. Yu, Phys. Rev. **102**, 670 (1956).
[50] G. Rassmann and O. Henkel, phys. stat. sol. **1**, 517 (1961).
[51] W. D. Doyle, J. E. Rudisill, and S. Shtrikman, J. appl. Phys. **32**, 1785 (1961); J. Phys. Soc. Japan **17**, Suppl. B-I, 567 (1962).
[52] G. A. Bassett, J. W. Menter, and D. W. Pashley, in: Structure and Properties of Thin Films, Ed. C. A. Neugebauer, J. B. Newkirk, and D. A. Vermilyea, Wiley, New York 1959 (p. 24).
[53] C. Kittel, Rev. mod. Phys. **21**, 541 (1949).
[54] D. S. Rodbell, in: Growth and Perfection of Crystals, Ed. R. H. Doremus, B. W. Roberts, and D. Turnbull, Wiley, New York 1958 (p. 247).

[55] F. E. LUBORSKY and C. R. MORELOCK, J. appl. Phys. **35**, 2055 (1964).
[56] F. E. LUBORSKY and T. O. PAINE, J. appl. Phys. **31**, 66 S (1960).
[57] I. S. JACOBS and F. E. LUBORSKY, J. appl. Phys. **28**, 467 (1957).
[58] F. E. LUBORSKY, E. F. FULLAM, and D. HALLGREN, J. appl. Phys. **29**, 989 (1958).
[59] A. ARROTT and J. E. GOLDMAN, Electr. Manufacturer **1**, 109 (1959).
[60] F. E. LUBORSKY, J. appl. Phys. **34**, 1706 (1963).
[61] F. E. LUBORSKY and C. R. MORELOCK, Proc. Internat. Conf. Magnetism, Nottingham, 1964 (p. 763).
[62] L. F. BATES and D. H. MARTIN, Proc. Phys. Soc. (London) **B 69**, 145 (1956).
[63] W. D. NIX and R. A. HUGGINS, Phys. Rev. **121**, 1038 (1961).
R. CAREY and E. D. ISAAC, Brit. J. appl. Phys. **15**, 551 (1964).
[64] W. F. BROWN JR., J. appl. Phys. **33**, 3022 (1962); ref. [1], p. 72.
[65] R. W. DEBLOIS and C. P. BEAN, J. appl. Phys. **30**, 225 S (1959).
[66] R. W. DEBLOIS, J. appl. Phys. **32**, 1561 (1961).
[67] D. HADFIELD (Ed.), Permanent Magnets and Magnetism, Iliffe Books Ltd., London 1962 (p. 101).
[68] C. KOOY and U. ENZ, Philips Res. Rep. **15**, 7 (1960).
[69] R. M. BOZORTH, Ferromagnetism, van Nostrand Comp., Princeton 1951 (p. 575).
W. C. ELLIS, H. J. WILLIAMS, and R. C. SHERWOOD, J. appl. Phys. **28**, 1215 (1957).
[70] G. W. RATHENAU, J. SMIT, and A. L. STUYTS, Z. Phys. **133**, 250 (1952).
[71] A. AHARONI, Phys. Rev. **119**, 127 (1960); J. appl. Phys. **32**, 245 S (1961).
[72] R. CLAD, doctoral thesis (third cycle), The University of Strasbourg, 1962.
[73] G. HEIMKE, Z. angew. Phys. **15**, 271 (1962); **17**, 181 (1964).
[74] W. A. J. J. VELGE and J. K. DE VOS, J. appl. Phys. **34**, 3568 (1963).
[75] E. F. KNELLER and F. E. LUBORSKY, J. appl. Phys. **34**, 656 (1963).
[76] W. F. BROWN JR., Phys. Rev. **130**, 1677 (1963).
[77] A. AHARONI, J. appl. Phys. **33**, 1324 (1962).
[78] F. D. STACEY, Proc. Phys. Soc. (London) **73**, 136 (1959).
[79] A. AHARONI, Phys. Rev. **135**, A 447 (1964).
[80] W. F. BROWN JR., in: Fluctuation Phenomena in Solids, Ed. R. E. BURGESS, Academic Press, New York 1965 (p. 59 and 72).
[81] W. F. BROWN JR., J. Phys. Radium **20**, 101 (1959); Ref. [1], p. 71.
[82] W. F. BROWN JR., Rev. mod. Phys. **17**, 15 (1945).
[83] H. L. STADLER and P. L. ZACHMANIDIS, J. appl. Phys. **35**, 2625 (1964).
[84] C. A. FOWLER, JR., E. M. FRYER, and D. TREVES, J. appl. Phys. **32**, 296 S (1961).
[85] G. BATE, D. E. SPELIOTIS, and J. R. MORRISON, J. appl. Phys. **35**, 972 (1964).
[86] L. F. BATES and D. J. CRAIK, J. Phys. Soc. Japan **17**, Suppl. B-I, 535 (1962).
[87] R. W. DEBLOIS, Rev. sci. Instrum. **32**, 816 (1961).
[88] R. W. DEBLOIS, J. appl. Phys. **36**, 1647 (1965).
[89] R. W. DEBLOIS, private communication.
[90] C. GUILLAUD, J. Recherches C.N.R.S., No. 9 (1949); J. Phys. Radium **12**, 492 (1951).
[91] A. A. HIRSCH, Proc. Intermag. Conf., Washington, D.C., 1965, paper No. 5.4.
[92] J. S. SHUR, E. W. SHTOLZ, and W. I. MARGOLINA, Zh. eksper. teor. Fiz. **38**, 46 (1960).


*(Received October 18, 1965)*

# *Original Papers*

phys. stat. sol. **16**, 43 (1966)

*Institute of Semiconductors, Academy of Sciences of the Ukrainian SSR, Kiev*

# Electrical Pinch in Elastically Deformed Germanium

By

E. I. Rashba, V. A. Romanov, I. I. Boiko, and I. P. Zhadko

The phenomena accompanying the flow of an electric current through uniaxially compressed plates of intrinsic Ge are investigated. The plates are cut with definite orientations relative to the main crystallographic axes so that the directions of the current and field do not coincide. Under these conditions transverse gradients of the nonequilibrium carrier concentration arise in specimens with long carrier lifetimes, this results in a modulation of the conductivity, emergence of rectifying characteristics, etc. In strong fields the carriers and current are concentrated in a narrow layer close to one of the surfaces, the carriers being almost completely removed from the rest of the specimen (the electrical pinch in an electron–hole plasma). The redistribution of carriers across the sample and the current-voltage characteristics are investigated experimentally for these pinch conditions. For fields of about 100 V/cm an increase and decrease of concentration is observed, giving an extremely nonlinear current-voltage characteristic if opposite plate surfaces have different surface recombination velocities. A rectification ratio of about 50 may be obtained in this way. Under electrical pinch conditions the dependence of the resistance of the specimen on compression exceeds the ordinary piezoresistance by an order of magnitude.

Систематически исследованы, теоретически и экспериментально, явления, сопровождающие прохождение электрического тока через пластинки из собственного Ge, подвергнутого одноосному сжатию. Пластинки вырезались под углом к главным кристаллографическим осям с тем, чтобы направления электрического тока и поля не совпадали между собой. В этих условиях в образцах с большим временем жизни неравновесных носителей возникают поперечные градиенты их концентраций; это приводит к модуляции проводимости образцов, появлению выпрямляющих характеристик и т. д. В сильных полях высокие плотности носителей и токов концентрируются в узком слое у одной из поверхностей; в то же время из всей остальной части образца носители почти полностью удаляются (электрический пинч в электронно-дырочной плазме). Экспериментально исследованы распределение носителей по толщине образца и вольтамперные характеристики в условиях пинча. При полях около 100 в/см наблюдались возрастание и понижение концентраций примерно на порядок, а при различии скоростей поверхностной рекомбинации на поверхностях пластинки — резко нелинейные вольт-амперные характеристики с коэффициентом выпрямления около 50. Зависимость сопротивления образца от сжатия в условиях электрического пинча по порядку величины превышает обычное пьезосопротивление.

## 1. Introduction

It was shown [1] that the spatial distribution of electrons and holes in a macroscopically anisotropic bipolar semiconductor deviates sufficiently from the equilibrium distribution if an electrical current flows through the crystal. Really, the densities of electron and hole currents ($j_n$ and $j_p$) in an anisotropic crystal are directed under some angles to the total current $j = j_n + j_p$. Boundary conditions for $j_n$ and $j_p$ may be satisfied only if transverse gradients in the concentration of electrons and holes ($n$ and $p$) exist. The spatial distribution of current depends essentially on the crystal shape, on its orientation with respect to the crystallographic axes, and on the volume and surface recombination rate.

In high electric fields $E$ extensive depletion regions ($n, p \ll p_i$) and narrow accumulation regions ($n, p \gg p_i$) may arise. The total number of carriers in a specimen may vary considerably under these conditions but everywhere $n \approx p$ holds as before because of quasi-neutrality. The sharp inhomogeneity in the concentration distribution results in a redistribution of the current density across the specimen; under certain conditions the whole current practically may be concentrated in a narrow channel close to one of the surfaces.

This effect of carrier and current concentration in a narrow channel caused by conductivity anisotropy can be called electrical pinch in an electron–hole semiconductor plasma.

Under pinch conditions the specimen resistance considerably differs from the equilibrium one and the current-voltage characteristic must be nonlinear and is of rectifying type for the nonsymmetrical specimen. We have already reported briefly the observation of such characteristics in elastically deformed Ge [2].

In the present paper we derive the theory of the effect and analyse its results; an experimental procedure is also described, which allows the observation of phenomena connected with redistribution of carrier concentration throughout the specimen for various parameter values of the system and various forms of the current-voltage characteristics. Finally the theory is compared with the experimental data.

In view of the strong nonlinearity of the system the initial equations may contain significant inaccuracies which are primarily due to the dependence of the recombination rate on the injection level. Therefore, we did not compare quantitatively our theory with the experimental data, but paid principal attention to qualitative tendencies and the dependence of the physical process on the parameters of the system and also on approximate values of the investigated quantities. A satisfactory agreement has been obtained of the experimental data with the theoretical predictions.

Although the anisotropy of the intrinsic conductivity, $\sigma_{xy}/\sigma_{yy}$, has been only about 1% in our experiments (for compressive stress of $10^3$ kg/cm$^2$), a strongly pronounced electrical pinch occurred with a change of the concentrations by approximately one order of magnitude from their equilibrium values and with the overfall of the concentrations by two orders; the rectification ratio being some tens for fields of about 100 V/cm.

The occurrence of carrier concentration gradients is closely connected with the existence of circulation currents within anisotropic crystals; their influence on the thermal e.m.f. was investigated in [3] and on the photo e.m.f. in [4]. There is also a definite analogy between the phenomena considered here and the peculiarities of Hall-effect and magnetoresistance found in bipolar semiconductors [5]. It is worth noting that in some aspects our investigation, especially the consideration of the phenomenon in a strong electric field, is more complete than that reported in [5].

## 2. Theory

We consider a bipolar semiconductor with electron and hole conductivity tensors $\boldsymbol{\sigma}_n$ and $\boldsymbol{\sigma}_p$, which are anisotropic in the $xy$ plane, $Ox$ and $Oy$ are axes which do not coincide with the main axes of these tensors, $Oz$ has the same direction as one of the main axes. For a specimen of slab geometry with thickness $2\,d$ ($-d \leqq y \leqq d$) and an electrical current flowing in $Ox$ direction, a transverse electric field $E_y$ will arise because $\sigma_{xy} \neq 0$ (here $\boldsymbol{\sigma} = \boldsymbol{\sigma}_n + \boldsymbol{\sigma}_p$). This field is of such magnitude that $j_y$ (the density of the transverse current) will be zero. However, the currents $j_{ny}$ and $j_{py}$ will not vanish because $\boldsymbol{\sigma}_n$ and $\boldsymbol{\sigma}_p$ have different anisotropies, therefore electron and hole currents and consequently concentration gradients of $n$ and $p$ will arise in the $Oy$ direction.

It is evident that $n$, $p$, $E_y$, and $j_x$ depend only on $y$, besides $j_y = 0$ and $E_x = \text{const}$ (the latter follows from rot $E = 0$). From the diffusion equations for an intrinsic quasi-neutral semiconductor ($n \approx p$) and monomolecular recombination [1] follows

$$\frac{\partial p}{\partial t} = D\,\frac{\partial^2 p}{\partial y^2} + \frac{a\,e\,D}{2\,k\,T}\,E_x\,\frac{\partial p}{\partial y} - \frac{p - p_i}{\tau}\,, \tag{1}$$

where

$$D = \frac{2\,D_{nyy}\,D_{pyy}}{D_{nyy} + D_{pyy}}\,, \qquad a = \frac{u_{nxy}}{u_{nyy}} - \frac{u_{pxy}}{u_{pyy}}\,. \tag{2}$$

Formula (1) is a generalization of the bipolar diffusion equation. The field term appears here because of the anisotropy of the semiconductor and is formally equivalent to the occurrence of a homogeneous electric field $-\frac{a}{2}\,E_x$ directed along $E_y$. The condition $j_y = 0$ is reduced to

$$e\,(D_{nyy} - D_{pyy})\,\frac{\partial p}{\partial y} + \sigma_{yx}\,E_x + \sigma_{yy}\,E_y = 0\,. \tag{3}$$

Because of the dependence of $p(y)$ on $y$ the components $\sigma_{ij}$ depend also on $y$. Using (3) one can represent the formula for the hole current as

$$j_{py} = -\,e\,D\,\frac{\partial p}{\partial y} - \frac{a\,e^2\,D}{2\,k\,T}\,p\,E_x\,, \tag{4}$$

$$j_x = j_{nx} + j_{px} = e\,a\,D\,\frac{\partial p}{\partial y} + e\,u^*\,p\,E_x\,, \tag{5a}$$

$$J_x = \int_{-d}^{d} j_x(y)\,\mathrm{d}y = e\,a\,D\,[p(d) - p(-\,d)] + e\,u^*\,E_x \int_{-d}^{d} p(y)\,\mathrm{d}y\,. \tag{5b}$$

Here

$$u^* = \frac{1}{e\,p}\,\frac{\sigma_{xx}\,\sigma_{yy} - \sigma_{xy}^2}{\sigma_{yy}}\,. \tag{6}$$

The spatial distribution of carriers will be determined from equations (1) and (4). Since these equations differ from analogous ones for isotropic specimen only by the presence of terms proportional to $a$, this parameter will further be of particular interest. As it is seen from (3) it equals to the difference of the relative anisotropies of the electron and hole conductivity tensors.

Introducing the surface recombination velocities $s_1$ and $s_2$ for the surfaces $y = -\,d$ and $y = +\,d$, we can write down by means of (4) the usual boundary conditions for $p(y)$ and solve equation (1). Then one obtains for the stationary case

$$p(y) = p_i + \frac{p_i \gamma}{\xi(\gamma)} \left\{ \exp\left(\alpha_1 \frac{y}{L}\right) [(S_- + 2\alpha_1) \sinh(\alpha_2 \beta) - S_+ \cosh(\alpha_2 \beta)] - \right.$$
$$\left. - \exp\left(\alpha_2 \frac{y}{L}\right) [(S_- + 2\alpha_2) \sinh(\alpha_1 \beta) - S_+ \cosh(\alpha_1 \beta)] \right\}, \quad (7)$$

where

$$\alpha_{1,2} = -\gamma \pm \sqrt{1 + \gamma^2}, \quad \gamma = \frac{a}{4} \frac{e L}{k T} E_x, \quad \beta = \frac{d}{L}, \quad L = \sqrt{D \tau}, \quad (8)$$

$$S_{1,2} = \frac{s_{1,2} L}{D}, \qquad S_\pm = S_1 \pm S_2, \quad (9)$$

$$\xi(\gamma) = [1 + S_1 S_2 + \gamma S_-] \sinh(2\beta\sqrt{1+\gamma^2}) + \sqrt{1+\gamma^2}\, S_+ \cosh(2\beta\sqrt{1+\gamma^2}). \quad (10)$$

It was mentioned that $\frac{a}{2} E_x$ has (apart of the sign) the meaning of an effective field, measured in units $E_L = k\,T/e\,L$, and $L/\alpha_i$ $(i = 1, 2)$ are the increased and reduced diffusion lengths. As it is seen from (7) these two lengths determine the transverse distribution of carriers. Pinch is displayed most strikingly as soon as the reduced length is considerably less than the specimen dimensions, the greater part of carriers is then concentrated at a distance of several reduced diffusion lengths from the surface. From (5b) and (7) one obtains

$$J_x = 2\, d\, E_x\, \Sigma(\gamma)\,; \quad \Sigma(\gamma) = \sigma_i^* \left\{1 - \frac{1}{\beta\,\xi}\left(\gamma\,\zeta + \frac{a^2}{4}\frac{D}{D^*}\eta\right)\right\}. \quad (11)$$

Here

$$\eta(\gamma) = 2\sqrt{1+\gamma^2}\,[\cosh(2\beta\sqrt{1+\gamma^2}) - \cosh(2\beta\gamma)] + S_+ \sinh(2\beta\sqrt{1+\gamma^2})\,; \quad (12)$$

$$\zeta(\gamma) = S_- \sqrt{1+\gamma^2}\,[\cosh(2\beta\sqrt{1+\gamma^2}) - \cosh(2\beta\gamma)] +$$
$$+ S_+ [\gamma \sinh(2\beta\sqrt{1+\gamma^2}) - \sqrt{1+\gamma^2} \sinh(2\beta\gamma)]\,, \quad (13)$$

and $\sigma_i^* = e\,u^*\,p_i$, $D^* = \frac{k\,T}{e} u^*$. For $S_+ = S_- = 0$ formulae (7) and (11) coincide with the results of [1]. Here $\Sigma(\gamma)$ has the meaning of an effective specific electroconductivity and $\sigma_i^*$ is the equilibrium electroconductivity. They coincide only for $\beta \to \infty$ or $S_1, S_2 \to \infty$. In the other cases $\Sigma$ differs from $\sigma_i^*$ even for $E_x \to 0$. In this connection it is of interest to note that the second term on the right side of (1) which describes an anisotropy effect, is squared by $E_x$ and therefore the difference between $\Sigma(0)$ and $\sigma_i^*$ is completely due to a change in the boundary condition arising from the last term in (4), linear in $E_x$.

Formulae (7) and (11) are very complicate and therefore we investigate in the next section their asymptotic form and show the results of numerical calculations for some cases.

### 3. Analysis of the Formulae

From (5a, b), (7), and (11) we find

$$\left.\begin{aligned} p(y, \gamma, S_1, S_2) &= p(-y, -\gamma, S_2, S_1)\,, \\ j_x(y, \gamma, S_1, S_2) &= -j_x(-y, -\gamma, S_2, S_1)\,, \\ J_x(\gamma, S_1, S_2) &= -J_x(-\gamma, S_2, S_1)\,. \end{aligned}\right\} \quad (14)$$

For $S_1 = S_2$ the spatial distributions of $p(y)$ and $|j_x(y)|$ have mirror symmetry (relative to $y = 0$) for opposite signs of the field. This symmetry is destroyed for $S_1 \neq S_2$, and this results in an asymmetric current-voltage characteristic. From (7) follows that the carrier concentration is a monotonous function of $y$. With increasing $y$ the function $p(y)$ decreases for positive $\gamma$ and increases for negative $\gamma$.

For the most interesting case of strong fields, when the quantities $1/\gamma$ and $e^{-|\gamma|}$ are sufficiently small, all formulae may be simplified.

With $S_1, S_2 \neq 0, \infty$ the condition of a strong field can be written as

$$|\gamma| \operatorname{Max} \{\beta, \beta^{-1}; S_{1,2}, S_{1,2}^{-1}\} \gg 1 . \tag{15}$$

If $\gamma > 0$ the current carriers are concentrated close to the surface $y = -d$ at a distance of about one reduced diffusion length $L/2\gamma$.

$$\left.\begin{aligned} p(y, \gamma) &\approx p_i \frac{2\beta + S_+}{S_1} \exp\left(-2\gamma \frac{y+d}{L}\right), \\ j_x &\approx \frac{4 e \gamma p_i}{a L} (D_n^* + D_p^*) \frac{2\beta + S_+}{S_1} \exp\left(-2\gamma \frac{y+d}{L}\right). \end{aligned}\right\} \tag{16}$$

Here

$$D_q^* = \frac{k T}{e} \left(u_{q x x} - \frac{u_{q x y}^2}{u_{q y y}}\right); \qquad q = n, p .$$

The accumulation region ($p \geqq p_i$) is very narrow, its limits are $-d \leqq y \leqq y_1$ where $y_1 \approx -d + \frac{L}{2\gamma} \ln\left(\frac{S_+ + 2\beta}{S_1}\right)$. It is seen from (16) that $p$ and $j_x$ are inversely proportional to $S_1$ on the boundary $y = -d$ when $S_1$ is small; $p(-d, \gamma)$ saturates at large $\gamma$. The occurrence of the length $L/2\gamma$ in the exponent points out to considerable gradients of carrier concentration and current density.

The depletion region ($p \leqq p_i$) covers the most part of the crystal ($y_1 \leqq y \leqq d$), the dependence of $p$ and $j_x$ on the coordinate in this region is given in the main by the increased diffusion length $2\gamma L$. In this region we have with exception of the closest neighbourhood to the point $y_1$

$$p(y, \gamma) \approx p_i \frac{S_2 + \dfrac{d-y}{L}}{2\gamma}, \qquad j_x(y, \gamma) \approx \frac{2 e D^* p_i}{a L}\left[S_2 + \frac{d-y}{L}\right], \tag{17}$$

i.e. the carrier concentration is inversely proportional to the field, but the current density is field independent. The boundary values of $p$ and $j_x$ for $y = d$ are proportional to $S_2$.

For high fields the total current tends to the field-independent saturation value

$$J_x \approx \frac{4 e p_i \beta D^*}{S_1 a}\left[S_1 (S_2 + \beta) + \frac{S_+ + 2\beta}{2\beta D^*} (D_n^* + D_p^*)\right]. \tag{18}$$

The saturation is extremely distinctive for the electrical pinch because the total number of carriers in the specimen decreases then as $1/\gamma$. If the last term dominates in (18) almost the whole current concentrates in a channel below the surface. The case $\gamma < 0$ can be treated by means of (14).

Special cases arise when the surface recombination velocities are zero or infinite (actually $\ll |\gamma|$ or $\gg |\gamma|$). Consideration of these cases helps to understand the influence of surface recombination velocities on carrier distribution under pinch condition. The high-field condition now has the form

$$|\gamma| \operatorname{Max}(\beta, \beta^{-1}) \gg 1 .$$

For $S_1 = S_2 = 0$ and $\gamma > 0$ we find for the accumulation region $\left(-d \leqq y \leqq \leqq -d + \frac{L}{2\gamma} \ln(4\beta\gamma)\right)$

$$\left.\begin{aligned} p(y,\gamma) &\approx 4\beta\gamma\, p_i \exp\left(-2\gamma\frac{d+y}{L}\right), \\ j_x(y,\gamma) &\approx 16\frac{e p_i \beta}{a L}(D_n^* + D_p^*)\,\gamma^2 \exp\left(-2\gamma\frac{d+y}{L}\right), \end{aligned}\right\} \tag{19}$$

i.e. the carrier concentration on the boundary $y = -d$ increases linear with field and the current density depends quadratically on field. In a depletion region $\left(-d + \frac{L}{2\gamma}\ln(4\beta\gamma) < y \leqq d\right)$

$$p(y,\gamma) \approx p_i \frac{d-y}{2\gamma L} + \frac{p_i}{4\gamma^2}, \qquad j_x(y,\gamma) \approx \frac{2eD^*}{aL} p_i \frac{d-y}{L} + \frac{e D^* p_i}{a L \gamma}, \tag{20}$$

i.e. on the boundary $y = d$ the carrier concentration is proportional to $E_x^{-2}$, and the current density is proportional to $E_x^{-1}$; in the most part of the specimen the current density practically does not depend on field.

For $S_2 = \infty$ and $\gamma > 0$ the whole crystal volume is enriched with current carriers, and we have

$$p(y,\gamma) \approx p_i \left\{1 + \frac{2\gamma}{S_1 + \frac{1}{2\gamma}} \exp\left(-2\gamma\frac{d+y}{L}\right)\right\}, \tag{21}$$

$$\left.\begin{aligned} j_x(y,\gamma) &\approx \frac{4D^* e\gamma p_i}{aL}\left\{1 + \frac{D_n^* + D_p^*}{D^*}\,\frac{2\gamma}{S_1 + \frac{1}{2\gamma}} \exp\left(-2\gamma\frac{d+y}{L}\right)\right\}, \\ J_x &\approx \frac{8\gamma e p_i}{a}\left\{\beta D^* + (D_n^* + D_p^*)\frac{\gamma}{2\gamma S_1 + 1}\right\}. \end{aligned}\right\} \tag{22}$$

The carrier concentration on the boundary $y = -d$ and the total current increase quadratically with field if $2 S_1 \gamma \ll 1$ and linearly if $2 S_1 \gamma \gg 1$. For $\gamma < 0$ the crystal volume is greatly depleted with current carriers. At distances of about one reduced diffusion length one obtains close to $y = +d$

$$\left.\begin{aligned} p(y,\gamma) &\approx p_i \frac{S_1 + 2|\gamma| \exp\left(2|\gamma|\frac{y-d}{L}\right)}{S_1 + 2|\gamma|}, \\ j_x(y,\gamma) &\approx \frac{-4 e p_i |\gamma|}{aL}\,\frac{D^* S_1 + 2|\gamma|(D_n^* + D_p^*)\exp\left(2|\gamma|\frac{y-d}{L}\right)}{S_1 + 2|\gamma|}. \end{aligned}\right\} \tag{23}$$

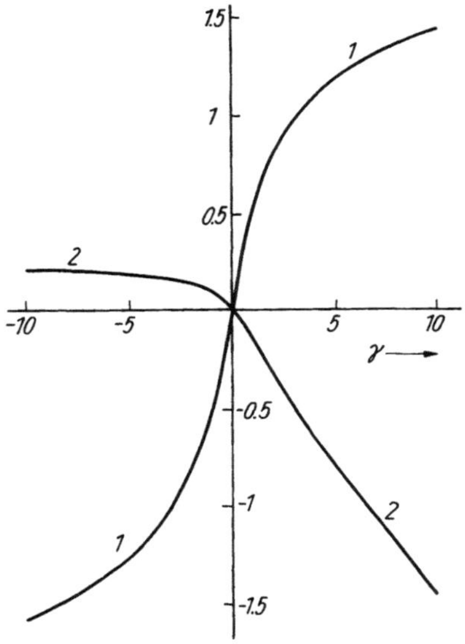


Fig. 1. Boundary concentrations vs. dimensionless field $\gamma$ for $\beta = 1$, $S_1 = 0.5$, $S_2 = 50$. 1 — lg $[p(-d, \gamma)/p_i]$; 2 — 10 lg $[p(+d, \gamma)/p_i]$

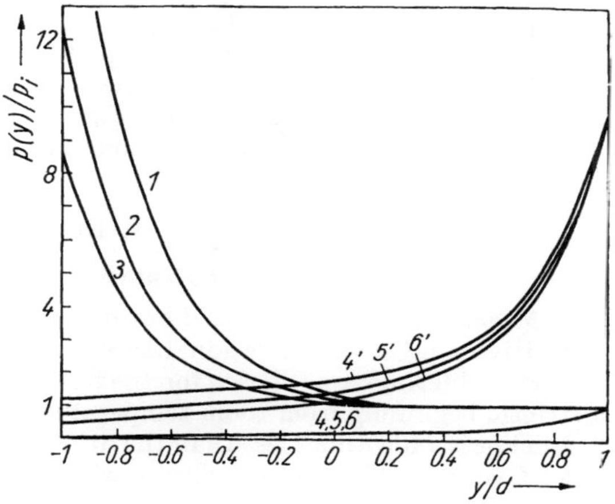


Fig. 2

Fig. 2. Transverse distribution of carriers for $\beta = 0.45$, $|\gamma| = 4.2$, $S_2 = \infty$
Accumulation regime ($\gamma > 0$): 1 — $S_1 = 0.3$; 2 — $S_1 = 0.6$; 3 — $S_1 = 1.0$.
Depletion regime ($\gamma < 0$): 4 — $S_1 = 0$; 5 — $S_1 = 0.6$; 6 — $S_1 = 1.0$.
In the last three dependences (curves (4′), (5′), and (6′)) the ordinate is multiplied by ten

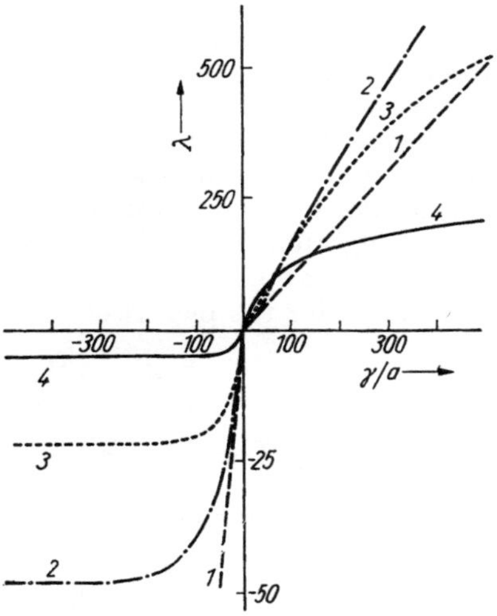


Fig. 3. Dimensionless current $\lambda = \dfrac{\gamma}{a} \dfrac{\Sigma(\gamma)}{\sigma_i^*}$ vs. dimensionless field $\dfrac{\gamma}{a}$ ($S_1 = 0.5$, $S_2 = 50$, $\beta = 1$). 1 — $a = 0$; 2 — $a = 0.0195$; 3 — $a = 0.0455$; 4 — $a = 0.195$

In the remaining part of the crystal

$$\left.\begin{aligned} p(y, \gamma) &\approx p_i \frac{S_1 + \dfrac{d + y}{L} + \dfrac{1}{2\,|\gamma|}}{S_1 + 2\,|\gamma|}, \\ j_x(y, \gamma) &\approx -4\,\frac{e\,D^*\,p_i\,|\gamma|}{a\,L}\,\frac{S_1 + \dfrac{d + y}{L} + \dfrac{1}{2\,|\gamma|}}{S_1 + 2\,|\gamma|}. \end{aligned}\right\} \tag{24}$$

The total current

$$J_x(\gamma) \approx -\frac{4\,e\,D^*\,p_i\,\beta}{a}\,\frac{2\,|\gamma|}{S_1 + 2\,|\gamma|}\left[\frac{D_n^* + D_p^*}{2\,\beta\,D^*} + S_1 + \beta\right] \tag{25}$$

is proportional to the field if $|\gamma| \ll S_1$, and reaches saturation if $|\gamma| \gg S_1$. For $S_1 = S_2 = \infty$ $p(y, \gamma)$ is always equal to $p_i$.

Fig. 1 shows the boundary carrier concentrations as a function of the dimensionless field (formula (7)), and Fig. 2 and 3 show the spatial distribution of carriers and the current-voltage characteristics, respectively, which are given by formula (11).

## 4. Experimental

### *4.1 Orientation of the specimens*

Measurements were carried out on Ge specimens of thin slab geometry. High-resitivity p-Ge with a residual acceptor concentration of $\approx 10^{12}$ cm$^{-3}$ and high carrier lifetime was chosen for the following reasons: i) The specimens were intrinsic beginning from room temperature; ii) it was easy to prepare plates with thickness a few times less than the diffusion length; iii) it was easy to change the surface recombination velocity in wide limits by means of mechanical treatment or etching. Anisotropy of conductivity was created by uniaxial compression in the press and was relatively small in comparison with that of naturally anisotropic materials. Nevertheless, this method possessed some advantages for the observation of pinch development with increasing deformation, and for a check of absence of analogous phenomena in isotropic materials without deformation, the latter was important for eliminating of boundary effects due to contacts and the influence of anisotropy connected with heating of electrons. Since the value of the parameter $a$ determines the deviation of carrier concentration from equilibrium one, it is of importance to calculate the orientation of a slab, with respect to the crystallographic axes $X$, $Y$, $Z$, which corresponds to the maximum of $|a|$.

In the linear approximation for the components of the tension tensor $X_{\alpha\beta}$ the usual formula for the piezoresistance may be written as

$$\Delta_{\alpha\beta} \equiv \frac{\delta\varrho_{\alpha\beta}}{\varrho_i} = \Pi_{\alpha\beta,\alpha'\beta'}\, X_{\alpha'\beta'} \,. \tag{26}$$

It follows from $\dfrac{\delta\varrho_{\alpha\beta}}{\varrho_i} = -\dfrac{\delta\sigma_{\alpha\beta}}{\sigma_i}$ and from comparison of (26) with (2) that for uniaxial compression along the axis $Ox$, when $X_{\alpha\beta} = -X\,\delta_{\alpha\beta}\,\delta_{\alpha x}$, the coefficient $a$ is given by

$$a = \Delta_{xy}^{(p)} - \Delta_{xy}^{(n)} = -\left(\Pi_{xy,xx}^{(p)} - \Pi_{xy,xx}^{(n)}\right) X \,. \tag{27}$$

The upper indices p and n refer to monopolar hole and electron materials, respectively. The components of the piezoresistance tensor may be expressed by their values in the main crystallographic axes, namely

$$\left.\begin{aligned} \Pi_{xy,xx} &= (\Pi_{11} - \Pi_{12} - \Pi_{44})\,(l_1^3\, l_2 + m_1^3\, m_2 + n_1^3\, n_2)\,; \\ \Pi_{11} &= \Pi_{XX,XX},\, \Pi_{12} = \Pi_{XX,YY},\, \Pi_{44} = 2\,\Pi_{XY,XY}\,. \end{aligned}\right\} \tag{28}$$

Here $(l_1, m_1, n_1)$ and $(l_2, m_2, n_2)$ are the direction cosines of the axes $Ox$ and $Oy$. The extreme value of the trigonometric polynomial in the second bracket of the right part of (28) is equal to $\pm$ 0.281 [6]. The positive value corresponds to the following orientation

$$\begin{pmatrix} l_1 & m_1 & n_1 \\ l_2 & m_2 & n_2 \\ l_3 & m_3 & n_3 \end{pmatrix} = \begin{pmatrix} 0.305 & 0.305 & 0.902 \\ -0.638 & -0.638 & 0.432 \\ 0.707 & -0.707 & 0 \end{pmatrix} \tag{29}$$

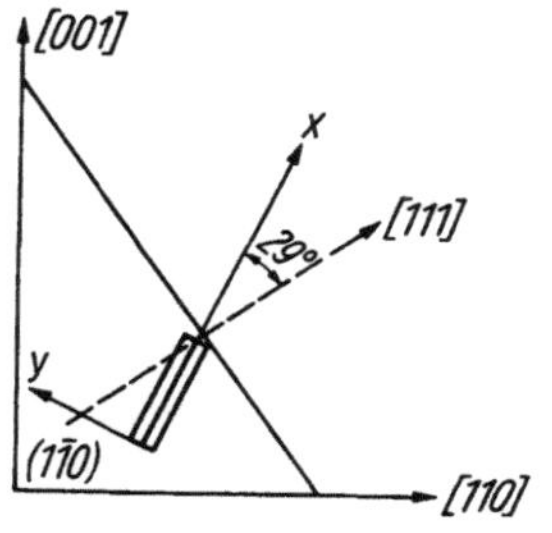


Fig. 4. Orientation of the specimen with respect to the crystallographic axes

as shown in Fig. 4. The axis $Oz$, directed along $[1\bar{1}0]$ is one of main axes of the tensors $\boldsymbol{\sigma}_n$ and $\boldsymbol{\sigma}_p$. All measurements described below were carried out with the mentioned orientation of the Ge plates.

The coefficients $\Pi_{\alpha\beta}$ are reported in [7] for 300 °K. Their combination $\Pi_{11} - \Pi_{12} - \Pi_{44}$ is equal to $+1.36 \times 10^{-4}$ cm²/kg for electrons and $-1.02 \times 10^{-4}$ cm²/kg for holes. Therefore in agreement with (27)

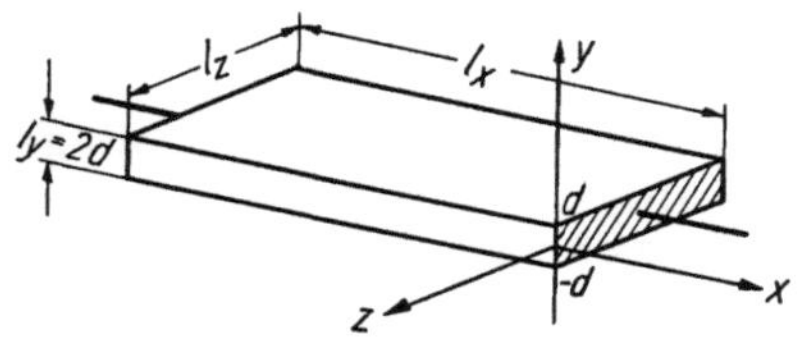


$$a = 0.67 \times 10^{-4} \frac{\text{cm}^2}{\text{kg}} \cdot X\,. \tag{30}$$

If the temperature rises $a$ decreases approximately like $1/T$. Similarly we may write

$$\frac{\delta\sigma_{xy}}{\sigma_i} = 0.15 \times 10^{-4} \frac{\text{cm}^2}{\text{kg}} \cdot X\,. \tag{31}$$

The numerical coefficient in (30) is greater than in (31) because the contributions of electrons and holes sum up in the former case and partly cancel out in the latter case.

### *4.2 Experimental procedure*

Our measurements covered the current–voltage characteristics of deformed specimens, the transverse distribution of concentration as a function of a compression $X$ and electric field $E_x$, and kinetic phenomena of the current.

The block-diagram of the experimental arrangement is shown in Fig. 5. Voltage pulses (amplitude up to 200 V, duration $\Delta t = 0.2$ to 2 ms, and repetition frequency 8.6 Hz) were applied to the electrodes (on the $y\,z$ planes) of the specimen which had undergone an uniaxial compression with stress up to $2 \times 10^3$ kg/cm² along the $Ox$ direction (see Fig. 4). Pulse fields were employed in order to avoid a heating of the specimen. The spatial distribution of carriers in the crystal was investigated by means of the usual infrared spot method. A globar was employed as infrared source, its image was focused on a slit (0.15 × 0.4 mm). The specimen with polished $x\,y$ sides was shifted relative to the slit at a rate of 0.1 mm per min. The radiation passing through the specimen in $Oz$ direction was modulated owing to a modulation of carrier concentration by the pulse field. Bolometer and narrow-band amplifier (band width equal to 0.1 Hz) registered the first harmonic of the signal $\Delta I$, corresponding to the resonance frequency of the amplifier (8.6 Hz). The distribution $\Delta I(y)$ was normalized to a distribution $I_0(y)$ measured at $E_x = 0$ when the incident infrared radiation was modulated by rectangular pulses of the same length as in the $\Delta I$ measurements. The experimental value $\Delta I/I_0$ is proportional to the mean concentration of nonequilibrium carriers at the place of the spot. The arrangement also permitted to measure the phase of the signal and thus to determine whether depletion or accumulation of

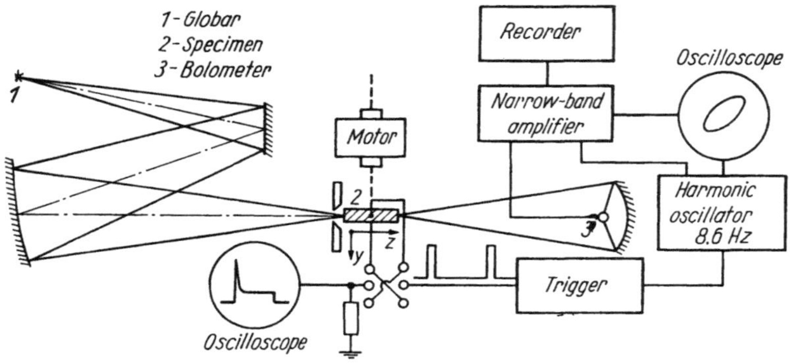


Fig. 5. Block-diagram of the experimental arrangement

carriers took place. The current amplitude of the steady-state current–voltage characteristics was measured with an oscilloscope after the current had become stationary. It was possible to determine the rise time of current by photographing the oscillogram.

The specimen temperature was established in the range 15 to 70 °C by means of a special thermostat.

## 5. Experimental Results and Discussion

### *5.1 Transverse distribution of nonequilibrium carriers*

Fig. 6 shows the transverse distributions of carrier concentrations in deformed specimens for different values of $E_x$, $X$, and $s_{1,2}$. Two kinds of surface treatments were employed: etching in perhydrole ($s \approx 10^2$ cm/s) and rough grinding ($s \approx 10^4$ cm/s). According to (7) the distribution $p(y)$ depends only on the product $X\,E_x \sim \gamma$, and this gives the possibility to rule over the distribution by field $E_x$ as well as by stress $X$. Such conclusion is confirmed experimentally as seen from the similarity of Fig. 6a and b.

It is of interest to compare the curves of Fig. 6a with the theoretical ones of Fig. 2, which are shown for close value of $\beta$; $\gamma = 5.4$ corresponding to a field 30 V/cm and $X = 1500$ kg/cm². It is seen that the general shape of carrier redistribution throughout the specimen is similar in both figures. From the curves in Fig. 6b follows that at 30 V/cm the predominant part of carriers is already drown out from the major part of the specimen. An analysis of the curves shows that the concentration decreases at least by one order of magnitude. In the accumulation range the concentration increases almost by one order, but its spatial distribution is more smooth than the theoretical one (see Fig. 2).

The especially pronounced phenomenon of pinch arises in the symmetrical case when both surfaces posses small $s$-values (see Fig. 6c). One may see the appearance of the narrow accumulation region and a wide almost depleted region; the concentration varies by two orders across the specimen.

Therefore the fundamental conclusions of the theory are experimentally confirmed. The phenomenon of pinch turned out to be pronounced, although the anisotropy of $\sigma_{ij}$ was only about 1% for $X \approx 10^3$ kg/cm², as seen from (31). This is because the concentration distribution $p(y)$ (formulae (7) and (8)) depends only on the combination $\gamma \sim a\,L\,E_x$, and the smallness of $a$ is compensated by a large $L$.

However, there are some deviations from the theory. The curves 4 and 5 in Fig. 6c which correspond to the largest fields, differ from others by the occur-

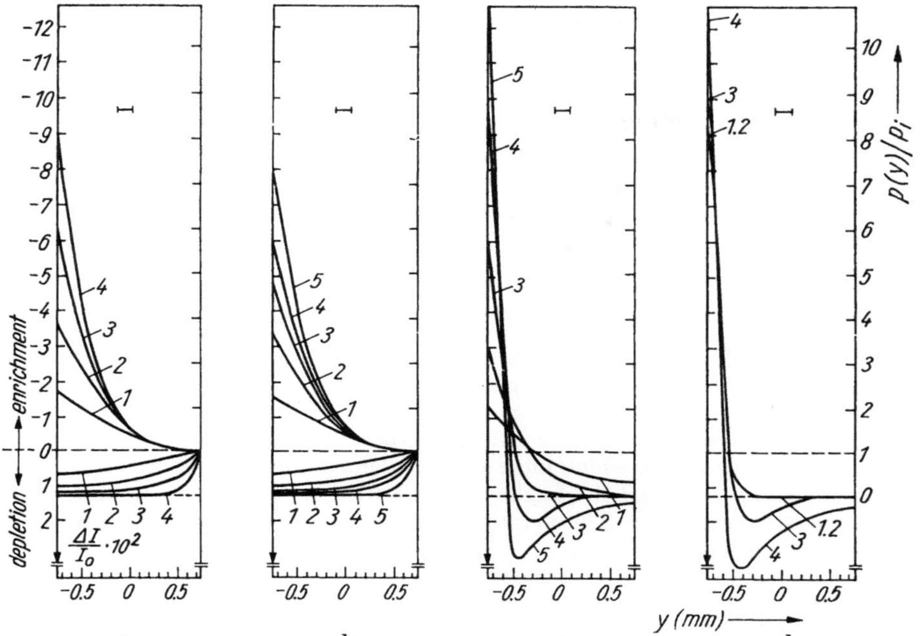


Fig. 6. Change in transmission (left scale) and transverse distribution of carrier concentration (right scale). $l_x = 15$ mm, $l_y = 1.5$ mm, $l_z = 8$ mm; $L \approx 0.2$ cm; $\beta \approx 0.4$; $t_0 = 50$ °C

a) $s_1 \approx 10^2$ cm/s, $s_2 \approx 10^4$ cm/s, $E_x = \pm 40$ V/cm, $\Delta t = 1{,}6$ ms. 1 — $X = 200$ kg/cm²; 2 — $X = 400$ kg/cm²; 3 — $X = 800$ kg/cm²; 4 — $X = 1200$ kg/cm².

b) $s_1 \approx 10^2$ cm/s, $s_2 \approx 10^4$ cm/s, $X = 1500$ kg/cm², $\Delta t = 1{,}6$ ms. 1 — $E_x = \pm 5$ V/cm; 2 — $E_x = \pm 10$ V/cm; 3 — $E_x = \pm 15$ V/cm; 4 — $E_x = \pm 20$ V/cm; 5 — $E_x = \pm 30$ V/cm.

c) $s_1 \approx s_2 \approx 10^2$ cm/s, $X = 1500$ kg/cm², $\Delta t = 1{,}6$ ms. 1 — $E_x = 20$ V/cm; 2 — $E_x = 40$ V/cm; 3 — $E_x = 60$ V/cm; 4 — $E_x = 85$ V/cm; 5 — $E_x = 100$ V/cm.

d) $s_1 \approx s_2 \approx 10^2$ cm/s, $X = 1500$ kg/cm², $E_x = 100$ V/cm. 1 — $\Delta t = 0.4$ ms; 2 — $\Delta t = 0.8$ ms; 3 — $\Delta t = 1.2$ ms; 4 — $\Delta t = 1.8$ ms.

The horizontal segment in the upper part of the figures corresponds to the width of the infra-red spot

rence of distinctive minima within the specimen. It is of interst to consider the distribution of the $\Delta I$ phase over the specimen. Before the occurrence of the minima ($E_x < 80$ V/cm) the phase is nearly constant in the accumulation and depletion regions and abruptly changes almost by $\pi$ close to $\Delta I = 0$. In the region of the minima the phase changes by more than $\pi/4$ over a wide length which indicates the occurrence of inertial processes. Fig 6d shows that these processes are caused by inhomogeneous heating of the specimen under pinch condition and by thermoconductivity; the minima vanish if the heating is diminished by reducing $\Delta t$. It shall be noted here that for $\Delta t = 1.6$ ms and $E_x = 100$ V/cm we observed a stationary transverse temperature difference of about 10 degrees.

In our specimens $\tau$ decreased with increasing $p$ and therefore the spatial distribution of concentration became more smooth. The experimental inaccuracy, caused by the width of the light spot, became considerable in sections of steepest $p(y)$ change.

Processes connected with the finite length of the specimen are to be mentioned. The transverse distribution of concentration sets up under the influence of a transverse diffusion and a transverse drift (with an effective field $\frac{a}{2} E_x$ — see

(1)). On the contrary, the carrier injection and extraction, which disturb this distribution, are caused by the longitudinal drift due to the effective fields $\left(\frac{a}{2}\right)^2 E_x$ and $\frac{n-p}{n+p} E_x$, which is connected with the fact that the specimen is anisotropic and somewhat extrinsic, respectively. As was shown experimentally, the longitudinal drift in the thin specimens disturbed only the concentration distribution at distances $\approx 2\,d$ from the contacts, the concentration distribution in the remaining part of the sample did not depend on $x$.

Finally, the carrier distribution turns out to be sensitive to the conductivity inhomogeneity of the specimen.

### *5.2 Current-voltage characteristics*

Fig. 7 shows the experimental current–voltage characteristics for symmetric and nonsymmetric surface treatments; it is convenient to compare these characteristics with the theoretical ones shown in Fig. 3. As it is seen from Fig. 7 the main theoretical conclusions are confirmed experimentally: the strong nonlinearity of the characteristics which increase with deformation, the occurrence of rectification for $s_1 \neq s_2$, and the tendency to saturation which is most pronounced for the reverse branches. The maximum rectification ratio observed for a field $E_x = 100$ V/cm was about 50.

The lack of full saturation of the reverse branches (depletion regime) is partly caused by heating — the declination decreased by some tens per cent when $\Delta t$ decreased up to 0.2 ms. The residual declination is accounted for by a small hole concentration ($\approx 10^{12}$ cm$^{-3}$) in the depletion region of the specimen, due to the presence of some acceptors. The direct branches of the current-voltage characteristics (accumulation range) differ from the theoretical ones more significantly because of intensive heating, which cannot be diminished by decreasing $\Delta t$ because a large time is required for setting up of the diffusion-drift equilibrium.

Fig. 8 represents the theoretical and experimental current-tension characteristics for comparable parameter values: It is seen that they are in satisfactory agreement with each other. It is worth while to emphasize that under pinch condition $dJ_x/dX$ is considerably larger than for the ordinary piezoeffect, particularly for $X \approx 10^3$ kg/cm$^2$. These quantities differ by two orders of magnitude.

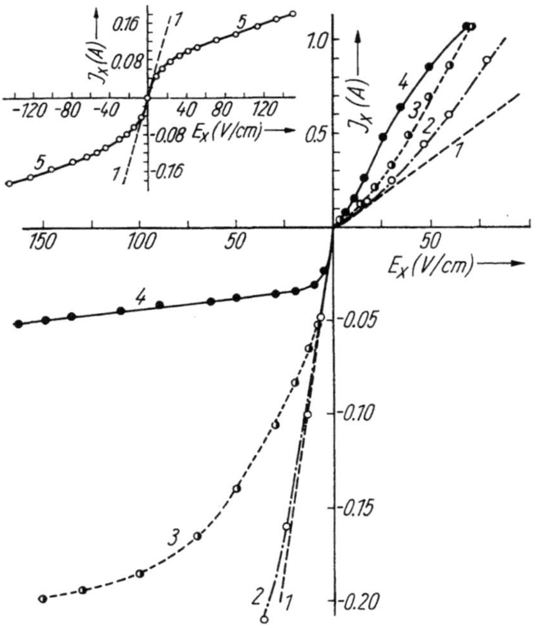


Fig. 7. Current-voltage characteristics for the same specimen as in Fig. 6.
1 — Nondeformed specimen ($X = 0$).
Rectifying characteristics:
2 — $X = 100$ kg/cm$^2$; 3 — $X = 300$ kg/cm$^2$; 4 — $X = 1500$ kg/cm$^2$; 5 — Nonrectifying characteristic of the specimen for $s_1 \approx s_2 \approx 10^2$ cm/s, $X = 1500$ kg/cm$^2$

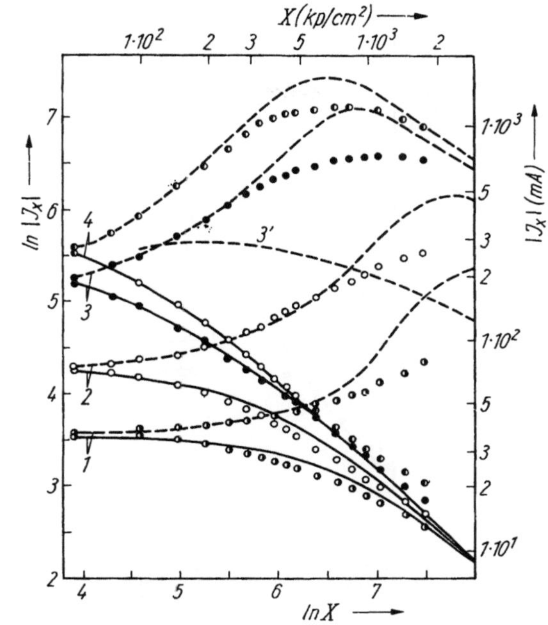


Fig. 8. Current-tension characteristics.

Points — experimental values for a specimen with $l_x = 1.5$ cm, $l_y = 0.075$ cm, $l_z = 0.4$ cm; $L \approx 0.2$ cm ($\beta \approx 0.2$); $t_0 = 60$ °C; $s_1 \approx 10^2$ cm/s, $s_2 \approx 10^4$ cm/s. The curves were calculated according to (11) and (30) for $\beta = 0.1$, $S_1 = 0.5$, $S_2 = 50$. Curve (3') was calculated for $\beta = 1$ and $E_x = 100$ V/cm. Solid curves — direct branches of the characteristics. Dashed curves — reverse branches of the characteristics.

1 — $E_x = \pm$ 20 V/cm;
2 — $E_x = \pm$ 40 V/cm;
3 — $E_x = \pm$ 100 V/cm;
4 — $E_x = \pm$ 140 V/cm

### *5.3 Kinetics of the current*

Because of strong nonlinearity the kinetics of current rise under pinch conditions is very complicate. For a plate with $s_1 = 0$, $s_2 = \infty$ a solution of equation (1) with the initial condition $p(y, t = 0) = p_i$ was obtained, which describes the setting up of the concentration after the field has been applied. Here we represent the main results of the investigation of this solution. In weak fields ($|\gamma| \ll 1$) the transverse diffusion dominates and the current rise follows a law which is nearly exponential. The effective time is

$$\tau_{\text{eff}} = \frac{\tau}{1 + \left(\frac{\pi}{4\beta}\right)^2}. \tag{32}$$

For $\beta \gg 1$ it coincides with $\tau$ and for $\beta \ll 1$ with the ordinary lifetime of carriers in a thin plate.

In strong fields ($|\gamma|\,\beta$, $|\gamma|\,\beta^{-1} \gg 1$) the transverse drift dominates and for the depletion range the rise time is equal to the time required to draw the carriers across the specimen to the side with $s_2 = \infty$

$$\tau_{\text{eff}} = \frac{\tau\beta}{|\gamma|} = \frac{2d}{\left|\frac{a}{2}\frac{e\,D}{k\,T}E_x\right|}. \tag{33}$$

The current rise follows now a linear law. In the accumulation range $\tau_{\text{eff}} \to \tau$; though transverse drift dominates here $\tau_{\text{eff}}(E_x)$ differs from the previous case, it increases because the increasing number of nonequilibrium carriers in the plate must be accumulated.

Fig. 9 represents the experimental data. Agreement of theory and experiment is seen in the depletion regime for the pulse shape and the dependence of $\tau_{\text{eff}}$ on $E_x$; the value of the current at $t = 0$ on the oscillogram corresponds to the initial resistance of the sample, the decrease of current to the stationary value is caused by carrier extraction. For the accumulation regime $\tau_{\text{eff}}$ at first increases with $E_x$ in agreement with theory, and then begins to decrease. The latter fact is caused

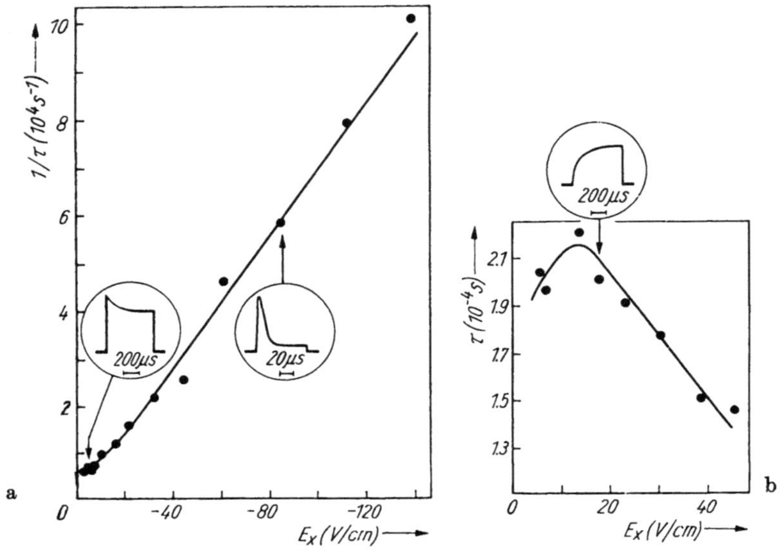


Fig. 9. Rise time of current vs. electric field $E_x$ for the same specimen as in Fig. 6, but for $t_0 = 20$ °C. a) Depletion regime; b) accumulation regime. Oscillograms of the current rise are shown by circles for the values of the electric field indicated by arrows

by (a) the slowing down of the increase rate of the concentration with rise of $\gamma$, which is connected with $s_1 \neq 0$ and begins from $\gamma \sim \frac{1}{2 s_1}$ (see (21)), and (b) by the decrease of the volume lifetime with rise of the injection level.

### *Acknowledgements*

The authors are indebted to Prof. V. E. Lashkarev for his continuous interest, L. V. Babkov and his collaborators for supplying us with Ge crystals, and Yu. A. Bondarenko for considerable technical assistance.

### References

[1] E. I. Rashba, Fiz. tverd. Tela **6**, 3247 (1964).
[2] I. I. Boiko, I. P. Zhadko, E. I. Rashba, and V. A. Romanov, Fiz. tverd. Tela **7**, 2239 (1965).
[3] A. G. Samoilovich and L. L. Korenblit, Fiz. tverd. Tela **3**, 2054 (1961).
[4] W. van Rosbroeck and W. G. Pfann, J. appl. Phys. **33**, 2304 (1962).
I. K. Kikoin and S. D. Lazarev, Zh. eksper. teor. Fiz. **47**, 780 (1964).
I. P. Zhadko, E. I. Rashba, V. A. Romanov, I. M. Stakhira, and K. D. Tovstyuk, Fiz. tverd. Tela **7**, 1777 (1965).
[5] H. Welker, Z. Naturf. **6a**, 184 (1951).
A. I. Anselm, Zh. tekh. Fiz. **22**, 1146 (1952).
L. Landauer and J. Swanson, Phys. Rev. **91**, 555 (1953).
G. E. Pikus, Zh. tekh. Fiz. **26**, 22 (1956); **26**, 36 (1956).
V. P. Zhuze, G. E. Pikus, and O. V. Sorokin, Zh. tekh. Fiz. **27**, 1167 (1957).
M. Bernard, J. Phys. Chem. Solids **8**, 332 (1959).
[6] W. G. Pfann and R. N. Thurston, J. appl. Phys. **32**, 2008 (1961).
[7] C. S. Smith, Phys. Rev. **94**, 42 (1954).

*(Received January 7, 1966)*

phys. stat. sol. **16**, 57 (1966)

*Institute of Physics, Polish Academy of Sciences, Warsaw*

# Distribution Function of Hot Carriers

By

E. KAMIENIECKI

From the general, time-dependent, Boltzmann equation for a non-degenerate electron (or hole) gas, the integral form of the distribution function of hot carriers is obtained. Assuming that the scattering of carriers is due mainly to optical modes, the distribution function in the "zero-order approximation" is derived and numerical calculations performed. The distribution function of hot electrons is found to be strongly anisotropic with a plateau and small local minimum in the direction opposite to that of the electric field.

Aus der allgemeinen zeitabhängigen Boltzmanngleichung für ein nichtentartetes Elektronen- oder Löchergas wird die Integralform der Verteilungsfunktion heißer Ladungsträger erhalten. Unter der Annahme, daß die Streuung der Ladungsträger hauptsächlich am optischen Zweig erfolgt, werden die Verteilungsfunktion in der ,,nullten Näherung" abgeleitet und numerische Berechnungen durchgeführt. Die Verteilungsfunktion heißer Elektronen ist stark anisotrop mit einem Plateau und einem schwachen örtlichen Minimum in entgegengesetzter Richtung zum elektrischen Feld.

To find the distribution function for hot electrons (holes) we start from the general, time-dependent Boltzmann equation

$$\left[\frac{\partial}{\partial t} + \boldsymbol{F}\,\nabla_{\varrho}\right] f(\varrho, t) = $$
$$= \int [P(\varrho', \varrho)\, f(\varrho', t)\,(1 - f(\varrho, t)) - P(\varrho, \varrho')\, f(\varrho, t)\,(1 - f(\varrho', t))]\, \mathrm{d}^3\varrho' \,, \tag{1}$$

where $f(\varrho, t)$ is the probability of finding an electron at the phase space point $\varrho(\boldsymbol{k}, \boldsymbol{r})$ (where $\boldsymbol{k}$ is the wave vector of a carrier and $\boldsymbol{r}$ is its position) at time $t$; $P(\varrho, \varrho')$ is the probability per unit time that a carrier makes a transition from point $\varrho = (\boldsymbol{k}, \boldsymbol{r})$ into point $\varrho' = (\boldsymbol{k}', \boldsymbol{r}')$. According to the dynamical law of motion we have for an electron [1]

$$\boldsymbol{F} = \dot{\varrho} = (\dot{\boldsymbol{k}}, \dot{\boldsymbol{r}}) = \left(-\frac{e}{\hbar}\,[\vec{\mathscr{E}}(\boldsymbol{r}, t) + \boldsymbol{v}(\boldsymbol{k}) \times \vec{\mathscr{B}}(\boldsymbol{r}, t)]\,,\ \boldsymbol{v}(\boldsymbol{k})\right), \tag{2}$$

where $\vec{\mathscr{E}}$ and $\vec{\mathscr{B}}$ are the electric and the magnetic fields, respectively. $\boldsymbol{v}$ is the velocity of the electron. In a system of coordinates moving along the trajectory given by the equation

$$\mathrm{d}\varrho = \boldsymbol{F}\,\mathrm{d}t$$

equation (1) has the form

$$\frac{\mathrm{d}}{\mathrm{d}t} f(\varrho, t) =$$
$$= \int [P(\varrho', \varrho)\, f(\varrho', t)\,(1 - f(\varrho, t)) - P(\varrho, \varrho')\, f(\varrho, t)\,(1 - f(\varrho', t))]\, \mathrm{d}^3\varrho' \,, \tag{3}$$

where the distribution function $f$ changes only via collisions. For a nondegenerate electron gas $f \ll 1$ and

$$\frac{\mathrm{d}}{\mathrm{d}t} f(\varrho, t) = \int [P(\varrho', \varrho)\, f(\varrho', t) - P(\varrho, \varrho')\, f(\varrho, t)]\, \mathrm{d}^3\varrho' \,. \tag{4}$$

Denoting

$$\frac{1}{\tau_1} = \int P(\varrho, \varrho')\, d^3\varrho' , \tag{5}$$

$$\frac{\bar{f}}{\tau_2} = \int f(\varrho'', t)\, P(\varrho'', \varrho)\, d^3\varrho'' , \tag{6}$$

equation (4) passes into an ordinary differential equation[1])

$$\frac{d}{dt} f = -\frac{f}{\tau_1} + \frac{\bar{f}}{\tau_2}$$

which may be integrated [1] in the time interval $t' \div t_0$ yielding

$$f(\varrho_{t_0}, t_0) = f(\varrho_{t'}, t') \exp\left(-\int_{t'}^{t_0} \frac{dt'''}{\tau_1(\varrho_{t'''})}\right) + \\ + \int_{t'}^{t_0} \frac{\bar{f}(\varrho_{t''}, t'')}{\tau_2(\varrho_{t''})} \exp\left(-\int_{t''}^{t_0} \frac{dt'''}{\tau_1(\varrho_{t'''})}\right) dt'' . \tag{7}$$

The distribution function looses its dependence on the initial condition if we set $t'$ equal to minus infinity:

$$f(\varrho_{t_0}, t_0) = \int_{-\infty}^{t_0} \frac{\bar{f}(\varrho_t, t)}{\tau_2(\varrho_t)} \exp\left(-\int_t^{t_0} \frac{ds}{\tau_1(\varrho_s)}\right) dt . \tag{8}$$

For a homogeneous crystal and homogeneous fields equations (5), (6), and (8) pass into

$$f(\boldsymbol{k}_{t_0}, t_0) = \int_{-\infty}^{t_0} \frac{\bar{f}(\boldsymbol{k}_t, t)}{\tau_2(\boldsymbol{k}_t)} \exp\left(-\int_t^{t_0} \frac{ds}{\tau_1(\boldsymbol{k}_s)}\right) dt , \tag{9}$$

$$\frac{1}{\tau_1} = \int P(\boldsymbol{k}, \boldsymbol{k}')\, d^3k' , \tag{10}$$

$$\frac{\bar{f}}{\tau_2} = \int f(\boldsymbol{k}'', t)\, P(\boldsymbol{k}'', \boldsymbol{k})\, d^3k'' . \tag{11}$$

According to conventional time-dependent perturbation theory the probability of a transition per unit time is

$$P_{if} = \frac{2\pi}{\hbar} |M_{if}|^2\, \delta(E_f - E_i) , \tag{12}$$

where $M_{if}$ is the matrix element for scattering from the initial state $\langle i|$ into the final state $|f\rangle$; $E_i$ and $E_f$ are the initial and the final energies of the whole system. For a spherical and parabolic band with an effective mass $m^*$ equations (10) and (11) can be rewritten in spherical coordinates making use of equation (12):

$$\frac{1}{\tau_1} = \frac{2\pi}{\hbar} \left(\frac{m^*}{\hbar^2}\right)^{3/2} \int |M(\boldsymbol{k}, \boldsymbol{k}')|^2\, \delta\,(E_f - E_i) \sqrt{2\, E_{k'}}\, dE_{k'}\, d\Omega' , \tag{13}$$

$$\frac{\bar{f}}{\tau_2} = \frac{2\pi}{\hbar} \left(\frac{m^*}{\hbar^2}\right)^{3/2} \int f(E_{k''}, \Omega'', t)\, |M(\boldsymbol{k}'', \boldsymbol{k})|^2\, \delta\,(E_f - E_i) \sqrt{2\, E_{k''}}\, dE_{k''}\, d\Omega'' , \tag{14}$$

where $d\Omega$ is an element of the solid angle.

---

[1]) Analogous calculations for the Boltzmann equation in the relaxation-time approximation was performed by MacCallum [1], Budd [2], and Tavernier [3].

As it is well known hot carriers in Ge and Si with energies exceeding that of the optical modes[2]), interact mainly with these phonons. Furthermore the process of phonon emission will be dominant.

In order to avoid the effect of impact ionisation we restrict our considerations to electrons with small energies (less then a few eV). The relations between energies of the states $\boldsymbol{k}$, $\boldsymbol{k}'$, $\boldsymbol{k}''$ in the case of optical phonon emission are

$$E_{k''} = E_k + \hbar\,\omega_{\boldsymbol{q}}\,, \qquad E_k = E_{k'} + \hbar\,\omega_{\boldsymbol{q}}\,.$$

Substituting these relations into (13) and (14) we get

$$\frac{1}{\tau_1} = \frac{2\pi}{\hbar}\left(\frac{m^*}{\hbar^2}\right)^{3/2} \int |M(\boldsymbol{k}, \boldsymbol{k}')|^2\, \delta\,(E_{k'} + \hbar\,\omega_{\boldsymbol{q}} - E_k)\,\sqrt{2\,E_{k'}}\;\mathrm{d}E_{k'}\,\mathrm{d}\Omega\,', \tag{15}$$

$$\frac{\bar{f}}{\tau_2} = \frac{2\pi}{\hbar}\left(\frac{m^*}{\hbar^2}\right)^{3/2} \int f(E_{k''}, \Omega'', t)\; |M(\boldsymbol{k}'', \boldsymbol{k})|^2\, \delta\,(E_{k''} - E_k - \hbar\,\omega_{\boldsymbol{q}})\,\sqrt{2\,E_{k''}}\;\mathrm{d}E_{k''}\,\mathrm{d}\Omega''\,. \tag{16}$$

If the maximal value of the wave vector of an optical phonon is larger (or equal) than the maximum vector connecting the points of the surfaces of the energies $E_k$ and $E_{k'}$ (or $E_{k''}$) in $\boldsymbol{k}$-space, then the integration over $\boldsymbol{k}'$ (or $\boldsymbol{k}''$) is taken over all directions. We shall restrict our considerations to such energies of a carrier that satisfy this condition.

According to Harrison's [6] consideration for germanium and silicon type nonpolar crystals, optical modes always give the zero-order contribution to the scattering, i.e. the contribution independent on the wave vector of carriers. In this zero-order approximation the integrals of (15) and (16) are

$$\frac{1}{\tau_1} = \frac{4\pi^2}{\hbar}\left(\frac{2\,m^*}{\hbar^2}\right)^{3/2} M_0^2\,\sqrt{E - \hbar\,\omega_{\boldsymbol{q}}}\,, \tag{17}$$

$$\frac{\bar{f}}{\tau_2} = \frac{4\pi^2}{\hbar}\left(\frac{2\,m^*}{\hbar^2}\right)^{3/2} M_0^2\,\sqrt{E + \hbar\,\omega_{\boldsymbol{q}}}\;\frac{1}{4\pi}\int f\,(E + \hbar\,\omega_{\boldsymbol{q}}, \Omega'', t)\,\mathrm{d}\Omega''\,. \tag{18}$$

We can also put

$$\frac{1}{\tau_2} = \frac{4\pi^2}{\hbar}\left(\frac{2\,m^*}{\hbar^2}\right)^{3/2} M_0^2\,\sqrt{E + \hbar\,\omega_{\boldsymbol{q}}}\,, \tag{19}$$

$$\bar{f} = \frac{1}{4\pi}\int f\,(E + \hbar\,\omega_{\boldsymbol{q}}, \Omega'', t)\,\mathrm{d}\Omega''\,. \tag{20}$$

For carriers with energies much greater than the phonon energy i.e. for $\frac{\hbar\,\omega_{\boldsymbol{q}}}{E} \ll 1$

$$\bar{f} \approx \frac{1}{4\pi}\int f(E, \Omega'', t)\,\mathrm{d}\Omega'' \tag{21}$$

and

$$\frac{1}{\tau_1} \approx \frac{1}{\tau_2} \approx \frac{1}{\tau} = \left(\frac{m^*}{2}\right)^{-1/2} \frac{1}{\lambda}\sqrt{E}\,,$$

where "the mean free path" $\lambda = \hbar^4/8\,\pi^2\,m^{*2}\,M_0^2$. As it is seen from (21), $\bar{f}$ is the isotropic part of the distribution function.

---

[2]) The highest energy for optical modes $\hbar\,\omega_{\boldsymbol{q}}$ in Ge is 0.037 eV [4] and in Si 0.063 eV [5].

The distribution function (9) for these high-energy carriers becomes analogous to that suggested by Chambers [7] for the time-relaxation approximation:

$$f_{t_0} = \int\limits_{-\infty}^{t_0} \frac{1}{\tau_t} \bar{f}_t \exp\left(-\int\limits_t^{t_0} \frac{ds}{\tau_s}\right) dt$$

or after integration over parts:

$$f_{t_0} = \bar{f}_{t_0} - \int\limits_{-\infty}^{t_0} \frac{d\bar{f}_t}{dt} \exp\left(-\int\limits_t^{t_0} \frac{ds}{\tau_s}\right) dt\,. \tag{22}$$

The integral term in equation (22) is the anisotropic part of the distribution function.

If only a constant homogeneous electric field is applied to a homogeneous isotropic crystal, the electron gas will be in steady-state and the distribution function will have the axial symmetry $f(\boldsymbol{k}) = f(k, \varphi)$ where $\varphi$ is the angle between the direction of the electric field $\vec{\mathcal{E}}$ and the wave vector of the carrier $\boldsymbol{k}$. Assuming the isotropic part of the distribution function to be Maxwellian with an effective "carrier temperature" $T_e$ i.e.

$$\bar{f}_t = C \exp\left(-\frac{E - \Delta E_t}{W}\right),$$

where $W = \mathrm{k} T_e$, k the Boltzmann constant and $C = h^3\, N/2\,(2\,\pi\, m^*\, W)^{3/2}$, equation (22) can be written after [8], introducing the new variables

$$x = \frac{e\,\mathcal{E}}{m^{*1/2}}(t_0 - s) \quad \text{and} \quad y = \frac{e\,\mathcal{E}}{m^{*1/2}}(t_0 - t)\,,$$

in the form

$$f(E) = C\left\{\exp\left(-\frac{E}{W}\right) - \frac{1}{W}\int\limits_0^{\infty} dy\,(y + \sqrt{2\,E}\cos\varphi) \times \right.$$
$$\left. \times \exp\left[-\left(\frac{E - \Delta E_y}{W} + \frac{\sqrt{2}}{e\,\mathcal{E}\,\lambda}\int\limits_0^y \sqrt{E - \Delta E_x}\, dx\right)\right]\right\}, \tag{23}$$

where the energy gained by an electron in the electric field $\vec{\mathcal{E}}$ in the time interval $(t_0 - t)$ is

$$\Delta E_t = -\,e\int\limits_t^{t_0} \vec{\mathcal{E}}\,\boldsymbol{v}(s)\, ds = -\left(\frac{1}{2}\,y^2 + \sqrt{2\,E}\,y\cos\varphi\right).$$

If we put the new parameters

$$W' = \alpha\, W\,, \qquad E' = \alpha\, E\,, \qquad (e\,\mathcal{E}\,\lambda)' = \alpha(e\,\mathcal{E}\,\lambda)$$

into (23), the expression within the outer brackets remains invariable, and then

$$f_{W',(e\mathcal{E}\lambda)'}(E') = \frac{1}{\alpha^{3/2}} f_{W,(e\mathcal{E}\lambda)}(E)\,.$$

Because of this invariancy, the calculation of the distribution function as a function of $E$ and $\varphi$ (Fig. 1, 2, and 3) has been performed for one value of $e\,\mathcal{E}\,\lambda$ (1 eV) and several values of $W$ (0.025, 0.125, 0.2, and 0.625 eV).

Equation (23) and numerical date displayed on Fig. 1, 2, and 3 are valid, of course, for electrons (for holes we should take the opposite direction of $\boldsymbol{F}$) with energy much greater than that of optical modes. For carriers with an energy

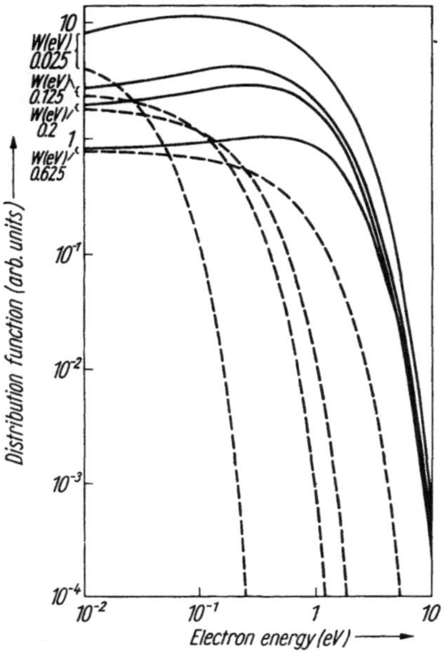


**Fig. 1.** Distribution function vs. energy of carriers for $e \mathcal{E} \lambda = 1$ eV; $W = 0.025$, 0.125, 0.2, and 0.625 eV; $\varphi = 0$ and 180°

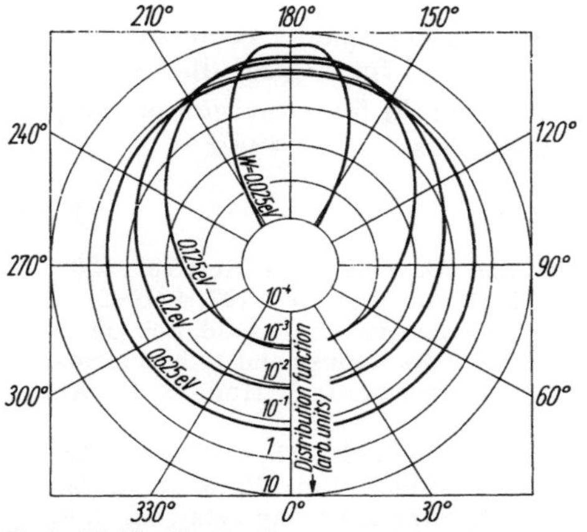


**Fig. 2.** Distribution function vs. angle for $E = 1$ eV; $W = 0.025$, 0.125, 0.2, and 0.625 eV

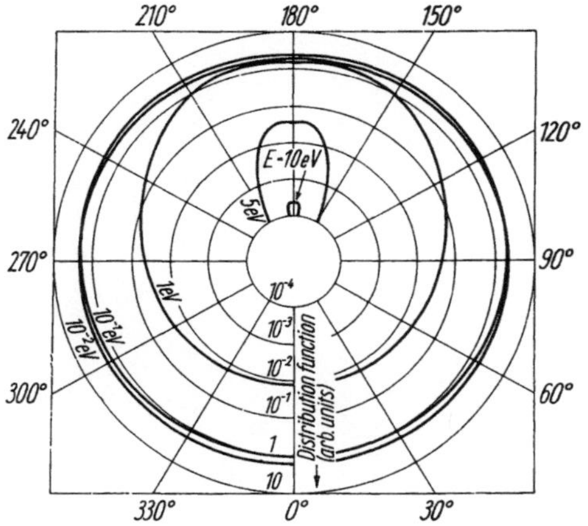


**Fig. 3.** Distribution function vs. angle for $W = 0.2$ eV; $E = 0.01$, 0.1, 1, 5, and 10 eV

slightly exceeding the optical mode energy two different relaxation times, $\tau_1$ from (17) and $\tau_2$ from (19), and $\bar{f}$ from (20) should be accounted for in (9). For energies smaller than that of optical modes the dominant contribution to lattice scattering is from acoustical phonon interactions [9].

As it is seen from Fig. 2 and 3 the distribution function of high energy carriers for a strong field is strongly anisotropic. This anisotropy increases with the decrease of energy (Fig. 3) and temperature (Fig. 2) of carriers at constant value of electric field strength. As it is seen from Figs. 2 and 3 the distribution function of electrons is flat with a small local minimum in the direction opposite to that of the electric field. The distribution of holes is similar in the direction of electric field. This effect is due to an increase in the relaxation time

$$\frac{1}{\tau_t} = \left(\frac{m^*}{2}\right)^{-1/2} \frac{1}{\lambda} \sqrt{E - \Delta E_t}$$

with angle $\varphi$ (from 0 to 180°).

The author thanks Professor L. Sosnowski, Doc. J. Kołodziejczak and Dr. J. Blinowski for helpful discussions and remarks. He is also obliged to Mgr. S. Żukotyński for a revision of the paper and Mr. R. Kowalczyk for numerical calculations.

## References

[1] C. MACCALLUM, Phys. Rev. **132**, 930 (1963).
[2] H. BUDD, Phys. Rev. **127**, 4 (1962); C. R. Acad. Sci. (France) **256**, 615 (1963).
[3] J. TAVERNIER, C. R. Acad. Sci. (France) **255**, 1868 (1962).
[4] B. N. BROCKHOUSE and P. K. IYENGAR, Phys. Rev. **111**, 747 (1958).
[5] H. PALEVSKY, D. J. HUGHES, W. KLEJ, and E. TUNKELO, Phys. Rev. Letters **2**, 258 (1959).
[6] W. A. HARRISON, Phys. Rev. **104**, 1281 (1956).
[7] R. CHAMBERS, Proc. Phys. Soc. **A 65**, 458 (1952).
[8] E. KAMIENIECKI, phys. stat. sol. **6**, 877 (1964).
[9] W. E. PINSON and R. BRAY, Phys. Rev. **136 A**, 1449 (1964).

*(Received April 25, 1966)*

phys. stat. sol. 16, 63 (1966)

*Department of Quantum Chemistry,*
*Academy of Sciences of the Moldavian SSR, Kishinev*

# Jahn-Teller Distortions and Inversion Splitting for Transition Metal Complexes and Local Centres in Crystals with Degenerate Electronic T-Terms

By

I. B. Bersuker and B. G. Vekhter

It is shown that the Jahn-Teller equivalent configurations of a threefold degenerate electronic T-term do not seem to correspond to pure tetragonal or pure trigonal distortions as expected from the work of Opik and Pryce [10]. The conclusions of the latter work appear to be valid only to first approximation. The distortions of tetragonal minima under trigonal displacements are obtained. This shows that inversion splitting does occur as predicted earlier [1], in contrary to the statements of Ham [9]. The results of the theory are in agreement with the last experiments by Sturge [8].

In der Arbeit wird gezeigt, daß die dem Jahn-Teller-Effekt äquivalenten Konfigurationen eines dreifach entarteten elektronischen T-Terms nicht, wie es aus früheren Arbeiten von Opik und Pryce [10] erwartet wurde, mit reinen tetragonalen oder trigonalen Verzerrungen übereinstimmen, und daß die Schlußfolgerungen dieser Arbeiten nur in erster Näherung gültig zu sein scheinen. In der vorliegenden Arbeit werden die Verzerrungen der tetragonalen Minima unter trigonalen Verschiebungen erhalten. In diesem Fall tritt Inversionsaufspaltung ein, wie es schon früher [1] vorausgesagt wurde, im Gegensatz zu den Schlußfolgerungen von Ham [9]. Die Ergebnisse der Theorie befinden sich in Übereinstimmung mit den letzten Experimenten von Sturge [8].

## 1. Introduction

In an earlier paper one of us [1] proposed a theory of the so-called "inversion" splitting of electronic-vibrational levels in transition metal complexes with degenerate electronic states of E- and T-types. This splitting appears when in accordance with the Jahn-Teller theorem the system possesses several equivalent minima of the adiabatic potential with large barriers between them.[1]) The idea about inversion states allows us to explain some interesting features of ESR spectra of the systems under consideration [2, 3], in particular the temperature transition in the $g$-factor and hyperfine structure parameters [2], and to predict such manifestations of the effect as the resonance ultrasonic absorption [4][2]) and the heat capacity and heat conductivity anomalies of the appropriate crystals [7].

[1]) The inversion splitting having values from small fractions to several tens of 1 $cm^{-1}$ are not to be confused with the Jahn-Teller splitting, the value of which has the order of 100 to 1000 $cm^{-1}$. Neither can it be called "the dynamic Jahn-Teller effect" because by this term one usually understands the case when the barriers between the equilibrium configurations are small and there is a complete coupling of electronic and nuclear motions, while inversion splitting describes the dynamic situation in the case of high barriers and quasistationary equilibrium configurations.

[2]) Recent experimental works [5, 6] may be probably interpreted as the confirmation of such absorption. However, the authors of [5] connecting the observed ultrasonic absorption with the Jahn-Teller effect consequences, don't refer to [4], in which the possibility of the specific ultrasonic absorption in such systems was shown for the first time.

The obtained results naturally concern also local centres of the crystal, which together with the nearest environment don't differ in the main from complexes.

Recently Sturge [8] observed directly $T_2$-state inversion splitting in the "no-phonon" line of the $^4A_2 \rightarrow {}^4T_2$ transition of $V^{2+}$ in MgO at low temperatures. This splitting has a value of about 40 $cm^{-1}$ and its behaviour under the influence of uniaxial stress and temperature shows it to be the consequence of the interaction of three equilibrium configurations in the tetragonal distorted cubic environment of the $V^{2+}$ ion (without any alternative explanation). These results are in full agreement with the predictions of the theory [1] for the excited $T_2$-state. However, the author of [5] believes that the splitting observed by him is not connected with the inversion splitting predicted in [1] and tries another explanation of the phenomenon (still unsuccesfully as he affirms himself). He refers to Ham [9] who stated that our results [1] for the $T_2$-state and tetragonal distortions are incorrect, because in this case the inversion splitting is impossible.

Taking into account the complexity of the task for systems with several minima of the adiabatic potential and the increasing interest to the phenomenon in connection with new stimulant experiments we give here a more rigorous deduction of inversion splitting in systems with degenerate T-terms. We show that the well-known work by Opik and Pryce [10] on the equilibrium Jahn-Teller configurations have, generally speaking, a restricted application and Ham's conclusion about the absence of inversion splitting based on the results of this work is incorrect. Besides, we show that the experiments of Sturge are consistent with the proposed inversion splitting theory.

## 2. Trigonal Distortions of Tetragonal Minima

We consider an octahedral system with degenerate electronic T-terms. In this case the Hamiltonian of the system contains terms linear in tetragonal distortions of $e_g$-type ($Q_2$ and $Q_3$) and in trigonal distortions of $t_{2g}$-type ($Q_4$, $Q_5$, and $Q_6$)

$$V' = A\,(l_2\,Q_2 + l_3\,Q_3) + B\,(t_4\,Q_4 + t_5\,Q_5 + t_6\,Q_6)\,. \tag{1}$$

The constants $A$ and $B$ and the matrixes $l$ and $t$ are well known [1, 11]. The terms $V'$ are essential for the Jahn-Teller effect. Since only two matrixes $l_2$ and $l_3$ of the five ones appearing in (1) commute, the physical situation is characterized by a complicate coupling between the electronic and nuclear motions. This circumstance doesn't allow the direct full investigation of the surface of the potential energy as functions of all five coordinates $Q_2$ to $Q_6$.

Studying the stationary points of this surface Opik and Pryce [10] have shown that its minima correspond either to pure tetragonal (only the coordinates $Q_2$ and $Q_3$ are displaced) or to pure trigonal distortions (only the coordinates $Q_4$, $Q_5$, and $Q_6$ are displaced). The former is realized if $\frac{A^2}{K_a} > \frac{4}{3}\frac{B^2}{K_b}$, and the latter in the opposite case ($K_a$ and $K_b$ are the force constants for tetragonal and trigonal displacements, respectively). In the case of tetragonal minima under consideration the potential surface in $Q_2\,Q_3$ space has the form of three paraboloids crossing one another without interaction. That is why Ham [9] concluded about the impossibility of transitions between these paraboloids which give rise to the inversion splitting.

However, Opik and Pryce's conclusions are not true in the general case. Indeed, the results of Opik and Pryce are based upon a representation of the wave function as a linear combination of the three T-term basis functions with constant coefficients $a_i$. In a more precise examination these coefficients depend on the nuclear coordinates $Q_\alpha$. When $a_i = a_i(Q_\alpha)$ the conclusion about the pure tetragonal or pure trigonal minima do not hold any longer (see the solution in [10]). The coefficients $a_i$ are constant only in the first approximation (in zero approximation for the wave function) and therefore the conclusions of Opik and Pryce hold only in this approximation. By the way the authors of [10] themselves mentioned it casually.

Rejecting this limitation we look for eigenfunctions of the Hamiltonian in the degenerate case in the form $\sum \psi_i(q, Q)\, \chi^{(i)}(Q)$ where $\psi_i$ are the electron wave functions of the degenerate term and $\chi^{(i)}(Q)$ are some coefficients (which Opik and Pryce suppose to be constant near by the stationary points but are in fact functions of nuclear displacements). For these coefficients one may obtain the following equations in the case of threefold degenerate T-terms:

$$\begin{Vmatrix} \sum_{i=2}^{6} H_0(Q_i) - A\,Q_3 - \varepsilon & B\,Q_4 & B\,Q_5 \\ B\,Q_4 & \sum_{i=2}^{6} H_0(Q_i) + \frac{1}{2} A\,(Q_3 + \sqrt{3}Q_2) - \varepsilon & B\,Q_6 \\ B\,Q_5 & B\,Q_6 & \sum_{i=2}^{6} H_0(Q_i) + \frac{1}{2} A\,(Q_3 - \sqrt{3}\,Q_2) - \varepsilon \end{Vmatrix} \begin{Vmatrix} \chi^{(1)}(Q) \\ \chi^{(2)}(Q) \\ \chi^{(3)}(Q) \end{Vmatrix} = 0\,, \qquad (2)$$

where $H_0(Q)$ is the Hamiltonian of the harmonic oscillator.

The pure tetragonal minima are seen to be realized if and only if $B = 0$, that is in the absence of the influence of trigonal displacements. Then, as it was mentioned earlier, the potential surface in $Q_2\, Q_3$ space takes the form of three displaced paraboloids which don't interact with one another. For each of them the displaced coordinates of normal vibrations are given by the expressions

$$\left.\begin{aligned} Q_2^{(1)} &= Q_2\,, & Q_3^{(1)} &= Q_3 - Q^0\,, \\ Q_2^{(2)} &= \frac{1}{2} Q_2 + \frac{\sqrt{3}}{2} Q_3\,, & Q_3^{(2)} &= -\frac{1}{2} Q_3 + \frac{\sqrt{3}}{2} Q_2 - Q^0\,, \\ Q_2^{(3)} &= -\frac{1}{2} Q_2 + \frac{\sqrt{3}}{2} Q_3\,, & Q_3^{(3)} &= -\frac{1}{2} Q_3 - \frac{\sqrt{3}}{2} Q_2 - Q^0\,, \end{aligned}\right\} \qquad (3)$$

where $Q^0 = A/K_\alpha$.

We assume that in zero approximation such pure tetragonal minima are realized, and consider how they are changing when interactions with trigonal displacements are taken into account. The latter displacements in equation (2) perturb the trigonal coordinate functions only and therefore one may separate the variables and write $\chi^{(i)}(Q)$ as the product $\chi^{(i)}_{tt}(Q_2 Q_3)\, \chi^{(i)}_{tr}(Q_4 Q_5 Q_6)$. Multiplying on the left side each term of (2) by the tetragonal functions of the corresponding minimum and integrating over the tetragonal coordinates we obtain the following system for the trigonal functions in tetragonal minima:

$$\begin{Vmatrix} \sum_{i=4}^{6} H_0(Q_i) - \varepsilon & \gamma\, B\, Q_4 & \gamma\, B\, Q_5 \\ \gamma\, B\, Q_4 & \sum_{i=4}^{6} H_0(Q_i) - \varepsilon & \gamma\, B\, Q_6 \\ \gamma\, B\, Q_5 & \gamma\, B\, Q_6 & \sum_{i=4}^{6} H_0(Q_i) - \varepsilon \end{Vmatrix} \begin{Vmatrix} \chi^{(1)}_{tr} \\ \chi^{(2)}_{tr} \\ \chi^{(3)}_{tr} \end{Vmatrix} = 0 \,. \tag{4}$$

Here $\gamma$ denotes the earlier introduced [1] overlap integral of the tetragonal oscillatory functions of different minima:

$$\gamma = \int \mathrm{d}Q_2\, \mathrm{d}Q_3\, \chi^{(i)}_{tt}\,(Q_2 Q_3)\, \chi^{(j)}_{tt}(Q_2 Q_3)\,; \quad i \neq j\,. \tag{5}$$

For the ground oscillatory state we have $\gamma_0 = \exp\left\{-\frac{3}{4}\frac{A^2}{K_a}\cdot\frac{1}{\hbar\,\omega}\right\}$, where $\hbar\,\omega$ is the quantum of the tetragonal oscillations. Since for the deep minima unter consideration $\frac{A^2}{K_a} \gg \hbar\,\omega$ and therefore $\gamma \ll 1$, the system (4) may be solved by an iteration procedure. In zero approximation $\chi^{(1)}_{tr}$, $\chi^{(2)}_{tr}$, and $\chi^{(3)}_{tr}$ coincide and are functions of undisplaced harmonic oscillators. In the next approximation for the first-order terms ($\sim \gamma$) we substitute in the first of the equations (4) $\chi^{(2)}_{tr}$ and $\chi^{(3)}_{tr}$ by $\chi^{(1)}_{tr}$, and so on. Then we obtain

$$\left.\begin{aligned} \left[\sum_{i=4}^{6} H_0(Q_i) + \gamma\, B\, Q_4 + \gamma\, B\, Q_5 - \varepsilon\right] \chi^{(1)}_{tr} &= 0\,, \\ \left[\sum_{i=4}^{6} H_0(Q_i) + \gamma\, B\, Q_4 + \gamma\, B\, Q_6 - \varepsilon\right] \chi^{(2)}_{tr} &= 0\,, \\ \left[\sum_{i=4}^{6} H_0(Q_i) + \gamma\, B\, Q_5 + \gamma\, B\, Q_6 - \varepsilon\right] \chi^{(3)}_{tr} &= 0\,. \end{aligned}\right\} \tag{6}$$

The solutions of these equations are also harmonic oscillator functions for which the equilibrium positions of two from three trigonal coordinates in each tetragonal minima are displaced by $\gamma\, B/K_b$. Thus, when the interaction with both tetragonal and trigonal displacements is taken into account, the equilibrium configurations become no more pure tetragonal. However, as the overlap integral $\gamma$ is small the displacements of the trigonal coordinates are much smaller than displacements of the tetragonal ones and we can speak about pseudotetragonal configurations.

It must be noticed that it is impossible to obtain a result proportional to $\gamma \sim \mathrm{e}^{-1/\alpha}$ where $\alpha = \frac{\hbar\,\omega}{A^2/K_a} \ll 1$ by means of common perturbation treatment (see the analogous situation in the theory of superconductivity [12]).

### 3. The Inversion Splitting — Discussions

Using the total electronic-vibrational functions of distinct distorted configurations, for which both tetragonal and trigonal coordinates are displaced [1], we obtain the following expression for the Hamiltonian matrix of the system (the energy of the distorted configuration is considered to be zero)

$$\begin{vmatrix} -\varepsilon & u & u \\ u & -\varepsilon & u \\ u & u & -\varepsilon \end{vmatrix} = 0 , \tag{7}$$

where $u = -\gamma^2 B^2/K_b$. This is the same form of the effective Hamiltonian matrix which Sturge had to assume for the explanation of the observed splitting [8]. As it is clear from the preceding such a form of the Hamiltonian is essentially due to the inversion splitting effect.

The separation $\delta$ between inversion levels (singlet ($\varepsilon = +2\,u$) and doublet ($\varepsilon = -u$)) is easily obtained from equation (4),

$$\delta = 3\,|u| = \frac{9}{2}\,\Delta E\,\gamma^2 , \tag{8}$$

where $\Delta E = \frac{2}{3}\,B^2/K_b$ is the Jahn-Teller splitting of the degenerate electronic T-term in the trigonal minimum. If we accept $\gamma = 0.1$ and $\Delta E = 1000\ \text{cm}^{-1}$ according to Sturge, we obtain $\delta = 45\ \text{cm}^{-1}$. This is in good agreement with the observed value $\delta \approx 40\ \text{cm}^{-1}$. However, in contrast with the results of [8] the ground state in our calculation is the singlet and not the doublet state ($u < 0$).

The magnetic nature of the observed transitions is easily understood by taking into account the spin-orbital coupling. Splittings of the inversion levels caused by the latter have the order of magnitude of $\lambda\,\gamma$ [3, 14], where $\lambda$ is the spin-orbital constant. Since for $V^{2+}$ $\lambda = 120\ \text{cm}^{-1}$ this extra splitting has the same order as the resolution of the monochromator used by Sturge ($\approx 10\ \text{cm}^{-1}$) and consequently it couldn't be observed.

The authors wish to express their gratitude to Dr. M. D. Sturge (Bell Telephone Laboratories) for sending his work [8] prior to publication.

### References

[1] I. B. Bersuker, Zh. eksper. teor. Fiz. **43**, 1315 (1962) or Soviet Phys. — J. exper. theor. Phys. **16**, 933 (1963).

[2] I. B. Bersuker, Zh. eksper. teor. Fiz. **44**, 1239 (1963) or Soviet Phys. — J. exper. theor. Phys. **17**, 836 (1963).
I. B. Bersuker, S. S. Budnikov, B. G. Vekhter, and B. I. Chinik, Fiz. tverd. Tela **6**, 2583 (1964).

[3] I. B. Bersuker and B. G. Vekhter, Fiz. tverd. Tela **5**, 2432 (1963) or Soviet Phys. — Solid State Phys. **5**, 1772 (1964); Fiz. tverd. Tela **7**, 1231 (1965).

[4] I. B. Bersuker, Zh. eksper. teor. Fiz. **44**, 1577 (1963) or Soviet Phys. — J. exper. theor. Phys. **17**, 1060 (1963); Fiz. tverd. Tela **6**, 436 (1964).

[5] E. M. Gyorge, M. D. Sturge, D. B. Fraser, and R. C. LeCraw, Phys. Rev. Letters **15**, 19 (1965).

[6] I. S. Ciccarello, R. Arst, and K. Dransfeld, Phys. Rev. **138**, A 934 (1965).

[7] B. G. Vekhter and I. B. Bersuker, Fiz. tverd. Tela **7**, 1569 (1965).

[8] M. D. Sturge, Phys. Rev. **140**, A 880 (1965).
[9] F. S. Ham, Phys. Rev. **138**, A 1727 (1965).
[10] U. Opik and M. H. L. Pryce, Proc. Roy. Soc. **A 238**, 425 (1957).
[11] W. Moffitt and W. Thorson, Phys. Rev. **108**, 1251 (1957).
[12] N. N. Bogolyubov, V. V. Tolmachov, and D. V. Shirkov, Novii metod v teorii sverkh-provodimosti, AN SSSR, Moscow 1958 (p. 3); english translation: New Method in the Theory of Superconductivity, Acad. Sci. USSR, Dubna 1958.
[13] I. B. Bersuker and B. G. Vekhter, Trudy Komissii po Spektroskopii **3**, 520 (1965).
[14] I. B. Bersuker, Litov. Fiz. Sbornik **3**, 327 (1963).

*(Received March 26, 1966)*

phys. stat. sol. **16**, 69 (1966)

*Faculty of Physics, University of Bucharest*

# Some Kinetic Phenomena in Impure Ionic Semiconductors

By

C. N. Plăviţu

A criterion is given for finding the regions where one of the various scattering mechanisms is dominant. The phonon drag contribution to some of the kinetic coefficients is only slightly modified when ionized impurity scattering is taken into account.

Es wird ein Kriterium zur Bestimmung der Bereiche mit einem vorherrschenden Streumechanismus gegeben. Im Falle von Streuung an ionisierten Störstellen ändert sich der Beitrag des Phononendrag zu einigen kinetischen Koeffizienten nur wenig.

## 1. Introduction

We consider an ionic semiconductor of cubic symmetry with donor impurities incorporated into the lattice. For the calculation of the kinetic coefficients we use the approximate solutions of the system of kinetic Boltzmann equation for the distribution functions of carriers and acoustic and optical phonons [1 to 3]. The solutions were obtained under the same simplifying conditions as in [3].

The acoustic and optical phonon drag effect is taken into account in the same manner as in [3]. In the present paper we take the electric field $\boldsymbol{E}$ and the temperature gradient $\nabla T$ as generalized external forces. Further we consider the scattering of carriers by ionized impurities and acoustic and optical phonons. The relaxation time approximation is considered to be valid for every scattering mechanism.

By an exact evaluation of the integrals entering in the expressions for the kinetic coefficients we found the regions where the various scattering mechanisms of carriers are dominant.

Besides, the impurity-induced change in the contributions of the acoustic and optical phonon drag effect to the kinetic coefficients will be treated by a scattering mechanism. As in [3] we consider a high and a low temperature range.

## 2. The High Temperature Range

We use the expressions for the kinetic coefficients which are reported in [3]:

$$\left.\begin{aligned}
&\sigma = -\,a^{(0)}\,, \\
&\alpha_{\mathrm{e}} = \frac{1}{T}\left[\frac{a^{(1)}}{a^{(0)}} - \frac{\mu}{e}\right], \qquad && \varkappa_{\mathrm{e}} = \frac{1}{T}\left[\frac{a^{(1)}\,a^{(1)}}{a^{(0)}} - a^{(2)}\right], \\
&\alpha_{\mathrm{f}}^{(1)} = \frac{1}{T}\,\frac{a_{(1)}^{(0)}}{a^{(0)}}\,, && \varkappa_{\mathrm{f}}^{(1)} = \frac{1}{T}\left[\frac{a^{(1)}\,a_{(1)}^{(0)}}{a^{(0)}} - a_{(1)}^{(1)}\right], \\
&\alpha_{\mathrm{f}}^{(2)} = \frac{1}{T}\,\frac{a_{(2)}^{(0)}}{a^{(0)}}\,, && \varkappa_{\mathrm{f}}^{(2)} = \frac{1}{T}\left[\frac{a^{(1)}\,a_{(2)}^{(0)}}{a^{(0)}} - a_{(2)}^{(1)}\right],
\end{aligned}\right\} \qquad (2.1)$$

where $\sigma$ is the electric conductivity, $\alpha$ the coefficient of the thermoelectric power, and $\varkappa$ the thermal conductivity. The indices (e) and (f) refer to carrier and phonon contributions to the kinetic coefficients, respectively. The indices (1) and (2) of $\alpha$ and $\varkappa$ distinguish the contributions of the acoustic and optical phonon drag parts, respectively, $\mu$ is the chemical potential of carriers and $T$ the temperature. The quantities involved in (2.1) are given by

$$\left.\begin{aligned}
a^{(i)} &= \frac{e^{2-i}\hbar^2}{3\pi^2 m^2} I^{(i)}, && (i = 0, 1, 2),\\
a_{(1)}^{(j)} &= \frac{2^{2-n} e^{1-j} (\hbar v_0)^2 \tau_{0\mathrm{f}}^{(1)}}{3\pi^2 m (4-n)} K_{(1)}^{(j)}, && (n = 0, 2); \quad (j = 0, 1),\\
a_{(2)}^{(j)} &= \frac{2^{3-n'} e^{1-j} R \hbar^2 \omega_0 \tau_{0\mathrm{f}}^{(2)}}{3\pi^2 m (2-n')} K_{(2)}^{(j)}, && (n' = 0, 1); \quad (j = 0, 1),\\
a_{(2)}^{(j)} &= \frac{4 e^{1-j} R \hbar^2 \omega_0 \tau_{0\mathrm{f}}^{(2)}}{3\pi^2 m} K_{(2)}^{(j)}, && (n' = 2); \quad (j = 0, 1),
\end{aligned}\right\} \tag{2.2}$$

where $e$ is the charge of carriers, $m$ their effective mass, $\hbar$ is the Planck constant divided by $2\pi$, $v_0$ the velocity of acoustic phonons, $\omega_0$ the maximum angular frequency of optical phonons, and $R$ is a scalar which enters in the expression for the angular frequency of optical phonons.

The phonon relaxation times $\tau_\mathrm{f}^{(s)}$, $(s = 1, 2)$ depend, according to [4], in the following manner on the wave vector $\boldsymbol{q}$

$$\tau_\mathrm{f}^{(1)} = \tau_{0\mathrm{f}}^{(1)} q^{-n}, \qquad \tau_\mathrm{f}^{(2)} = \tau_{0\mathrm{f}}^{(2)} q^{-n'}, \tag{2.3}$$

where $n = 0, 2$ and $n' = 0, 1, 2$ for crystals of cubic structure [4] ($n = 0$ when the interaction of acoustic phonons with the surface of the crystals dominates). The quantities $I^{(i)}$ and $K_{(s)}^{(j)}$ $(s = 1, 2)$ are given by

$$\left.\begin{aligned}
I^{(i)} &= \int_0^\infty \tau\, \varepsilon^i k^4 \frac{\partial f_0}{\partial \varepsilon} \mathrm{d}k,\\
K_{(1)}^{(j)} &= \int_0^\infty \frac{\tau}{\tau^{(1)}} \varepsilon^j k^{4-n} \frac{\partial f_0}{\partial \varepsilon} \mathrm{d}k, && (n = 0, 2),\\
K_{(2)}^{(j)} &= \int_0^\infty \frac{\tau}{\tau^{(2)}} \varepsilon^j k^{4-n'} \frac{\partial f_0}{\partial \varepsilon} \mathrm{d}k, && (n' = 0, 1),\\
K_{(2)}^{(j)} &= \int_0^\infty \frac{\tau}{\tau^{(2)}} \varepsilon^j k^2 \ln \frac{2k}{\varkappa'} \frac{\partial f_0}{\partial \varepsilon} \mathrm{d}k, && (n' = 2),
\end{aligned}\right\} \tag{2.4}$$

where $\varepsilon$ is the energy of charge carriers, $f_0$ the equilibrium distribution function of charge carriers, $\boldsymbol{k}$ their wave vector, $\varkappa' = \left(\frac{2 m \omega_0}{\hbar}\right)^{1/2}$, and $\tau^{(1)} = \tau_0^{(1)}/k$ and $\tau^{(2)} = \tau_0^{(2)} k$ the relaxation times for interactions between carriers and acoustic and optical phonons, respectively, [1, 2]; $\tau$ is the total relaxation time.

In the present case we have

$$\left.\begin{aligned}
\frac{1}{\tau} &= \frac{1}{\tau^{(1)}} + \frac{1}{\tau^{(2)}} + \frac{1}{\tau^{(3)}}, & \tau &= \frac{\tau_0^{(3)} k^3}{b k^4 + a k^2 + 1},\\
\frac{\tau}{\tau^{(1)}} &= \frac{b k^4}{b k^4 + a k^2 + 1}, & \frac{\tau}{\tau^{(2)}} &= \frac{a k^2}{b k^4 + a k^2 + 1},
\end{aligned}\right\} \tag{2.5}$$

where $\tau^{(3)} = \tau_0^{(3)} k^3$ is the relaxation time for interaction between carriers and ionized impurities [1] and $b = \frac{\tau_0^{(3)}}{\tau_0^{(1)}}$, $a = \frac{\tau_0^{(3)}}{\tau_0^{(2)}}$. To compute the integrals (2.4) we have to take into account the expressions (2.5). Further we assume $\varepsilon = \frac{\hbar^2 k^2}{2m}$ (i. e. an isotropic band structure).

$$\left.\begin{aligned}
I^{(i)} &= \left(\frac{\hbar^2}{2m}\right)^i \frac{\tau_0^{(3)} e^{\frac{\mu}{k_0 T}}}{2 k_0 T} \sum_{\alpha=1}^{2} A_\alpha x_\alpha^{4+i} \left[\sum_{r=0}^{2+i} \frac{\Gamma(3+i-r)}{\gamma_\alpha'^{\,3+i-r}} - e^{-\gamma_\alpha'} \mathrm{Ei}(\gamma_\alpha')\right],\\
K_{(1)}^{(j)} &= \left(\frac{\hbar^2}{2m}\right)^j \frac{b\, e^{\frac{\mu}{k_0 T}}}{2 k_0 T} \sum_{\alpha=1}^{4} B_\alpha k_\alpha^{9+2j-2p} \times\\
&\times \left\{\sum_{r=0}^{3+j-p}\left[\frac{\Gamma(4+j-p-r)}{\gamma_\alpha^{4+j-p-r}} + \frac{\Gamma\left(\frac{7}{2}+j-p-r\right)}{\gamma_\alpha^{\frac{7}{2}+j-p-r}}\right] - \right.\\
&\left. - \left[e^{-\gamma_\alpha} \mathrm{Ei}(\gamma_\alpha) + \frac{\sqrt{\pi}(-\gamma_\alpha)^{\frac{1}{4}} e^{-\frac{\gamma_\alpha}{2}}}{\gamma_\alpha^{1/2}} W^{(-\gamma_\alpha)}_{-\frac{1}{4},-\frac{1}{4}}\right]\right\};\\
&\qquad (n = 2p;\ p = 0, 1),\\
K_{(2)}^{(j)} &= \left(\frac{\hbar^2}{2m}\right)^j \frac{a\, e^{\frac{\mu}{k_0 T}}}{2 k_0 T} \sum_{\alpha=1}^{4} B_\alpha k_\alpha^{7+2j} \left\{\sum_{r=0}^{2+j}\left[\frac{\Gamma(3+j-r)}{\gamma_\alpha^{3+j-r}} + \frac{\Gamma\left(\frac{5}{2}+j-r\right)}{\gamma_\alpha^{\frac{5}{2}+j-r}}\right] - \right.\\
&\left. - \left[e^{-\gamma_\alpha} \mathrm{Ei}(\gamma_\alpha) + \frac{\sqrt{\pi}(-\gamma_\alpha)^{\frac{1}{4}} e^{-\frac{\gamma_\alpha}{2}}}{\gamma_\alpha^{1/2}} W^{(-\gamma_\alpha)}_{-\frac{1}{4},-\frac{1}{4}}\right]\right\}; \qquad (n' = 0),\\
K_{(2)}^{(j)} &= \left(\frac{\hbar^2}{2m}\right)^j \frac{a\, e^{\frac{\mu}{k_0 T}}}{2 k_0 T} \sum_{\alpha=1}^{4} B_\alpha k_\alpha^{6+2j} \left\{\sum_{r=0}^{2+j} \frac{\Gamma\left(\frac{5}{2}+j-r\right)}{\gamma^{\frac{5}{2}+j-r}} + \right.\\
&\left. + \sum_{r=0}^{1+j} \frac{\Gamma(2+j-r)}{\gamma_\alpha^{2+j-r}} - \left[e^{-\gamma_\alpha} \mathrm{Ei}(\gamma_\alpha) + \frac{\sqrt{\pi}(-\gamma_\alpha)^{\frac{1}{4}} e^{-\frac{\gamma_\alpha}{2}}}{\gamma_\alpha^{1/2}} W^{(-\gamma_\alpha)}_{-\frac{1}{4},\frac{1}{4}}\right]\right\};\\
&\qquad (n' = 1),
\end{aligned}\right\} \quad (2.6)$$

where $\Gamma(x)$ is the Euler function, $\mathrm{Ei}(x)$ the exponential integral function, $W_{\mu,\nu}(x)$ the Whittaker functions [5], and $x_\alpha$ and $k_\alpha$ are solutions of

$$z^2 + a z + b = 0; \quad \left(z = \frac{1}{x} = \frac{1}{k^2}\right), \quad y^4 + a y^2 + b = 0; \quad \left(y = \frac{1}{k}\right). \quad (2.7)$$

$A_\alpha$ and $B_\alpha$ are the coefficients of a decomposition of the integrands into partial fractions. The arguments of $I^{(i)}$ is $\gamma_\alpha' = \beta x_\alpha$ and that of $K_{(s)}^{(j)}$ is $\gamma_\alpha = \beta k_\alpha^2$ with $\beta = \frac{\hbar^2}{2 m k_0 T}$, where $k_0$ is the Boltzmann constant.

The integrals $K^{(j)}_{(2)}$ have a singularity for $n' = 2$ which must be eliminated in the same way as in [3]. In the following we will give the approximate expressions used in the paper.

To establish the ranges where the different scattering mechanisms predominate the values of $D = (a/2)^2 - b$ in (2.7) have to be known which depend on the type of semiconductor and the temperature. Numerical estimates indicate that $\frac{a^2}{4b} \sim \frac{10^{67}\, \varepsilon'^2 Z^4 v_0}{N_i\, \omega_0^4} T$, where $\varepsilon'$ is the dielectric constant, $Z$ the ionic charge, and $N_i$ the concentration of impurities. Taking the following parameters $\varepsilon' \approx 10$, $v_0 \approx 10^5$ cm/s, $\omega_0 \approx 10^{13}$ rad/s, $N_i \approx 10^{15}$ cm$^{-3}$ we find $a^2/4b \approx 10^{12}\, T$, i.e. $D > 0$ in the high temperature range, and thus $D \approx (a/2)^2$ to a good approximation.

Under these conditions the first equation (2.7) has only one and the second equation two solutions which differ from zero. The relations (2.6) become

$$
\left.
\begin{aligned}
I^{(i)} &= \left(\frac{\hbar^2}{2m}\right)^i \frac{\tau_0^{(3)} e^{\frac{\mu}{k_0 T}}}{2 k_0 T} x_1^{4+i} \left[\sum_{r=0}^{2+i} \frac{\Gamma(3+i-r)}{\gamma_1'^{3+i-r}} - e^{-\gamma_1'} \mathrm{E_i}(\gamma_1')\right],\\
K^{(j)}_{(1)} &= \left(\frac{\hbar^2}{2m}\right)^j \frac{b\, e^{\frac{\mu}{k_0 T}}}{2 k_0 T} \sum_{\alpha=1}^{2} B'_\alpha k_\alpha^{9+2j-2p} \times\\
&\times \left\{\sum_{r=0}^{3+j-p}\left[\frac{\Gamma(4+j-p-r)}{\gamma_\alpha^{4+j-p-r}} + \frac{\Gamma\left(\frac{7}{2}+j-p-r\right)}{\gamma_\alpha^{\frac{7}{2}+j-p-r}}\right] -\right.\\
&\left.- \left[e^{-\gamma_\alpha} \mathrm{Ei}(\gamma_\alpha) + \frac{\sqrt{\pi}(-\gamma_\alpha)^{\frac{1}{4}} e^{-\frac{\gamma_\alpha}{2}}}{\gamma_\alpha^{1/2}} W^{(-\gamma_\alpha)}_{-\frac{1}{4},-\frac{1}{4}}\right]\right\};\\
&\qquad\qquad (n = 2p,\ p = 0, 1),\\
K^{(j)}_{(2)} &= \left(\frac{\hbar^2}{2m}\right)^j \frac{a\, e^{\frac{\mu}{k_0 T}}}{2 k_0 T} \sum_{\alpha=1}^{2} B'_\alpha k_\alpha^{7+2j} \left\{\sum_{r=0}^{2+j}\left[\frac{\Gamma(3+j-r)}{\gamma_\alpha^{3+j-r}} + \frac{\Gamma\left(\frac{5}{2}+j-r\right)}{\gamma_\alpha^{\frac{5}{2}+j-r}}\right] -\right.\\
&\left.- \left[e^{-\gamma_\alpha} \mathrm{Ei}(\gamma_\alpha) + \frac{\sqrt{\pi}(-\gamma_\alpha)^{\frac{1}{4}} e^{-\frac{\gamma_\alpha}{2}}}{\gamma_\alpha^{1/2}} W^{(-\gamma_\alpha)}_{-\frac{1}{4},-\frac{1}{4}}\right]\right\}; \qquad (n' = 0),\\
K^{(j)}_{(2)} &= \left(\frac{\hbar^2}{2m}\right)^j \frac{a\, e^{\frac{\mu}{k_0 T}}}{2 k_0 T} \sum_{\alpha=1}^{2} B'_\alpha k_\alpha^{6+2j} \left\{\sum_{r=0}^{2+j} \frac{\Gamma\left(\frac{5}{2}+j-r\right)}{\gamma_\alpha^{\frac{5}{2}+j-r}} +\right.\\
&\left.+ \sum_{r=0}^{1+j} \frac{\Gamma(2+j-r)}{\gamma_\alpha^{2+j-r}} - \left[e^{-\gamma_\alpha} \mathrm{Ei}(\gamma_\alpha) + \frac{\sqrt{\pi}(-\gamma_\alpha)^{\frac{1}{4}} e^{-\frac{\gamma_\alpha}{2}}}{\gamma_\alpha^{1/2}} W^{(-\gamma_\alpha)}_{-\frac{1}{4},-\frac{1}{4}}\right]\right\};\\
&\qquad\qquad (n' = 1).
\end{aligned}
\right\} \quad (2.8)
$$

The absolute values of the parameter $\gamma_1'$ and $\gamma_\alpha$ are equal to $\beta/a \approx 10^{-4}/T^2$, i.e. they are small in the high temperature range. Then the expressions (2.8) are approximately given by

$$\left.\begin{aligned}
I^{(i)} &\approx -\left(\frac{\hbar^2}{2\,m}\right)^i \frac{\tau_0^{(2)}\,e^{\frac{\mu}{k_0\,T}}}{2\,k_0\,T}\,\frac{\Gamma\,(3+i)}{\beta^{3+i}}\,,\\
K_{(1)}^{(j)} &\approx -\left(\frac{\hbar^2}{2\,m}\right)^j \frac{\tau_0^{(2)}\,e^{\frac{\mu}{k_0\,T}}}{\tau_0^{(1)}\,2\,k_0\,T}\,\frac{\Gamma\left(\frac{7}{2}+j-p\right)}{\beta^{\frac{7}{2}+j-p}}\,; \qquad (n = 2\,p;\; p = 0, 1)\,,\\
K_{(2)}^{(j)} &\approx -\left(\frac{\hbar^2}{2\,m}\right)^j \frac{e^{\frac{\mu}{k_0\,T}}}{2\,k_0\,T}\,\frac{\Gamma\left(\frac{5}{2}+j\right)}{\beta^{\frac{5}{2}+j}}\,; \qquad (n' = 0)\,,\\
K_{(2)}^{(j)} &\approx -\left(\frac{\hbar^2}{2\,m}\right)^j \frac{e^{\frac{\mu}{k_0\,T}}}{2\,k_0\,T}\,\frac{\Gamma\,(2+j)}{\beta^{2+j}}\,; \qquad (n' = 1)\,,\\
K_{(2)}^{(j)} &\approx -\left(\frac{\hbar^2}{2\,m}\right)^j \frac{e^{\frac{\mu}{k_0\,T}}}{4\,k_0\,T}\,\frac{\Gamma\left(\frac{3}{2}+j\right)}{\beta^{\frac{3}{2}+j}}\left(c^{(j)} + \ln\frac{k_0\,T}{\hbar\,\omega_0}\right);\\
&\qquad (n' = 2),\; (c^{(0)} = 1.42,\; c^{(1)} = 2)\,,
\end{aligned}\right\} \tag{2.9}$$

where use has been made of the following relations:

$$\sum_{\alpha=1}^{2} B_\alpha' = 1\,, \qquad \sum_{\alpha=1}^{2} B_\alpha'\,k_\alpha = 0\,, \qquad \sum_{\alpha=1}^{2} B_\alpha'\,k_\alpha^2 = -\frac{1}{a}\,. \tag{2.10}$$

From (2.9), (2.2), and (2.1) we obtain the known expressions for the kinetic coefficients in the case when scattering by optical phonons is predominant at high temperatures [3]. In the case $n' = 1$ (which was not studied in [3]) we get

$$\alpha_f^{(2)} = \frac{16\,R\,\hbar\,\omega_0\,\tau_{0f}^{(2)}}{3\,(2\,\pi\,m\,k_0)^{1/2}\,T^{3/2}\,u_2}\,, \qquad \varkappa_f^{(2)} = -\frac{k_0\,\alpha_f^{(2)}}{e}\,\sigma\,T\,, \tag{2.11}$$

where $u_2$ is the mobility of charge carriers interacting with optical phonons.

When $D = 0$ or $D < 0$ all solutions of equation (2.7) are different from zero, and then the integrals (2.4) are given by the expressions (2.6). It must be noted that the quantities $A_\alpha$ in these expressions satisfy also the relations (2.10), whereas the quantities $B_\alpha$ satisfy the relations

$$\left.\begin{aligned}
&\sum_{\alpha=1}^{4} B_\alpha = 1\,,\\
&\sum_{\alpha=1}^{4} B_\alpha\,k_\alpha = \sum_{\alpha=1}^{4} B_\alpha\,k_\alpha^2 = \sum_{\alpha=1}^{4} B_\alpha\,k_\alpha^3 = 0\,,\\
&\sum_{\alpha=1}^{4} B_\alpha\,k_\alpha^4 = -\frac{1}{b}\,.
\end{aligned}\right\} \tag{2.12}$$

If the parameters $|\gamma'_\alpha|$ and $|\gamma_\alpha|$ have large or small values so that asymptotic expressions or power series can be used for the functions $\mathrm{Ei}(x)$ and $W_{\mu,\nu}(x)$ then, by taking into account (2.10) and (2.12), approximate expressions can be obtained for the integrals (2.6) which will lead to kinetic coefficients for the following scattering mechanisms:

In the first case (i.e. $|\gamma'_\alpha|$ and $|\gamma_\alpha|$ are large) the expressions for the kinetic coefficients are valid when scattering of carriers by ionized impurities predominates. In the second case (i. e. $|\gamma'_\alpha|$ and $|\gamma_\alpha|$ are small) the expressions for the kinetic coefficients are valid when scattering of carriers by acoustic phonons is predominant.

In the high temperature range the conditions $D = 0$ and $D < 0$ are practically not satisfied for an ionic semiconductor even when the dielectric constant and the acoustic phonon velocity are smaller than usually, the maximum angular frequency of optical phonons is higher than usually and the impurities are completely ionized (the semocinductor is assumed to be non-degenerate, $N_\mathrm{i} \approx \approx 10^{17}\ \mathrm{cm}^{-3}$). Indeed, an approximate numerical estimate shows that $a^2/4\,b \approx \approx 10^2\,T$ (if $\varepsilon' \approx 1$, $v_0 \approx 10^4$ cm/s, $\omega_0 \approx 10^{14}$ rad/s, $N_\mathrm{i} \approx 10^{17}\ \mathrm{cm}^{-3}$), i.e. the inequality $D > 0$ holds and the parameters $\gamma'_\alpha$ and $\gamma_\alpha$ remain small ($|\gamma'_\alpha| = = |\gamma_\alpha| \approx 10^2/T^2$).

In conclusion, in impure semiconductors scattering of carriers by optical phonons prevails in the high temperature range.

### 3. The Low Temperature Range

For a calculation of the kinetic coefficients in this range we use again the expressions (2.1) where $a^{(i)}$ and $a^{(j)}_{(1)}$ have the same form as in (2.2). For $a^{(j)}_{(2)}$ we have [3]

$$\left.\begin{aligned} a^{(j)}_{(2)} &= \frac{24\, e^{1-j}\, R^2\, k_0\, T\, \tau^{(2)}_{0\,\mathrm{f}}\, \mathrm{e}^{\frac{\hbar\omega_0}{k_0 T}}}{3\,\pi^2} K^{(j)}_{(2)}\,; & (n' = 0)\,,\\ a^{(j)}_{(2)} &= \frac{8\, e^{1-j}\, R^2\, k_0\, T\, \hbar\, \tau^{(2)}_{0\,\mathrm{f}}}{3\,\pi^2\, m} \left(\frac{2\,m}{\hbar\,\omega_0}\right)^{\frac{1}{2}} \mathrm{e}^{\frac{\hbar\omega_0}{k_0 T}} K^{(j)}_{(2)}\,; & (n' = 1)\,,\\ a^{(j)}_{(2)} &= \frac{4\, e^{1-j}\, R^2\, k_0\, T\, \hbar\, \tau^{(2)}_{0\,\mathrm{f}}\, \mathrm{e}^{\frac{\hbar\omega_0}{k_0 T}}}{3\,\pi^2\, m\, \omega_0} K^{(j)}_{(2)}\,; & (n' = 2)\,, \end{aligned}\right\} \tag{3.1}$$

where

$$K^{(j)}_{(2)} = \int\limits_0^\infty \frac{\tau}{\tau^{(2)}}\, \varepsilon^j\, k^4\, \frac{\partial f_0}{\partial \varepsilon}\, \mathrm{d}k\,.$$

The relations (2.5) become

$$\tau = \frac{\tau^{(3)}_0\, k^3}{b\, k^4 + a\, k^3 + 1}\,, \qquad \frac{\tau}{\tau^{(1)}} = \frac{b\, k^4}{b\, k^4 + a\, k^3 + 1}\,, \qquad \frac{\tau}{\tau^{(2)}} = \frac{a\, k^3}{b\, k^4 + a\, k^3 + 1}\,, \tag{3.2}$$

where $a$ and $b$ have the same meaning as in the first part of this paper; $\tau^{(3)}_0$ and $\tau^{(1)}_0$ are given by the same expressions as in the high temperature range, and $\tau^{(2)}_0$ is given by the known expression in the low temperature range [1, 2].

The dependence of the phonon relaxation time $\tau^{(s)}_\mathrm{f}$ on the wave vector is the same as in the high temperature range [4]. Taking into account these remarks and computing the integrals which are involved in $a^{(i)}$, $a^{(j)}_{(1)}$, and $a^{(j)}_{(2)}$ we obtain

$$\left.\begin{aligned}
I^{(i)} &= \left(\frac{\hbar^2}{2m}\right)^i \frac{\tau_0^{(3)}\, e^{\frac{\mu}{k_0 T}}}{2\,k_0\,T} \sum_{\alpha=1}^{4} A_\alpha\, k_\alpha^{8+2i} \Bigg\{ \sum_{r=0}^{3+1} \frac{\Gamma\left(\frac{7}{2}+i-r\right)}{\gamma_\alpha^{\frac{7}{2}+i-r}} + \sum_{r=0}^{2+i} \frac{\Gamma(3+i-r)}{\gamma_\alpha^{3+i-r}} - \\
&\quad - \left[ e^{-\gamma_\alpha}\,\mathrm{Ei}(\gamma_\alpha) + \frac{\sqrt{\pi}\,(-\gamma_\alpha)^{\frac{1}{4}}\, e^{-\frac{\gamma_\alpha}{2}}}{\gamma_\alpha^{1/2}}\, W^{(-\gamma_\alpha)}_{-\frac{1}{4},-\frac{1}{4}} \right] \Bigg\}, \\
K_{(1)}^{(j)} &= \left(\frac{\hbar^2}{2m}\right)^j \frac{b\, e^{\frac{\mu}{k_0 T}}}{2\,k_0\,T} \sum_{\alpha=1}^{4} A_\alpha\, k_\alpha^{9+2j-2p} \times \\
&\quad \times \Bigg\{ \sum_{r=0}^{3+j-p} \left[ \frac{\Gamma\left(\frac{7}{2}+j-p-r\right)}{\gamma_\alpha^{\frac{7}{2}+j-p-r}} + \frac{\Gamma(4+j-p-r)}{\gamma_\alpha^{4+j-p-r}} \right] - \\
&\quad - \left[ e^{-\gamma_\alpha}\,\mathrm{Ei}(\gamma_\alpha) + \frac{\sqrt{\pi}\,(-\gamma_\alpha)^{\frac{1}{4}}\, e^{-\frac{\gamma_\alpha}{2}}}{\gamma_\alpha^{1/2}}\, W^{(-\gamma_\alpha)}_{-\frac{1}{4},-\frac{1}{4}} \right] \Bigg\};\ (n = 2p,\ p = 0.1), \\
K_{(2)}^{(j)} &= \left(\frac{\hbar^2}{2m}\right)^j \frac{a\, e^{\frac{\mu}{k_0 T}}}{2\,k_0\,T} \sum_{\alpha=1}^{4} A_\alpha\, k_\alpha^{8+2j} \times \\
&\quad \times \Bigg\{ \sum_{r=0}^{3+j} \frac{\Gamma\left(\frac{7}{2}+j-r\right)}{\gamma^{\frac{7}{2}+j-r}} + \sum_{r=0}^{2+j} \frac{\Gamma(3+j-r)}{\gamma_\alpha^{3+j-r}} - \\
&\quad - \left[ e^{-\gamma_\alpha}\,\mathrm{Ei}(\gamma_\alpha) + \frac{\sqrt{\pi}\,(-\gamma_\alpha)^{\frac{1}{4}}\, e^{-\frac{\gamma_\alpha}{2}}}{\gamma_\alpha^{1/2}}\, W^{(-\gamma_\alpha)}_{-\frac{1}{4},-\frac{1}{4}} \right] \Bigg\};\ (n = 0, 1, 2).
\end{aligned}\right\} \tag{3.3}$$

The coefficients of the decomposition of the integrands into partial fractions, $A_\alpha$, satisfy the relations (2.12), $k_\alpha$ are the solutions of the following equation:

$$x^4 + a\,x + b = 0\,; \qquad \left(x = \frac{1}{k}\right). \tag{3.4}$$

To establish the ranges in which the various carrier scattering mechanisms are predominant we must know the values of the discriminator $D$ of equation (3.4). The discriminator $D = \frac{a^4}{4} - \left(\frac{4\,b}{3}\right)^3$ depends on the type of semiconductor and its temperature. Approximate numerical estimates indicate that in this case $\frac{a^4}{b^3} \sim \frac{10^{82}\,\varepsilon'^2\,Z^8\,v_0^6}{\omega_0^6\,N_i\,T^3}\, e^{-\frac{4\,\hbar\omega_0}{k_0 T}}$. For a semiconductor with the parameter values $\varepsilon' \sim 10$, $v_0 \sim 10^5$ cm/s, $\omega_0 \sim 10^{13}$ rad/s, $N_i \sim 10^{15}\ \mathrm{cm}^{-3}$ we find that the ratio $a^4/b^3$ varies between $\sim 10^{13}$ and $\sim 10^3$ for a temperature variation from the upper limit of the low temperature range ($\sim 10^2$ °K) to about 12 °K. (In these conditions $D > 0$.) At about 10 °K the ratio $a^4/b^3$ is of the order unity, i.e. $D = 0$. For lower temperatures (between 10 and 1 °K) this ratio varies between values of order unity and $\sim 10^{-154}$. Therefore we have at low temperatures $D < 0$.

When the impurity concentration is large (but the semiconductor remains still non-degenerate) then the temperature corresponding to $D \approx 0$ is higher than in the first case.

It must be mentioned that the ranges of $D$-variation change essentially with maximum angular frequency. For instance, if $\omega_0 \approx 10^{12}$ rad/s the ratio $a^4/b^3$ varies between $\sim 10^{20}$ and $\sim 10^9$ in the whole low temperature range $10^2$ to 1 °K. This means that $D > 0$. If $\omega_0 \approx 10^{14}$ rad/s, the ratio $a^4/b^3$ is under $10^{-5}$ (i.e. $D < 0$).

In the following we will consider the three ranges of $D$ separately.

a) $D < 0$

In this case we can write to a very good approximation $D \approx \left(\frac{a^2}{2}\right)^2$, then equation (3.4) has three solutions differing from zero. The integrals $I^{(i)}$ and $K^{(j)}_{(s)}$ are given by (3.3) with $\alpha = 1, 2, 3$ and the coefficients $A_\alpha$ satisfy the relations

$$\sum_{\alpha=1}^{3} A_\alpha = 1\,, \qquad \sum_{\alpha=1}^{3} A_\alpha\, k_\alpha = \sum_{\alpha=1}^{3} A_\alpha\, k_\alpha^2 = 0\,, \qquad \sum_{\alpha=1}^{3} A_\alpha\, k_\alpha^3 = -\frac{1}{a}\,. \tag{3.5}$$

The absolute values of $\gamma_\alpha = \beta\, k_\alpha^2$ are of the same order of magnitude $\left(\text{i.e. } \approx \frac{\beta}{a^{2/3}}\right)$ and vary, in the temperature range corresponding to $D < 0$, between $\approx 10^{-3}$ and order unity for $N_i \approx 10^{15}$ cm$^{-3}$ or between $\approx 10^{-1}$ and $\approx 10^2$ when the impurity concentration increases ($N_i \approx 10^{17}$ cm$^{-3}$). If the maximum angular frequency of the optical phonons is small ($\omega_0 \approx 10^{12}$ rad/s) the parameters $|\gamma_\alpha|$ vary from $\approx 10^{-5}$ to order unity (for $N_i \approx 10^{15}$ cm$^{-3}$) or from $\approx 10^{-3}$ to $\approx 10^2$ (for $N_i \approx 10^{17}$ cm$^{-3}$). When the parameters $|\gamma_\alpha|$ are small (at the upper limit of the low temperature range) a power series can be used for the functions $\mathrm{Ei}(x)$ and $\mathrm{W}_{\mu,\nu}(x)$ [5]. With the help of (3.3) and (3.5) we can now compute the kinetic coefficients and obtain the known result that in this case scattering of carriers by optical phonons prevails at low temperature [3].

When the parameters $|\gamma_\alpha|$ are large (at the lower limit of the low temperature range) asymptotic expressions can be used for $\mathrm{Ei}(x)$ and $\mathrm{W}_{\mu,\nu}(x)$ [5] which, by using (3.5), are of the following from

$$\left.\begin{aligned} I^{(i)} &\approx -\left(\frac{\hbar^2}{2\,m}\right)^i \frac{\tau_0^{(3)}\, \mathrm{e}^{\frac{\mu}{k_0 T}}}{2\,k_0\,T}\, \frac{\Gamma(4+i)}{\beta^{4+1}}\,, \\ K^{(j)}_{(1)} &\approx -\left(\frac{\hbar^2}{2\,m}\right)^j \frac{b\, \mathrm{e}^{\frac{\mu}{k_0 T}}}{2\,k_0\,T}\, \frac{\Gamma\left(\frac{9}{2}+j-p\right)}{\beta^{\frac{9}{2}+j-p}}\,; \qquad (n = 2\,p;\quad p = 0, 1)\,, \\ K^{(j)}_{(2)} &\approx -\left(\frac{\hbar^2}{2\,m}\right)^j \frac{a\, \mathrm{e}^{\frac{\mu}{k_0 T}}}{2\,k_0\,T}\, \frac{\Gamma(4+j)}{\beta^{4+j}}\,; \qquad (n' = 0, 1, 2)\,. \end{aligned}\right\} \tag{3.6}$$

The kinetic coefficients can be again computed by using (3.6) together with (2.1), (2.2), and (3.1). For the contribution of carriers to the kinetic coefficients we obtain the known expressions for scattering of carriers by ionized impurites. The contribution of phonon drag to the kinetic coefficients is given by the following expressions:

$$\left.\begin{aligned}
\alpha_f^{(1)} &\approx \frac{2^{2-2p}\, v_0^2\, \tau_{0\,f}^{(1)}}{9\sqrt{\pi}\,(4-2p)\,T\,u_1}\cdot\frac{\Gamma\left(\frac{9}{2}-p\right)}{\beta^{-p}}\,,\\
\varkappa_f^{(1)} &\approx -\frac{k_0\,\alpha_f^{(1)}\,\sigma\cdot T}{e}\left[4-\frac{\Gamma\left(\frac{11}{2}-p\right)}{\Gamma\left(\frac{9}{2}-p\right)}\right]; \qquad (p=0,1)\,,\\
\alpha_f^{(2)} &= \frac{24\,R^2\,m\,k_0\,\tau_{0\,f}^{(2)}\,e^{\frac{\hbar\omega_0}{k_0T}}}{\hbar^2\,u_2}\,; \qquad \varkappa_f^{(2)}=0\,, \quad (n'=0)\,,\\
\alpha_f^{(2)} &= \frac{8\,R^2\,k_0\,\tau_{0\,f}^{(2)}}{\hbar\,u_2}\left(\frac{2\,m}{\hbar\,\omega_0}\right)^{\frac{1}{2}} e^{\frac{\hbar\omega_0}{k_0T}}\,; \qquad \varkappa_f^{(2)}=0\,, \quad (n'=1)\,,\\
\alpha_f^{(2)} &= \frac{4\,R^2\,k_0\,\tau_{0\,f}^{(2)}\,e^{\frac{\hbar\omega_0}{k_0T}}}{\hbar\,\omega_0\,u_2}\,; \qquad \varkappa_f^{(2)}=0\,, \quad (n'=2)\,,
\end{aligned}\right\} \tag{3.7}$$

where $u_1$ and $u_2$ are the mobilties of charge carriers interacting with acoustic and optical phonons, respectively.

It is seen that $\alpha_f^{(1)}$ and $\varkappa_f^{(1)}$ differ only by numerical factors from the corresponding expressions in the case of carrier scattering by optical or acoustic phonons. The quantities $\alpha_f^{(2)}$ and $\varkappa_f^{(2)}$ are the same irrespective of the scattering mechanism. Hence, the contribution of phonon drag to the kinetic coefficients is not essential influenced by the presence of impurities in a semiconductor.

b) $D = 0$

Now all solutions of equation (3.4) are different from zero, and this means that the integral expressions for $I^{(i)}$ and $K_{(s)}^{(j)}$ are given by (3.3). The absolute values of the parameters $\gamma_\alpha = \beta\, k_\alpha^2$ are of the same order of magnitude, i.e. $|\gamma_\alpha| \approx \beta\,\frac{b}{a^2}$ which is of order unity when $N_i \approx 10^{15}\ \mathrm{cm}^{-3}$ and of the order $\approx 10^2$ when $N_i \approx 10^{17}\ \mathrm{cm}^{-3}$.

When the parameters $|\gamma_\alpha|$ are large (high impurity concentration) the asymptotic expressions of $I^{(i)}$ and $K_{(s)}^{(j)}$ are given by (3.6), where use has been made of the fact that the coefficients $A_\alpha$ satisfy (2.2). Hence, scattering of carriers by impurities prevails, as it was shown before.

If the parameters $|\gamma_\alpha|$ are small (low impurity concentration) power series can be used for the functions $I^{(i)}$ and $K_{(s)}^{(j)}$. Then we find the same expressions for the kinetic coefficients as in the case of carrier scattering by acoustic phonons [3].

c) $D < 0$

Then all solutions of equation (3.4) are different from zero, and the functions $I^{(i)}$ and $K_{(s)}^{(j)}$ are given by (3.3). The absolute values of $\gamma_\alpha$ are of the same order of magnitude $\left(|\gamma_\alpha| \approx \frac{\beta}{b^{1/2}}\right)$.

When $\omega_0 \approx 10^{13}$ rad/s and $T$ varies between 10 and 1 °K then $\frac{\beta}{b^{1/2}}$ varies between $\approx 1$ and $10^2$ (for $N_i \approx 10^{15}\ \text{cm}^{-3}$) and between $\approx 10$ and $\approx 10^3$ (for $N_i \approx 10^{17}\ \text{cm}^{-3}$). If $\omega_0 \approx 10^{14}$ rad/s then the ratio $\frac{\beta}{b^{1/2}}$ varies from $\approx 10^{-1}$ to $\approx 10^2$ (for $N_i \approx 10^{15}\ \text{cm}^{-3}$) and from $\approx 1$ to $\approx 10^3$ (for $N_i \approx 10^{17}\ \text{cm}^{-3}$) in the whole low temperature range $10^2$ to 1 °K. Hence, in this range scattering of carriers by ionized impurities prevails too, whereas for small $|\gamma_a|$ scattering of carriers by acoustic phonons prevails. It is found that in every range of $D$ two carrier scattering mechanisms occur whose nature depends on the temperature and type of semiconductor. Further it should be stressed that, depending on type of semiconductor, the optical and acoustic phonon scattering mechanisms can be completely neglected throughout the low temperature range. For instance, if $\omega_0 \approx 10^{12}$ rad/s the acoustic phonon scattering can be completely neglected, and for $\omega_0 \approx 10^{14}$ rad/s the same holds for optical phonon scattering. The impurity scattering mechanism can be completely neglected throughout the low temperature range if the semiconductor is pure.

The discussions about the ranges in which different carrier scattering mechanisms prevail have to include considerations about the dielectric constant $\varepsilon'$ and acoustic phonon velocity $v_0$ for different semiconductors. But these are omitted because $D$ depends strongly (exponentially) on temperature and maximum angular frequency of optical phonons.

Finally, it must be noted that the results of this paper about the ranges in which different carrier mechanisms prevail, are only orienting ones which is due to the simplifying assumptions involved.

## References

[1] A. I. ANSELM, Introduction to the Theory of Semiconductors, Fizmatgiz, 1962 (in Russian).
[2] I. M. TSIDILKOVSKII, Thermomagnetic Phenomena in Semiconductors, Fizmatgiz, 1962 (in Russian).
[3] C. N. PLĂVIŢU, phys. stat. sol. **12**, 265 (1965).
[4] C. HERRING, Phys. Rev. **95**, 954 (1954).
[5] I. M. RYSHIK and I. S. GRADSTEIN, Tables of Series, Products, and Integrals, VEB Deutscher Verlag der Wissenschaften, Berlin 1957.

*(Received March 31, 1966)*

phys. stat. sol. **16**, 79 (1966)

*Physikalisches Institut der Universität Freiburg i. Br. (a)*
*and Institut für Elektrowerkstoffe, Freiburg i. Br. (b)*

# Comments on Paraelectric Resonance

By

P. Sauer (a), O. Schirmer (b), and J. Schneider (a)

The conditions under which paraelectric resonance can be observed are investigated. Paraelectric resonance is defined as a resonant transition between energy levels formed by a site-degenerate electric dipole in a crystal. Possible applications of the present theory may be found in systems of the type KCl: $OH^-$. In these systems the ground state of the impurity ion has a sixfold site degeneracy which is removed by the tunnel effect and the electric field. Low temperature anomalies of the static dielectric constant caused by the tunnel effect are also discussed.

Wir untersuchen die Bedingungen, unter denen paraelektrische Resonanz beobachtet werden kann. Paraelektrische Resonanz ist dabei definiert als resonanter Übergang zwischen den Niveaus, die ein mehrfach lageentarteter elektrischer Dipol in einem Kristall besitzt. Die Theorie wird dargestellt an einem System des Typs KCl: $OH^-$, dessen Grundzustand sechsfach lageentartet ist. Tunneleffekt und äußeres elektrisches Feld können diese Entartung aufheben. Außerdem werden Anomalien der statischen Dielektrizitätskonstante diskutiert, die bei tiefen Temperaturen infolge des Tunneleffektes zu erwarten sind.

## 1. Introduction

A molecular impurity occupying a defined lattice site in a cubic host crystal has in general the choice between different, but energetically identical, equilibrium orientations. If the impurity ion possesses a permanent electric dipole moment, this orientational or site degeneracy can be lifted by an externally applied electric field.

As a practical example, we will consider $OH^-$ molecular ions in KCl. Kuhn and Lüty [1] have demonstrated that (i) the dipoles are oriented along the six $C_4$-axes of the host crystal; (ii) the population of the dipoles in the six equivalent sites can be changed by an externally applied electric field; (iii) the $OH^-$ centre possesses an effective electric dipole moment of $\mu = 3.94$ Debye.

We will now ask for the Stark splitting of the sixfold degenerate ground state, caused by an externally applied electric field $\boldsymbol{E}$. For $\boldsymbol{E} \| [100]$, there is one type of dipole oriented parallel to $\boldsymbol{E}$, its energy being increased[1]) by the amount $+\mu E$. Consequently, dipoles which point opposite to $\boldsymbol{E}$ are energetically lowered by $-\mu E$. Finally, there are four types of dipoles pointing normal to $\boldsymbol{E}$ which experience no change of their potential energy (see Fig. 1).

One is lead to ask whether it may be possible to induce paraelectric resonance transitions between these Stark levels, which should occur at a frequency $\nu = \mu E/h$ or $\nu = 2\mu E/h$. For externally applied electric fields of only a few kV/cm in strength, the energy splitting of the Stark levels corresponds already

[1]) Throughout this paper, we assign the positive energies $W$ to the more stable states.

to microwave frequencies. The possibility of observing paraelectric resonance had already been visualized previously by Känzig [2], Kuhn and Lüty [3], and Shepherd and Feher [4].

In this paper, we will give a short outline of the theory of paraelectric resonance. We will discuss the theory on the typical example of a sixfold site degenerate system in cubic symmetry, as KCl:$OH^-$. As a first step, we will discuss the splitting of the sixfold degenerate ground state which results from the tunnel effect. Next, we will investigate the additional splitting of the levels by the Stark effect. It will turn out that a finite transition probability in paraelectric resonance is intimately related with the tunnel effect.

Observation of paraelectric resonance has very recently been reported by Dreyfus and Bron [5] for the system KCl:$OH^-$. However, we feel that the experimental data given are not yet sufficient to justify a direct comparison with the theoretical predictions to be presented in this paper.

## 2. Paraelectric Resonance in a Sixfold Site Degenerate System

### *2.1 Tunnel effect and Stark effect*

The zero-point librational and translational motion of the $OH^-$ dipole will induce tunneling between its six equilibrium sites. Consequently, the sixfold ground state will be split, in close analogy to the inversion splitting of ammonia. The energies are obtained from the characteristic equation

$$\det |H_{ij} - W S_{ij}| = 0 \,, \tag{1}$$

where $i, j = 1, 2 \ldots 6$. To simplify the discussion, we will neglect the overlap integrals, $S_{ij} = \delta_{ij}$. This assumption is justified in the weak coupling limit, i.e. for small tunneling rates. The tunnel effect is caused by two types of interactions[2])

$$H_{14} = H_{25} = H_{36} = \gamma$$

and

$$H_{ij} = \beta$$

for all other combinations of $i$ and $j$. Here, $\gamma$ and $\beta$ represent the interaction between opposite and neighbouring sites, respectively (see Fig. 1). The sixfold

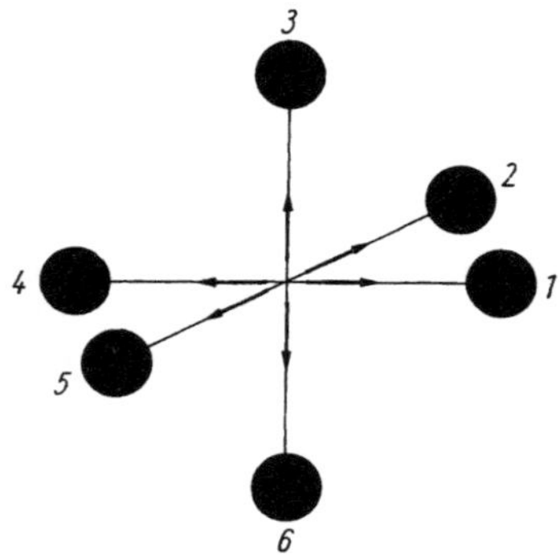


Fig. 1. Site degeneracy of the $OH^-$ dipole in KCl

[2]) It is apparent that this scheme corresponds closely to a LCAO-MO description of the electronic energy states of molecules.

degenerate ground state is found to split into a singlet $\Gamma_1^+$, a triplet $\Gamma_4^-$, and a doublet $\Gamma_3^+$ [3]),

$$\left.\begin{aligned} W(\Gamma_1^+) &= +4\beta + \gamma\,, \\ W(\Gamma_4^-) &= \qquad\; -\gamma\,, \\ W(\Gamma_3^+) &= -2\beta + \gamma\,. \end{aligned}\right\} \tag{2}$$

The matrix elements $H_{ii}$ do not cause a splitting of the ground state and will therefore be omitted. The eigenfunctions are

$$\left.\begin{aligned} &\Gamma_1^+: && \frac{1}{\sqrt{6}}(\varphi_1 + \varphi_2 + \varphi_3 + \varphi_4 + \varphi_5 + \varphi_6)\,, \\ &\Gamma_4^-: && \frac{1}{\sqrt{2}}(\varphi_1 - \varphi_4)\,, \\ & && \frac{1}{\sqrt{2}}(\varphi_2 - \varphi_5)\,, \\ & && \frac{1}{\sqrt{2}}(\varphi_3 - \varphi_6)\,, \\ &\Gamma_3^+: && \frac{1}{2}(\varphi_1 - \varphi_2 + \varphi_4 - \varphi_5)\,, \\ & && \frac{1}{\sqrt{12}}(2\,\varphi_3 + 2\,\varphi_6 - \varphi_1 - \varphi_2 - \varphi_4 - \varphi_5)\,. \end{aligned}\right\} \tag{3}$$

The same splitting is also found for quite different cases which have in common only a sixfold degenerate ground state in octahedral symmetry, as, possibly, $Li^+$ ions in KCl [7, 8] or the shallow donors in silicon [9]. The occurrence of a tunnel splitting in KCl:$OH^-$ was first visualized by Känzig and coworkers [10].

Next, we will investigate the additional splitting of the energy levels of KCl:$OH^-$ which is induced by an externally applied electric field. In Section 1, we had already mentioned the high-field solutions, for $\boldsymbol{E}\|[100]$, which will be approximated when the electric dipole energy is much greater than the energy splitting resulting from the tunnel effect. In the presence of tunneling, the Stark effect is readily taken into account by augmenting the diagonal elements in the characteristic equation (1) by the potential energies $\boldsymbol{\mu}\cdot\boldsymbol{E}$ appropriate to the different orientations of the dipole $\boldsymbol{\mu}$ with respect to $\boldsymbol{E}$. For $\boldsymbol{E}\|[100]$, e.g., $H_{11}$ and $H_{44}$ are augmented by $\pm\,\mu\,E$. The electric field will lower the symmetry of the centre from $O_h$ to $C_{4v}$, and the states are further split. The positions of the energy levels, calculated in this way, are listed below.

$$\left.\begin{aligned} &\boldsymbol{E}\|[100]\,,\ C_{4v}: \\ &\{\mu^2 E^2 + \gamma^2 - W(\Gamma_1)^2\}\,\{2\beta + \gamma - W(\Gamma_1)\} - 8\beta^2\,\{\gamma + W(\Gamma_1)\} = 0\,, \\ &\qquad W(\Gamma_3) = -2\beta + \gamma\,, \\ &\qquad W(\Gamma_5) = -\gamma\,, \end{aligned}\right\} \tag{4}$$

[3]) Throughout this work, we will label the irreducible representations following the nomenclature given in the monography of Koster et al. which is exceptionally appropriate for the problem under study (see reference [6]).

$\boldsymbol{E}||[111]$, $C_{3v}$:

$$\left.\begin{aligned} W(\Gamma_1) &= +2\beta \pm \sqrt{(2\beta+\gamma)^2 + \frac{\mu^2 E^2}{3}}\,, \\ W(\Gamma_3) &= -\beta \pm \sqrt{(\beta-\gamma)^2 + \frac{\mu^2 E^2}{3}}\,, \end{aligned}\right\} \tag{5}$$

$\boldsymbol{E}||[110]$, $C_{2v}$:

$$\left.\begin{aligned} &\{4\beta+\gamma-W(\Gamma_1)\}\{\gamma+W(\Gamma_1)\}\{2\beta-\gamma+W(\Gamma_1)\} - \\ &\qquad - \{\gamma - W(\Gamma_1)\}\frac{\mu^2 E^2}{2} = 0\,, \\ &W(\Gamma_2) = -\gamma\,, \\ &W(\Gamma_4) = -\beta \pm \sqrt{(\beta-\gamma)^2 + \frac{\mu^2 E^2}{2}}\,. \end{aligned}\right\} \tag{6}$$

The Stark effect is illustrated in Fig. 2a and b for the two limiting cases $\gamma = 0$ and $\beta = 0$, where tunneling occurs only between neighbouring sites, $\gamma = 0$, or between opposite sites, $\beta = 0$. There is some similarity with a Breit-Rabi diagram which describes the transition from the zero-field energy splitting to the Paschen-Back region, for an unpaired electron interacting with a nuclear spin and with an externally applied magnetic field.

### *2.2 Selection rules for paraelectric resonance*

In the absence of an external static electric field, electric dipole transitions can be induced by an oscillating electric field $\boldsymbol{E}_{\mathrm{hf}}$ between states of opposite parity. The transition moments do not depend on the orientation of $\boldsymbol{E}_{\mathrm{hf}}$ with respect to the crystalline axes and are calculated from the zero-field eigenfunctions, as given by (3),

$$\Gamma_1^+ - \Gamma_4^-:\quad \frac{1}{\sqrt{3}}\,\mu\,,$$

$$\Gamma_4^- - \Gamma_3^+:\quad \sqrt{\frac{2}{3}}\,\mu\,.$$

The situation is quite different in the high-field regime, where $\mu E \gg \beta, \gamma$. Here, we are looking for transitions which involve a change of the potential energy of the dipole in the static electric field, in close analogy to paramagnetic resonance. The transition moments are now calculated in the following way: We first take the high-field eigenfunctions appropriate to the symmetry of the problem, e.g. those for $C_{4v}$ if the static field is along [100]. These eigenfunctions are correct in the absence of tunneling. Consequently, the intensity of a paraelectric resonance transition, calculated from these basis functions, is zero. However, as soon as a finite tunneling rate is admitted, the exchange interaction $H_{ij}$ weakly perturbs the eigenfunctions and paraelectric resonance transitions appear. The transition moments are of the order of $\beta/E$ and $\gamma/E$, being much smaller than those in zero static field.

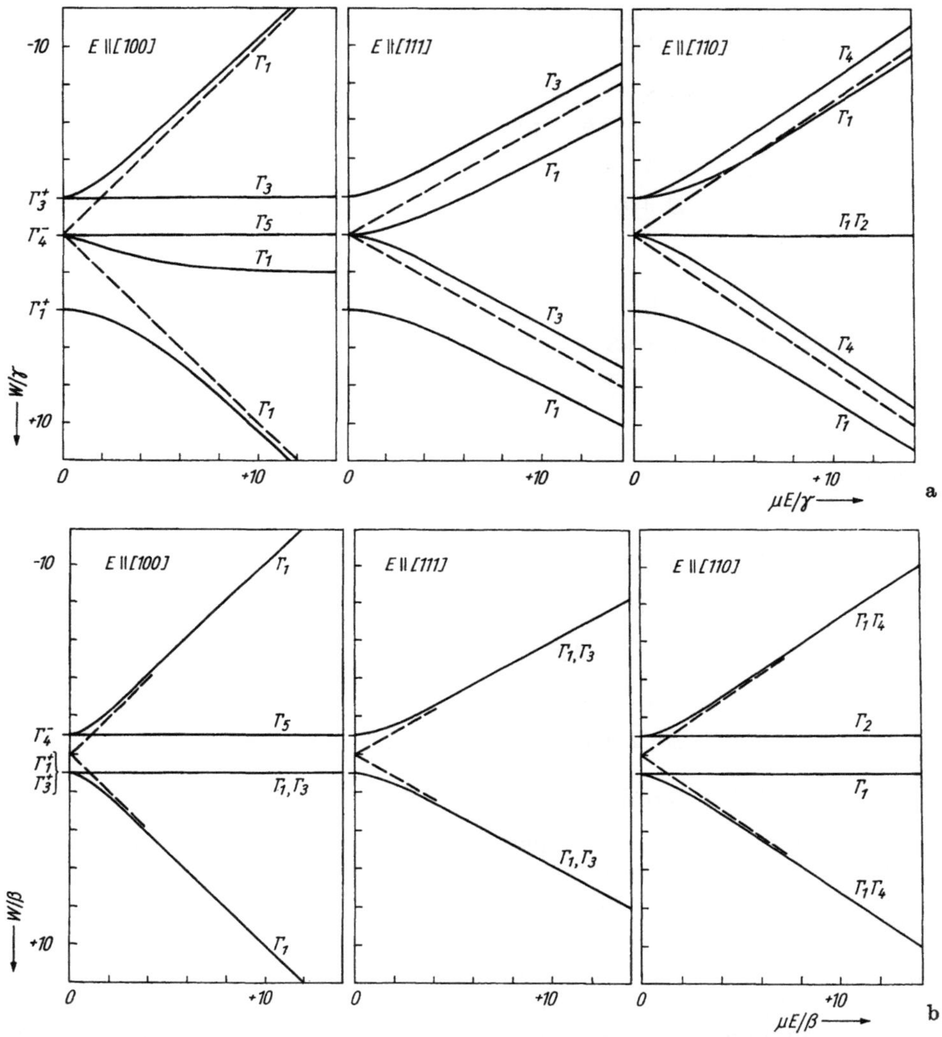


Fig. 2. Stark effect of a sixfold site degenerate state for the case that tunneling occurs only a) between neighbouring sites, $\gamma = 0$, or b) between opposite sites, $\beta = 0$. The energy values are plotted in normalized units, $W/\beta$ or $W/\gamma$, against the Stark field in units of $\mu E/\beta$ or $\mu E/\gamma$. The broken lines indicate the Stark splitting in the classical limit, $\gamma = 0$ or $\beta = 0$. In Fig. 2a the ordinate should read $W/\beta$, the abscissa $\mu E/\beta$, in Fig. 2b the ordinate should read $W/\gamma$, the abscissa $\mu E/\gamma$.

In our discussion, we have sofar purposely neglected off-diagonal electric transition moments since their energetical contribution, $\langle \varphi_i | \mu E | \varphi_j \rangle$, is much smaller than that of the matrix element given above, $\langle \varphi_i | H | \varphi_j \rangle = \beta, \gamma$. The reason is that we only consider systems where the Stark splitting is very small as compared with the librational or vibrational splitting. In other words, the externally applied electric field is assumed to be much weaker than the crystalline electric fields, locking the dipole in its equilibrium position.

A summary of the energies, polarisations, and transition moments of the paraelectric resonance transitions is compiled in the following tables.

$\boldsymbol{E}||[100]$

| | | | |
|---|---|---|---|
| $\Gamma_1$—$\Gamma_1$: | $\mu E \pm (2\beta + \gamma)$ | $\parallel$ | $2\beta/E$ |
| $\Gamma_1$—$\Gamma_1$: | $2\mu E$ | $\parallel$ | $\gamma/E$ |
| $\Gamma_1$—$\Gamma_5$: | $\mu E \pm \gamma$ | $\perp$ | $\sqrt{2}\,\beta/E$ |

$\boldsymbol{E}||[111]$

| | | | |
|---|---|---|---|
| $\Gamma_1$—$\Gamma_1$: | $2\mu E/\sqrt{3}$ | $\parallel$ | $(2\beta + \gamma)/E$ |
| $\Gamma_3$—$\Gamma_3$: | $2\mu E/\sqrt{3}$ | $\parallel$ | $\sqrt{2}\,(\beta - \gamma)/E$ |
| $\Gamma_3$—$\Gamma_3$: | $2\mu E/\sqrt{3}$ | $\perp$ | $(\beta - \gamma)/E$ |
| $\Gamma_1$—$\Gamma_3$: | $2\mu E/\sqrt{3} \pm 3\beta$ | $\perp$ | $(\beta + 2\gamma)/2E$ |

$\boldsymbol{E}||[110]$

| | | | |
|---|---|---|---|
| $\Gamma_1$—$\Gamma_1$: | $\sqrt{2}\,\mu E$ | [110] | $(\beta + \gamma)/E$ |
| $\Gamma_1$—$\Gamma_1$: | $\mu E/\sqrt{2} \pm (\beta - \gamma)$ | [110] | $2\beta/E$ |
| $\Gamma_4$—$\Gamma_4$: | $\sqrt{2}\,\mu E$ | [110] | $(\beta - \gamma)/E$ |
| $\Gamma_1$—$\Gamma_4$: | $\sqrt{2}\,\mu E \pm 2\beta$ | $[1\bar{1}0]$ | $\gamma/E$ |
| $\Gamma_1$—$\Gamma_4$: | $\mu E/\sqrt{2} \pm (\beta + \gamma)$ | $[1\bar{1}0]$ | $2\beta/E$ |
| $\Gamma_1$—$\Gamma_2$: | $\mu E/\sqrt{2} \pm (\beta + \gamma)$ | [001] | $2\sqrt{2}\,\beta/E$ |

### 3. The Transition from Hindered Rotation to Libration

The motion of a molecular impurity in a solid, involving its angular degrees of freedom, can be discussed from two different points of view, hindered rotation or libration. It is our feeling that dipolar impurities in alkali halides are rather tightly locked in their equilibrium positions by the lattice forces. Therefore, the ground state of the dipolar ions will be probably better described in terms of libration. In contrast, the crystalline forces acting on molecular impurities in van der Waals lattices, as rare gas matrices, are sufficiently weak to favour hindered rotation. In this section, we want to show that there is a continuous transition between these two limiting cases, keeping in mind the example of the $OH^-$ dipole in KCl.

The energy states of a freely rotating diatomic molecule are classified by its angular momentum $J$:

$$W(J) = \frac{\hbar^2}{2\,\Theta} J\,(J + 1)\,,$$

where $\Theta$ is the moment of inertia. Each level contains $2J + 1$ sublevels which are degenerate in the absence of an external perturbation. If such a rotator is built into a crystal, this degeneracy will be lifted in part and the motion will become hindered. The splitting can be discussed by the methods of crystal field theory developed for paramagnetic ions. Here, we consider only the pertur-

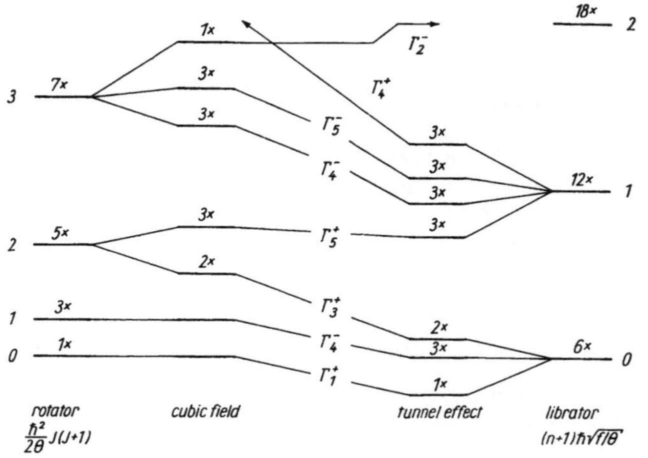


Fig. 3. Transition from hindered rotation to libration, as discussed for the $OH^-$ dipole in KCl. The relative positions of the librational levels with respect to the rotational levels are not drawn to scale

bations of the $J = 0, 1, 2$ states of the rotator by a cubic crystalline field. For this case, the left part of Fig. 3 gives all the information which can be derived from symmetry alone. Since we consider a molecule on a sixfold coordinated anion site, the doublet $\Gamma_3^+$ of the $J = 2$ state has been placed below the triplet, $\Gamma_5^+$, and, consequently, the singlet $\Gamma_2^-$ of the $J = 3$ state above the two triplets $\Gamma_4^-$ and $\Gamma_5^-$.

Quantitative values for the energy levels of a diatomic hindered rotator in $O_h$ symmetry have been obtained as early as 1936 by Devonshire [11]. On going to higher crystalline potentials, his energy levels are found to be grouped into sets of closely spaced terms. We would like to show that these sets mouth into the librational levels of the molecule. The tunnel splitting of these levels can be treated by LCAO theory, as in Section 2.1 of our paper.

If the potential becomes so high that tunneling is negligible, we are left with the equidistant energy states of the librational oscillator,

$$W(n) = (n + 1)\,\hbar\sqrt{\frac{f}{\Theta}}\,,$$

where $f$ is a force constant. The degeneracy of these librational levels is $6(n+1)$-fold, which can be understood by the following argument: In the ground state, the libration of the molecule in the two-dimensional potential is described by a non-degenerate wave function. Multiplication by the sixfold site degeneracy gives the multiplicity six for the ground state. The first excited state, separated from the ground state by one librational quantum $\hbar\sqrt{f/\Theta}$, is twelvefold degenerate, since there are two possibilities to accomodate one quantum in a two-dimensional oscillator. The extension of this argument to the higher levels is straightforward.

For a discussion of the tunnel splitting of the excited librational states, we must know their irreducible representations according to the point group $O_h$. This can be achieved in the following way: The characters of the several symmetry operations which connect the six librator sites are derived by inspection of the six-dimensional transformation matrices, i.e. only those librators lying on the symmetry element considered have diagonal elements and contribute to the character. Sofar, this is the the same technique, as used to construct molecular orbitals. In the excited states, however, the diagonal elements must be replaced by the characters for the specific symmetry operations of the overtones of a two-dimensional oscillator [12]. A summation of all diagonal elements leads to the the reducible character; the reduction into irreducible representations is then straightforward.

It then turns out that the librational states are split by the tunnel effect exactly into those symmetry types which we found for the sets of approaching terms in Devonshire's paper.

## 4. Dielectric Constant

The classical expression for the paraelectric susceptibility of a dilute system of non-interacting electric dipoles is

$$\chi = \frac{N \mu^2}{3 k T}, \tag{7}$$

where $N$ is the number of electric dipole moments $\mu$ per $cm^3$. Deviations from equation (7) are expected to occur at sufficiently low temperatures, when the thermal energy $k T$ becomes comparable with the energy splitting resulting from the tunnel effect.[4] We will therefore recalculate $\chi$ from the general expression[1])

$$\chi = \frac{N}{E} \frac{\sum_{i=1}^{6} \frac{\partial W_i}{\partial E} \cdot e^{W_i/k T}}{\sum_{i=1}^{6} e^{W_i/k T}} \tag{8}$$

taking into account the tunnel splitting of the sixfold degenerate ground state. The quadratic Stark effect of the tunnel-split energy levels in a weak [5]) electric field is calculated by second order perturbation theory, or by specializing equations (4), (5), and (6) to weak electric fields.

With the abbreviations $W(\Gamma_1^+) = W_1$, $W(\Gamma_1^+) - W(\Gamma_4^-) = W_{14} = -W_{41}$, etc., we then obtain

$$\chi = \frac{2}{3} N \mu^2 \frac{\frac{1}{W_{14}} (e^{W_1/k T} - e^{W_4/k T}) + \frac{2}{W_{43}} (e^{W_4/k T} - e^{W_3/k T})}{e^{W_1/k T} + 3 \cdot e^{W_4/k T} + 2 \cdot e^{W_3/k T}}. \tag{9}$$

In the high temperature limit, $k T \gg W_i$, this specializes to the classical relation (7). At very low temperatures, all dipoles are in the state $\Gamma_1^+$, and

$$\chi = N \mu^2 \frac{2}{3} \frac{1}{W(\Gamma_1^+) - W(\Gamma_4^-)}. \tag{10}$$

---

[4]) Additional low temperature anomalies for $\chi$ will occur when $k T$ becomes comparable with the mean dipole–dipole interaction energy, which increases with the dipole concentration $N$ (see [10]). We will consider only sufficiently diluted systems where these cooperative effects can be neglected.

[5]) Under normal experimental conditions, the electric field applied on the sample to measure $\chi$ will not exceed a few V/cm. Therefore, the Stark energy will be much smaller than the zero-field splitting.

This case is entirely analogous to that of van Vleck temperature-independent paramagnetism. The occurrence of low temperature anomalies in $\chi$, which result from the tunnel effect, was also recently mentioned by Bosshard, Dreyfus, and Känzig [13][6]).

Känzig, Hart, and Roberts [10] have measured the dielectric constant of KCl doped with $OH^-$ in very low concentration ($N = 4 \times 10^{16}$ cm$^{-3}$). No deviation from the $1/T$ dependence of $\chi$ was observed, down to 0.3 °K. This sets an upper limit of 0.3 °K for the tunnel splitting of the ground state.

The breakdown of the $1/T$ law of the paraelectric susceptibility, at temperatures where $kT$ becomes comparable with the tunnel splitting, also sets a lower limit for the ultimate temperature attainable by the technique of paraelectric cooling [3, 4].

## 5. Conclusion

The phenomenon of paraelectric resonance has been discussed on the typical example of a sixfold site degenerate system, as KCl:$OH^-$. For simplicity, we have considered only systems where the tunnel splitting of the sixfold ground state, as well as the electric dipole energy, are very small as compared with the energy of libration or vibration of the impurity centre in the host crystal. In this limit, the number of parameters required to describe the energy states and eigenfunctions, can be greatly reduced and the system becomes accessible to straightforward physical interpretation.

Härtel and Lüty [14] have shown that, by virtue of their elastic tensor moment, the $OH^-$ ions can also be oriented by an externally applied mechanical stress. The effect of uniaxial stress on the energy levels of a sixfold site degenerate system, split by exchange interaction, has been already treated by Wilson and Feher [9] for the isomorphous case of the shallow donor in silicon, and we can refer the reader to their work.

Our discussion of paraelectric resonance did not consider the effects of paraelectric relaxation, resulting from the interaction of the dipole with the lattice phonons. This dipole-lattice relaxation will determine part of the width of a paraelectric resonance line. In addition, random crystalline strains and electric fields will contribute to the linewidth.

***Acknowledgements***

We are greatly indebted to Dr. T. L. Estle of Texas Instruments for critical comments on the manuscript and for pointing out a serious error in its original form. We also wish to thank Prof. H. C. Wolf, Stuttgart, for bringing to our attention the phenomenon of paraelectric resonance. Thanks are also due to Prof. S. Flügge for his interest in this work.

## References

[1] U. Kuhn and F. Lüty, Solid State Commun. **2**, 281 (1964).
[2] W. Känzig, unpublished.
[3] U. Kuhn and F. Lüty, Solid State Commun. **3**, 31 (1965).
[4] I. Shepherd and G. Feher, Phys. Rev. Letters **15**, 194 (1965).

[6]) The expression for the low temperature susceptibility, as given in this paper, contains a typographical error (W. Känzig, private communication).

[5] W. E. Bron and R. W. Dreyfus, Phys Rev. Letters **16**, 165 (1966).
[6] G. F. Koster, J. O. Dimmock, R. G. Wheeler, and H. Statz, Properties of the Thirty-Two Point Groups, M.I.T. Press, Cambridge (Mass.) 1963.
[7] G. Lombardo and R. O. Pohl, Phys. Rev. Letters **15**, 291 (1965).
[8] H. S. Sack and M. C. Moriarty, Solid State Commun. **3**, 93 (1965).
[9] D. K. Wilson and G. Feher, Phys. Rev. **124**, 1068 (1961).
[10] W. Känzig, H. R. Hart, Jr., and S. Roberts, Phys. Rev. Letters **13**, 543 (1964).
[11] A. F. Devonshire, Proc. Roy. Soc. **A 153**, 601 (1936).
[12] M. Tinkham, Group Theory and Quantum Mechanics, McGraw-Hill Publ. Comp. 1964 (p. 246).
[13] U. Bosshard, R. W. Dreyfus, and W. Känzig, Phys. kondens. Materie **4**, 254 (1965).
[14] H. Härtel and F. Lüty, phys. stat. sol. **12**, 347 (1965).

*(Received April 1, 1966)*

*Note added in proof:*

Feher, Shepherd and Shore, Phys. Rev. Letters 16, 500 (1966), have observed paraelectric resonance of $OH^-$ dipoles in KCl at 35 GHz. The spectra were interpreted under the assumption that tunneling occurs only between neighbouring sites, i. e. $\gamma = 0$.

phys. stat. sol. **16**, 89 (1966)

*Institute of Physics, Czechoslovak Academy of Sciences, Prague*

# Etch Pits at Dislocations in High-Purity Iron Single Crystals

By

B. Šesták and S. Kadečková

A description is given of the etching (in a solution of 2% nital) of both fresh and grown-in dislocations on polished (100) surfaces of iron single crystals. The single crystals contain 25 to 50 ppm of carbon, 40 ppm of oxygen, 2 ppm of nitrogen, and other impurities totalling 15 ppm, and are prepared by the strain annealing method from Johnson and Matthey spectrographic iron.

Es wird das Ätzen (mit einer 2%igen Nitallösung) von neuen sowie beim Wachstum entstandener Versetzungen auf polierten (100) Flächen von Eiseneinkristallen beschrieben. Die Einkristalle enthalten 25 bis 50 ppm Kohlenstoff, 40 ppm Sauerstoff, 2 ppm Stickstoff und andere Verunreinigungen mit einer Gesamtkonzentration von 15 ppm. Die Präparation erfolgte durch die „Strain-anneal" Methode aus spektrographischem Eisen von Johnson und Matthey.

## 1. Introduction

An etching technique which reliably reveals dislocation sites in crystals by surface pits is very important for studying dislocations in specimens with densities of dislocations not exceeding $10^9$ to $10^{10}$ cm$^{-2}$. This technique has the advantage that the dislocations can be revealed at the surface of bulk specimens.

Boswell [1], Gorsuch [2], Seemann and Benedict [3], Schmidtmann, Klärner, and Schenck [4], Hahn [5], and recently Takeuchi [6] have described the etching of dislocations in iron. All these authors report, however, that before the etching it is necessary to cool down slowly the crystals from a temperature of 750 to 850 °C or age them at a lower temperature, e.g. 150 °C. This means that the methods which they describe allow only "aged" dislocations to be etched. Without prior decorating, dislocations have been successfully etched on natural {100} growing faces of iron whiskers using 2% nital or 4% picral [7], or a mixture of both [8]. So far it has been the only case of etching fresh dislocations in body-centred cubic metals. The report about the etching of fresh dislocations in tungsten single crystals (Schadler [9]) does not seem to be quite conclusive (see also Berlec [10]).

It has been the aim of the present work to etch dislocations in single crystals of high-purity iron. The method of etching "aged" dislocations, recently published by Takeuchi [6], was verified. On the single crystals which were prepared by recrystallization and which contained the impurities listed below it was possible to etch dislocations after ageing at 150 °C, as well as dislocations forming a substructure. On the same crystals, after annealing for 120 h at 860 °C in wet hydrogen, we failed to etch any dislocations. This method, therefore, does not appear to be sufficiently reliable, for the result of etching obviously depends on the amount of impurities and on their distribution in the crystals.

We have found, however, that on the (100) surface of these iron single crystals, both annealed and as-grown, it is possible to etch fresh and grown-in dislocations by using a solution of 2% nital, similarly as with the whiskers. Etching in a mixture of 2% nital and 4% picral does not yield reliable results.

## 2. Experimental Technique

The etching of dislocations was studied on single crystals prepared by the strain anneal method from Johnson and Matthey spectrographic iron [11, 12]. The crystals contained 25 to 50 ppm carbon, 40 ppm oxygen, 2 ppm nitrogen, and other elements totalling 15 ppm (as determined by spectrographic analysis). Samples having their surface parallel to the (100) plane were cut by the electro-spark technique from the single crystals, 4 mm in diameter. The deformed layer, approximately 1.2 mm thick, which was formed during the cutting and determined by etching dislocations, was removed by grinding and chemical polishing in the following solution [6]: 5 cm$^3$ HF (50%) + 80 cm$^3$ $H_2O_2$ (30%) + 15 cm$^3$ $H_2O$. After being rinsed in alcohol, the polished surface was etched in a 2% solution of nitric acid in alcohol (nital) 30 to 90 s, rinsed in water, then in alcohol, and dried in an air stream. The surface of the specimen was watched in a Zeiss-Neophot metallograph with oblique illumination.

## 3. Results and Characteristics of Etching

The described method was found capable of producing well observable etch pits only on the (100) surface. Their shape depends on the magnitude of the deviation of the surface from the (100) plane. When the deviation is greater than about 4°, the pits become asymmetric (elongated in one direction) and it becomes more difficult to distinguish them from the background. Three kinds of etch pits can be observed on a well oriented surface: a) pits having the form of a cone with a sharp vertex, marked by $A_1$ and $A_2$ in Fig. 1, b) a small number of pits of the same shape but of markedly larger size (Fig. 1, B), c) shallow pits with a flat bottom (Fig. 1, C).

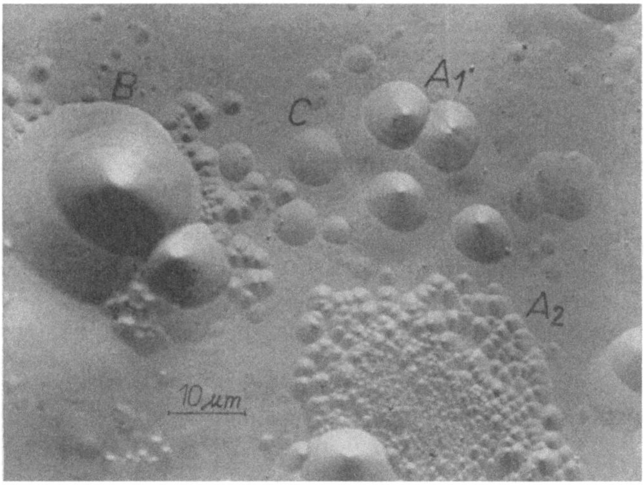


Fig. 1. (100) surface of iron single crystal after etching 30 s in 2% nital. $A_1$ — etch pits at grown-in dislocations, $A_2$ — etch pits at fresh dislocations

A more detailed study of the etched (100) surface revealed that the pits of the first group, marked by A, exhibit the following properties:

1. After repeated etching, the pits increase in size, remain sharp-bottomed, and do not change their position. When gradually polishing away thin layers parallel to the surface, these pits are found to follow distinctly a line defect in the crystal.

2. In as-grown and annealed crystals these pits arrange themselves in a manner which is characteristic of the dislocation substructure of crystals, e.g. rows of pits (Fig. 2).

3. They appear in places where the crystal was plastically deformed. An example is shown in Fig. 1, in which dense clusters of pits occur in places where the surface was previously scratched. Fig. 3 shows etch pits originating from a scratch made on the crystal surface by a steel scriber having a spherical tip of $r \approx 0.2$ mm and loaded with 5 g. Fig. 3a shows a surface with a scratched line after etching, Fig. 3b the same surface after polishing away a layer of approximately 12 μm and etching. Fig. 3c shows the surface after repeatedly polishing away approximately 40 μm, and Fig. 3d the same surface after polishing away approximately 30 μm. The specimen was etched immediately after scratching at room temperature.

We believe that these facts prove sufficiently that the etch pits reveal dislocations. Experiments have shown that they form both at fresh dislocations (Fig. 1 and 3) and at grown-in dislocations (i.e. at aged dislocations, Fig. 2), in as-grown and annealed crystals.

The small number of markedly larger sharp-bottomed pits (Fig. 1, B) have the same properties as the etch pits A. When the etching time is increased, the pits remain sharp-bottomed. After polishing and repeated etching their position is nearly the same. We believe that these pits form either at dislocations with large Burgers vectors or at dislocations along which a greater number of impurities has precipitated. Shallow flat-bottomed pits can be observed in a great number especially after prolonged etching. As the etching time is increased the

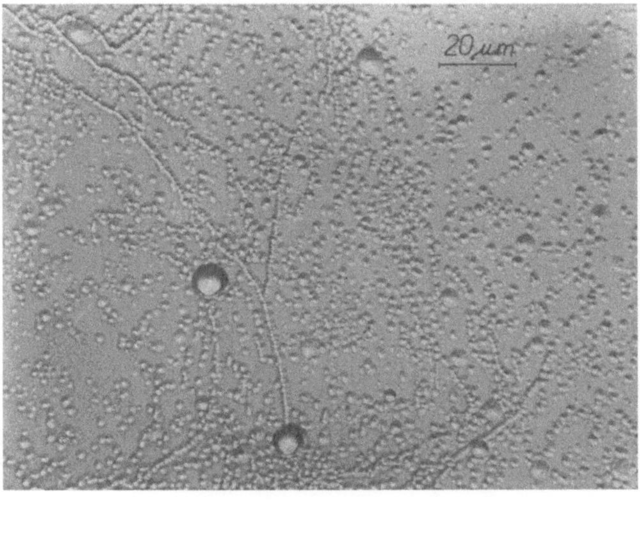


Fig. 2. (100) surface of as-grown iron single crystal after 30 s etching in 2% nital

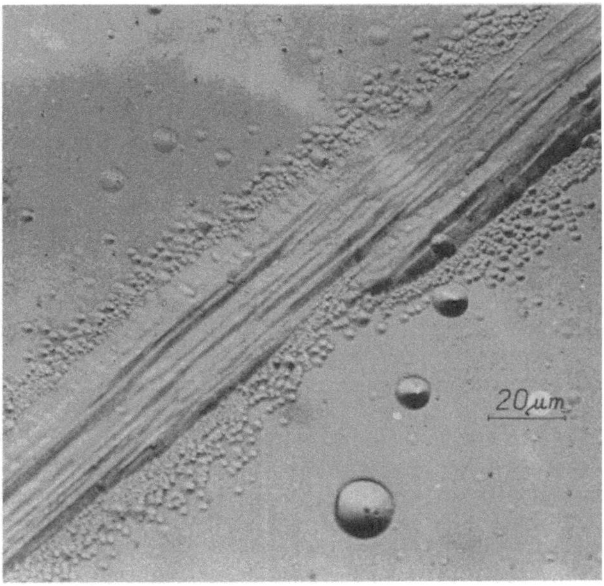


a

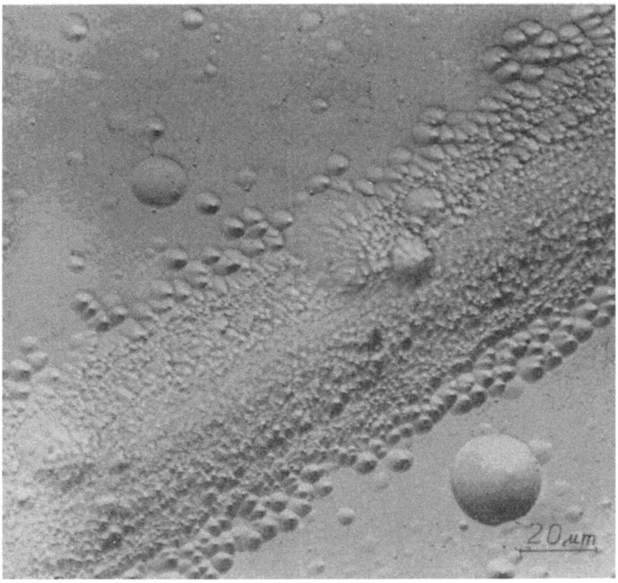


b

Fig. 3. (100) surface of annealed iron single crystal deformed by scratching, after etching in 2% nital.
a) Original crystal surface after scratching, etched for 30 s;
b) layer of approximately 12 µm polished away, etched for 60 s;

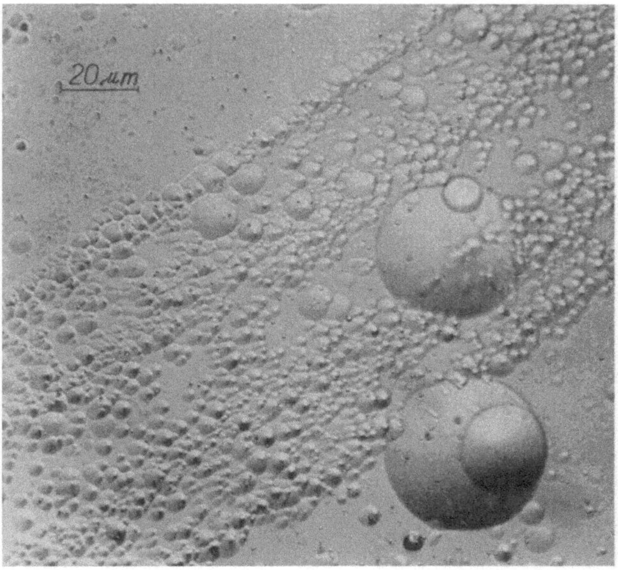


c

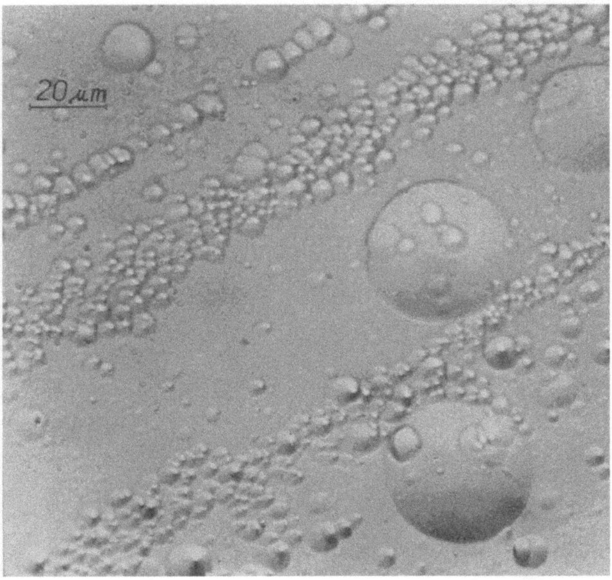


d

c) another layer of 40 μm polished away, etched for 60 s;
d) another layer of 30 μm polished away, etched for 60 s

shallow pits increase in size, and after polishing and repeated etching they appear at random. In our opinion these pits are due to defects other than dislocations at which the dissolution of the crystal nucleates.

## 4. Conclusion

It has been shown that it is possible to reveal the dislocation sites on a (100) surface of single crystals of high-purity iron by etching them in 2% nital. The etch pits are formed both at fresh dislocations and grown-in ones.

The authors express their gratitude to Dr. F. Kroupa and Dr. Z. Hauptman for remarks on the manuscript.

## References

[1] F. W. C. Boswell, Metal Progr. **72**, 92 (1957).
[2] P. D. Gorsuch, Research on the Behaviour of Nearly Perfect Crystals, General Electric Res. Lab. Progr. Rep. No. **3**, October 1959.
[3] H. J. Seemann and V. Benedict, Arch. Eisenhüttenwesen **33**, 177 (1962).
[4] E. Schmidtmann and H. F. Klärner, Arch. Eisenhüttenwesen **32**, 769 (1961).
[5] G. T. Hahn, Trans. AIME **224**, 395 (1962).
[6] T. Takeuchi, J. Phys. Soc. Japan **20**, 942 (1965).
[7] R. V. Coleman, J. appl. Phys. **29**, 1487 (1958).
[8] I. P. Kushnir, Kristallografiya **7**, 247 (1962).
[9] H. W. Schadler, General Electric Res. Lab. Rep. No. 61 RL—2683 M, April 1961.
[10] I. Berlec, J. appl. Phys. **33**, 197 (1962).
[11] S. Kadečková and B. Šesták, to be published.
[12] B. Šesták and S. Kadečková, International Symposium „Reinststoffe in Wissenschaft und Technik", Dresden 1965, in press.

*(Received April 1, 1966)*

phys. stat. sol. **16**, 95 (1966)

*Inorganic Materials Research Division, Lawrence Radiation Laboratory, and Department of Mineral Technology, College of Engineering, University of California, Berkeley*

# An Absolute Determination of the Extrinsic and Intrinsic Stacking Fault Energies in Ag–In Alloys

By

P. C. J. GALLAGHER

Observations have been made of extrinsic-intrinsic node pairs, and of a hitherto unreported fault pair. It is shown that a simple estimate of the ratio of extrinsic to intrinsic stacking fault energy ($\gamma_e/\gamma_i$) cannot be made from the inscribed radii of node pairs, and that earlier qualitative results that $\gamma_c/\gamma_i > 2$, or more, are invalid. The simple geometry of the new fault pair facilitates a straight forward absolute determination of both $\gamma_e$ and $\gamma_i$. Contrary to published results, $\gamma_e$ and $\gamma_i$ have been found to be approximately equal. For the present alloys, $\gamma_e/\gamma_i = 1.09 \pm 0.1$ (for the electron–atom ratio $e/a = 1.15$) and $\gamma_e/\gamma_i = 1.03 \pm 0.1$ (for $e/a = 1.23$). Good absolute agreement has been obtained between $\gamma_i$ determined from the fault pairs, and from the observation of extended three-fold nodes. It is concluded that extrinsic faults are rarely observed because of a high impedance to their formation arising from the co-operative glide which is necessary, but that, once formed, the extrinsic fault energy is closely the same as the intrinsic.

Des noeuds doubles extrinsèques–intrinsèques et une paire de fautes jusqu'ici non décrite, ont été observés. On montre qu'une estimation simple du rapport des énergies de faute extrinsèque et intrinsèque ($\gamma_e/\gamma_i$) ne peut être déduite des rayons des circles inscrits dans les noeuds doubles, et que les précédents résultats qualitatifs indiquant un rapport $\gamma_e/\gamma_i > 2$ ou plus ne sont pas valables. La simple géométrie de la nouvelle paire de fautes facilite la détermination rigoureuse de $\gamma_e$ et $\gamma_i$. Contrairement aux résultats publiés, $\gamma_e$ et $\gamma_i$ sont trouvés sensiblement égales. Pour les présents alliages, $\gamma_e/\gamma_i = 1,09 \pm 0,1$ (pour $e/a = 1,15$) and $\gamma_e/\gamma_i = 1,03 \pm 0,1$ (pour $e/a = 1,23$). Un bon accord est trouvé entre les valeurs de $\gamma_i$ obtenues à partir des paires de fautes ou de l'observation des noeuds triples étendus. On conclut que les fautes extrinsèques sont rarement observées à cause de leur formation difficile qui nécessite un glissement coopératif, mais qu'une fois formées elles ont une énergie pratiquement égale à celle des fautes intrinsèques.

## 1. Introduction

Previous work has shown that both intrinsic and extrinsic faults are present in f.c.c. alloys of low stacking fault energy. Estimates of the relative magnitudes of the intrinsic and extrinsic stacking fault energies ($\gamma_i$ and $\gamma_e$ respectively) have been made from examination of node pairs in Au-Sn [1] and Ag-Sn [2] and from a statistical treatment of the occurrence of stacking fault tetrahedra in Ni-70 at% Co [3]. Only tentative conclusions could be reached, but the ratio of $\gamma_e$ to $\gamma_i$ was variously estimated as from 2 to 4.5 or more. Work by Booker and Brown [4] suggests that the result that $\gamma_e \approx \gamma_i$ in silicon [5] may be spurious due to an anomalous diffraction effect.

In the present work silver-indium alloys have been examined in the electron microscope, their composition ranging from pure silver to Ag-12.5 wt% In (elec-

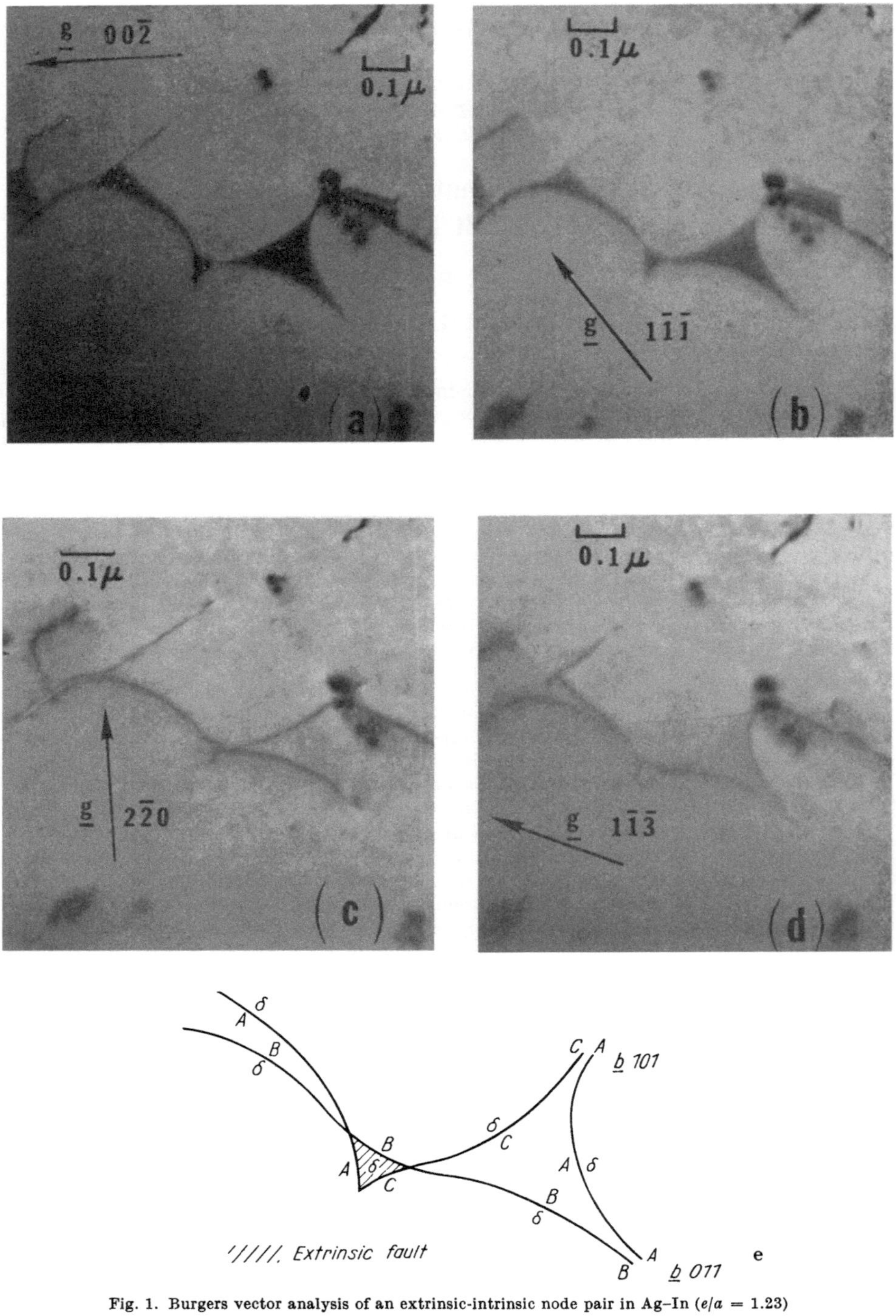


Fig. 1. Burgers vector analysis of an extrinsic-intrinsic node pair in Ag–In ($e/a$ = 1.23)

tron-atom ratio $e/a = 1.23$). Attention has been focused particularly upon the two alloys with lowest stacking fault energy ($e/a = 1.15$ and $1.23$). The following section covers observations which have been made on extrinsic-intrinsic node pairs while subsequent sections describe an interesting new extrinsic-intrinsic fault configuration which provides a very advantageous method for determining the absolute magnitudes of both $\gamma_e$ and $\gamma_i$.

## 2. Intrinsic and Extrinsic Node Pairs

Fig. 1 shows a good example of an extrinsic-intrinsic node pair. Several reflections were used so that the Burgers vector assignment (Fig. 1e) could be made with certainty. The notation used to describe the Burgers vectors is that of Thompson [6], and throughout this paper $\delta$ refers to the $(11\bar{1})$ plane, and BA = [011], CB = $[1\bar{1}0]$, and CA = [101].

Previous workers have drawn qualitative conclusions from fault configurations such as that in Fig. 1, by assuming that the ratio of the inscribed radius of one node to the other in the pair was approximately equal to the ratio of the extrinsic to the intrinsic stacking fault energy. This criterion applied to the node pair in Fig. 1 would suggest $\gamma_e/\gamma_i \approx 3$, but, as shown later, such a comparison is in fact invalid.

Fig. 2a contains an extrinsic-intrinsic node pair in an alloy for which $e/a = 1.10$, and $\gamma_i = 16$ erg/cm². A number of similar examples have been observed in this alloy. Fig. 2b shows at A what is almost certainly a node pair in pure silver to which oxygen was added. Similar configurations have also been observed in pure silver, for which $\gamma_i \approx 20$ erg/cm². Thus, the presence of extrinsic faults is by no means confined to materials of exceptionally low stacking fault energy.

Fig. 3 contains a number of interesting features. The network in the central picture area contains only extended nodes establishing that intrinsic and extrinsic faulting has occurred. Further examples of extrinsic-intrinsic node pairs are indicated at points A, B, C, while extrinsic and intrinsic faults are also present at D, E, and F. In several areas (e.g., F) an offset of the stacking fault fringes is

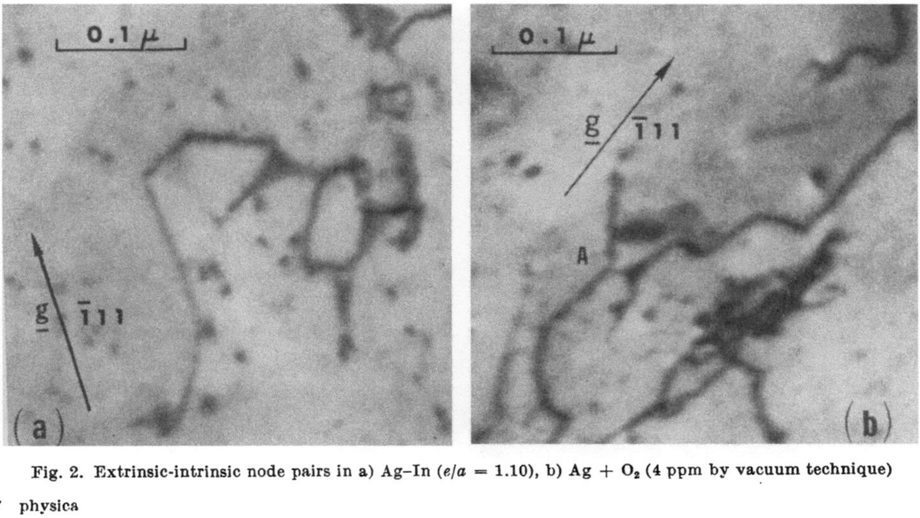


Fig. 2. Extrinsic-intrinsic node pairs in a) Ag–In ($e/a = 1.10$), b) Ag + $O_2$ (4 ppm by vacuum technique)

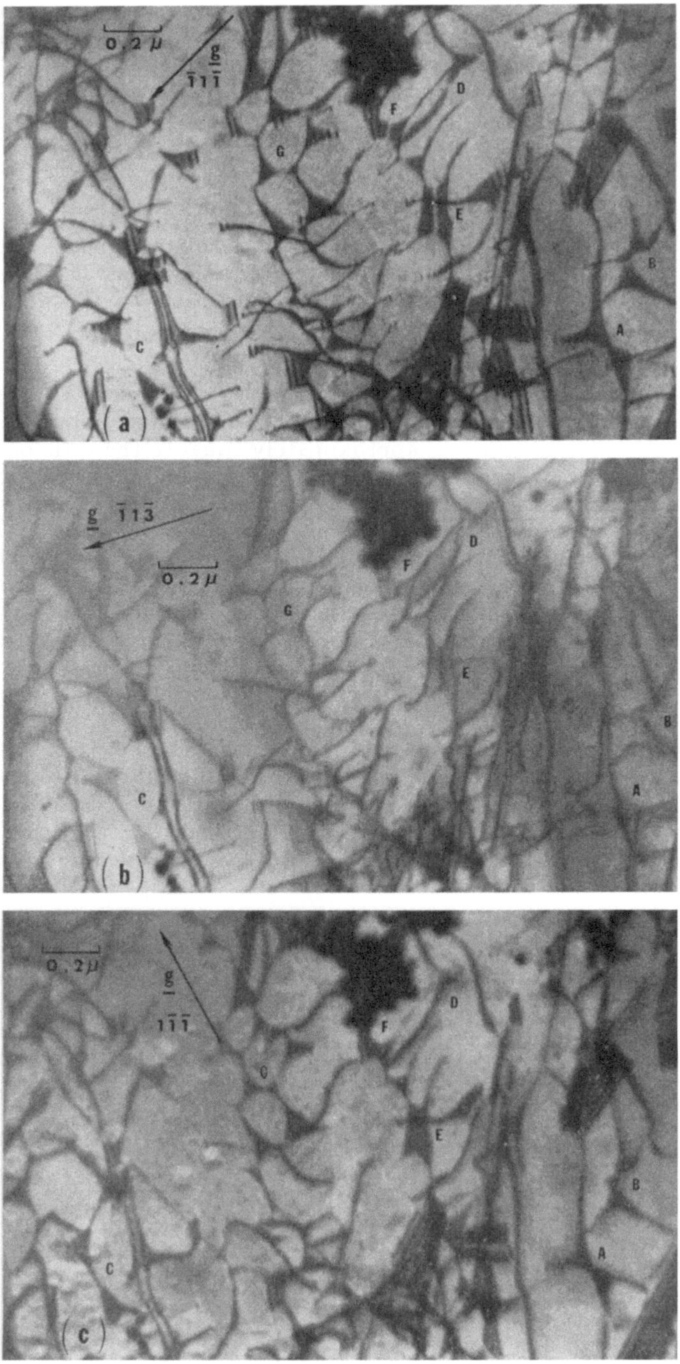


Fig. 3. Extrinsic and intrinsic node pairs, and fault pair (E) in Ag–In ($e/a = 1.23$)

noticeable where intrinsic and extrinsic faults are adjacent, as expected from contrast theory due to the phase shift of $2\pi/3$. It is emphasized that in these cases, as also throughout this paper, the extrinsic and intrinsic faults share a common dislocation line, and that effects from overlapping faults do not arise. A comparison of Fig. 3a and b clearly reveals the presence of a cross-linking partial dislocation dividing the intrinsic and extrinsic faulted region for many of the node pairs. The presence of such a link is particularly clear in Fig. 3b at points A, B, D, and F. The striking formation at point E, resembling an elongated node pair, is an example of the extrinsic-intrinsic fault configuration which is to be discussed in detail in the latter part of this paper.

Of the extrinsic-intrinsic node pairs in the figure, those at A and C each contain sensibly equal areas of extrinsic and intrinsic fault. By contrast, the node pairs at B and G consist of a considerably larger fraction of intrinsic fault. From Fig. 1, and particularly in Fig. 8, point P, we have evidence that the ratio of the inscribed radii in an intrinsic-extrinsic node pair can be as large as 5 or more, whereas in this same alloy previous work [7] has shown that for a sample of 40 single intrinsic nodes the standard deviation about the mean of a single measurement of the inscribed radius ($w$) was only $0.25\,\overline{w}$.

All the evidence points to the conclusion that the resultant configuration in an extended node pair depends very strongly upon factors other than the relative magnitudes of $\gamma_e$ and $\gamma_i$. The evidence is substantiated by the results obtained in the latter part of this work.

The most important factor in determining the equilibrium configuration of the node pair is the elastic energy which results from the cross-over of the partial dislocations emerging from the extrinsically faulted node, particularly when $\gamma_e$ and $\gamma_i$ are small.

This elastic energy depends sensitively upon the angle at which the partials cross, since the repulsive force decreases markedly as the angle increases. Considerable distortion of the node pair is to be expected in cases for which the cross-over energy may thereby be reduced, especially when $\gamma_e$ and $\gamma_i$ are closely equal.

The observed shape of such pairs is in complete agreement with the above, in that the partials bounding the extrinsic fault are straight, or even have the reverse curvature to that found in intrinsic nodes, so as to maintain as large a cross-over angle as possible (cf. Fig. 3, regions A, B, C, F, G).

Fig. 4 illustrates three node pairs, for which let us consider that $\gamma_e = \gamma_i$, and that both are low. Without making detailed calculations it can be surmised that

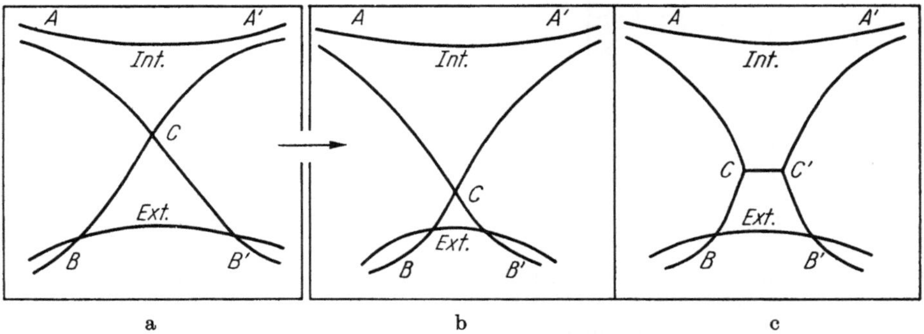


Fig. 4. Extrinsic-intrinsic node pair configurations. a) Unstable, b) and c) stable

configuration (a) is unstable. By moving point C toward BB′ (Fig. 4b) the total faulted area remains substantially the same, but the energy arising from the dislocation cross-overs at B and B′ is reduced. For this reason too, the partials may take the reverse curvature in order to increase the angle of cross-over at points B and B′, despite the increase in extrinsically faulted area which results. The formation of a cross-linking partial dislocation (CC′ in Fig. 4c) also reduces the total energy for the same reason, despite the increased faulted area and dislocation line length.

We may conclude, therefore, that even for $\gamma_e = \gamma_i$ only in rare cases will a node pair contain equal areas of intrinsic and extrinsic fault. Although by no means intractable, the theoretical analysis of extrinsic-intrinsic node pair configurations would involve a careful consideration of the character of each of the emerging dislocations, while, in addition, the network in region P (Fig. 8) clearly indicates that the constraints imposed by the surrounding dislocations are also of great importance.

It is concluded, therefore, that an examination of node pairs can lead to meaningful values of $\gamma_i$ and $\gamma_e$ only following a considerable expenditure of effort. The argument illustrated by Fig. 4 suggests that the extrinsically faulted area will almost invariably be smaller than one would expect from the relative magnitudes of $\gamma_e$ and $\gamma_i$, due to the strong dislocation interactions involved.

## 3. Extrinsic-Intrinsic Fault Pairs

The nature of the extrinsic-intrinsic fault pair at point E in Fig. 3 is most clearly appreciated from Fig. 3b. The observed magnitudes of $d_i$ and $d_e$ (defined in Fig. 5) vary in the same sense as the angle of projection is altered by tilting the specimen, establishing that the extrinsic and intrinsic faulted areas lie in the same plane.[1]) It has been established from pictures under different diffracting conditions that the three long, sensibly parallel partial dislocations have the same Burgers vector. Particular care has been taken to establish this fact without any doubt, and as many as six different reflections $(\bar{1}1\bar{1}, \bar{1}11, 002, 2\bar{2}0, \bar{1}13, 1\bar{1}3)$ were used to photograph particular fault pairs. Fig. 5 illustrates the fault pair

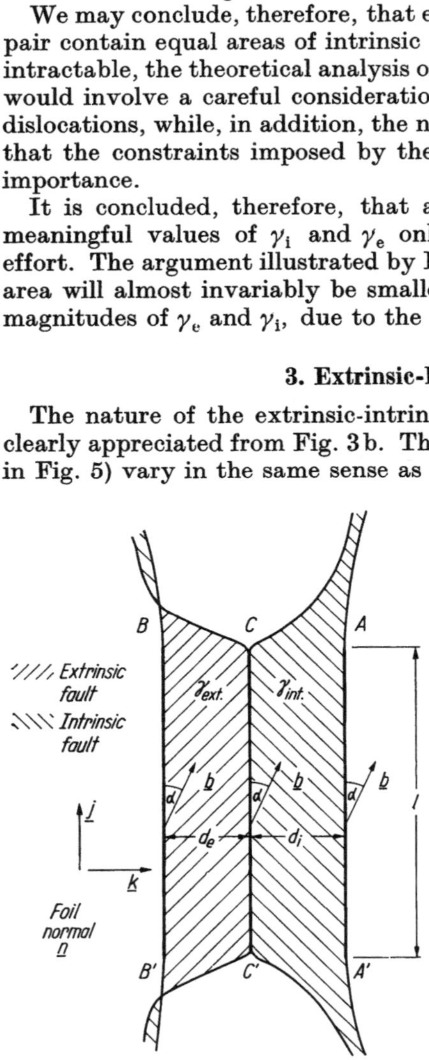


Fig. 5. Extrinsic-intrinsic fault pair

[1]) The configuration is not, of course, strictly co-planar, in that the dislocations lie on a pair of adjacent planes as is clear in Fig. 13. The configuration is planar in the sense that all the partial dislocations bounding the fault pair lie in the $(1\bar{1}1)$ plane.

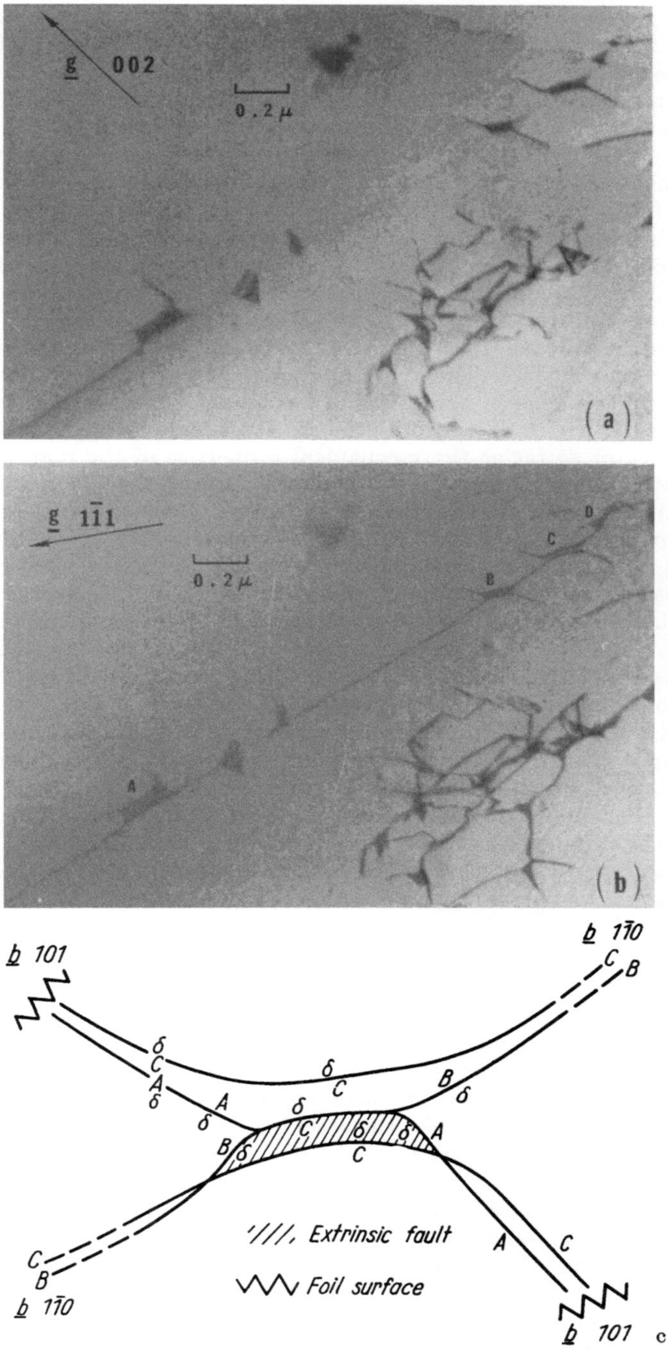


Fig. 6. a) and b) Fault pairs in Ag–In ($e/a$ = 1.15); c) Burgers vector assignment for fault pair B

configuration diagramatically. The fault pair in Fig. 5 is lettered analogously to the node pair in Fig. 4c to emphasize that the fault pair is merely a node pair containing a cross-linking dislocation which has been elongated in the direction defined by the unit vector $\boldsymbol{j}$ in Fig. 5. A number of advantages stem from the changes in geometry between the configurations of Fig. 4c and 5.

For sufficiently long fault pairs the equilibrium intrinsic and extrinsic fault widths can be calculated solely by allowing for the repulsive forces between the parallel partials (BB′, CC′, and AA′ in Fig. 5) and the attractive force which results from the faulted strips; the fact that the three parallel partials have the same Burgers vector and character $\alpha$ greatly facilitates the calculation, and eliminates the need for corrections to the measurements due to image displacements arising from the diffraction conditions. The simplicity of the configuration makes it feasible to use anisotropic elasticity theory in calculating the stacking fault energy in metals. The above advantages have also been noted [8] in work on three fold ribbons in nonmetals of hexagonal crystal structure.

In addition to their use in determining the stacking fault energy, fault pairs may also be important so far as the mechanical properties of the material are concerned in that subsequent glide in the plane containing the faults could be impeded. Fig. 6 illustrates the interaction of a long screw dislocation with a number of other dislocations of different Burgers vectors. The Burgers vector assignment in Fig. 6c is readily accomplished from Fig. 6a and b; Fig. 6a, in particular, makes it clear that the long screw dislocation intersects the other dislocations, rather than curving away to the surface as one might conclude from Fig. 6b. (A number of anomalous contrast effects are contained in the figures in this paper; also in earlier work [1, 2]. These will be described below.)

It should be noted that the fault configuration illustrated in Fig. 6c (and diagrammatically in Fig. 5) arises from the intersection of dislocations whose long range interaction is repulsive. A more detailed description of the formation process and the factors which favor it will be published shortly.

It is of interest to compare the fault pairs at B and C in Fig. 6b with the node pair at point D. The difference between the intrinsic and extrinsic faulted areas in the node pair is very marked, whereas the fault pairs (formed from dislocations with the same Burgers vectors as the node pair) contain much more closely equal areas of extrinsic and intrinsic fault. At point A we have an example of a fault pair which, due to external constraints, is not suitable for measurement. Only fault pairs with closely parallel partials have been measured, since the theory which will be used applies only to such cases.

Fig. 7 illustrates complex intrinsic-extrinsic faults which are analysed in detail in Fig. 7e and f. While the faults are completely unsuitable for a determination of $\gamma_e$ and $\gamma_i$ the figure nevertheless contains very useful information. Once again, the lateral displacement of the stacking fault fringes makes it clear that intrinsic and extrinsic faults are present. It was suggested in the previous paragraph that fault pairs may represent a relatively sessile configuration. Nevertheless, a comparison of Fig. 7d with 7a shows that glide of complex intrinsic-extrinsic faults can take place, in this case simply as a result of heating due to the electron beam. It should be noted, however, that the fault pairs in Fig. 7 are of a less stable nature than those in the previous figures. The dislocations forming them have formed junctions locally, but have not intersected. The configurations are apparently prevented from collapsing by the forces arising from the proximity of the foil surfaces.

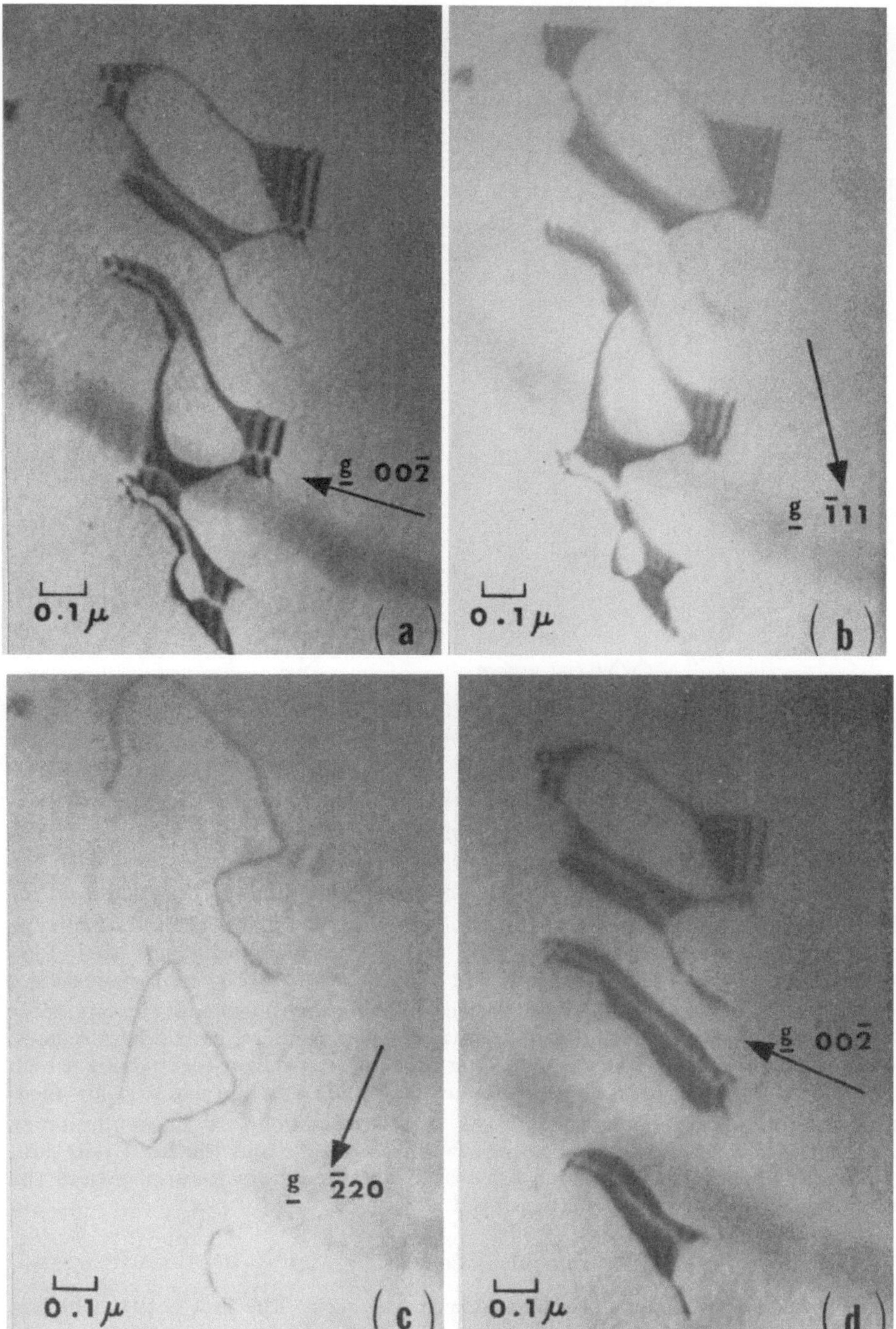


Fig. 7. a), b), and c) Complex extrinsic-intrinsic faults; d) same after glide

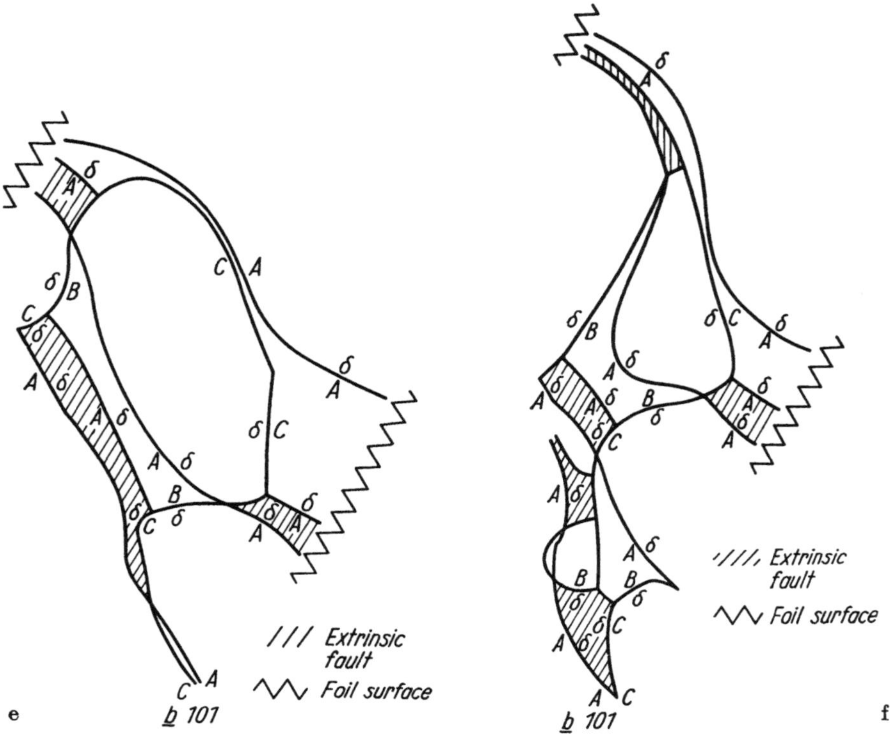


Fig. 7. e) and f) Burgers vector assignments

Confirmation of the assignment of identical Burgers vectors to the long partials in a fault pair is evident in Fig. 7c taken with the $\bar{2}20$ reflection. It has been found most convenient to measure the fault pairs photographed with the $\bar{2}20$ reflection so that only the partials are in contrast. The foil normal used for the majority of the pictures is [110], which means that $\bar{2}20$ is the only 220 type reflection available, and that one third of the fault pairs will have their long partials out of constrast (e.g., as in Fig. 7c). Fortunately good pictures have been obtained using the $1\bar{1}\bar{3}$ or $1\bar{1}3$ reflections by which means all three possible Burgers vectors can be brought into contrast. Furthermore, as will be discussed below with reference to Fig. 10, as long as care is taken in recognizing which particular diffracting conditions are operative the fault pairs can also be measured when photographed with $1\bar{1}1$, $\bar{1}11$, or 002 reflections. Before going on to present the theoretical relationship between $\gamma_i$ and $\gamma_e$, and the fault pair configuration, and the results which have been obtained from its application, the anomalous contrast effects are described.

## 4. Anomalous Contrast Effects

Two types of anomalous effect have been observed. The first is that which is present in most of the photographs in this paper, and is particularly clearly visible for a number of diffracting conditions in Fig. 7a, b, d and Fig. 8a, b, c. It concerns the contrast observed at the central long partial, that which divides the extrinsic and intrinsic faulted areas. The same effect occurs for the cross-linking dislocation in node pairs, and may be seen in the previous work [1, 2] and in a large number of instances in Fig. 3 and 8 in the present work.

One expects, since all three long partials have the same Burgers vector, that these three dislocations would be either in or out of contrast together. However, it is quite clear that this is not the case. In Fig. 7a, c and 8a a line of no contrast separates the intrinsic and extrinsic faults in all cases, except at Q in Fig. 8a. In all these cases (except Q) the outer two partials having the same Burgers vector as the central partial (or cross-link in the case of node pairs) are in contrast.

In Fig. 7b and 8c the central partial (or cross-link in node pairs) is in contrast, while the outer partials are out of contrast (except at Q). These effects can be very confusing in making Burgers vector assignments. The effects at Q are reversed from those elsewhere simply because the cross-linking dislocation has a different Burgers vector from those in the other faults. One may state the general rules:

1. A Shockley partial which would normally be in contrast for a given reflection appears out of contrast if it separates regions of extrinsic and intrinsic faulting which are also in contrast.

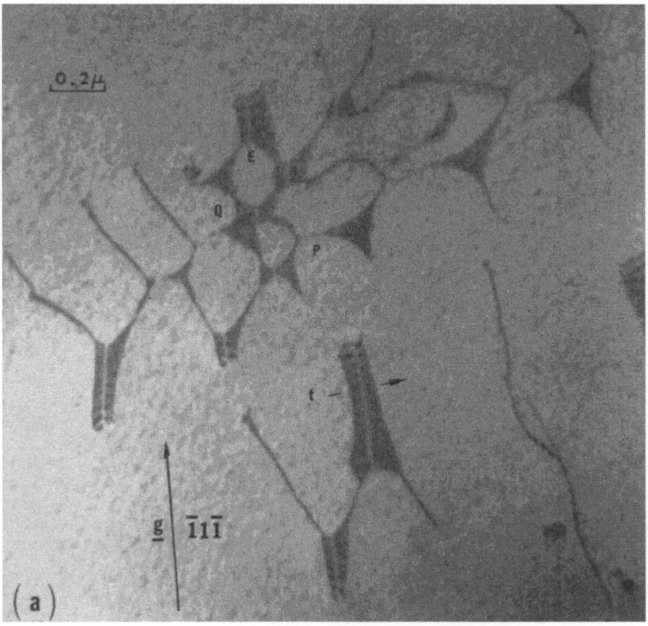


Fig. 8a

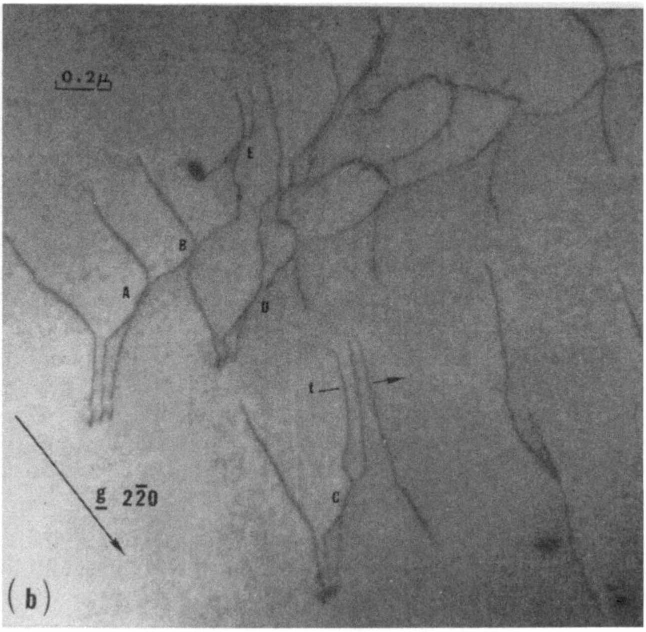


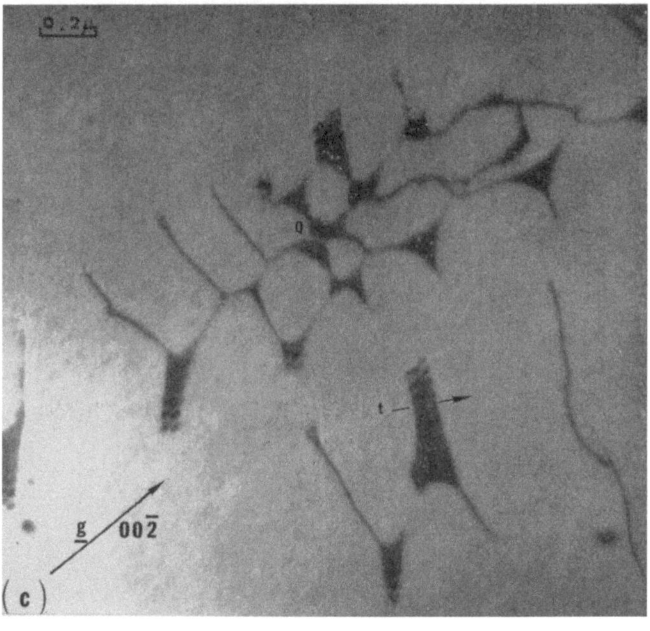


Fig. 8. a), b), and c) Examples of fault pairs in Ag–In ($e/a$ = 1.23); d) analysis of lower fault pairs

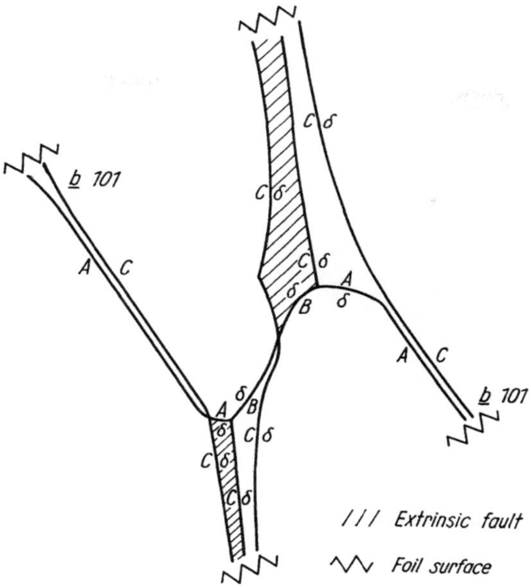


Fig. 8d

2. Conversely, a Shockley partial which would normally be out of contrast for a given reflection appears in contrast if it separates regions of extrinsic and intrinsic faulting which are also in contrast.

Further examples of these effects can be seen in Fig. 3, and, of course, in Fig. 6b where the line of no contrast makes the dislocations appear not to cross. Detailed calculations on these effects are in progress and will be reported at a later date.

The second anomalous effect is visible under diffracting conditions which show only the partial dislocations (i.e., $2\bar{2}0$, $\bar{1}1\bar{3}$, or $1\bar{1}3$ in the present work). In Fig. 8b there are numerous points, e.g., A, B, C, D, E, at which partial dislocations cross over one another at the join of an extrinsic to an intrinsic fault. Only at point A, however, it is clear that a cross-over has occurred. In the other cases the contrast suggests, anomalously, that the dislocations remain parallel but do not cross.

A further particularly clear example of this is shown in Fig. 9. In both these bright-field pictures the magnitude of $s$, the deviation from the Bragg position, was positive and close to that normally used for optimum contrast. In Fig. 9a with $\boldsymbol{g} = 2\bar{2}0$, anomalous contrast is observed at points A and B where the partial dislocations do, in fact, cross over. With $\boldsymbol{g} = \bar{2}20$ in Fig. 9b, the dislocation cross-over at point A is now clearly visible, while the contrast at B remains anomalous. That similar dislocation interactions in closely adjacent sites should give different contrast indicates a very sensitive dependence upon the magnitude of $s$.

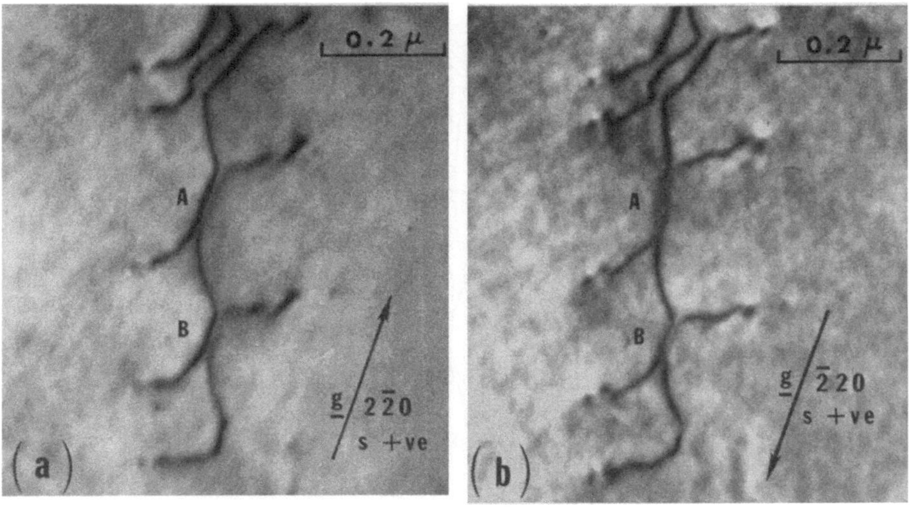


Fig. 9. Anomalous contrast effects at partial dislocation cross-overs

## 5. Fault Pairs and the Extrinsic and Intrinsic Stacking Fault Energy

### *5.1 Theory*

In the present paper the theoretical treatment is limited to isotropic elasticity. More comprehensive experimental results are being obtained in order to test the predictions of anisotropic theory. The work of Teutonico [9] provides a valuable basis for the necessary calculations.

With the notation of Fig. 5, the force between the parallel dislocations of character $\alpha$ is obtained by resolving the partial Burgers vector $\boldsymbol{b}$ in the $\boldsymbol{j}$ and $\boldsymbol{k}$ directions. The repulsive force per unit length ($F$) between two dislocations distance $d$ apart is

$$F = \frac{\mu b^2}{2 \pi d}\left(\cos^2 \alpha + \frac{\sin^2 \alpha}{(1-\nu)}\right). \tag{1}$$

For the case of an infinitely long fault pair, the equilibrium configuration is described by the following equations:

$$\gamma_i = \frac{\mu b^2}{2 \pi}\left(\frac{1}{d_i} + \frac{1}{(d_i + d_e)}\right)\left[\cos^2 \alpha + \frac{\sin^2 \alpha}{(1-\nu)}\right], \tag{2}$$

$$\gamma_e = \frac{\mu b^2}{2 \pi}\left(\frac{1}{d_e} + \frac{1}{(d_i + d_e)}\right)\left[\cos^2 \alpha + \frac{\sin^2 \alpha}{(1-\nu)}\right]. \tag{3}$$

We may also write

$$\gamma_e - \gamma_i = \frac{\mu b^2}{2 \pi}\left(\frac{1}{d_e} - \frac{1}{d_i}\right)\left[\cos^2 \alpha + \frac{\sin^2 \alpha}{(1-\nu)}\right] \equiv \frac{\mu b^2}{2 \pi}\left(\frac{1}{d_e} - \frac{1}{d_i}\right) f(\alpha). \tag{4}$$

Essentially identical equations apply to the threefold ribbons found in non-metals [8]. The magnitude of $f(\alpha)$ varies from unity in the case of a screw fault ($\alpha = 0$) to $f(\alpha) = 1.92$ at $\alpha = 90°$ in the present alloys for which $\nu \approx 0.48$.

The extent to which equations (2) and (3) are applicable to the observed fault pairs depends upon a number of factors. It is clearly desirable that the partials should be parallel and that the configuration should be as symmetrical as possible about a line parallel to $\boldsymbol{k}$ through the middle of the fault pair (cf. Fig. 5). For a fault pair satisfying these conditions, e.g. at B, Fig. 6, the main failing of equations (2) and (3) is that their application will lead to an underestimate of $\gamma_i$ and an overestimate of $\gamma_e$. The longer the fault pair, of course, the smaller is this discrepancy which arises from the strong forces involved in the dislocation cross-overs at B and B′. Estimates of the accuracy of (2) and (3) in their present application are made later in the text.

The most commonly observed fault pair configuration is that which at one end terminates in the foil surface. Numerous examples are contained in Fig. 8. Measurements have been made in the regions where the three partials are most nearly parallel, which is usually about one third of the way along the fault pair from the foil surface. Care must also be taken to avoid measuring fault pairs which are contained in a complex network of dislocations, e.g. at E in Fig. 8. It is interesting that in some, but not all of the fault pairs the curvature of one of the long bounding partials changes suddenly locally, e.g. at E in Fig. 3, at C in Fig. 6b, and near C in Fig. 8b. This appears to arise from the presence of a strong local solute impedance.

A valuable aid in establishing the absolute accuracy of the determinations of $\gamma_i$ and $\gamma_e$ made from the fault pairs is a comparison with $\gamma_i$ obtained from applying the recently published node theories [10 to 12] to the extended intrinsic nodes which have been observed. Particularly useful is an area such as that in Fig. 8 where good examples of nodes and fault pairs are in close proximity.

### 5.2 *Experimental results*

The magnitudes of $\gamma_i$ and $\gamma_e$ have been determined from an examination of eleven fault pairs in the alloy with $e/a = 1.15$ and from nine fault pairs in the alloy with $e/a = 1.23$. The correct Burgers vector was readily assigned to the long partials in all cases so that the character of the fault pair, $\alpha$, could be obtained. The foil normal unit vector, $\boldsymbol{n}$, was generally in the [110] direction. From trace analysis, using the 110 standard projection, the direction of the unit vector, $\boldsymbol{k}$, could be ascribed in terms of crystallographic indices applying the condition that it lies in the $(11\bar{1})$ plane. The measured values of $d$ could thus be corrected for inclination in the foil since

$$d_{\text{meas}} = d \sin \varphi\,, \qquad \text{with } \cos \varphi = \boldsymbol{k} \cdot \boldsymbol{n}\,. \tag{5}$$

The maximum correction is by 22% corresponding to $\boldsymbol{k}$ in the [112] direction, while for $\boldsymbol{k}$ along $[1\bar{1}0]$ no correction is necessary. In all the fault pairs examined it was possible to recognize the characteristic shape illustrated in Fig. 5. The partial dislocations bounding the ends of one part of the fault pair curve away gradually toward the long partial (intrinsic fault), while on the other side the end partials cross the long partial as close to the fault pair as possible, and a constriction may frequently be seen (extrinsic fault). Dark-field pictures

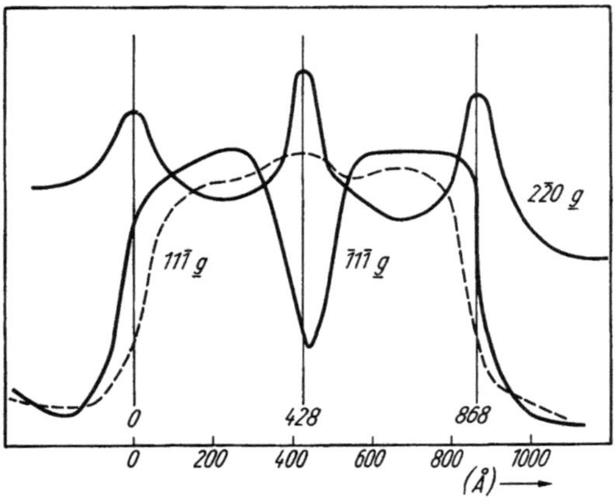


Fig. 10. Microdensitometer traces from traverse $t$ in Fig. 8

have also been taken, and an established technique [13] for distinguishing between intrinsic and extrinsic faults has been followed to confirm that the assignments have, in fact, been made correctly.

Fig. 10 contains microdensitometer traces from the traverse $t$ in Fig. 8a, b, c. Similar sets of traces have been obtained from other faults which had been

Table 1

$\gamma_e$ and $\gamma_i$ from fault pair data in Ag–In alloys, $e/a = 1.15$ and $1.23$

| $e/a = 1.15$ | | | | $e/a = 1.23$ | | | |
|---|---|---|---|---|---|---|---|
| $d_e$ (Å) | $d_i$ (Å) | Character $\alpha$ (deg) | $f(\alpha)$ | $d_e$ (Å) | $d_i$ (Å) | Character $\alpha$ (deg) | $f(\alpha)$ |
| 244 | 238 | 14 | 1.05 | 395 | 393 | 23 | 1.15 |
| 268 | 255 | 10 | 1.03 | 351 | 417 | 14 | 1.06 |
| 352 | 372 | 8 | 1.01 | 510 | 511 | 6 | 0.99 |
| 304 | 342 | 34 | 1.28 | 375 | 390 | 9 | 1.02 |
| 197 | 250 | 15 | 1.06 | 409 | 589 | 38 | 1.35 |
| 250 | 249 | 26 | 1.20 | 375 | 370 | 30 | 1.22 |
| 204 | 237 | 6 | 1.00 | 511 | 454 | 34 | 1.30 |
| 144 | 200 | 6 | 1.00 | 325 | 325 | 24 | 1.15 |
| 146 | 208 | 6 | 1.00 | 452 | 450 | 17 | 1.08 |
| 347 | 331 | 8 | 1.01 | | | | |
| 266 | 285 | 10 | 1.03 | | | | |

From which, with $\mu = 2.45 \times 10^{11}$ dyn/cm², $b = 1.67$ Å.
$\gamma_e = 7.4 \pm 0.6$ erg/cm² (deviation of single reading $\pm 1.8$ erg/cm²);
$\gamma_i = 6.8 \pm 0.4$ erg/cm² (deviation of single reading $\pm 1.2$ erg/cm²).
For $e/a = 1.15$, $\gamma_e/\gamma_i = 1.09 \pm 0.1$

With $\mu = 2.5 \times 10^{11}$ dyn/cm², $b = 1.67$ Å.
$\gamma_e = 4.7 \pm 0.3$ erg/cm² (deviation of single reading $\pm 0.8$ erg/cm²);
$\gamma_i = 4.55 \pm 0.3$ erg/cm² (deviation of single reading $\pm 0.8$ erg/cm²).
For $e/a = 1.23$, $\gamma_e/\gamma_i = 1.03 \pm 0.1$

examined with a number of different reflections. It is clear from Fig. 10 why the $2\bar{2}0$ reflection has been preferred, since the position of the partials is readily measured. Diffracting conditions which lead to the outer partials in contrast, and the center one out ($\boldsymbol{g}$ either $1\bar{1}\bar{1}$, $1\bar{1}1$, or 002, depending on $\boldsymbol{b}$) also enable $d_e$ and $d_i$ to be determined easily. In such cases a number of comparisons such as that of Fig. 10 show that the position of the outer partials under these reflecting conditions is where the intensity of the microdensitometer trace has dropped $20 \pm 10\%$ from its peak value. The intensity decreases so rapidly at the edges of the faults (see trace for $\boldsymbol{g} = \bar{1}1\bar{1}$ in Fig. 10) that this uncertainty corresponds to an error of only $\pm 3\%$ in $d$. Less satisfactory measurements can be made when the outer partials are out of contrast, and measurements have not been made from such plates.

Table 1 contains the results of the measurements and the values of $\gamma_e$ and $\gamma_i$ calculated therefrom using equations (2) and (3). From the table we see that all the fault pairs had characters within 40° of pure screw. For both alloys we find that, contrary to earlier qualitative results, the ratio $\gamma_e/\gamma_i$ is very close to unity. For the reasons mentioned earlier it is possible that $\gamma_e$ has been overestimated and $\gamma_i$ underestimated by the application of (2) and (3). In the following section two methods are employed to arrive at the magnitude of this effect, and a preliminary report is made of the effect of heat treatment on the fault pairs.

### *5.3 Checks on the absolute accuracy of $\gamma_e$ and $\gamma_i$*

#### *5.3.1 Variation of $d_i/d_e$ with $l/d_i$*

Fig. 11 shows the variation of $d_i/d_e$ with the ratio of the length of the center partial to the width of the intrinsic fault, $l/d_i$. We know that when $l/d_i \to \infty$ the equations (2) and (3) are completely valid in describing the fault pair. However, it is clear from Fig. 11 that even for relatively small values of $l/d_i$ the equations which have been used are a very good approximation. The influence of the forces arising from the end partials of the fault pair will cause a variation of $d_i/d_e$ as a function of $l/d_i$. From Fig. 11 we see that these forces have a negligible effect on the equilibrium value of $d_i/d_e$ as long as $l/d_i \gtrsim 2$. Valid conclusions may therefore be drawn using (2) and (3) even for short fault pairs, since the above condition, generally speaking, excludes only node pairs.

From the above we may conclude that the present measurements have led to very good absolute determinations of $\gamma_e$ and $\gamma_i$.

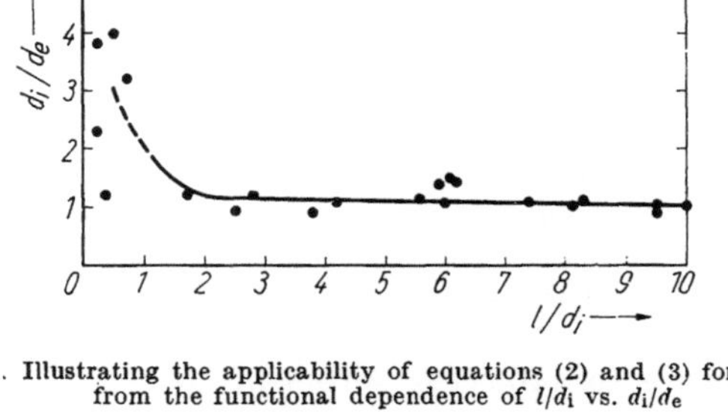


Fig. 11. Illustrating the applicability of equations (2) and (3) for $l/d_i > 2$, from the functional dependence of $l/d_i$ vs. $d_i/d_e$

Table 2

Comparison of $\gamma_i$ determined from fault pairs and from isolated intrinsic nodes

| | | $\gamma_i$ (erg/cm$^2$) | Standard deviation of the mean (erg/cm$^2$) | Standard deviation of a single reading (erg/cm$^2$) |
|---|---|---|---|---|
| $e/a = 1.23$ | Sample of 40 nodes [7] | 5.55 | 0.3 | 1.3 |
| | Nodes (3) close to fault pairs | 4.8 | | |
| | Nodes in Ag–Sn ($e/a = 1.23$) [14] | 4.9 | | |
| | Fault Pairs | 4.55 | 0.3 | 0.8 |
| $e/a = 1.15$ | Sample of 40 nodes | 8.7 | 0.5 | 2.5 |
| | Nodes (4) close to fault pairs | 9.2 | | |
| | Fault Pairs | 6.8 | 0.4 | 1.2 |

### *5.3.2 Comparison with $\gamma_i$ calculated from node data*

Table 2 compares the results obtained from measurements of the inscribed radius of extended intrinsic nodes with those from fault pairs. The node results have been calculated using the equations of Brown [10]. It has been shown [7] that for a given screw node configuration, applying the theory of Siems [11] leads to a value of $\gamma_i$ about 10% less than after Brown, while the theory of Jøssang et al. [12] leads to a value of $\gamma_i$ about 20% less than that calculated after Brown. A further uncertainty in the node theories arises from the term involving the cut-off radius of the dislocation core. A change by a factor of 2 in the magnitude of the core radius which is inserted in the equations alters $\gamma_i$ by $\approx$ 10%.

Bearing in mind these uncertainties in calculations made from node data we see from Table 2 that the absolute magnitudes of $\gamma_i$ determined from fault pairs are in rather good agreement with those obtained from nodes. This is particularly true for the alloy with $e/a = 1.23$, while even for $e/a = 1.15$ nodes and fault pairs in close proximity gave values of $\gamma_i$ which agreed within 25%.

### *5.3.3 High-temperature experiments*

One further possibility which must be examined before concluding that the present results establish that $\gamma_e \approx \gamma_i$, is that preferential segregation may take place to the extrinsic fault. This would have the effect of reducing the apparent extrinsic stacking fault energy. That preferential segregation should occur seems unlikely since the atomic displacements differ only slightly for the two types of fault.

Fig. 12 contains a sequence of pictures illustrating the appearance of extrinsic and intrinsic faults before, during, and after an annealing treatment in the electron microscope. The offset of the stacking fault fringes is clearly visible establishing that both extrinsic and intrinsic faulting has occurred.

It is interesting to note that at room temperature prior to anneal wide bands of both extrinsic and intrinsic fault are present, probably due to the trailing partial being held up while the rest of the configuration continues to glide. Raising the temperature to 140 °C is sufficient to activate movement of the

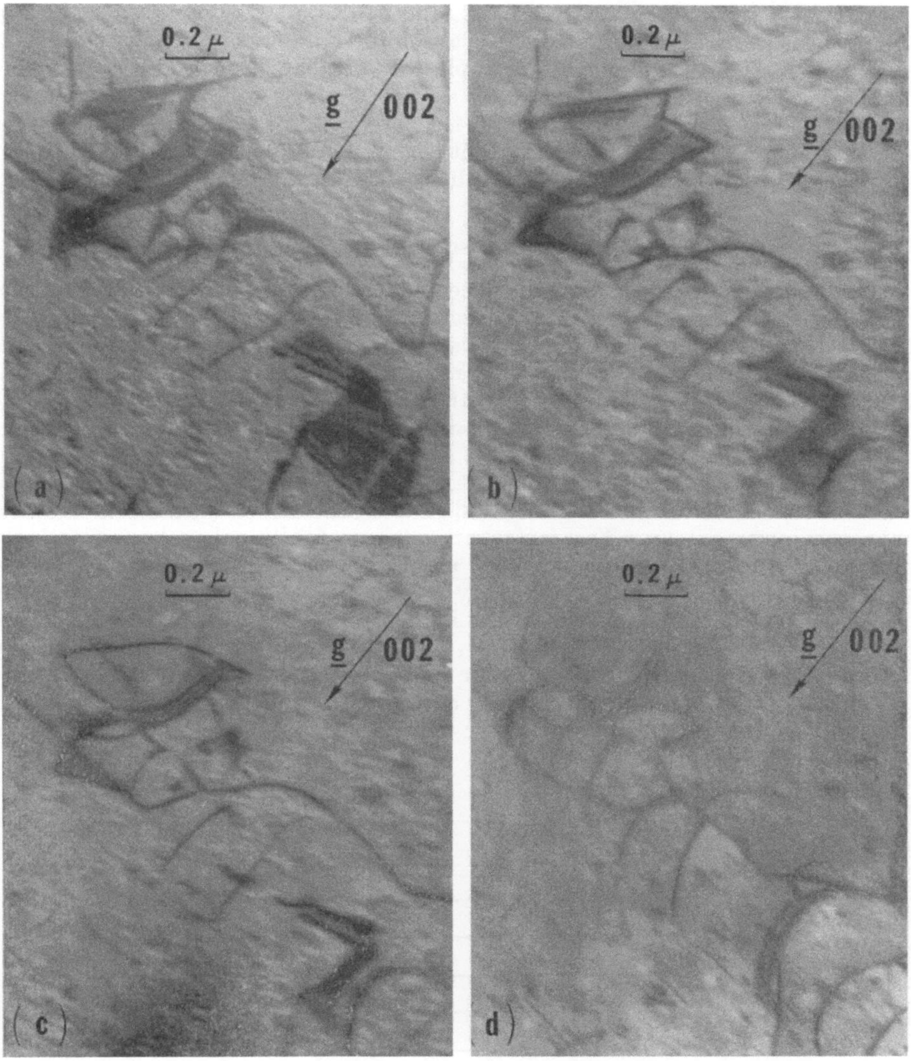


Fig. 12. The glide of partials bounding extrinsic-intrinsic fault pairs for the alloy with $e/a = 1.23$; a) at room temperature,b) and c) at 140 °C, d) subsequently at room temperature

partials, as shown in Fig. 12b, c. On cooling to room temperature (Fig. 12d), we find that the equilibrium configurations which have been reached are fault pairs for which $d_i$ and $d_e$ are closely equal.

The fact that in this case the same result has been obtained as for fault pairs observed at room temperature following room temperature deformation is strong evidence that preferential segregation does not occur.

It appears from Sections 5.3.1, 5.3.2, and 5.3.3 that the values of $\gamma_i$, $\gamma_e$ and $\gamma_e/\gamma_i$ which have been obtained have absolute significance, Fig. 11 making it clear that equations (2) and (3) are applicable without correction even to short fault pairs.

We are able to conclude, therefore, that the ratio of $\gamma_i$ to $\gamma_e$ in the present alloys is only slightly greater than unity, and there seems no reason to doubt that this result is of more general validity.

## 6. Discussion

Fig. 13 illustrates the planar displacements which arise from the formation of an extrinsic-intrinsic pair, the configurations being those first described by Frank [15]. As pointed out by Frank and Nicholas [16] for both the extrinsic and intrinsic faults there are two next nearest neighbor misfits of the type AA, BB, or CC from which one expects the major part of the fault energy to be derived. Alternatively the intrinsic and extrinsic faults may be regarded as one- and two-layer twins, respectively.

Throughout the earlier work on the nature of faults it was stated that no a priori reason could be seen to conclude which of the two types of fault would have the lower energy. The fact that extrinsic faulting has been relatively infrequently observed seems to be the main reason for the generally held view that the extrinsic stacking fault energy is higher than the intrinsic.

A more plausible explanation in the light of the present results is that generation of an extrinsic fault is difficult, but that once formed the fault energy per unit area is closely the same as for an intrinsic fault. In the latter respect the present experimental results are in accord with simple theoretical considerations in that the next nearest neighbor misfit energy is the same for both types of fault.

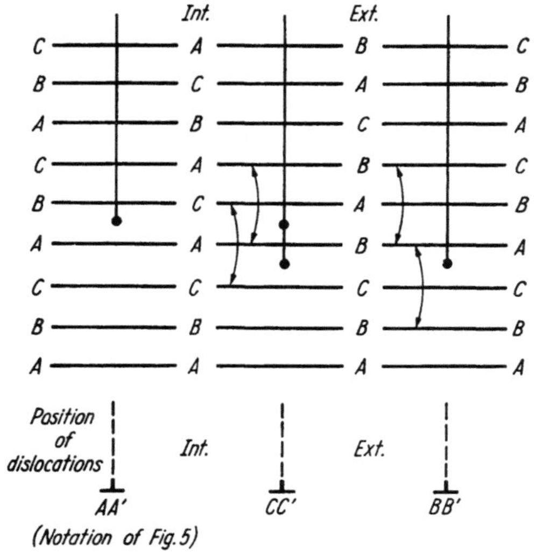


Fig. 13. Description of an extrinsic-intrinsic fault pair in terms of atomic planes, illustrating the second nearest neighbor misfits

## 7. Summary

1. Extrinsic-intrinsic node pairs are shown to be unsuitable for a straight forward determination of $\gamma_e$ and $\gamma_i$.

2. A new extrinsic-intrinsic fault configuration is described which possesses many advantages for the determination of $\gamma_e$ and $\gamma_i$.

3. In the two alloys examined, $\gamma_e/\gamma_i$ has been determined as $1.03 \pm 0.1$, and $1.09 \pm 0.1$. Comparison of $\gamma_i$ determined in this way with $\gamma_i$ from node data indicates that satisfactory absolute determinations of $\gamma_i$ and $\gamma_e$ can be obtained from the fault pairs, using the very simplest theoretical relations.

### *Acknowledgements*

Thanks are due to J. Washburn, P. Guyot, and G. Thomas for useful discussions. The work was supported by the United States Atomic Energy Commission through the Inorganic Materials Research Division of the Lawrence Radiation Laboratory, University of California, Berkeley, California.

## References

[1] M. H. Loretto, Phil. Mag. **10**, 467 (1964).
[2] L. K. Ives and A. W. Ruff, J. appl. Phys. **37**, 1831 (1966).
[3] M. H. Loretto, Phil. Mag. **11**, 125 (1965).
[4] G. R. Booker and L. M. Brown, Phil. Mag. **11**, 1315 (1965).
[5] E. Aerts, P. Delavignette, R. Siems, and S. Amelinckx, J. appl. Phys. **33**, 3078 (1962).
[6] N. Thompson, Proc. Phys. Soc. **B 66**, 481 (1953).
[7] P. C. J. Gallagher, J. appl. Phys. **37**, 1710 (1966).
[8] S. Amelinckx, Solid State Phys. Suppl. **6**, 1 (1964).
[9] L. J. Teutonico, Acta metall. **11**, 1283 (1963).
[10] L. M. Brown, Phil. Mag. **10**, 44 (1964).
[11] R. Siems, Disc. Faraday Soc. **38**, 42 (1964).
[12] T. Jøssang, M. J. Stowell, J. P. Hirth, and J. Lothe, Acta metall. **13**, 279 (1965).
[13] R. Gevers, A. Art, and S. Amelinckx, phys. stat. sol. **3**, 1563 (1963).
[14] A. W. Ruff, private communication, 1965.
[15] F. C. Frank, Phil. Mag. **42**, 809 (1951).
[16] F. C. Frank and J. F. Nicholas, Phil. Mag. **44**, 1213 (1953).

*(Received April 5, 1966)*

phys. stat. sol. **16**, 117 (1966)

*Tube Investments Research Laboratories, Hinxton Hall, Cambridge*

# The Migration of Double Positioning Boundaries During the Growth of Epitaxial (111) Gold Films

By

M. J. STOWELL and T. J. LAW

A study has been made of the migration of double positioning boundaries during continuous growth of (111) gold films inside an electron microscope. The boundary mobility is shown to be temperature dependent and the results indicate that double positioning boundary migration is a thermally activated process. The influence of rate of coalescence of deposit islands and of deposition rate on the size of double positioning domain structures in (111) gold films is discussed and the conclusions are extended to other multiply positioned and polycrystalline structures formed by vacuum evaporation.

Es wird die Wanderung von „double positioning" Zwillingsgrenzen während des kontinuierlichen Wachstums von (111)-Goldschichten innerhalb eines Elektronenmikroskops untersucht. Die Beweglichkeit der Grenzen erweist sich als temperaturabhängig, und die Ergebnisse zeigen, daß die Wanderung ein thermisch aktivierter Prozeß ist. Der Einfluß der Verschmelzungsrate der aufgedampften Inseln und der Verdampfungsrate auf die „double positioning" Domänenstruktur der (111)-Goldschichten wird diskutiert, und die Schlußfolgerungen werden auf ähnliche durch Vakuumaufdampfung erzeugte Strukturen ausgedehnt.

## 1. Introduction

It has been established that, when gold is deposited from the vapour phase onto $MoS_2$ cleavage flakes, two (111) orientations of the deposit are nucleated [1]. The deposit is said to be "doubly positioned" and the two "positions", which are epitaxially oriented, occur randomly on the substrate. During growth of the film, islands in the two orientations coalesce and a type of grain boundary is formed. We refer to this boundary as a "double positioning" (D.P.) boundary; it is, in essence, an incoherent twin boundary as the orientations it separates are twin related. In a continuous deposit, several hundreds of angstroms thick, the two positions are arranged in domains several μm across and it is clear that considerable changes in the double positioning structure have taken place during film growth. Some details of these structural modifications have already been reported [2, 3]. The aim of this paper is to present further experimental evidence on the mechanism by which double positioning structural changes take place, and to investigate the temperature dependence of D.P. boundary mobility.

Although the present study has been carried out on a (111) deposit, the results are relevant to the problem of thin film growth in general and we shall discuss the role of recrystallization and coalescence of deposit islands in determining the final structure of thin films.

## 2. Experimental

The experimental arrangement for thin film deposition inside our Siemens Elmiskop I electron microscope and details of the substrate preparation have been described previously [4]. In the current work, gold was deposited on thin molybdenite substrates whose temperature was varied in the range of 350 to 475 °C. The substrates were mounted in a split washer so that the substrate normal was inclined at about 20° to the incident electron beam. By observing areas where a (111) gold reflection was operative, the two deposit orientations were seen in black-white contrast [2]. The image was recorded from the electron microscope viewing screen on Kodak Tri-X 35 mm ciné film and the camera was run at 8 frames per second.

## 3. Experimental Observations and Measurements

### *3.1 General behaviour*

It has been mentioned in Section 1 and described in previous papers [2, 3, 4] that double positioning boundaries are formed as a result of the coalescence between deposit islands in different epitaxial orientations (see Fig. 5 in [2]). They are sometimes eliminated quite rapidly (in less than 1 s), especially when a small island coalesces with a much larger one. More commonly, the boundaries remain within the islands for periods of the order of 1 min and then move by migration parallel to the boundary normal. It is sometimes observed that migration commences very soon after coalescence has occurred with a second island, as shown in Fig. 1 and 2. It is not necessary, however, that such a coalescence precedes boundary migration, as is seen from Fig. 3. The boundaries move to reduce their total area and generally move completely out of an island in a very short time unless it is geometrically unfavourable for them to do so.

It is apparent from our observations that the boundaries, which are formed at the neck between coalescing islands, are initially prevented from moving by variations in surface topography. The time taken for the coalescence neck to grow sufficiently will depend on temperature island size and shape. Further, it is possible that a thermal groove could develop at the boundary, thus hindering its motion. However, the important feature is that the boundaries, on breakaway from "pinning" sources, move with measurable velocities and so it is possible to investigate the kinetics of D. P. boundary migration.

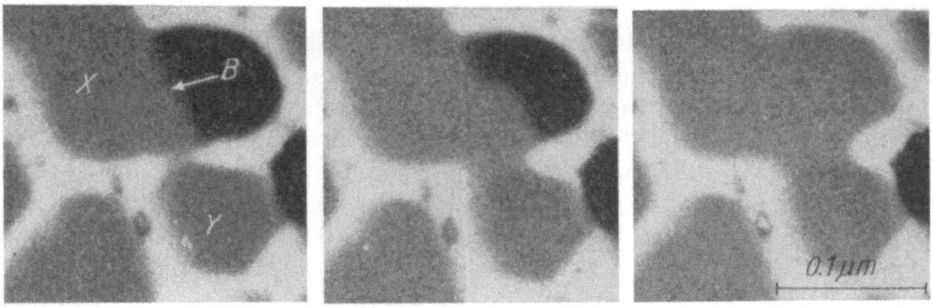


a b c

Fig. 1. Ciné-film sequence of the growth of Au on $MoS_2$ at 400 °C. Coalescence between gold islands X and Y in a) causes the D.P. boundary B to migrate ahead of the coalescence neck (b) and, subsequently, completely out of island X (c). Time scale: a) arbitrary zero, b) 0.125 s, c) 0.25 s

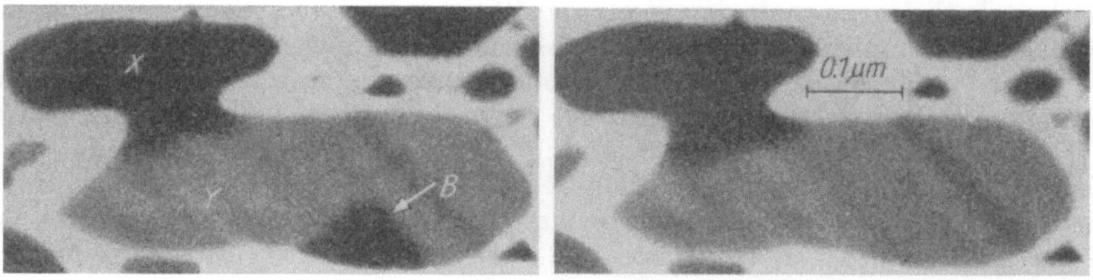


a b

Fig. 2. Elimination of the D.P. boundary B as a result of the coalescence between islands X and Y. a) Arbitrary zero, b) 0.25 s; substrate temperature 475 °C

i ii

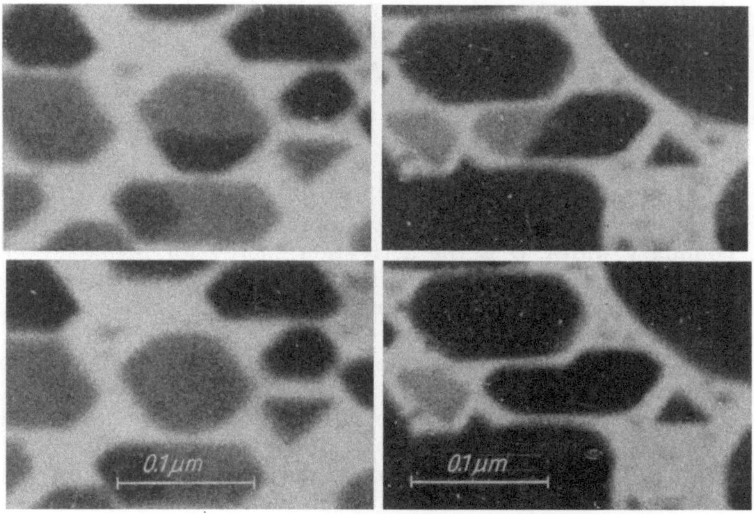


a a

b b

Fig. 3. Two examples of spontaneous D.P. boundary migration
i) Substrate temperature 475 °C, b) 0.125 s after a);
ii) substrate temperature 450 °C, b) 0.125 s after a)

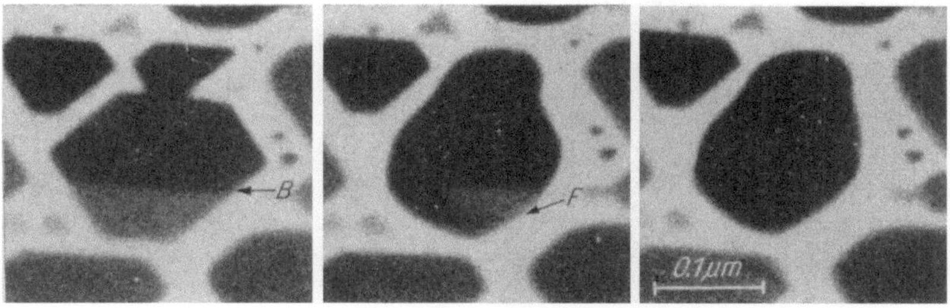


a b c

Fig. 4. Elimination of the D.P. boundary B by the motion of an irregular front, F. Substrate temperature 475 °C; a) arbitrary zero, b) 0.5 s, c) 0.75 s

Although the large majority of composite islands recrystallize in the manner just discussed, another mode of contrast change is occasionally observed. In this, an irregular black-white front moves across an island, usually at a rate rather slower than that observed for the normal manner of boundary migration. This is illustrated in the sequence of Fig. 4. It is not certain that this change of contrast signifies a change in orientation and more work is necessary on this point. However, the frequency of occurrence of this type of event is not high and we feel justified in concentrating on the more straightforward mode of orientation change.

### *3.2 Kinetics of D.P. boundary migration*

The observations detailed in Section 3.1 lead one to suspect that D.P. boundary migration might be a thermally activated process. A simplified theory for such a process was put forward by Turnbull [5] and from this it evolves that the velocity $V$ of a migrating boundary is related to the driving force $F$ (per atom of the boundary) by the relation

$$V = M \cdot F\,, \tag{1}$$

where $M$ is the mobility of a boundary atom, given by

$$M = \frac{\nu\, b^2}{k\,T} \exp\left(-\frac{Q}{k\,T}\right) = \frac{D_m}{k\,T}\,. \tag{2}$$

Here, $\nu$ is an atomic vibration frequency (approximately the Debye frequency), $b$ is the boundary width, $Q$ is the activation energy for boundary migration and $D_m$ has the form of a diffusion coefficient. In principle, it should be possible to measure D.P. boundary velocities from our ciné-films and, from details of island shapes, to evaluate driving forces. We have attempted to do this in order to investigate the temperature dependence of $M$ through the parameter $D_m$ in the range 350 to 475 °C.

In order to estimate the driving force $F$, it is necessary to make assumptions concerning deposit island shapes. In view of our previous studies [2], we consider the islands to be platelets with side faces normal to the substrate. Some possible sources of driving force have been considered by Jacobs et al. [2] and it was concluded that, in the system Au/$MoS_2$, the free energy of the boundary surface provided by far the major contribution, since contributions from free surface energy and deposit-substrate interfacial energy are small. On this basis, it is easy to derive an expression for $F$, as we now show.

Consider a plate-like island, shown schematically in Fig. 5, which contains a D.P. boundary of length $L(x)$, where the $x$-axis is normal to the boundary

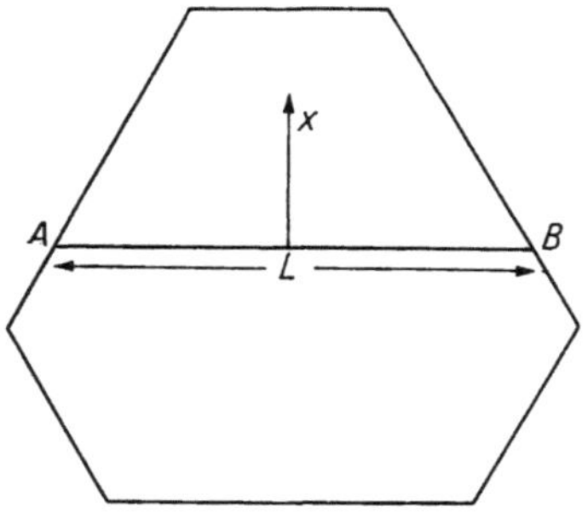


Fig. 5. Schematic representation of a deposit island containing a D.P. boundary AB of length $L$; the $x$-axis is normal to AB

plane. If the D.P. boundary surface free energy is $\sigma$, the driving force for migration (per atom of the boundary) is given by

$$F = \frac{-\sigma}{N\,L(x)}\,\frac{\mathrm{d}L(x)}{\mathrm{d}x}\,, \tag{3}$$

where $N$ is the number of atoms per unit area of the boundary. In most of the cases analysed, the sides of the islands were straight, as in Fig. 5, so that $\mathrm{d}L(x)/\mathrm{d}x$ was constant. In such cases, the time $t$ taken for a boundary to move a distance $x$ is, from equations (1), (2), and (3),

$$t = \frac{k\,T}{D_\mathrm{m}}\,\frac{N}{\sigma}\,\frac{\bar{L}}{\Delta L}\,x^2\,, \tag{4}$$

where $\bar{L}$ is the mean boundary length and $\Delta L$ is the change in boundary length.

In principle, it is possible to obtain $D_\mathrm{m}$ from measurements of D.P. boundary movement on the ciné-films. In practice, this has not proved possible. What is observed is that either a boundary is clearly present on one frame of an 8 frame/s ciné-film and absent on the next frame, or that a boundary moves during the exposure of one or more frames, giving rise to a blurred image of the boundary. In cases where a boundary has vanished between successive exposures, we can place an upper limit on the time $t$ in equation (4) and then obtain a lower limit $D_\mathrm{m}^{(-)}$ for $D_\mathrm{m}$. In cases where a boundary has migrated during exposure of two or more frames, we can also deduce an upper limit $D_\mathrm{m}^{(+)}$ for $D_\mathrm{m}$. Our technique was to obtain values of $D_\mathrm{m}^{(+)}$ and $D_\mathrm{m}^{(-)}$ and so to determine a range in which $D_\mathrm{m}$ must lie at temperatures of 350, 400, 450 and 475 °C.

In the evaluation of $D_\mathrm{m}^{(\pm)}$, we require a value for $\sigma$; this is unknown for gold. We used the relation $\sigma \approx 0.8\,\sigma_\mathrm{gb}$, where $\sigma_\mathrm{gb}$ is the mean high angle grain boundary energy [6] and data for $\sigma_\mathrm{gb}$ [7] to obtain an estimated value for $\sigma$ of 300 erg cm$^{-2}$. The value of $N$ in (4) was taken to be $9.8 \times 10^{14}$ cm$^{-2}$, the atom density in a gold {112} plane.

Equation (4) is only valid for plate-like islands for which $\mathrm{d}L(x)/\mathrm{d}x$ is constant. Islands having curved profiles were assumed to be platelets and the curved profiles were approximated by a series of linear segments to which equation (4) was applied.

The final results of the analysis of forty separate events are shown in Fig. 6, a log-linear plot of $D_\mathrm{m}$ against $1/T$. To simplify the graph, only the ranges in

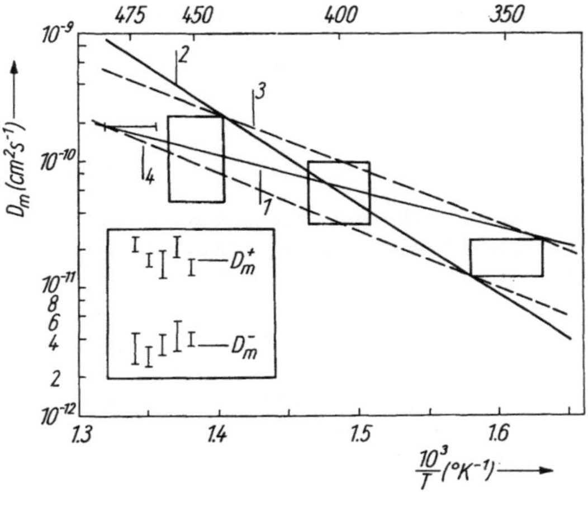


Fig. 6. Results of analysis of D.P. boundary migration; log $D_\mathrm{m}$ plotted against $1/T$. The inset illustrates the procedure for arriving at upper and lower limits for $D_\mathrm{m}$ from the experimental results

which $D_m$ must lie are shown. At 475 °C, no event was recorded which could lead to a value of $D_m^{(+)}$, so that the value shown is a lower limit to $D_m$. With each experimentally determined value of $D_m^{(\pm)}$ are associated errors arising from inaccuracies of measurement on the photographic prints ($\approx$ 20 to 30%) and from inaccuracies of temperature measurements. The latter errors were taken to be $\pm 10$ °C and are represented by the vertical sides of the "boxes" drawn in Fig. 6. These "boxes" define the regions in which $D_m$ must lie and their vertical extremities were arrived at as follows. The upper limit on $D_m$ is defined as the smallest upper extreme value of the $D_m^{(+)}$ determinations. The lower limit is the largest lower extreme value of $D_m^{(-)}$. The inset of Fig. 6 serves to illustrate these definitions; the plotted values of $D_m^{(\pm)}$ are hypothetical and have been staggered on the $1/T$ axis for the sake of clarity.

Assuming that $D_m$ obeys an Arrhenius relation $D_m = D_{m0} \exp(-Q/RT)$, where $R$ is the gas constant, it may be deduced from the lines of maximum and minimum slope in Fig. 6 (lines (1) and (2)) that $Q$ must lie between the extreme limits of 13 and 33 kcal/mol; the corresponding extreme limits on $D_{m0}$ are $1.3 \times 10^{-6}$ and 3.0 cm$^2$ s$^{-1}$, respectively. Within the temperature range 350 to 475 °C, lines (3) and (4) in Fig. 6 show that the numerical values of $D_m$ lie within the limits given by $D_m = 10^{-3.4\pm0.2} \exp(-21000/RT)$. The $Q$ value in this relation must *not* be taken to represent the activation energy for D.P. boundary migration, it is just a convenient parameter through which the range of possible $D_m$ values may be expressed.

The experimental errors arising out of measurements of photographic prints are quite large. This is due not only to inaccuracies of measurement but also due to the lack of clarity of the image at the edges of an island. This last point is of importance and is the major reason for not increasing the camera speed. Because of the loss of definition at island edges, no corrections were made for faceting of the islands.

It has been assumed that the D.P. boundaries extend from top to bottom of the deposit islands. We know that this is not always the case and that a coherent twin boundary parallel to the substrate can sometimes extend over part of an island [8]. Motion of the incoherent part of this D.P. boundary may either be assisted or retarded by a contribution to the driving force from the coherent twin boundary free energy $\gamma$, and to account for this we have to include a term $\pm \frac{\gamma}{N} \frac{1}{t}$ in equation (3), where $t$ is the depth of the incoherent boundary. This term will be small unless $t < 20$ Å. Unfortunately, we have no means of assessing the contribution of this term and we have ignored it in our calculations.

## 4. Discussion

Although the activation energy for D.P. boundary migration has not been determined with great accuracy, the results show that this quantity is considerably lower than that for volume self diffusion in gold (41.7 kcal/mol, see [9]). There is no theoretical model from which we may estimate an expected value of $Q$. It has been suggested (see [10]) that the activation energy for grain boundary migration may be similar to that for grain boundary diffusion, or that mobility should be controlled by atom-vacancy interchange, so that $Q$ should be close to the activation energy for vacancy migration in the bulk crystal. No values for grain boundary diffusion coefficients are known for gold, but silver has an activation energy for grain boundary diffusion of 20.2 to 21.5 kcal/mol [11]. The activation energy for vacancy migration in gold is about 19.5 kcal/mol (see [12]). From equation (2), it is seen that the simple theoretical value for $D_{m0}$ is $\nu b^2$. The Debye frequency for gold is about $3.5 \times 10^{12}$ s$^{-1}$ and if we take $b$ to be $(a/6)\langle 112\rangle = 1.67$ Å, we obtain $D_{m0} \approx 10^{-3}$ cm$^2$ s$^{-1}$. Our results for $D_m$ within the range of 350 to 475 °C are not at variance with the predictions above.

The observations detailed in Section 3.1 indicate that it is necessary for neck growth between coalescing islands in different orientations to be at least almost complete before boundary migration can commence. Further, a boundary may often remain within a composite island until another coalescence event occurs and disrupts the trend of the first island towards an equilibrium shape. It is, therefore, apparent that recrystallization in discontinuous thin films is intimately linked with the mechanism of coalescence and of island shape change. In view of the size of the islands concerned, shape changes in these deposits almost certainly take place by surface diffusion [4]. The shapes of coalescing islands have a marked effect on deciding whether a boundary can migrate soon after coalescence or whether it will be constrained to remain within the island for a long time; a preliminary discussion of this point has been presented by Pashley and Stowell [3]. A boundary which is constrained to remain in an island may well develop a thermal groove during the final stages of neck growth. In this case, a further coalescence, not necessarily close to the emergence of the boundary at the island surface, will cause large surface diffusion currents to be set up which destroy the previous (quasi-) equilibrium island shape and then allow the boundary to migrate. It is not possible at present to separate the roles of slow neck growth and thermal grooving on the pinning of D.P. boundaries but since it is frequently observed that boundary migration takes place without the assistance of a coalescence event, it is probable that thermal grooving is not very important during continuous growth at the temperatures used in this study.

It evolves from this discussion that two conditions for obtaining large double positioning domain sizes in our deposits are that both D.P. boundary mobility and surface diffusion must be sufficiently rapid to allow a boundary to be removed from an island before many further coalescences take place. However, the coalescence frequency is determined by the lateral growth rate of islands; this is governed by the deposition rate, which also has an influence on the initial density of islands. If the deposition rate is high, say $10^2$ to $10^3$ Å s$^{-1}$, a high density of nuclei would result and the early stages of coalescence between small islands would allow some boundary migration to take place. However, in view of the large rate of island growth, many coalescences will take place before the associated necks are able to be eliminated and a high density of D.P. boundaries will be "frozen" into the structure. Only after continuity of the film has been reached and surface irregularities are smoothed out will recrystallization be possible. One effect of increasing the deposition rate is, then, to delay the onset of recrystallization until a later stage of film growth is attained. We conclude that a slow rate of deposition is also an important condition to be satisfied in order to obtain large double positioning domains in (111) epitaxial films.

The importance of our observations and conclusions will now be discussed with reference to other deposit-substrate combinations. The effect of substrate temperature and deposition rate on the oriented growth of thin films has been demonstrated on a variety of materials, for example Ag/NaCl [13], Ge/$CaF_2$ [14], Si/Si [15] with the evaporation technique and with sputtered films [16]. All of the authors showed that, for a given substrate temperature, low deposition rates favour the production of single crystals and that, as the substrate temperature was decreased, the transition from a single crystal to a polycrystalline growth occurred for lower deposition rates. These results, which have largely remained unexplained can be accounted for qualitatively with the aid of the arguments presented above. For a complete explanation of the effects it would be necessary to have very much more information about orientation distributions at different stages of growth (including the nucleation stage) for a wide range of deposition rates at each temperature used. It is important to realize that the structure of a continuous film is dependent not only on the nucleating conditions but it is strongly dependent on recrystallization occurring during the growth process. This has already been demonstrated by Pashley et al. [4], who illustrated marked changes in the grain size of polycrystalline deposits with thickness at a fixed substrate temperature.

A most interesting and striking demonstration of the importance of recrystallization during thin film growth has been obtained by Matthews and Grünbaum [17] and by Matthews [18], who studied in detail the growth of gold on clean and contaminated NaCl substrates in ultra-high vacuum. In their work, the initial deposits contained mainly (001) nuclei with about 10% of (111) nuclei multiply positioned on the (001) substrate. The (111) islands, although initially small in number, controlled the orientation of the growing film when the initial nucleation density was low (on clean surfaces), but when the nucleation density was high (on contaminated surfaces), the (001) orientation predominated throughout the growth sequence. Some of the factors controlling growth recrystallization in these deposits have been discussed by Pashley and Stowell [3], and it was noted that both free and interfacial surface energies contributed towards the driving force for recrystallization. In order to obtain an epitaxial (001) deposit, one has to inhibit the preferred growth of the (111) orientations before continuity of the film is reached, since the contribution of the free surfaces to the driving force for grain boundary migration is predominant in a continuous film and favours the growth of (111) grains to produce a $\langle 111 \rangle$ textured deposit. Matthews and Grünbaum [17] produced a (001) film on a clean substrate by depositing at a high rate (2000 Å $s^{-1}$). Later, Matthews [19] obtained a high nucleus density by initially depositing 20 Å of gold at 1000 Å $s^{-1}$ and then grew the film to continuity at the low rate of 5 Å $s^{-1}$. A similar technique has been applied by Matthews [20] to the growth of multiply positioned iron on clean NaCl substrates. These results do not contradict our conclusions because, in this case, the initial deposit consisted (presumably) of a large number of oriented (001) islands; what is important is that the inhibition of recrystallization due both to deposition rate and nucleus density considerations has been clearly demonstrated. It must be borne in mind that a large amount of work has been carried out on growing f.c.c. metals in (111) orientation, which is the orientation favoured during recrystallization (assuming clean growth conditions). We suggest that further experiments along the lines of that performed by Matthews [20] would produce very interesting results because the structure produced by growth at a slow deposition rate should change with increased substrate temperature or decreased growth rate to give a $\langle 111 \rangle$ textured film.

## 5. Conclusions

The coarsening of double positioning structures in evaporated (111) gold films has been observed directly in the electron microscope. The temperature dependence of D.P. boundary mobility has been determined and the relation between coalescence phenomena and recrystallization during thin film growth has been examined. It is emphasized that these effects have a pronounced influence on the structure of continuous films. The initial, nucleated structure can be markedly changed during growth, the final structure often bearing little resemblance to the initial one.

*Acknowledgements*

We thank Dr. D. W. Pashley for many stimulating and encouraging discussions and Mr. A. G. Doggett for experimental assistance. This paper is published by permission of the Chairman of Tube Investments Limited.

## References

[1] R. Uyeda, Proc. Physico-Math. Soc. Japan **22**, 1023 (1940).
[2] M. H. Jacobs, D. W. Pashley, and M. J. Stowell, Phil. Mag. **13**, 129 (1966).
[3] D. W. Pashley and M. J. Stowell, J. Vac. Sci. Technol. (to be published).
[4] D. W. Pashley, M. J. Stowell, M. H. Jacobs, and T. J. Law, Phil. Mag. **10**, 127 (1964).
[5] D. Turnbull, Trans. AIME **191**, 3 (1951).
[6] R. L. Fullman, J. appl. Phys. **22**, 448 (1950).
[7] M. C. Inman and H. R. Tipler, Met. Rev. **8**, 105 (1963).
[8] M. H. Jacobs and M. J. Stowell, Phil. Mag. **11**, 591 (1965).
[9] D. Lazarus, Solid State Phys. **10**, 71 (1960).
[10] K. T. Aust and J. W. Rutter, in: Recovery and Recrystallization of Metals, Ed. L. Himmel, Interscience Publishers, 1963 (p. 131).
[11] R. E. Hoffman and D. Turnbull, J. appl. Phys. **22**, 1949 (1951).
[12] A. C. Damask and G. J. Dienes, Point Defects in Metals, Gordon and Breach, New York/London 1963.
[13] B. W. Sloope and C. O. Tiller, J. appl. Phys. **32**, 1331 (1961).
[14] B. W. Sloope and C. O. Tiller, J. appl. Phys. **36**, 3174 (1965).
[15] B. A. Unvala and G. R. Booker, Phil. Mag. **9**, 691 (1964).
[16] E. Krikorian, in: Single Crystal Films, Ed. M. H. Francombe and H. Sato, Pergamon Press, 1964 (p. 113).
[17] J. W. Matthews and E. Grünbaum, Phil. Mag. **11**, 1233 (1965).
[18] J. W. Matthews, Phil. Mag. **12**, 1143 (1965).
[19] J. W. Matthews, Appl. Phys. Letters **7**, 131 (1965).
[20] J. W. Matthews, Appl. Phys. Letters **7**, 255 (1965).

*(Received April 5, 1966)*

phys. stat. sol. **16**, 127 (1966)

*Zentralinstitut für Kernforschung der Deutschen Akademie der Wissenschaften, Rossendorf (a), Akademie-Werkstätten für Forschungsbedarf der Deutschen Akademie der Wissenschaften, Berlin (b), und II. Institut für Experimentalphysik der Technischen Universität Dresden (c)*

# Mößbauer-Untersuchungen an Mn- und Mg–Mn-Ferriten

Von

E. WIESER (a), W. MEISEL (b) und K. KLEINSTÜCK (a, c)

Die Mößbauer-Spektren von stöchiometrischem Mn-Ferrit sowie zwei Mg–Mn-Ferriten wurden bei Zimmertemperatur und bei der Temperatur des flüssigen Stickstoffs aufgenommen. Bei Zimmertemperatur wurde eine Überlagerung zweier Sechs-Linien-Spektren unterschiedlicher Intensität für alle drei Proben gefunden, die sich in den Tieftemperaturspektren nicht mehr nachweisen läßt. Die zwei überlagerten Spektren werden den $Fe^{3+}$-Ionen auf Tetraeder- und Oktaederplätzen zugeordnet. Für die beiden Mg–Mn-Ferrite ergibt sich dabei, daß das größere Magnetfeld am Kernort dem Untergitter mit dem kleineren mittleren magnetischen Moment entspricht. Zur Erklärung dieses Sachverhaltes werden Abweichungen von der kollinearen Néelschen Spinkonfiguration auf Oktaederplätzen diskutiert.

Mössbauer investigations of a stoichiometric Mn-ferrite and two Mg–Mn-ferrites are made at room and liquid nitrogen temperatures. At room temperature all samples show a superposition of two six-lines-spectra with different intensities. These spectra correspond to $Fe^{3+}$-ions on tetrahedral and octahedral sites. Hence it follows for the two Mg–Mn-ferrites, that the greater magnetic field at the iron nucleus corresponds to the sublattice having the smaller average magnetic moment. These results are explained as being due to deviations from collinear Néel spin-configurations on octahedral sites.

## 1. Problemstellung

Ferrimagnetische Metalloxyde mit Spinellstruktur besitzen als weichmagnetische Werkstoffe große Bedeutung. Zum Verständnis ihrer magnetischen Eigenschaften ist die Kenntnis der Wertigkeitsstufen, in denen die beteiligten Kationen auftreten, und der Verteilung der verschiedenen Ionenarten auf die beiden möglichen Typen von Gitterplätzen (Oktaeder- und Tetraederplätze) eine notwendige Voraussetzung. Bei manganhaltigen Ferriten erweist sich die Bestimmung dieser Größen als besonders schwierig.

Aussagen über die Verteilung der verschiedenen Kationen auf Oktaeder- und Tetraederplätze liefern Neutronenbeugungsuntersuchungen. Für das System $Mn_xFe_{3-x}O_4$ liegen mehrere derartige Arbeiten vor [1 bis 4]. Die Verteilungsparameter in Mg-Mn-Ferriten wurden von Nathans u. a. [5] sowie Kleinstück u. a. [6] bestimmt. Eindeutige Angaben über die Wertigkeiten der vorhandenen Kationen kann man aus den neutronographisch ermittelten mittleren magnetischen Momenten der Oktaeder- und Tetraederplätze nicht erhalten. Die Zahl der zu einer vollständigen Beschreibung der Wertigkeitsverhältnisse erforderlichen Parameter ist zu groß, da sowohl Fe als auch Mn in mehreren Wertigkeitsstufen vorliegen können ($Fe^{2+}$, $Fe^{3+}$; $Mn^{2+}$, $Mn^{3+}$, $Mn^{4+}$).

Beispielsweise für stöchiometrischen Mn-Ferrit sollte man die Zusammensetzung $Mn^{2+}Fe_2^{3+}O_4$ erwarten. Von Harrison u. a. [7] sowie Krupička und Závěta [8] wird jedoch $Mn^{3+}$ und eine äquivalente Menge $Fe^{2+}$ auf Oktaederplätzen angenommen, damit die Abweichung des bei 4,2 °K zu 4,6 $\mu_B$ bestimmten mittleren magnetischen Moments dieser Plätze von dem für 3 $d^5$-Ionen zu erwartenden Wert von 5 $\mu_B$ erklärt werden kann.

Die unterschiedlichen Verteilungsparameter, die Nathans u. a. [5] aus magnetischen Daten und Neutronenbeugungsmessungen sowie Kleinstück u. a. [6, 9] mit Hilfe einer kombinierten Neutronen- und Röntgenstrahlenbeugungsmethode an Mg-Mn-Ferriten erhielten, deuten darauf hin, daß auch bei diesen Substanzen entweder neben $Mn^{2+}$ und $Fe^{3+}$ noch andere Wertigkeitsstufen vorliegen oder die Anordnung der magnetischen Momente vom Néelschen Modell abweicht.

Zuverlässige chemische Verfahren zur Bestimmung der Wertigkeiten sind, wie in [6] näher angeführt, nicht bekannt. Es ist daher notwendig, nach physikalischen Methoden zu suchen, die zur Lösung des Wertigkeitsproblems in Ferriten beitragen können. Der Mößbauer-Effekt an $Fe^{57}$ liefert unterschiedliche Ergebnisse für $Fe^{2+}$- und $Fe^{3+}$-Ionen, weshalb Mößbauer-Messungen für die Untersuchung des Wertigkeitsproblems in Ferriten indiziert sind.

Die Werte für die Isomerieverschiebung in verschiedenen chemischen Substanzen liegen für $Fe^{2+}$ bei 1,1 bis 1,5 mm/s und für $Fe^{3+}$ bei 0,2 bis 0,7 mm/s (bezogen auf eine Quelle von $Co^{57}$ in natürlichem Fe) [10]. Dieser Unterschied wurde zum Nachweis beider Ionensorten in paramagnetischen Verbindungen ohne Zeemann-Aufspaltung der Spektren benutzt [11, 12]. Die magnetischen Felder am Kernort kann man in Substanzen mit geordneter Magnetstruktur ihrem Betrage nach dem mittleren magnetischen Moment der 3 d-Schale im betreffenden Untergitter als proportional annehmen. Die elektrische Quadrupolaufspaltung ist für $Fe^{2+}$ im allgemeinen größer als für $Fe^{3+}$. Da sie jedoch in starkem Maße strukturabhängig ist, ist sie zur Unterscheidung beider Ionen wenig geeignet.

Von Bauminger u. a. [13] sowie Ito u. a. [14] wurde das Verhalten von $Fe^{2+}$ und $Fe^{3+}$ in Oxyden, die Spinellstruktur besitzen, mit der Methode des Mößbauer-Effektes am Beispiel des Magnetits ($Fe_3O_4$) untersucht. Bei der Temperatur des flüssigen Stickstoffs wurde für $Fe^{2+}$-Ionen ein Feld von (470 $\pm$ 20) kOe [14] bzw. (450 $\pm$ 20) kOe [13] am Kernort gefunden. Für $Fe^{3+}$-Ionen beträgt das Feld (510 $\pm$ 20) kOe. Ito u. a. beobachteten außerdem eine beträchtliche Quadrupolaufspaltung von (1,9 $\pm$ 0,2) mm/s an den $Fe^{2+}$-Linien. Bei Zimmertemperatur tragen infolge des Elektronenaustausches zwischen $Fe^{2+}$ und $Fe^{3+}$ auf Oktaederplätzen alle Oktaederionen zu einer Liniengruppe bei, die einem inneren Feld am Kernort entspricht, das innerhalb der Fehlergrenzen mit dem der $Fe^{2+}$-Linien übereinstimmt. Aus Leitfähigkeitsmessungen folgt, daß bei allen Ferriten, die $Fe^{2+}$ und $Fe^{3+}$ nebeneinander enthalten, mit einem solchen Elektronenaustausch und demzufolge mit einer Verfälschung der Intensität der Mößbauer-Linien zu rechnen ist. Zur quantitativen Auswertung der Linienintensitäten sind daher Messungen bei tiefen Temperaturen erforderlich. (Im folgenden wird die Bezeichnung „$Fe^{2+}$" für Linien verwendet, die durch den Elektronenaustausch zwischen $Fe^{2+}$ und $Fe^{3+}$ beeinflußt werden.)

Am $MnFe_2O_4$ liegen Mößbauer-Untersuchungen von Hrynkiewicz u. a. [15] und Tanaka u. a. [16] vor. In der ersten Arbeit wird das auf 0 °K extrapolierte innere Magnetfeld mit (515 $\pm$ 20) kOe angegeben. In [16] wurde das Feld am Kernort bei Zimmertemperatur zu (425 $\pm$ 20) kOe bestimmt. Die Überlagerung eines zweiten Sechs-Linien-Spektrums wurde in beiden Arbeiten nicht beobachtet.

Belov u. a. [17] führten Mößbauer-Messungen an Mn- und Mg-Mn-Ferriten durch. Proben von der in der vorliegenden Arbeit untersuchten Zusammensetzung wurden nicht gemessen. Eine Überlagerung wurde auch in [17] nicht gefunden.

## 2. Experimentelles

Untersucht wurden $MnFe_2O_4$ (Probe 1) sowie zwei Mg-Mn-Ferrite der ungefähren Zusammensetzung $Mg_{0,5}Mn_{0,5}Fe_2O_4$ (Probe 2) und $Mg_{0,75}Mn_{0,5}Fe_{1,75}O_4$ (Probe 3). Angaben über die Probenherstellung, die genaue chemische Zusammensetzung, die magnetischen Eigenschaften und die Verteilung der Metalle auf Oktaeder- und Tetraederplätze finden sich in [6]. Die in dieser Arbeit untersuchten Proben sind mit den hier verwendeten identisch. Am $MnFe_2O_4$ wurden zusätzlich Messungen an einer Substanz durchgeführt, die auf etwa 80% $Fe^{57}$ angereichertes Fe enthielt.

Als $\gamma$-Strahlenquelle diente $Co^{57}$ in Platin. Die Messungen erfolgten mit statischem Absorber und bewegter Quelle. Angaben über die benutzte Apparatur finden sich in [18].

Das Flächengewicht der polykristallinen Proben betrug etwa 35 mg/cm². Vom Mn-Ferrit wurden außerdem zwei angereicherte Proben von 9 mg/cm² und 3 mg/cm² und eine Probe mit natürlichem Eisen von 18 mg/cm² Flächengewicht gemessen, um den Einfluß von Sättigungserscheinungen auf die beobachteten Linien verschiedener Intensität zu überprüfen. Da sich eine Abhängigkeit der Verhältnisse der Linienhöhen von der Präparatdicke bemerkbar machte, wurden Messungen an Ferritgemischen aus $NiFe_2O_4$ und $Fe_3O_4$ mit einem definierten Verhältnis „$Fe^{2+}$"-Ionen/$Fe^{3+}$-Ionen durchgeführt. Diese gestatten die Größe der Verfälschung der Intensitätsverhältnisse abzuschätzen. Die $Fe^{3+}$-Linien beider Substanzen fallen zusammen, da ihre Aufspaltungsparameter innerhalb der Fehlergrenzen übereinstimmen. Bei der Berechnung des theoretischen „$Fe^{2+}$"/$Fe^{3+}$-Verhältnisses wurde berücksichtigt, daß alle Oktaederionen des Magnetits zur „$Fe^{2+}$"-Linie beitragen.

Bei einem Präparat von 35 mg/cm² Flächengewicht wird ein Intensitätsverhältnis von 3 um etwa 20% zu klein gemessen. Diese Messungen dienten gleichzeitig zur Prüfung der Nachweisempfindlichkeit für geringe $Fe^{2+}$-Mengen. Es ergab sich, daß sich die „$Fe^{2+}$"-Linie bei einem „$Fe^{2+}$"/$Fe^{3+}$-Verhältnis von 1:10 unter den gegebenen experimentellen Bedingungen noch gut erkennen läßt.

## 3. Ergebnisse

In Tabelle 1 sind die am Mn-Ferrit und an den beiden Mg-Mn-Ferriten bei Zimmertemperatur und bei der Temperatur des flüssigen Stickstoffs erhaltenen Ergebnisse zusammengestellt. Die am $MnFe_2O_4$ bei anderen Probendicken gemessenen Größen für die Isomerieverschiebung und die inneren Felder stimmen innerhalb der Fehlergrenzen mit den in der Tabelle angeführten überein.

Die bei Zimmertemperatur aufgenommenen Spektren aller drei Proben (Fig. 1a, 2a und 3a) zeigen Linienformen, die sich als Überlagerungen eines Spektrums höherer und eines solchen geringerer Intensität mit unterschiedlicher magnetischer Aufspaltung und Isomerieverschiebung deuten lassen. Die Spektren, die zum größeren Magnetfeld am Kernort gehören, wurden mit dem Index 1 versehen. Beim Mn-Ferrit besitzt dieses Spektrum die geringere Inten-

Tabelle 1

Werte für die Magnetfelder am Kernort ($H$), die Schwerpunktsverschiebungen der Spektren ($S$) und die Intensitätsverhältnisse ($V$) der zwei sich bei Zimmertemperatur überlagernden Spektren der Proben 1, 2 und 3

| Präparat | $H_1$ ($10^5$ Oe) | $H_2$ ($10^5$ Oe) | $S_1$ (mm/s) | $S_2$ (mm/s) | $V$ | Temperatur (°K) |
|---|---|---|---|---|---|---|
| Probe 1 (3 mg/cm²) ($MnFe_2O_4$) | 4,70 ±0,25 | 4,50 ±0,25 | 0 ±0,15 | 0,05 ±0,15 | 0,30 ±0,10 | 293 |
| Probe 1 (5 mg/cm²) ($MnFe_2O_4$) | 5,10 ±0,20 | — — | 0,11 ±0,10 | — — | — — | 77 |
| Probe 2 (35 mg/cm²) ($Mn_{0,5}Mg_{0,5}Fe_2O_4$) | 4,75 ±0,25 | 4,40 ±0,25 | −0,02 ±0,15 | 0,04 ±0,15 | 2,8±1 | 293 |
| Probe 2 (35 mg/cm²) ($Mn_{0,5}Mg_{0,5}Fe_2O_4$) | 5,10 ±0,20 | — | 0,12 ±0,10 | — | — | 77 |
| Probe 3 (35 mg/cm²) ($Mn_{0,5}Mg_{0,75}Fe_{1,75}O_4$) | 4,50 ±0,25 | 4,0 ±0,25 | −0,03 ±0,15 | −0,01 ±0,15 | 3,1±1 | 293 |
| Probe 3 (35 mg/cm²) ($Mn_{0,5}Mg_{0,75}Fe_{1,75}O_4$) | 5,10 ±0,20 | — | 0,10 ±0,10 | — | — | 77 |

sität, während bei beiden Mg-Mn-Ferriten die stärkeren Linien dem größeren Feld entsprechen. Das in der Tabelle 1 angeführte Intensitätsverhältnis $V$ ist als das Verhältnis der Linienhöhen $Y_1/Y_2$ definiert. Es wurde aus den am besten aufgelösten Liniengruppen bestimmt. Diese waren für Probe 1 die beiden Linien Nummer 1 und für die Proben 2 und 3 die Linien 1 und 6, wenn man die sechs Linien eines Spektrums, von negativen Geschwindigkeiten des höchsten Betrages beginnend, durchnumeriert. Als positiv sind die Geschwindigkeiten definiert, bei denen sich die Quelle in Richtung auf den Absorber bewegt.

Bei der Temperatur des flüssigen Stickstoffs sind die Linien der drei Proben etwa symmetrisch und bedeutend schmaler als bei Zimmertemperatur (Fig. 1b, 2b und 3b). Diese Messungen lassen sich daher jeweils als ein einziges Sechs-Linien-Spektrum interpretieren. Allerdings weisen die gegenüber den Nachbarlinien verbreiterten Linien 2 und 6 (Probe 1) bzw. 1 und 6 (Probe 2 und 3) darauf hin, daß auch hier noch eine Überlagerung vorliegt, die jedoch nicht mehr aufgelöst werden kann.

## 4. Diskussion der Ergebnisse

Zuerst soll die als Ausgangspunkt der Arbeit gestellte Frage nach dem *Vorhandensein von* $Fe^{2+}$ näher untersucht werden.

Die Spektren aller drei Proben zeigen bei Zimmertemperatur Linien, deren zugehöriges inneres Feld mit dem bei $Fe_3O_4$ für die „$Fe^{2+}$"-Linie gefundenen ziemlich gut übereinstimmt (Fig. 1a, 2a und 3a). In den Tieftemperaturspek-

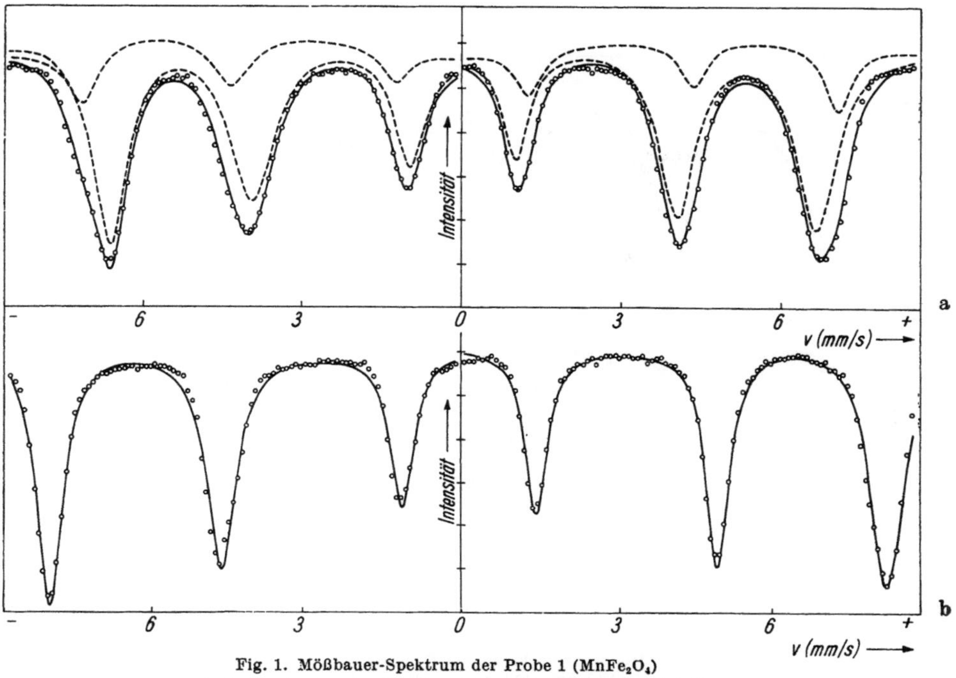


Fig. 1. Mößbauer-Spektrum der Probe 1 ($MnFe_2O_4$)
a) Zimmertemperatur, b) Temperatur des flüssigen Stickstoffs

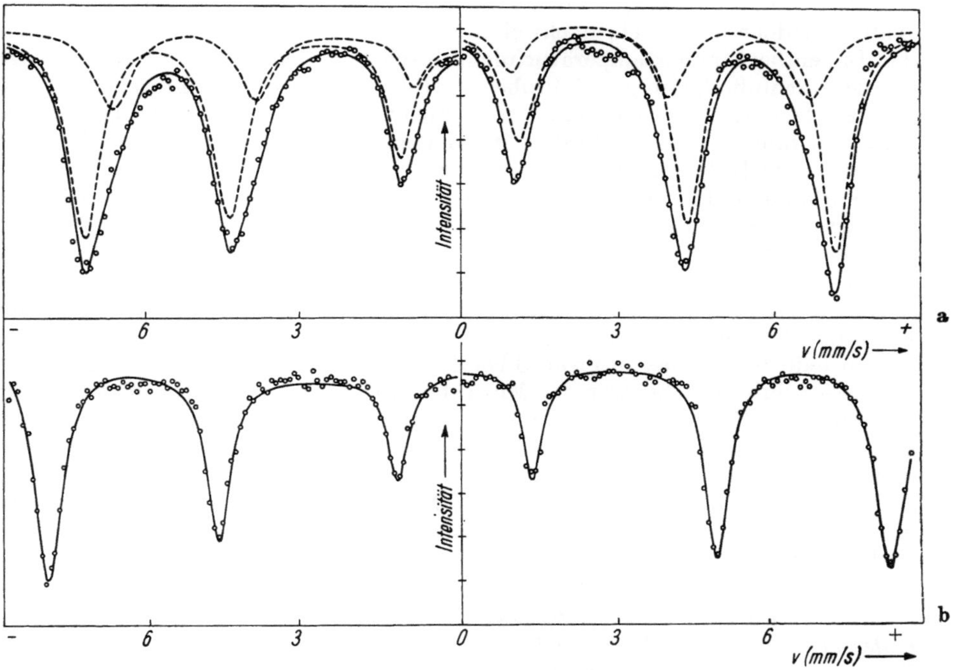


Fig. 2. Mößbauer-Spektrum der Probe 2 ($Mn_{0,5}Mg_{0,5}Fe_2O_4$)
a) Zimmertemperatur, b) Temperatur des flüssigen Stickstoffs

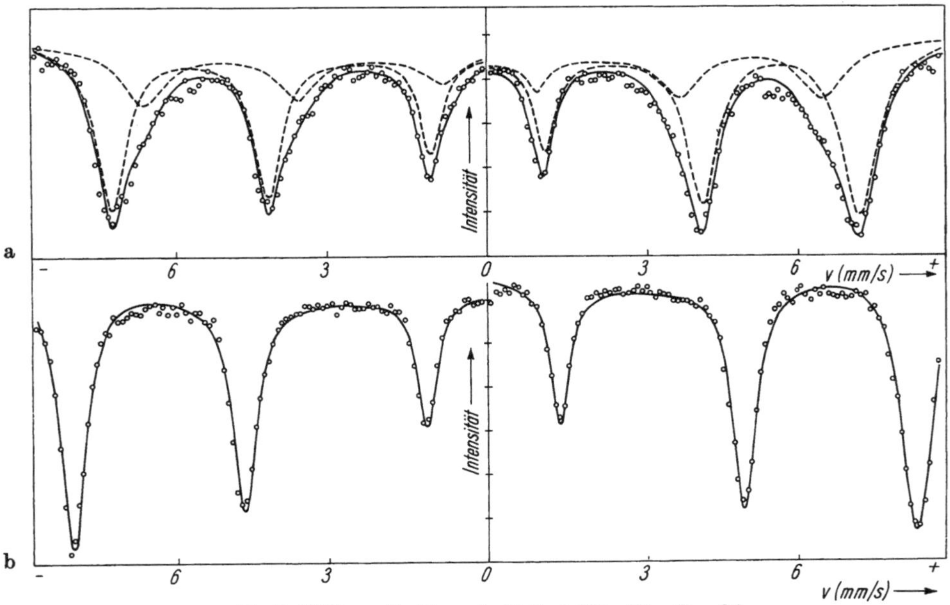


Fig. 3. Mößbauer-Spektrum der Probe 3 ($Mn_{0,5}Mg_{0,75}Fe_{1,75}O_4$)
a) Zimmertemperatur, b) Temperatur des flüssigen Stickstoffs

tren (Fig. 1b, 2b und 3b) läßt sich das Spektrum mit der geringeren Aufspaltung jedoch nicht mehr nachweisen. Das bedeutet, daß sich die entsprechenden Linien bei Zimmertemperatur nicht durch $Fe^{2+}$ deuten lassen. Am klarsten ist diese Schlußfolgerung bei Probe 1. Hier liegen die Intensitätsverhältnisse so, daß fast alle Oktaederionen zum Spektrum mit dem schwächeren Feld beitragen müssen. Dies erfordert in Analogie zu dem Befund beim Magnetit, daß etwa die Hälfte aller Oktaederplätze mit $Fe^{2+}$-Ionen besetzt sein müßte. Das Vorhandensein einer derartigen Menge $Fe^{2+}$ läßt sich auf Grund des Tieftemperaturspektrums mit Sicherheit ausschließen. Bei den Proben 2 und 3 sind die Intensitätsverhältnisse nicht so eindeutig, wenn man berücksichtigt, daß bei der Temperatur des flüssigen Stickstoffs ein eventueller Elektronenaustausch gegenüber Zimmertemperatur abgeschwächt sein müßte. Gegen eine Deutung der Spektren mit geringerer Zeeman-Aufspaltung durch „$Fe^{2+}$" spricht jedoch noch das an Probe 3 bei Zimmertemperatur gemessene Spektrum. Bei dieser Substanz beträgt der Mn-Anteil an den zwei Oktaederplätzen eines Moleküls 0,32 [6]. Nach der chemischen Analyse sind in einem Molekül mindestens 0,27 Atome $Mn^{3+}$ enthalten. Nimmt man an, daß alles Mn auf Oktaederplätzen dreiwertig ist, so folgt aus der Stöchiometrie der Verbindung die sehr geringe Menge von 0,05 Atomen $Fe^{2+}$ pro Molekül. Man dürfte daher für Probe 3 nur einen sehr kleinen Beitrag durch eine „$Fe^{2+}$"-Linie erwarten. Tatsächlich sind die Intensitäten der Spektren mit den geringeren inneren Magnetfeldern für Probe 2 und 3 jedoch etwa gleich stark.

*Zusammenfassend kann festgestellt werden, daß mit den hier vorliegenden Untersuchungen kein* $Fe^{2+}$ *nachgewiesen werden konnte. Das heißt, die bei Zimmertemperatur gefundenen Linienaufspaltungen müssen eine andere Ursache haben.*

Es sei ausdrücklich bemerkt, daß dieses Ergebnis nicht ausschließt, daß in den untersuchten Proben geringe Mengen $Fe^{2+}$ enthalten sind. Da über das Wesen des Elektronenaustausches zwischen $Fe^{2+}$ und $Fe^{3+}$ nur wenig bekannt ist, muß im Extremfall damit gerechnet werden, daß eine geringe Menge zweiwertigen Eisens bei Mittelung über die für den Mößbauer-Effekt charakteristische Wechselwirkungszeit ($10^{-7}$ s) nur eine kleine Abnahme des inneren Feldes für alle Oktaeder-Eisenionen bewirkt. Ein plötzliches Aufhören des Elektronenaustausches mit sinkender Temperatur wie beim $Fe_3O_4$ kann hier nicht erwartet werden, da dieses beim Magnetit mit einer Phasenumwandlung des Gitters verknüpft ist. Bei den in dieser Arbeit betrachteten Proben ist ein derartiger Umwandlungspunkt nicht bekannt. Man kann lediglich eine zunehmende Behinderung des Elektronenaustausches mit abnehmender Temperatur annehmen. In diesem Sinne lassen sich die Verbreiterungen der Linien 6 und 2 bzw. 6 und 1 der Tieftemperaturspektren im Vergleich zu den benachbarten Linien als Hinweis auf $Fe^{2+}$ deuten.

Zur *Erklärung der Überlagerung zweier Spektren* bei Zimmertemperatur bei allen drei Proben sollen zwei Möglichkeiten diskutiert werden:

1. Man kann die Spektren, die dem kleineren inneren Magnetfeld entsprechen, $Fe^{3+}$- und $Fe^{2+}$-Ionen zuordnen, die in Elektronenaustausch miteinander stehen. Um zu erklären, daß die dazu notwendigen $Fe^{2+}$-Ionen bei Zimmertemperatur vorhanden sind und bei Stickstofftemperatur nicht, müßte man annehmen, daß $Fe^{2+}$ durch thermische Anregung nach der Beziehung $Fe^{3+} + Mn^{2+} \leftrightarrows Fe^{2+} + Mn^{3+} - E$ entsteht. Aus den bei $MnFe_2O_4$ (Probe 1) gemessenen Linienintensitäten folgt mit dieser Vorstellung, daß bei Zimmertemperatur ein großer Teil des Mangans auf Tetraederplätzen dreiwertig vorliegt, während sich auf Oktaederplätzen eine äquivalente Menge $Fe^{2+}$ befindet. Die geringere Intensität der „$Fe^{2+}$"-Linien bei den Proben 2 und 3 wäre in diesem Modell zu verstehen, da die Hälfte des Mangans durch Magnesium ersetzt ist. Gegen diese Interpretation sprechen allerdings die Ergebnisse von Leitfähigkeitsmessungen. Man müßte für $MnFe_2O_4$ bei einem hohen Anteil von $Fe^{2+}$ auf Oktaederplätzen einen spezifischen Widerstand von etwa $10^{-2}$ $\Omega$ cm erwarten, wie er für $Fe_3O_4$ gefunden wurde. In [7] wurde jedoch ein spezifischer Widerstand von 10 $\Omega$ cm festgestellt, der sich nach Oxydation einer wahrscheinlich in der Probe enthaltenen geringen Menge $Fe^{2+}$ auf $10^4$ $\Omega$ cm erhöhte. Lotgering [19] bestimmte aus Leitfähigkeitsmessungen den Wert für $E$ in der obigen Gleichung für die thermische Erzeugung von $Fe^{2+}$ zu $E = 0{,}3$ eV. Mit dieser Größe erhält man bei Zimmertemperatur nicht die zur Erklärung des von uns gemessenen Intensitätsverhältnisses der Mößbauer-Linien notwendige $Fe^{2+}$-Menge.

2. Die zweite Möglichkeit der Interpretation besteht darin, *die beiden unterschiedlichen inneren Magnetfelder den* $Fe^{3+}$*-Ionen auf Oktaeder- und Tetraederplätzen zuzuordnen.* Das Spektrum mit der geringeren Intensität muß dann jeweils zu den Tetraederplätzen gehören, da bei allen drei Proben sich auf diesen weniger Fe befindet als auf Oktaederplätzen (siehe [1] und [6]). Die gemessenen Intensitätsverhältnisse sind mit diesem Modell bei Berücksichtigung der großen Fehlergrenzen für die Proben 2 und 3 zu verstehen, während das an Probe 1 gemessene Intensitätsverhältnis zu groß ist. Hastings und Corliss [1] stellten fest, daß für $MnFe_2O_4$ die mittleren magnetischen Momente von Oktaeder- und Tetraederplätzen bei 4,2 °K gleich groß sind und 4,6 $\mu_B$ betragen. Da die Brillouin-Funktionen der Untergittermagnetisierung für kleine Werte von $T/T_c$

($T_c \approx 600$ °K) sehr flach verlaufen, ist anzunehmen, daß sich die Momente der beiden Platzgruppen auch bei der Temperatur des flüssigen Stickstoffs nur wenig unterscheiden. Für die Proben 2 und 3 ist ein ähnliches Verhalten zu erwarten. Somit ist auch das Verschwinden der Linienaufspaltung bei tiefen Temperaturen verständlich.

Für $MnFe_2O_4$ stimmt das Verhältnis der inneren Felder für Oktaeder- und Tetraederplätze mit dem der mittleren magnetischen Momente annähernd überein $\left(\frac{H_{\text{Tetr.}}}{H_{\text{Okt.}}} = 1{,}04;\ \frac{\mu_{\text{Tetr.}}}{\mu_{\text{Okt.}}} = 1{,}15\right)$.

*Für die Proben 2 und 3 sind die neutronographisch ermittelten mittleren magnetischen Momente der Tetraederplätze* ebenfalls *größer als die der Oktaederplätze.* Diese Aussage gilt auch dann, wenn man den Einfluß der unmagnetischen Mg-Ionen auf die mittleren magnetischen Momente eliminiert. *Den Tetraederplätzen entspricht jedoch das kleinere Magnetfeld am Kernort.* Bei den beiden Mg-Mn-Ferriten würde in diesem Modell die Proportionalität zwischen innerem Magnetfeld am Kernort und magnetischem Moment der Hülle für die beiden Untergitter verletzt. Die Erklärung könnte darin liegen, daß bei beiden Proben *auf Oktaederplätzen Abweichungen vom kollinearen Néelschen Modell der Spinkonfiguration in der von Yafet und Kittel* [20] *vorgeschlagenen Art* vorliegen. Da diese Yafet-Kittel-Ordnung neutronographisch nicht nachgewiesen werden konnte, muß man annehmen, daß sie sich entsprechend den von Anderson [21] für kubische Spinelle vorgebrachten Argumenten nicht als Fern-, sondern als Nahordnung ausbildet. Die neutronographisch gemessenen Oktaedermomente entsprechen dann nur der Resultierenden der wahren Ionenmomente, die Winkel miteinander bilden. Dieser Winkel der Ionenmomente mit der Resultierenden läßt sich abschätzen, wenn man annimmt, daß das Verhältnis der Magnetfelder am Kernort dem Verhältnis der Ionenmomente entspricht und daß das neutronographisch bestimmte Tetraedermoment unverfälscht ist (für die Proben 2 und 3 ergibt sich der Wert $\varphi \approx 37°$). Die erste Annahme ist sicher nur angenähert richtig, da die inneren Felder auch Beiträge enthalten, die dem Moment der 3d-Schale nicht proportional sind [22]. Aus den Ergebnissen in [17] folgt jedoch, daß bei Mg-Mn-Ferriten der dem 3d-Moment proportionale, negative Anteil dominiert.

Unser Dank gebührt Herrn Dr. P. Kleinert für die Herstellung der Präparate und Herrn Dr. R. Perthel für wertvolle Diskussionen. Herrn Dipl.-Math. R. Hammer danken wir für die Unterstützung bei der Auswertung der Spektren mit einem Rechenautomaten.

## Literatur

[1] J. H. Hastings und L. M. Corliss, Phys. Rev. **104**, 328 (1956).

[2] J. S. Nozik und I. I. Yamzin, Kristallografiya (SSSR) **6**, 923 (1961).

[3] I. I. Yamzin, N. V. Belov und J. S. Nozik, J. Phys. Soc. Japan, Suppl. B-III **7**, 55 (1962).

[4] S. Pickart und H. A. Alperin, J. Phys. Soc. Japan, Suppl. B-III **7**, 57 (1962).

[5] R. Nathans, S. Pickart, S. E. Harrison und C. J. Kriessmann, Proc. Inst. electr. Eng. **B 104**, Suppl. 5, 217 (1957).

[6] K. Kleinstück, E. Wieser, P. Kleinert und R. Perthel, phys. stat. sol. 8, 271 (1965).

[7] F. W. Harrison, W. P. Osmond und R. W. Teale, Phys. Rev. **106**, 865 (1957).

[8] S. Krupička und K. Závěta, Czech. J. Phys. **9**, 324 (1959).

[9] E. Wieser und K. Kleinstück, Z. Krist. **121**, 14 (1965).
[10] P. P. Brady, J. F. Duncan und K. F. Mots, Proc. Roy. Soc. **287**, **343** (1965).
[11] G. Shirane, D. E. Cox, W. J. Takei und S. L. Ruby, J. Phys. Soc. Japan **17**, 1598 (1962).
[12] G. Shirane, D. E. Cox und S. L. Ruby, Phys. Rev. **125**, 1158 (1962).
[13] R. Bauminger, S. G. Cohen, A. Marinov, S. Ofer und E. Segal, Phys. Rev. **122**, 1447 (1961).
[14] A. Ito, K. Ono und Y. Ishikawa, J. Phys. Soc. Japan **18**, 1465 (1963).
[15] A. Z. Hrynkiewicz, D. S. Kulgawczuk und K. Tomala, Acta phys. Polon. **28**, 423 (1965).
[16] M. Tanaka, T. Moriguchi und Y. Aiyama, J. Phys. Soc. Japan **18**, 1091 (1963).
[17] V. F. Belov, M. N. Devisheva, I. S. Zheludev, E. F. Makarov, R. A. Stukan und V. A. Trukhtanov, Fiz. tverd. Tela **6**, **3435** (1964).
[18] W. Meisel und J. Jacob, Kernenergie **9**, 113 (1966).
[19] F. K. Lotgering, J. Phys. Chem. Solids **25**, 95 (1964).
[20] Y. Yafet und C. Kittel, Phys. Rev. **87**, 290 (1952).
[21] P. W. Anderson, Phys. Rev. **102**, 1008 (1956).
[22] R. Watson und A. J. Freeman, Phys. Rev. **123**, 2027 (1961).

*(Received April 5, 1966)*

phys. stat. sol. 16, 137 (1966)

*Institute of Physics of the Academy of the R.S.R. Bucharest*

# Colour Centres in X-Irradiated NaCl:Ag and NaCl:(Ag + Ca) Crystals

By

V. TOPA and S. HARAGEA

A study is made of colour centres induced by X-irradiation at room temperature in NaCl:Ag and NaCl:(Ag + Ca). It is shown that the induced bands in NaCl:Ag are strongly dependent on the concentration of Ag as well as on the irradiation time and also that the B-, C-, and D-bands have a complex structure. Three components of the B-band, four of the C-band and two of the D-band are found at liquid nitrogen temperature. Optical bleaching in the F-light shows that only a component of the B-, C-, and D-bands is definitely electronic. It is also noted that only the optical bleaching of the F-band in NaCl:Ag results in a simple B-band much like that obtained by electrolytic colouring. The incorporation of both Ca and Ag in NaCl and subsequent X-irradiation lead to very intensive B- and D-bands which are shifted to shorter wavelengths compared with those bands in crystals containing only Ag. The optical bleaching experiments on these crystals suggest that colour centres containing both Ca and Ag are probably formed, as in the case of similar crystals electrolytically coloured.

In NaCl:Ag und NaCl:(Ag + Ca) werden Untersuchungen an Farbzentren durchgeführt, die durch Röntgenbestrahlung bei Zimmertemperatur erzeugt werden. Die induzierten Banden in NaCl werden stark von der Ag-Konzentration und der Bestrahlungsdauer beeinflußt, und die B-, C- und D-Banden zeigen eine komplizierte Struktur. Drei Komponenten der B-Bande, vier der C-Bande und zwei der D-Bande werden bei der Temperatur des flüssigen Stickstoffs gefunden. Optisches Bleichen im F-Licht zeigt, daß nur eine Komponente der B-, C- und D-Banden eindeutig elektronisch ist. Auch führt nur optisches Bleichen der F-Bande in NaCl:Ag zu einer einfachen B-Bande genau wie bei elektrolytischer Verfärbung. Der Einbau von Ca und Ag in NaCl mit darauffolgender Röntgenbestrahlung ergibt intensive B- und D-Banden, die im Vergleich zu Kristallen, die nur Ag enthalten, nach kürzeren Wellenlängen verschoben sind. Die Experimente an diesen Kristallen weisen darauf hin, daß hauptsächlich Farbzentren gebildet werden, die sowohl Ca als auch Ag enthalten, wie im Falle ähnlicher Kristalle, die elektrolytisch verfärbt werden.

## 1. Introduction

In the last years the study of alkali halide crystals containing heavy metals has expanded very rapidly.

This is due, on the one hand, to technical applications, and, on the other hand, to an attempt to explain the luminescence of these crystals (subjected to ionizing radiation) on the basis of models of colour centres including these impurities. The problem of colour centres including impurities is interesting and constitutes a point of future interest.

Many studies have been devoted to Ag doped crystals. Kats [1] and independently Etzel and Schulman [2] showed that Ag doped alkali halide crystals X-irradiated at room temperature present additional absorption bands called A-, B-, C-, D-, and E-bands in [2]. These authors [1, 2] proposed models for colour centres responsible for these bands. Van der Vorst et al. [3], Ishiguro et al. [4], and Eppler and Driekammer [5] developed and proposed other models. Quite recently Parfianovich et al. [6] reported new data especially on the D-band which they found to be double in KCl:Ag and NaCl:Ag crystals.

Schmidt [7] showed that X-irradiation of Ag doped NaCl crystals at low temperatures does not lead to B-, C-, and D-bands but to other new bands (at 360 and 660 nm) which they assigned to transitions of the $V_K$ centre. Delbecq et al. [8] concluded from ESR spectra and optical absorption spectra measurements in polarized light that not only $V_K$ but also additional $Ag^0$ and $Ag^{++}$ centres are substitutionally formed during X-irradiation at low temperatures of Ag doped KCl crystals.

The aim of the present works is to investigate the production and bleaching of colour centres in NaCl:Ag and NaCl:(Ag + Ca) crystals after X-irradiation at room temperature.

The influence of various concentrations of Ag and Ca ions upon the induced absorption bands was followed in crystals with the same concentration of Ag but different amounts of Ca. Simultaneous introduction of Ag and Ca was performed in order to obtain V-type centres due to an enhancement of positive ion vacancies and, at the same time, to produce absorption bands associated with centres including both Ag and Ca as in [9].

In order to obtain a suitable resolution of the absorption spectra, all measurements were carried out at liquid nitrogen temperature.

## 2. Experimental Procedure

Some pure and Ag, Ca, and (Ag + Ca) doped NaCl and KCl single crystals were repeatedly recrystallized by the Kyropolous method in air. The majority of both pure and Ag and (Ag + Ca) doped NaCl crystals were recrystallized from the same crystals but in an inert atmosphere.

The impurities were incorporated as AgCl and CaCl p.a. The concentration of Ag and Ca was chemically determined; ir and uv absorption spectra of all crystals were recorded in order to detect eventual impurities. Except of small traces of $PO_4^-$ and $OH^-$ no other additional bands were found in crystals 3 cm thick.

The measurements were made by the use of the UR-10 Zeiss spectrophotometer in the infrared spectral region and of the universal Zeiss and EPS-2 Hitachi spectrophotometers in the visible and ultraviolet spectral regions.

The crystals were coloured at room temperature by X-irradiation of a tungsten target operating at 150 kV and 4 mA. The absorption spectra at liquid nitrogen temperature were performed in a glass cryostat having quartz windows and glass-covar-copper seals.

## 3. Results

### *3.1 Uncoloured crystals*

To observe the influence of Ca upon colour centres in Ag doped crystals, NaCl and KCl crystals were grown containing a large amount of Ag (1 mol% KCl and 0.6 mol% NaCl in the melt) and different amounts of Ca ($7.5 \times 10^{-3}$, $7.5 \times 10^{-2}$, and $7.5 \times 10^{-1}$ mol% in the melt).

Unexpectedly, only NaCl:(Ag + Ca) crystals present additional bands besides the three bands associated with the Ag ion. The spectrophotometer used in our experiments allowed the observation of only three maxima among those found recently by Fussgaenger et al. [10].

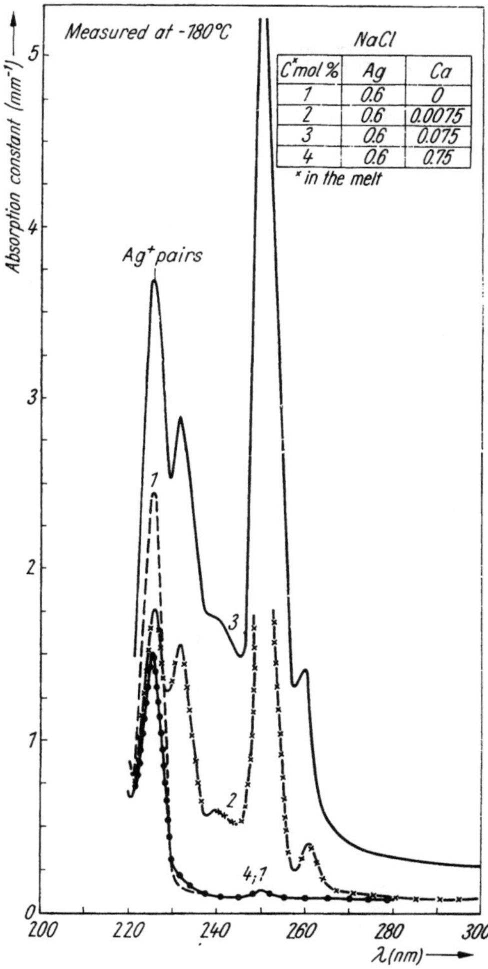


Fig. 1a. Absorption spectra of NaCl: Ag and NaCl: (Ag+Ca) crystals

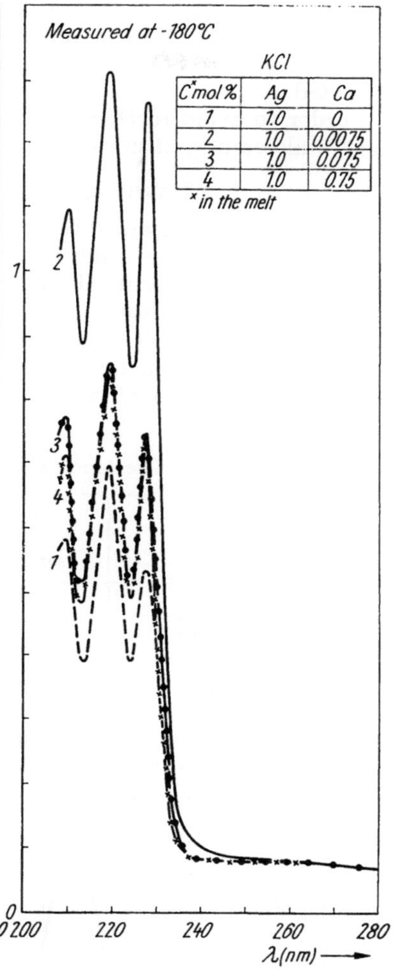


Fig. 1b. Absorption spectra of KCl: Ag and KCl: (Ag+Ca) crystals

Fig. 1a represents the absorption spectra of a set of NaCl crystals. They suggest that while the crystals without Ca incorporation as well as those containing a large amount of Ca (0.75 mol% in the melt) present no additional bands besides the bands associated with Ag ions (curves 1 and 4), the crystals containing a smaller Ca concentration (0.075 and 0.0075 mol% in the melt) present four absorption bands with peaks at 232, 240, 251, and 260 nm (curves 2 and 3). The band at 251 nm is a very strong one. Unlike NaCl crystals, KCl: (Ag + Ca) crystals present no bands besides the three bands associated with Ag ions.

If similar NaCl crystals are grown in an inert atmosphere, the additional bands appear no more, the 251 nm band is present, but it is very weak. The presence of these bands may be possibly explained by the existence of an Ag, Ca, and O combination.

The curves in Fig. 1 show that the bands associated to Ag ions are not affected by the presence of Ca. The intensity of the Ag ion band varies from crystal to crystal since there is no homogeneity in Ag repartition in the four crystals investigated.

In order to avoid further complications due to the additional bands most of the crystals employed were grown in an inert atmosphere.

### *3.2 X-irradiated crystals at room temperature*

When pure and Ag and (Ag + Ca) doped KCl and NaCl crystals are X-irradiated under the same conditions, the behaviour of the two salts is not alike. Fig. 2a and b represent absorption spectra of KCl and NaCl, respectively. Unlike NaCl:Ag and NaCl:(Ag + Ca) (Fig. 2b, curves 2 and 3), the absorption spectra of KCl:Ag and KCl:(Ag + Ca) are very much alike. We think this quite different behaviour to be due to differences in concentration of both Ag and Ca in the two sets of crystals (Table 1).

Table 1

The concentrations of Ag and Ca in mol%

| | In the melt | | In the crystal | |
|---|---|---|---|---|
| | Ag | Ca | Ag | Ca |
| NaCl | 0.6 | 0 | 0.33 | 0 |
| NaCl | 0.6 | 0.75 | 0.33 | 0.3 |
| KCl | 1.0 | 0 | 0.040 | 0 |
| KCl | 1.0 | 0.75 | 0.045 | 0.07 |

The presence of Ca produces not only a large increase in some of the induced bands, but also causes their shift towards shorter wavelengths (those close to the B- and D-bands). The absorption spectrum of NaCl:Ag performed at room

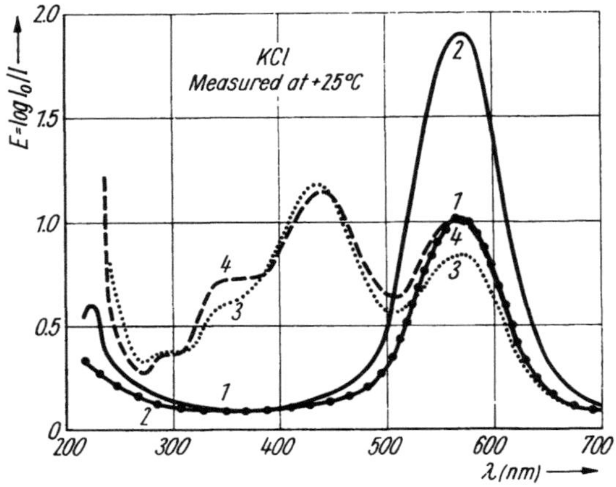


Fig. 2a. Absorption spectra of KCl crystals after X-irradiation for 30 min
1 — KCl; 2 — KCl:Ca ($C_{Ca} = 7.5 \times 10^{-1}$ mol%); 3 — KCl:Ag ($C_{Ag} = 1$ mol%); 4 — KCl:(Ag+Ca) $C_{Ag} = 1$ mol%, $C_{Ca} = 7.5 \times 10^{-1}$ mol%)
(The crystals were grown in open air, and the above concentrations refer to the melt)

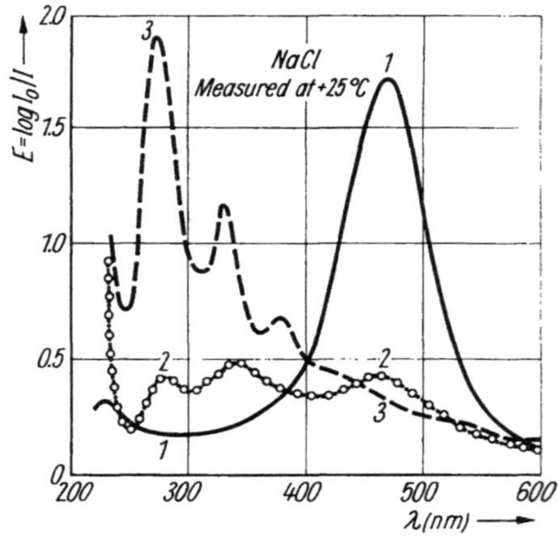


Fig. 2b. Absorption spectra of NaCl crystals after X-irradiation for 30 min
1 — NaCl; 2 — NaCl:Ag ($C_{Ag} = 6.0 \times 10^{-1}$ mol%); 3 — NaCl:(Ag+Ca) ($C_{Ag} = 6 \times 10^{-1}$ mol%, $C_{Ca} = 7.5 \times 10^{-1}$ mol%)
(The crystals were grown in open air, and the above concentrations refer to the melt)

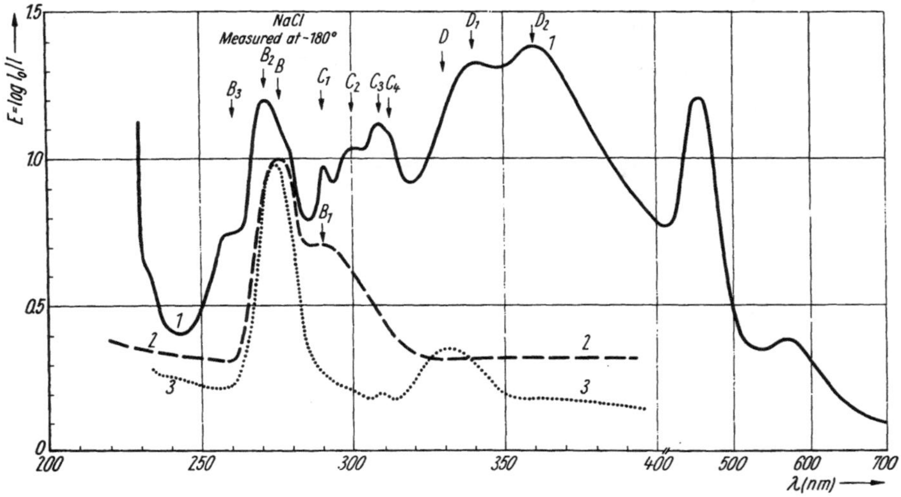


Fig. 3. Absorption spectra of NaCl:Ag and NaCl:(Ag + Ca)
1 – X-irradiated NaCl: Ag ($C_{Ag}$ = 0.33 mol%);
2 – Electrolytically coloured NaCl:(Ag + Ca) ($C_{Ag} \approx 2 \times 10^{-4}$ mol%, $C_{Ca}$ = 0.3 mol%);
3 – Curve 2 in Fig.6

temperature by the use of a recording spectrophotometer illustrated a complex structure of the induced bands.

Curve 1 in Fig. 3 represents the absorption spectrum of X-irradiated NaCl: Ag crystals performed at liquid nitrogen temperature. As may be seen, the B-band is a triplet (B, $B_2$, $B_3$) and the C-band is formed at least of four components ($C_1$, $C_2$, $C_3$, $C_4$). The D-band has two or even three components. A large band at about 580 nm is observed on the long-wavelength side of the F-band.

Fig. 4 represents the absorption curves of four crystals containing different amounts of Ag. The induced absorption bands are strongly dependent on the concentration of Ag ions.

Actually, with a smaller amount of Ag the induced bands correspond to those reported by Etzel and Schulman [2]. In the case of high concentration (1.2 mol% in the crystal), the B-band does not appear and the group of the C-bands is predominant — expecially the $C_3$-band. The $D_2$-band is also much increased. In addition, two new bands, hardly noticeable in the crystal with 0.15 mol% are formed at 259 and 560 nm.

A comparison of curve 3 in Fig. 4 and curve 1 in Fig. 3 shows that a good resolution for the C-bands is obtained for a given concentration and time of irradiation.

Additional experiments performed on crystals containing only Ag prove that the developement of the C-group bands at a given concentration depends upon the time of irradiation.

Fig. 5 shows the influence of Ca upon the induced bands arising in irradiated crystals containing the same amount of Ag (0.15 mol%) but different amounts of Ca. Different thickness and times of irradiation were used for comparing the position of the considered bands.

As the curves show an increase in the concentration of Ca leads to a shift of the peaks of the B- and D-bands towards shorter wavelengths and to a bleaching of the C-bands. At the same time a noticeable band arises at 370 nm in the

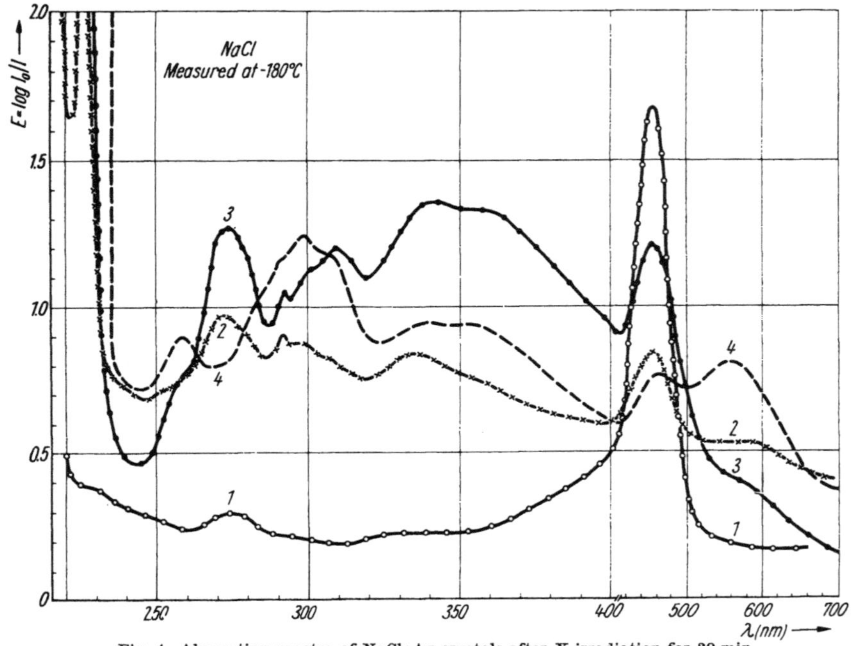


**Fig. 4.** Absorption spectra of NaCl:Ag crystals after X-irradiation for 30 min
1 — NaCl:Ag ($C_{Ag}$ = 0.03 mol%); 2 — NaCl:Ag ($C_{Ag}$ = 0.15 mol%);
3 — NaCl:Ag ($C_{Ag}$ = 0.33 mol%); 4 — NaCl:Ag ($C_{Ag}$ = 1.2 mol%)

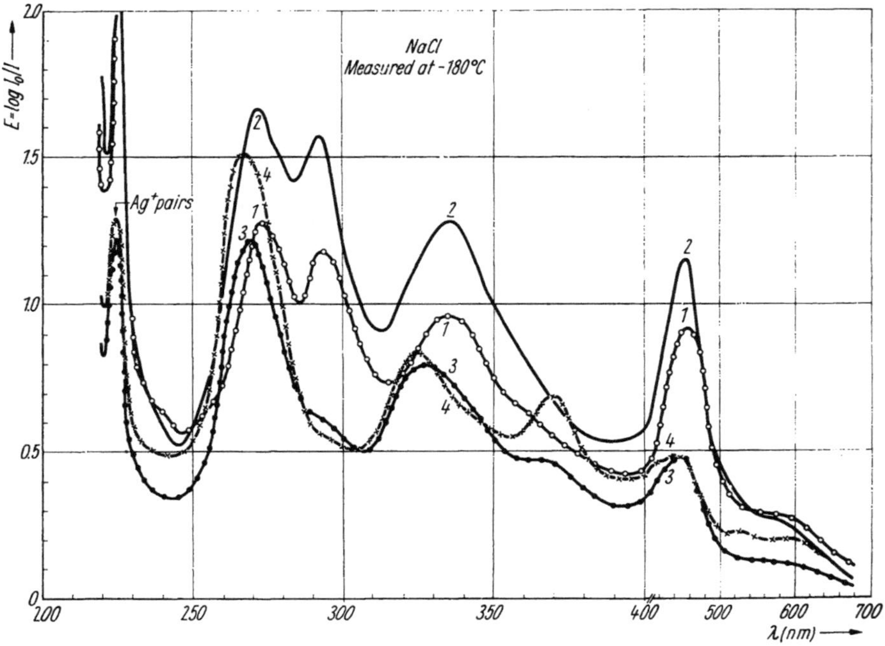


**Fig. 5.** Absorption spectra of NaCl:Ag and NaCl:(Ag+Ca) crystals after X-irradiation
1 — NaCl:Ag ($C_{Ag}$ = 0.15 mol%, $d$ = 6.53 mm, irradiated for 45 min);
2 — NaCl:(Ag+Ca) ($C_{Ag}$ = 0.15 mol%, $C_{Ca}$ = 0.005 mol%, $d$ = 6.53 mm, irradiated for 30 min);
3 — NaCl:(Ag+Ca) ($C_{Ag}$ = 0.15 mol%, $C_{Ca}$ = 0.043 mol%, $d$ = 3.00 mm, irradiated for 15 min);
4 — NaCl:(Ag+Ca) ($C_{Ag}$ = 0.15 mol%, $C_{Ca}$ = 0.205 mol%, $d$ = 3.07 mm, irradiated for 12 min)

two last crystals. The crystal containing the maximum amount of Ca presents a band close to the F-band and another band at 520 nm. A comparison of the curves in Fig. 4 and 5 indicates that incorporation of Ca results in an appreciable change in the induced bands arising in X-irradiated crystals.

Unlike NaCl:Ag in NaCl:(Ag + Ca) crystals containing a large amount of Ca the induced bands increase directly proportional to the time of irradiation.

An attempt to produce changes in the above bands was made by optical bleaching of the colour centres, but the irradiation in the B-band had no effect both at room and liquid nitrogen temperature under our experimental conditions.[1]) If a high Ag concentration is introduced a very weak F-band appears, so that the irradiation in the F-band is not efficient. In order to follow these changes crystals containing a small amount of Ag were employed. For comparison two crystals were taken which had the same Ag concentration but only one of them contained a large amount of Ca. The results are given in Fig. 6. Curve 1 represents the crystal previously X-irradiated.

After measuring the absorption spectrum, the crystal was irradiated at room temperature by means of an incandescent lamp with light of wavelength $\lambda >$ $> 410$ nm. Curve 2 represents the absorption spectrum following this irradiation. The initial crystal previously X-irradiated shows weak "B"- and "D"-bands and a prominent F-band. After irradiating with $\lambda > 410$ nm, the B-band at

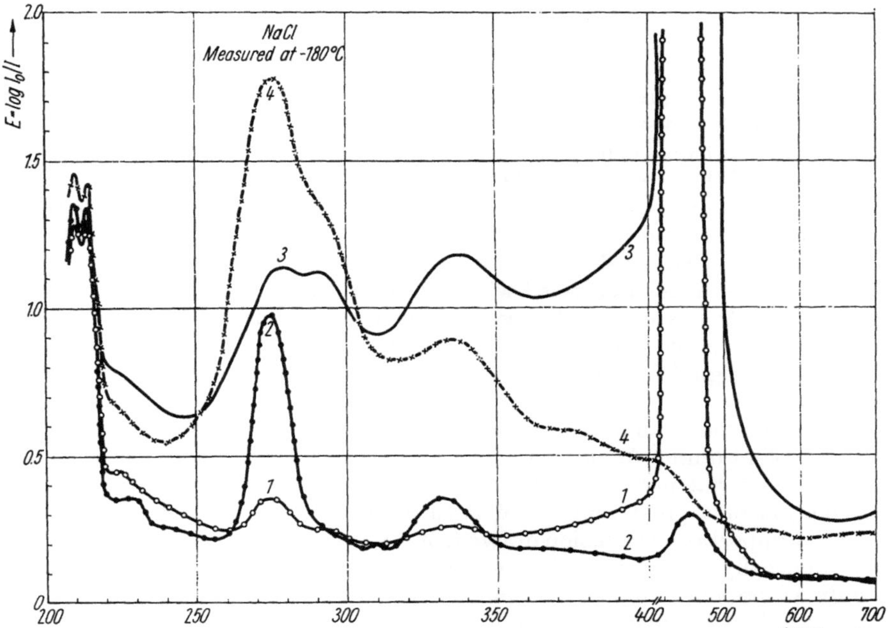


Fig. 6. Absorption spectra of NaCl: Ag and NaCl: (Ag + Ca) crystals
1 — X-irradiated NaCl: Ag ($C_{Ag} = 0.03$ mol%);
2 — Crystal 1 after irradiation with light of wavelength $\lambda > 410$ nm;
3 — X-irradiated NaCl: (Ag + Ca) ($C_{Ag} = 0.03$ mol%, $C_{Ca} = 0.205$ mol%);
4 — Crystal 3 after irradiation with light of wavelength $\lambda > 410$ nm

[1]) The crystal was irradiated for two hours in the spectrophotometer having a 2 mm slit.

274 nm is much increased, and a weak band at 310 nm and a large one at 332 nm appear.

The B-band appears now to have a simple structure. An increase in the B-band is also observed after similar treatments of NaCl:(Ag + Ca) crystals, but this time another band is formed which is marked as a shoulder superposed on this band. The height of the initial D-band is decreased and it changes its maximum position; simultaneously weak bands at 370 and 420 nm appear similar to those obtained by X-irradiation of the crystals containing 0.15 mol% Ag and 0.205 mol% Ca. As the F-band is very intensive, it is difficult to conclude whether they are caused by X-irradiation or appear as a result of bleaching of the F-band.

It is quite interesting to compare the absorption spectra of X-irradiated and electrolytically coloured crystals.

In a previous paper [9] we have shown that in electrolytically coloured NaCl:(Ag + Ca) crystals the B-band has a shoulder which we called $B_1$-band. For comparison, the absorption spectra of both electrolytically coloured NaCl:(Ag + Ca) and NaCl:Ag X-irradiated crystals were plotted in Fig. 3. The figure shows that the B-band formed in a NaCl:Ag crystal when irradiated in the F-band (Fig. 6, curve 2) has the same maximum position and halfwidth as that in an electrolytically coloured crystal, but it is only a component of the "B"-band arising usually in X-irradiated NaCl:Ag crystals containing large amounts of Ag.

## 4. Discussion

Measurement at liquid nitrogen temperature on crystals containing a large amount of Ag previously X-irradiated at room temperature prove that the induced absorption bands consist in a superposition of several neighbouring bands. The structure of these bands may be resolved partly at this temperature. These bands, especially those belonging to the C-group, depend strongly on both the concentration of Ag and time of irradiation. Optical bleaching experiments lead to the conclusion that most of the induced bands are V-type bands.

As regards the model associated with the centres responsible for the observed absorption bands it seems to be premature to make any assumption, as long as the model proposed for the B-band appears not to be correct. Brown [11] found no ESR signal due to the B-centre, and suggested that it is not a "one-electron" centre.

A discussion concerning the influence of Ca upon the induced bands formed in crystals containing Ag is more complicated. The numerous positive ion vacancies which balance the electric charges due to the divalent Ca ions enable the formation of a great number of both electronic and V-type centres.

The most important conclusion drawn from the above results is that the B-band in electrolytically coloured and X-irradiated crystals is not the same. In the latter case only a component of the "B"-band resembles to the B-band formed in the former case. It is interesting to note that only by F-light irradiation of a crystal containing X-ray induced bands it is possible to form a B-band identical with that formed by electrolytical colouring.

The F-light irradiation of X-ray coloured NaCl:(Ag + Ca) crystals leads to an increase in the B-band and a development of a shoulder on its long-wavelength side. The position of the maximum, the shape of the band, and the fact that this shoulder is formed only in crystals containing (Ag + Ca), suggest that it is identical with that observed in electrolytically coloured NaCl: (Ag + Ca) crystals. From the above data concerning the bands induced by X-irradiation of NaCl: (Ag + Ca) crystals one may conclude that in this case also Ca perturbed colour centres (B, D) appear. This assumption is supported by the fact that an increase in the Ca concentration is connected with a shift of the induced bands (B and D) towards shorter wavelengths, which affect only the electronic bands.

Photoconduction, fluorescence, and ESR studies will help us to obtain further information on phenomena occuring in Ag and (Ag + Ca) doped crystals.

### *Acknowledgements*

The authors would like to thank Mrs. M. Botez for her help in growing crystals and Mrs. G. Marinescu for chemical analysis. They are also grateful to Dr. V. Vasilescu for his amiability in placing the ESP-2 spectrophotometer at their disposal.

### References

[1] M. L. Kats, Dokl. Akad. Nauk SSSR **85**, 539 (1952); Zh. eksper. teor. Fiz. **23**, 720 (1952).
[2] H. W. Etzel and J. H. Schulman, J. chem. Phys. **22**, 1549 (1954).
[3] W. Maenhout van der Vorst and W. Dekeyser, Physica **23**, 903 (1957).
[4] M. Ishiguro, G. Okuro, and K. Kojima, Mem. Inst. Sci. Ind. Res., Osaka Univ. **15**, 1 (1958).
[5] R. A. Eppler, and N. G. Drickamer, J. chem. Phys. **32**, 1734 (1960).
[6] I. A. Parfianovich, E. I. Shuraleva, P. S. Ivanenko, and P. A. Babin, Izv. Akad. Nauk SSSR **24**, 427 (1965).
[7] H. J. Schmidt, Z. Phys. **166**, 447 (1962).
[8] C. J. Delbecq, W. Hayes, M. C. M. O'Brien, and P. M. Yuster, Proc. Roy. Soc. **A** 271, 243 (1963).
[9] V. Topa and A. Somlo, phys. stat. sol. **5**, K 13 (1964).
[10] K. Fussgaenger, W. Martienssen, and H. Bilz, phys. stat. sol. **12**, 383 (1965).
[11] F. C. Brown, private communication.

*(Received April 6, 1966)*

phys. stat. sol. **16**, 147 (1966)

*Institute of Solid State Physics, Czechoslovak Academy of Sciences, Prague*

# Excitonic Effects in the Interband Absorption of Semiconductors

By

B. VELICKÝ and J. SAK

Elliott's work on exciton effects at the direct absorption edge is extended to include other van Hove singularities and deviations from the effective mass approximation. The sum rule which replaces the $f$-sum rule in this case is derived. The Koster–Slater and Coulomb interactions are treated as limiting cases. Due to the electron-hole interaction, the $M_0$ and $M_1$ branch-points of $\varepsilon_2$ are enhanced and sharpened, whilst the $M_2$ and $M_3$ branch-points are weakened and smoothed out. Only the $M_0$ points generate (quasi) discrete levels; no evidence for these is found near the $M_1$ points. For a very prolate $M_1$ singularity, certain branch-points appear asymptotically below the critical energy.

Die Elliott'sche Arbeit über Exzitoneneffekte an der Absorptionskante für direkte Übergänge wird in der Richtung erweitert, daß andere van Hove Singularitäten und Abweichungen von der Näherung der effektiven Massen eingeschlossen werden. Die Summenregel, die in diesem Falle die $f$-Summenregel ersetzt, wird hergeleitet, und die Koster–Slater- und Coulomb-Wechselwirkung werden als Grenzfälle behandelt. Wegen der Elektron–Loch-Wechselwirkung werden die kritischen Punkte $M_0$ und $M_1$ in $\varepsilon_2$ stärker und schärfer, während die Punkte $M_2$ und $M_3$ abgeschwächt und geglättet werden. Nur die $M_0$ Punkte erzeugen (quasi) diskrete Energieniveaus, für solche Niveaus wird bei den $M_1$ Punkten kein Hinweis gefunden. Für eine sehr gestreckte $M_1$ Singularität erscheinen asymptotisch kritische Punkte unterhalb der kritischen Energie.

## 1. Dielectric Function

The essential features of the intrinsic absorption of covalent and weakly polar insulators can be explained by the presence of van Hove type singularities in the joint density of states [1]. For the interpretation of the fine structure it is necessary not only to refine the band model but also to account for correlation effects. In the neighbourhood of the fundamental absorption edge, the electron-hole interaction, giving rise to Wannier excitons, is most important. The theory of these effects as worked out by Elliott [2] was found to be in excellent agreement with experiments [3]. In the range of higher energies, other correlation effects become important, e.g. the Auger generation of pairs. After Phillips' proposal [4], even in the range far above the fundamental edge (up to about 10 eV from the edge) excitonic effects are dominant over other correlation effects (for applications, see [4, 5]). Neither a justification of the excitonic model nor its detailed mathematical treatment have been given so far. Here we are going to deal with the latter part of the problem.

We consider an ideal insulating cubic crystal with fixed ions. The volume of the crystal is $\Omega$, and all states are normalized to it. The electronic Hamiltonian $\mathcal{H} = \mathcal{H}_0 + V$ consists of the Bloch Hamiltonian $\mathcal{H}_0$ and the electron-hole interaction $V$ which is diagonal in the total quasi-momentum, in spin, and in the number of quasiparticles. The ground state of the crystal is of the Hartree type, and we write $\mathcal{H}\,|0\rangle = \mathcal{H}_0\,|0\rangle = E_0\,|0\rangle$. We denote by $|f\rangle = |c\,v\,\boldsymbol{k}\rangle$

a direct free electron-hole pair with an electron in the conduction band $c$ (with wave vector $\boldsymbol{k}$ and energy $E_c(\boldsymbol{k})$ and with a hole in the valence band $v$ (with wave vector $-\boldsymbol{k}$ and energy $-E_v(\boldsymbol{k})$). If the electron-hole interaction is taken into account we get two kinds of states, bound ones $|a\rangle$ with energies $E_a$ in the forbidden bands of the one-pair spectra of $\mathcal{H}_0$, and scattering states with a continuous spectrum identical with that of $\mathcal{H}_0$. In the range of the continuum, we shall use the basis of the outgoing states $|f\,+\rangle = |c\,v\,\boldsymbol{k}\,+\rangle$ defined by the equation

$$|f\,+\rangle = |f\rangle + (E_f - \mathcal{H} + i\,0)^{-1}\,V\,|f\rangle = |f\rangle + (E_f - \mathcal{H}_0 + i\,0)^{-1}\,V\,|f\,+\rangle\,. \quad (1)$$

The imaginary part of the dielectric function is given by the equation [1]

$$\varepsilon_2(\omega) = \frac{8\,\pi^2}{m^2\,\omega^2\,\Omega}\Big(\sum_a |\langle 0|p_z|a\rangle|^2\,\delta\,(E_a - E_0 - \omega) + $$
$$+ \sum_{c\,v\,\boldsymbol{k}} A(c\,v\,\boldsymbol{k})\,|\langle 0|p_z|c\,v\,\boldsymbol{k}\rangle|^2\,\delta\,\big(E_c(\boldsymbol{k}) - E_v(\boldsymbol{k}) - \omega\big)\Big)\,. \quad (2)$$

The two possible spin orientations are accounted for simply by a factor of 2, $p_z$ is the $z$-component of the total electronic momentum and the quantity

$$A(c\,v\,\boldsymbol{k}) = \left|\frac{\langle 0|\,p_z\,|f\,+\rangle}{\langle 0|\,p_z\,|f\rangle}\right|^2 \quad (3)$$

may be called amplifying factor. For $V \to 0$, the discrete levels disappear, and $A$ tends to unity, thus reducing equation (2) to the well-known formula for the dielectric function in the band approximation. The influence of the electron-hole interaction on the continuous part of intrinsic absorption is contained just in the factor $A$.

Thus, our problem is reduced to solving the scattering problem (1). On the basis of general principles of the scattering theory, the amplifying factor may be expected to have singularities of various nature (e.g. complex poles corresponding to resonances in the continuous spectrum). In the case of the optical absorption of atoms a detailed theory has been developed by Fano [8]. Phillips' idea is to apply Fano's theory to interband transitions [4, 5]. The complicated dependence of interband energies on the wave vector is believed to lead to some new effects, especially to resonances in the neighbourhood of saddle-points. Phillips assumes the existence of these saddle-point resonances and calls them hyperbolic excitons. To our knowledge, no detailed mathematical description of hyperbolic excitons exists. Another problem is the existence of singularities of other types, different from poles in the amplifying factor. The work of Elliott [2] shows that the form of the $M_0$ edge is drastically changed by the presence of an essential branching singularity at the direct edge energy.

Now, we briefly mention the problem of sum rules. The usual $f$-sum rule concerning the integration of the function $\omega\,\varepsilon_2(\omega)$ cannot be proved because the electron-hole interaction is not local [6]. We shall see below that the $f$-sum rule may really break down in the transition from the band model to models with electron-hole interaction. Thus, the deformation of the dielectric function which is due to excitonic effects cannot be explained simply by a redistribution of a fixed supply of oscillator strengths.

But we can deduce from (2) that for any form of electron-hole interaction (i.e. for any $V$ such that $V|0\rangle = 0$) the following equation holds:

$$\int_0^\infty \omega^2 \varepsilon_2(\omega)\, d\omega \equiv \frac{8\pi^2}{m^2 \Omega} \sum_f |\langle 0|\, p_z \,|f\rangle|^2 = \text{const}\,. \tag{4}$$

This is the sum rule governing our problem. This conclusion does not imply that the excitonic model is physically inconsistent, but merely that it describes correctly just the range of sufficiently low energies.

## 2. Koster-Slater Model

First we calculate $\varepsilon_2$ for the model defined by the following properties: 1. The matrix element of momentum, $\langle 0|p_z|c\, v\, \boldsymbol{k}\rangle$, is nearly constant within the Brillouin zone. 2. The electron-hole interaction is given by matrix elements $\langle c\, v\, \boldsymbol{k}|V|c'\, v'\, \boldsymbol{k}'\rangle = N^{-1} g$ where $N$ is the number of unit cells in the crystal and $g$ is the coupling constant of the interaction.

The second property makes this model quite analogous to the Koster-Slater model of impurity centres [7]. We take $g < 0$ (attractive forces) and small enough so that bound states may not appear. The electron-hole interaction of the Koster-Slater type is of extremely short-range nature, in contrast to the Coulomb forces. But there is an advantage that this problem is exactly soluble, and the result is expressed in relatively well known quantities. An easy calculation leads to the following expression for the dielectric constant:

$$\varepsilon_2(\omega) = |1 + g\, F(\omega)|^2\, \varepsilon_2^{(0)}(\omega)\,, \tag{5}$$

where $\varepsilon_2^{(0)}$ corresponds to the band approximation and

$$F(\omega) = -\frac{\Omega}{N} \int_{-\infty}^{\infty} \frac{d\xi}{\omega - \xi + i\,0}\, \varrho(\xi)\,, \tag{6}$$

where $\varrho(\omega)$ is the total joint density of states (the number of states per unit energy and unit volume)

$$\varrho(\omega) = (2\pi)^{-3} \sum_c \sum_v \int d^3\boldsymbol{k}\, \delta\big(\omega - E_c(\boldsymbol{k}) + E_v(\boldsymbol{k})\big)\,. \tag{7}$$

Hence, the knowledge of $\varrho(\omega)$ alone enables us to pass from $\varepsilon_2^{(0)}$ to $\varepsilon_2$. The analytic properties of the factor $|1 + g\,F|^{-2}$ are strictly given by the fact that the function $F(\omega)$ allows the integral representation (6) and, consequently, satisfies the dispersion relations. Only the region of non-zero density of states is of interest. The properties of the function $F$ may be quite intricated there. Thus, if $\varrho(\omega)$ possesses a van Hove singularity of the type $M_j$ at $\omega_j$, one has $F(z) \sim i^{j+1} (\omega_j - z)^{1/2} + \Phi(z)$, where $\Phi(z)$ is a function regular at $\omega_j$. The simplest way to see the influence of the factor $|1 + g\,F|^{-2} = |g|^{-2} \cdot |-g^{-1} - F|^{-2}$ is to plot the Argand diagram of the function $F(\omega)$ which yields the quantity $|-g^{-1} - F|^2$ directly (Fig. 1b). For a given $g$, the dashed circle separates the region of enhancement from that of reduction of the dielectric function $\varepsilon_2$ as compared to $\varepsilon_2^{(0)}$. The Argand diagram in Fig. 1b corresponds to the simplest possible band structure which generates the joint density of states presented in Fig. 1a. It is clear how any critical point in the density of states is connected

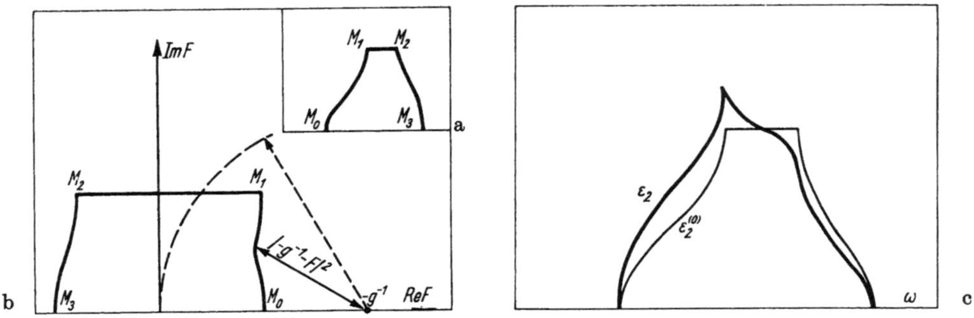


Fig. 1. a) Model density of states;
b) The corresponding Argand diagram according to equation (6);
c) The change in the dielectric constant induced by the Koster-Slater interaction

a b c

Fig. 2. Modifications of the dielectric function near combinations of critical points (in the Koster-Slater model)
a) $M_0 - M_1$, b) $M_1 - M_2$, c) $M_1 - M_1$

with a specific branching point of $F$. The corresponding changes in the dielectric function are drawn in Fig. 1c.

In Fig. 2 various examples are sketched how close combinations of van Hove singularities are deformed by the Koster-Slater interaction. It seems that most of the experimental data on the dielectric properties of semiconductors can be understood qualitatively if the band model is improved by the amplifying factor of the Koster-Slater type.

Some qualitative conclusions as to the effect of an attractive electron-hole interaction may be deduced from the above considerations:

1. Any critical point in the joint density of states is accompanied by a branching point in the amplifying factor.

2. Edges of the type $M_1$ are sharpened, those of the $M_2$ type are smeared out.

3. For bands with a shape similar to that of Fig. 1a, the dielectric function $\varepsilon_2(\omega)$ increases at low energies whereas it is reduced within the high-energy part of the band.[1])

[1]) Further comments may be the following:

4. The dip in Fig. 2c (antiresonance in Phillips' terminology) is of a rather accidental nature, corresponding to a loop in the Argand diagram. If a continuous change in $\varrho(\omega)$ occurs leading to the disappearance of the loop, the dip also disappears.

5. Fig. 1c illustrates how the $f$-sum rule may break down in an excitonic model: As we know a priori that the integral $\int \omega^2 (\varepsilon_2 - \varepsilon_2^{(0)})\, d\omega$ is equal to zero, the fact that the function $\varepsilon_2 - \varepsilon_2^{(0)}$ changes sign just once implies that the integral $\int \omega (\varepsilon_2 - \varepsilon_2^{(0)})\, d\omega$ is certainly different from zero.

A shortcoming of the Koster-Slater model is its failure in giving an adequate description of discrete and resonance levels. For this reason, we now consider a more realistic Coulomb-like electron–hole interaction.

## 3. Coulomb Interaction

### *3.1 Energy Band Extrema*

The Coulomb interaction is distinguished by two important properties:

1. It is weak when connecting Bloch pair states from different bands or distant regions in momentum space.

2. It is singularly strong for small momentum transfer processes within a single band.

According to the first property, we may expect that the effective mass approximation gives the main features of the excitonic effects correctly. The second property reflects a rather singular behaviour of the solutions of the effective mass equation. Near a stationary point of a single band, the effective equation reads

$$\left(\sum_i (2\,m_i)^{-1} \frac{\partial^2}{\partial x_i^2} + \frac{1}{\varepsilon r} + E\right) \Phi(x_1, x_2, x_3) = 0\,. \tag{8}$$

$\varepsilon$ is the static dielectric constant, $m_i$ are the principal effective masses, unknown are the Wannier amplitudes $\Phi$ and the energies $E$. Assuming again that the transition matrix element $\langle 0|\, p_z\, |c\, v\, \boldsymbol{k}\rangle$ varies slowly, and denoting by $|\Phi\rangle$ the state with the Wannier amplitude $\Phi$, we have

$$\langle 0|\, p_z\, |\Phi\rangle = \langle 0|\, p_z\, |c\, v\, \boldsymbol{k}_0\rangle\, \Phi(0, 0, 0)\,, \tag{9}$$

where the subscript 0 indicates the critical point under consideration. Thus, both the energies and the amplifying factor may be found from equation (8).

The solution of the problem formulated in this way was first found by Elliott [2] for the hydrogen-like case of an isotropic minimum ($m_i = m > 0$; $\Omega = 1$). His results read

$$E_n = -\frac{m}{2\,\varepsilon^2\, n^2}\,; \qquad |\Phi_n(0)|^2 = \frac{m^3}{\pi\, \varepsilon^3\, n^3} \quad \text{for bound states}\,, \tag{10}$$

$$\gamma = \pi(2\,\varepsilon^2\, m^{-1}\,[E - E_0])^{1/2}\,; \qquad |\Phi_E(0)|^2 = \frac{\gamma\, e^{\gamma}}{\sin h\, \gamma} \quad \text{in the continuum}\,. \tag{11}$$

The average intensity of the discrete part of absorption and the continuous part join continuously at the edge energy $E_0$.

The solutions of (8) for an isotropic band maximum are similar to those obtained in the repulsive Coulomb field with $m > 0$. Now, there exist no discrete levels, and the amplifying factor in the continuum range becomes

$$\gamma' = -\,\pi(2\,\varepsilon^2\, m^{-1}\,[E_3 - E])^{-1/2}\,; \qquad |\Phi_E(0)|^2 = \frac{\gamma'\, e^{\gamma'}}{\sin h\, \gamma'}\,. \tag{12}$$

The resulting $\varepsilon_2$ shapes are given in Fig. 3. Expressions (11) and (12) are easily recognized as Sommerfeld or Gamow factors. These results are in qualitative agreement with those obtained from the Koster-Slater model, but they are much more pronounced because of the singularity of the Coulomb potential. In particular, the $M_3$ singularity is weakened to such an extent that it should not be observable at all. In the case of anisotropic extrema, equation (8) is much more difficult to be solved. We shall mention this problem in the following section.

For band extrema overlapped by other parts of the band structure the effective mass considerations must be completed by taking into account the interband part of the electron-hole interaction. The total Hamiltonian is divided into its intraband and interband parts:

$$\left.\begin{aligned} \mathcal{H}_0 + V &= \mathcal{H}_m + U\,, \\ \mathcal{H}_m &= \mathcal{H}_0 + P_1\,V\,P_1 + P_2\,V\,P_2\,, \\ U &= P_1\,V\,P_2 + P_2\,V\,P_1\,. \end{aligned}\right\} \tag{13}$$

A scheme of the corresponding band model is represented in Fig. 4. The projectors $P_1$ and $P_2$ belong to the spaces of free pairs $v \to c_1$ and $v \to c_2$, respecti-

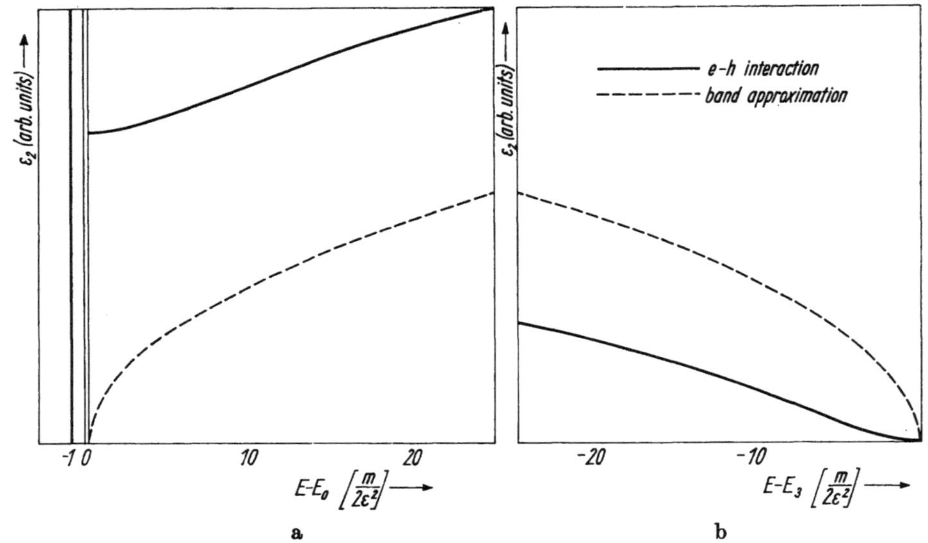


Fig. 3. a) Change in $\varepsilon_2$ near spherical interband edge due to Coulomb electron–hole interaction (after Elliott [2]); b) The same near a spherical $M_3$ point

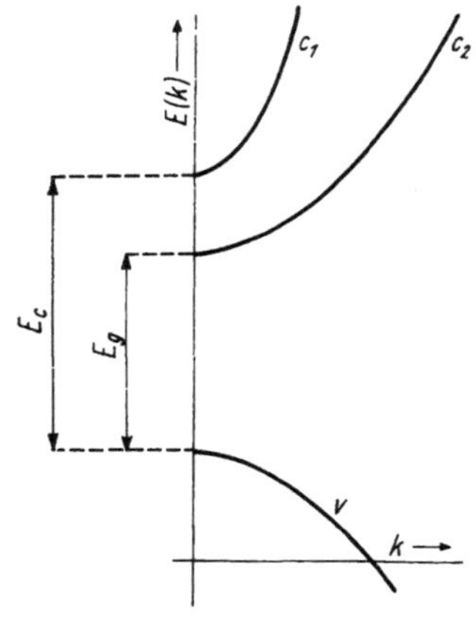


Fig. 4. Simplest band model leading to resonance effects

vely. We restrict ourselves to the space where $P_1 + P_2 = 1$. Both $\mathcal{H}_0$ and $\mathcal{H}$ have a purely continuous spectrum for energies near $E_c$. This is not the case with $H_m$, its $P_1$ block being equivalent to the effective Hamiltonian (8). The discrete levels of $\mathcal{H}_m$ decay when interband interaction is switched on, thus leading to resonances. We introduce outgoing distorted intraband-waves:

$$|c_i\, v\, \boldsymbol{k};\; m\rangle = |c_i\, v\, \boldsymbol{k}\rangle + (E_f - \mathcal{H}_m + i\,0)^{-1}\,(\mathcal{H}_m - \mathcal{H}_0)\,|c_i\, v\, \boldsymbol{k}\rangle\,. \qquad (14)$$

After some algebra [9] we get for the exact eigenfunctions of $\mathcal{H}$ as generated from the bands $P_1$ or $P_2$

$$|c_1\, v\, \boldsymbol{k} +\rangle =$$

$$= \left(1 + \left[1 + \frac{P_1\, U}{E_f - \mathcal{H}_m + i\,0}\right] \frac{P_2\, U}{E_f - \mathcal{H}_m - U \dfrac{P_1}{E_f - \mathcal{H}_m + i\,0} U + i\,0}\right) |c_1 v \boldsymbol{k};\, m\rangle\,, \qquad (15)$$

$$|c_2\, v\, \boldsymbol{k} +\rangle =$$

$$= \left(1 + \left[1 + \frac{P_2\, U}{E_f - \mathcal{H}_m + i\,0}\right] \frac{P_1\, U}{E_f - \mathcal{H}_m - U \dfrac{P_2}{E_f - \mathcal{H}_m + i\,0} U + i\,0}\right) |c_2 v \boldsymbol{k};\, m\rangle\,. \qquad (16)$$

The way in which expressions (15) and (16) lead to resonances in the dielectric constant may be illustrated by the simple example of an isolated resonance.[2])

We put $P_1 = |a\rangle$, $\langle a|\, \mathcal{H}_m |a\rangle = E_a|\, a\rangle$, $\langle a|\, a\rangle = 1$, $E_a < E_c$, and the expression (16) becomes

$$|c_2\, v\, \boldsymbol{k} +\rangle = |c_2\, v\, \boldsymbol{k}\;;\; m\rangle + |a\rangle\, \mathsf{G}_a(E_f + i\,0)\, \langle a|U|c_2\, v\, \boldsymbol{k}\;;\; m\rangle +$$
$$+ \frac{P_2}{E_f - \mathcal{H}_m + i\,0} U|a\rangle\, \mathsf{G}_a(E_f + i\,0)\, \langle a|U|c_2\, v\, \boldsymbol{k};\;\; m\rangle\,.$$

Here $\mathsf{G}_a(z)$ is the Green's function of the state $|a\rangle$. For its pole at $z = z_1 - i\, z_2$ in the lower half plane we obtain

$$\varepsilon_2(\omega) = \varepsilon_2^{(0)}(\omega) \left(1 + \frac{\alpha}{(\omega - z_1)^2 + z_2^2} + \frac{\beta\,(\omega - z_1)}{(\omega - z_1)^2 + z_2^2}\right).$$

Here, $\varepsilon_2^{(0)}$ denotes the effective-mass dielectric constant, the real parameters $\alpha, \beta$ are slowly varying functions of the energy and may be assumed to be constant within the region of the resonance, $|\omega - z_1| \lessapprox z_2$. We do not quote the rather clumsy expressions for these parameters and merely state the inequality $\beta^2 \leqq 4\,(\alpha + z_2^2)$, which is equivalent to the condition $\varepsilon_2 \geqq 0$. In the limiting case of equality, $\varepsilon_2$ vanishes for a single frequency and the resulting line shape is identical to that obtained by Fano [8]. But the crucial point about the theory of Fano is the spherical symmetry of the systems to which it applies (e.g. atoms). This is not generally valid in crystals and the theory contains two parameters even before the inhomogeneous broadening is taken into account [5].

---

[2]) Let us remark that the bound eigenstates of the operator enter equations (15) and (16) through $P_1$, as $P_1 = \Sigma\, |c_1\, v\, \boldsymbol{k};\, m\rangle \langle c_1\, v\, \boldsymbol{k};\, m| + \Pi$ where $\Pi$ is the projector on the bound states. In the limiting case $U \to 0$ we obtain $\Sigma\, |c_1\, v\, \boldsymbol{k} +\rangle \langle c_1\, v\, \boldsymbol{k} +| \to P_1 - \Pi$, $\Sigma\, |c_2\, v\, \boldsymbol{k} +\rangle\, \langle c_2\, v\, \boldsymbol{k} +| \to P_2 + \Pi$.

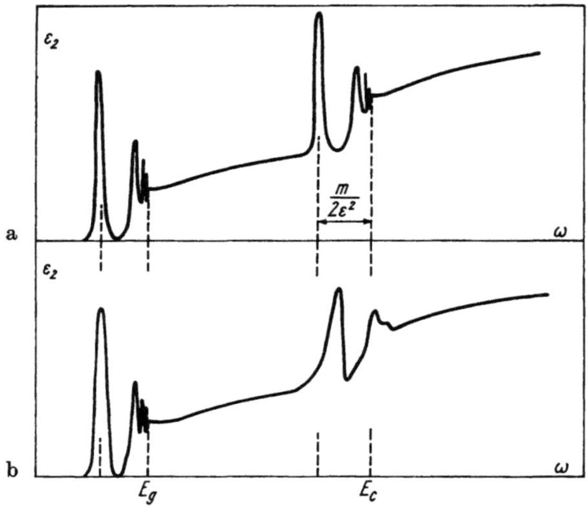


Fig. 5. Resonance effects in $\varepsilon_2$ for the band model of Fig. 4.
a) Effective mass approximation;
b) Resonance effects included

The effective-mass dielectric constant for the model of Fig. 4 is sketched in Fig. 5a. How do the resonance effects change this picture? Contributions from dressed transitions, both $v \to c_1$ and $v \to c_2$, must be taken into account.

1. The analytic structure of $\varepsilon\,(v \to c_2)$ is determined only by the Green's function of the $P_1$ space near the critical frequency. In $\varepsilon\,(v \to c_2)$ a "Rydberg series" of complex poles appears which lead to resonant line shapes with total integral strengths close to the unperturbed ones. Furthermore, a branching point is formed and it must be weak as it tends to zero with $U$.

2. The only singular point of $\varepsilon\,(v \to c_1)$ is at the energy $E_c$, and its character is mostly determined by the functions $|c_1\, v\, \boldsymbol{k};\, m\rangle$ (see equations (2), (15), and (16)). We conclude that interband resonance produces only a minor effect in $\varepsilon\,(v \to c_1)$. The positions of the resonant lines are shifted with respect to $E_c$ due to the real part of the Green's function of the $P_1$ space, whereas the singularity of $\varepsilon\,(v \to c_1)$ is not. In other words, the image of the complex Rydberg series may be shifted with respect to its "edge" as indicated in Fig. 5b. [3])

### *3.2 Saddle points*

Also in the case of critical points of the types $M_1$ or $M_2$ we may start with the effective mass approximation and then take into account resonance effects due to other bands. If we compare the situation at the extrema and saddle points (replace the higher minimum in Fig. 4 by a saddle point) we see that in the latter case, the $P_1$ continuum extends both downwards and upwards so that any artificially prepared bound state is degenerate with the scattering states of the "parent" band. The properties of the Coulomb interaction imply that the decay within the parent band will dominate over that to other bands. Hence, the residual interband interaction will be omitted in the following.

---

[3]) These conclusions do not depend on the physical nature of the spaces $P_1$, $P_2$. Probably in this way some observed phenomena may be explained, e.g. the difference between the $\Gamma$-excitons in annealed and unannealed solid Xe [5].

We shall inspect the effective equation in case of a saddle point of the $M_1$ type and put, without loss of generality, $m_1 = m_2 = m > 0$, $m_3 < 0$. The variables in the effective equation are not separable and we can say only little about its solutions. It is clear that the Hamiltonian has no discrete levels. Still, we might try to divide it into two parts, one of which would give discrete levels and the other would cause their decay. The former, diagonal part should have to contain states with positive kinetic energies. In $\boldsymbol{k}$-space these states correspond to the points outside the critical cone, forming a region there. Each of such regions corresponds to a Hamiltonian with a series of discrete levels. But the discrete spectrum becomes unstable on extending the region to the whole exterior of the critical cone, the levels falling down to $-\infty$. In this way, we get no resonances analogous to those near critical points of the type $M_0$.

But there is another possibility in the case of large absolute values of $m_3$. Then the effective equation contains the small parameter $m_3^{-1}$. For positive values of $m_3$, the critical point is a minimum and $\varepsilon_2$ will have discrete lines. We shall see that the lines are stable when $m_3^{-1} \to +0$ and we may expect that something of them will survive even after passing over to negative masses. So we may try to establish the existence of certain singularities of $\varepsilon_2$ as an asymptotic property for $m_3^{-1} \to -0$. The mathematical expression of this idea is to prove the existence of singularities in the adiabatic approximation.[4]) In the limiting case $m_3^{-1} = 0$ the effective equation becomes

$$\left[\frac{1}{2\,m}\left(\frac{\partial^2}{\partial x^2} + \frac{\partial^2}{\partial y^2}\right) + \frac{1}{\varepsilon\,(x^2 + y^2 + z^2)^{1/2}} + E\right] \Phi(x, y, z) = 0\,. \tag{17}$$

The eigenfunctions have the form

$$\Phi = \varphi_n(x, y; \zeta)\,\delta\,(z - \zeta)\,, \tag{18}$$

where the function $\varphi$ is the solution of the equation obtained from (17) if the variable $z$ is treated as a parameter with a fixed value $\zeta$, on which the functions $\varphi_n$ and the corresponding eigenvalues $\varepsilon_n(\zeta)$ depend. The corresponding amplifying factor is zero unless $\zeta = 0$. So we put $z = 0$ in (17), and obtain the Schrödinger equation for "two-dimensional hydrogen". For localized optically active states (see [11]),

$$\varepsilon_n(0) = -\frac{2\,m}{\varepsilon^2\,(2\,n - 1)^2}\,; \qquad |\varphi_n(0)|^2 = \frac{16\,m^3}{\pi\,\varepsilon^3\,(2\,n - 1)^3}\,. \tag{19}$$

In the continuum,

$$\gamma' = -\pi(2\,\varepsilon^2\,m^{-1}\,[E - E_1])^{-1/2}\,; \qquad |\varphi_E(0)|^2 = \frac{e^{-\gamma'}}{\cosh\gamma'}\,, \tag{20}$$

where the zero-point energy is at $E_1$. A sketch of the corresponding amplifying factor is given in Fig. 6a.[5])

The actual meaning of these calculations may be clarified only by comparison with a calculation for a finite mass $m_3$. This may be done in the adiabatic

---

[4]) A similar approximation is known from the theory of shallow impurity states [10].

[5]) This figure refers to the limiting case between $M_0$ and $M_1$ singularities. A similar problem for the "$M_2 - M_3$" singularity is that of repulsive two-dimensional Coulomb scattering. No bound states exist, and the interband absorption spectrum is reduced, due to the electron–hole interaction, by a factor the effect of which is visualized in Fig. 6b. Compare this result with that shown in Fig. 3b.

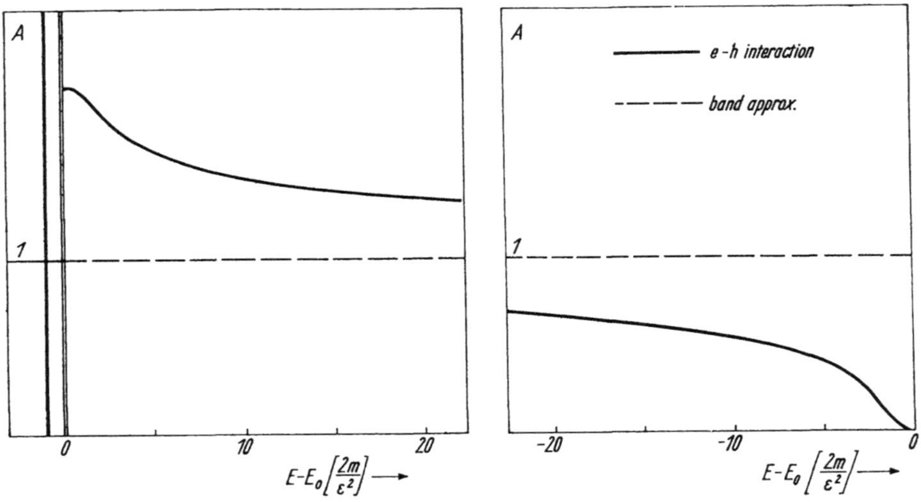


a b

Fig. 6. a) Amplifying factor near the "$M_0 - M_1$" critical point $|(m_3| = \infty)$;
b) The same near the "$M_2 - M_3$" singularity

approximation. We put

$$\Phi(x, y, z) = \varphi_n(x, y; z)\, \psi_{n\alpha}(z)\,, \tag{21}$$

where $\varphi$ is the solution of (17) and $\psi$ is to be found from the one-dimensional equation

$$\left(-\frac{1}{2\, m_3}\frac{\partial^2}{\partial z^2} + \varepsilon_n(z) - E_{n\alpha}\right)\psi_{n\alpha}(z) = 0 \tag{22}$$

which gives also the energy eigenvalue. For a given $\varphi$, with $\varphi(0) \neq 0$, we find the amplifying factor to be proportional to $|\psi(0)|^2$. For example, for $\varphi_1$, the ground state, we have estimated the form of $\varepsilon_1(z)$ by first-order perturbation theory (Fig. 7a). The function $\varepsilon_1(z)$ is the potential energy for the "particle" $m_3$ if $m_3 > 0$. Below $E_1$, the absorption corresponding to the functions $\varphi_1\, \psi_{1\alpha}$ is composed of a series of successively weaker lines. In the limiting case $m_3^{-1} \to +0$ these lines condense into a $\delta$-like singularity according to (18). For $m_3 < 0$, equation (22), being multiplied by $(-1)$ becomes that of scattering by a barrier (Fig. 7b) with energy $-E$. $\psi$ is chosen to be the outgoing wave again; $|\psi(0)|^2$ depends only slightly on energy in the region above the top of the barrier, but decreases exponentially below the top. The WKBJ estimate (which cannot be applied for energies close to the top of the barrier) gives the line shape shown in Fig. 8. The corresponding analytic expressions are

$E < -\dfrac{2\,m}{\varepsilon^2}$:

$$|\Phi_E(0)|^2 \approx |\varphi_1(0)|^2 \left(\frac{m_3}{m}\left(\frac{\varepsilon^2}{2\,m}E + 1\right)\right)^{-1/2}, \tag{23}$$

$E > -\dfrac{2\,m}{\varepsilon^2}$:

$$|\Phi_E(0)|^2 \approx |\varphi_1(0)|^2 \left(\frac{|m_3|}{m}\left(\frac{\varepsilon^2}{2\,m}E + 1\right)\right)^{-1/2} \mathrm{e}^{-\frac{1}{3}\sqrt{\frac{|m_3|}{m}}\left(\frac{\varepsilon^2}{2\,m}E + 1\right)^{3/2}} \tag{24}$$

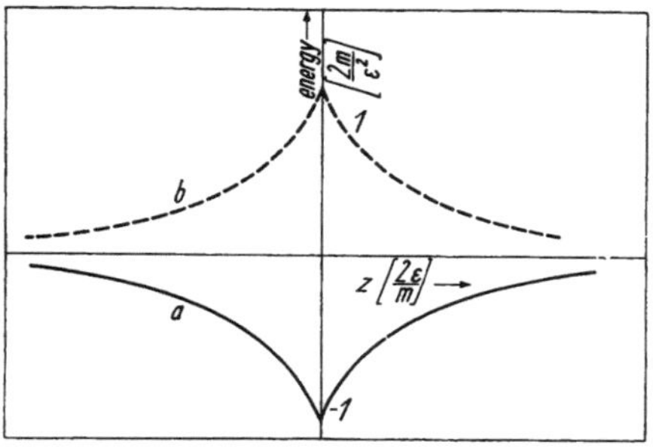


Fig. 7. Curve (a): The function $\varepsilon_1(z)$. Its form near $z = 0$ is given by the formula $\varepsilon_1(z) = -1+4\,|z|$. Curve (b): The function $-\varepsilon_1(z)$ is the barrier potential for the effective scattering problem in the case that $m_3$ is negative

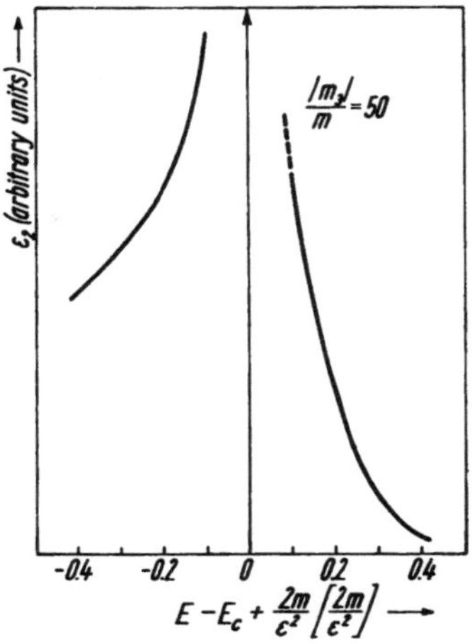


Fig. 8. "Exciton" line shape according to equations (23) and (24)

The singularity obtained is a branching point rather than a pole, and it is even not generated by the decay of a pole. If $|m_3|^{-1}$ increases, the adiabatic approximation fails to be valid, and the intensity of these singularities rapidly decreases. This reminds us once more of the asymptotic meaning of the results obtained.

***Acknowledgements***

We would like to express our thanks to Professor V. L. Bonch-Bruevich, Professor M. H. Cohen, Professor H. Ehrenreich and Professor J. C. Phillips for stimulating discussions during the Varenna Summer School 1965. We also thank Dr. E. Antončík and Professor J. Tauc for their kind interest in the work and for bringing several preprints to our attention.

## References

[1] J. Tauc, Progress in Semiconductors, Ed. A. F. Gibson, Vol. 9, Heywood, London 1965 (p. 87).
[2] R. J. Elliott, Phys. Rev. **108**, 1384 (1957).
[3] G. G. Macfarlane, T. P. McLean, J. E. Quarrington, and V. Roberts, Proc. Phys. Soc. **B 71**, 863 (1958).
[4] J. C. Phillips, The Fundamental Optical Spectra of Solids (preprint).
[5] K. P. Jain, Phys. Rev. **139**, A 544 (1965).
[6] R. Kubo, J. Phys. Soc. Japan **12**, 570 (1957).
[7] G. F. Koster and J. C. Slater, Phys. Rev. **96**, 1208 (1954).
[8] U. Fano, Phys. Rev. **124**, 1866 (1961).
[9] H. Feshbach, Ann. Phys. (U.S.A.) **5**, 357 (1958).
[10] W. Kohn and J. M. Luttinger, Phys. Rev. **98**, 915 (1955).
[11] S. Flügge and H. Marschall, Rechenmethoden der Quantentheorie, Springer-Verlag, 1952.

*(Received April 12, 1966)*

phys. stat. sol. **16**, 159 (1966)

*Physik-Department der Technischen Hochschule München*

# Delayed Fluorescence and Thermoluminescence of Anthracene Single Crystals at Very Low Temperatures

By

N. RIEHL and P. THOMA

The delayed fluorescence of anthracene single crystals of various origins is measured at 77 and 4.2 °K. At 4.2 °K the decay time of the delayed fluorescence is much larger than at 77 °K. The optical glow curves of the crystals are measured in the temperature range 4.2 to 77 °K after excitation with UV light. The nature of the traps responsible for the glow peaks is discussed.

Es wird die verzögerte Fluoreszenz von Anthrazeneinkristallen verschiedenen Ursprungs bei 77 und 4,2 °K gemessen. Bei 4,2 °K ist die Abklingzeit der verzögerten Fluoreszenz wesentlich größer als bei 77 °K. Die optischen Glowkurven der Kristalle wurden nach Anregung mit UV-Licht bei 4,2 °K zwischen 4,2 und 77 °K gemessen. Die Natur der für die Glowmaxima verantwortlichen Haftstellen wird diskutiert.

Previous papers on the electrical and optical properties of anthracene single crystals have shown a quasicontinuous and at least partially exponential distribution of hole traps [1, 2]. This quasicontinuous distribution was shown recently to consist of numerous resolvable, discrete trap levels [3, 4]. In two preceding publications we have reported on the observation of discrete low temperature trap-levels supposed to be electron traps [5, 6]. In this paper we report in more detail on measurements of the decay of luminescence (delayed fluorescence) and of the thermoluminescence of anthracene single crystals of different origin in the temperature range between 2.5 and 77 °K.

## 1. The Crystals

For these measurements four different types of crystals were used.

A) Monocrystalline platelets with surface area 0.5 to 2 cm², 10 to 80 μm thick, grown from the vapour phase.

B) Monocrystalline platelets, also grown from the vapour phase, with nearly the same dimensions. The raw substance, however, was aceotropically distilled and zone-refined 150 times. These crystals had a slightly stronger fluorescence than type A crystals.

C) Large single crystals grown from the melt which were kindly supplied by Dr. W. G. Schneider of NRC, Ottawa.

D) Large single crystals (Harshaw-scintillation grade crystals) 1″ ∅, 0.5 to 2 mm thick.

## 2. Experimental Details

For low temperature studies two different types of He cryostats were used making possible experiments either in high vacuum ($10^{-5}$ Torr) or in an atmosphere of He gas (at 840 Torr). The design of these cryostats is indicated in Fig. 1.

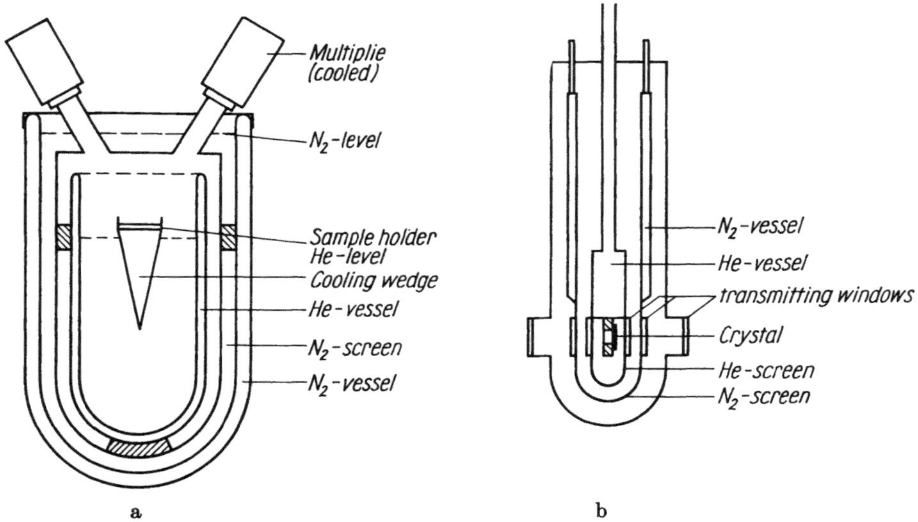


Fig. 1. He cryostats a) used for measurements at 840 Torr of a He atmosphere, and b) used for measurements in high vacuum

Photomultipliers of the types RCA 6655 A and 2020 were used for emission measurements in the blue-green range, Valvo type 150 CVP for the red range. Between multipliers and crystal either sets of filters or a Zeiss Monochromator were mounted.

Galvanometers and vibrating reed electrometers were used for the measurement of slow processes; cathode-ray oscilloscopes were applied to fast processes. Heating programs were used for the observation of glow curves such that a very linear time-temperature characteristic was obtained. These heating programs were checked by measuring the temperatures on both sides of sample crystals with Cu-Constantan thermocouples and, in the lowest temperature range, with carbon thermometers which had been calibrated previously against He or $H_2$ vapor pressure thermometers. The exciting light source was an Osram HBO-200 Hg high pressure lamp having many lines in its spectrum. Various bands of excitation light were selected with filters having an average bandwidth of 20 nm within a range of 345 to 420 nm. If not indicated otherwise excitation was applied for 5 min. To facilitate comparison of either the decay curves of crystal fluorescence or of glow curves of different crystals, equal sized crystals of types A and B were chosen and equal sized surface areas of type C and D crystals were left exposed. Fast and slow decay processes were observed in the following manner:

a) Fast processes: Observation of fluorescence and its decay with UV filter which opens at $\lambda = 390$ nm. Here an exact discrimination between fast and slow decay was not possible, because the photomultiplier saturated due to the high intensities of fluorescence during excitation. For more accurate measurements the excitation was attenuated with neutral filters. The voltage drop at the anode resistor of the multiplier was observed via cathode follower and CRT both having short time constants.

b) Slow processes: The slow decay processes were observed 4 ms after cessation of the excitation. Each shutter in the two optical paths needed 2 ms for motion from full opening to the complete closure.

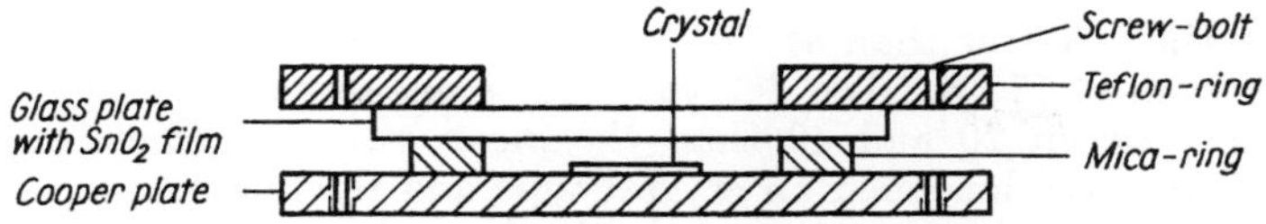


Fig. 2. Sample chamber used for studying the influence of electric fields on crystal properties

To investigate the effect of strong electric fields (between 0.5 and 20 kV/cm) on the decay characteristics and the glow emission different homogeneous dc fields were applied to crystals of types A, B, and D. Direct electrical contact with the crystal specimens was avoided by the arrangement shown in Fig. 2.

The distance between sample surfaces and the glass plates supporting the transparent electrodes made of evaporated $SnO_2$ films was at least 900 μm in the case of type A and B crystals and 100 μm in the case of type D crystals. During experiments with electric fields the cryostat was evacuated to better than $10^{-6}$ Torr.

## 3. Results

### *3.1 Decay measurements at 77 °K*

Here, the exciting wavelength range was restricted to the region between 349 and 370 nm. For the observation of delayed fluorescence the shutters at the cryostat windows were sequenced so that the multiplier shutter opened only after the shutter of the excitation path was fully closed. The decay curves were plotted beginning 5 ms after the closure of the excitation path was initiated. The short-lived fluorescence decays in a time $<10^{-6}$ s. For the delayed fluorescence of the type A, B, and D crystals, decay times were found in close correspondence with the values of Singh, Jones, Siebrand, Stoicheff, and Schneider [7] (quoted as SJSSS hereafter). Fig. 3 shows the averaged decay curves at 77 °K. The intensity of fluorescence of type B crystals was larger by a factor 3 than that of type A crystals while the decay times were nearly equal and had exponential characteristic. The decay times of type D crystals were markedly larger. No effect of electric fields on the decay times and on the intensities was found.

### *3.2 Decay measurements at 4.2 °K*

The temperature of the crystals was reduced to 4.2 °K and the measurements were performed as described above. Here, the short-lived fluorescence certainly

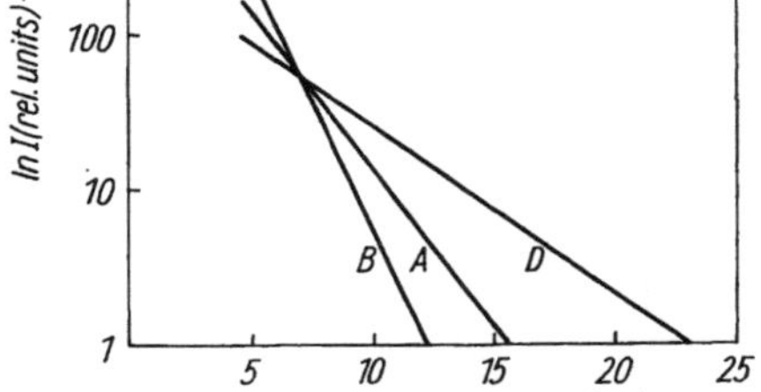


Fig. 3. Decay of delayed fluorescence of anthracene crystals (types A, B, and D) at 77 °K

decays in a time shorter than at 77 °K. The fluorescence intensities at 4.2 °K were larger than at 77 °K. The delayed fluorescence was decaying more slowly by a factor between 20 and 40 than the mean value of 12 ms measured at 77 °K. The increase depended on crystal type and thickness (Fig. 4).

It was not possible to measure exactly the dependence of the delayed fluorescence intensity on the exciting UV intensity $J_0$ because the curve within the first 5 ms had to be extrapolated. In the upper range of the available excitation intensities a relation approximating a square law ($J \approx J_0^2$) was observed. No delayed fluorescence was observable with excitation of wavelengths $> 400$ nm (wave numbers $< 25000$ cm$^{-1}$). Using a multiplier sensitive above 680 nm, a weak slowly decaying long wavelength emission with decay times of several seconds was observed. It can be interpreted as radiative transition from the lowest triplet level into the ground state (phosphorescence). At 77 °K no such emission was observable. Recently, by use of more sensitive multipliers we could confirm [8] that this emission is indeed the phosphorescence of solid anthracene. At 4.2 °K there was no observable effect of electric fields on any decay process.

### *3.3 Thermoluminescence measurements*

Helfrich and Mark [9] have studied the quasicontinuous hole trap distribution by means of space charge limited currents. The characteristics of space charge limited electron currents in anthracene [10] suggest a quasicontinuous electron trap distribution and therefore the absence of truly discrete traps. However, the thermoluminescence measurements reported here, electrical measurements of Thoma and Riehl [6], and results to be published in a following paper [11] prove the existence of centres, which are characterized by the essential properties of discrete traps known from anorganic phosphors. They might be a part of a quasicontinuous distribution with small but definite energy separation. It appears possible to interpret such discrete traps on the basis of the excited states of the anthracene molecule and of the crystal lattice. To measure thermoluminescence the different types of crystals were excited 5 min at 4.2 °K and in some cases at 2.5 °K with very intense UV light ($349 < \lambda_{exc} < 370$ nm). They were kept in darkness for 5 min after irradiation. Heating rates of 2, 3, 3.6 or 5.0 deg/min were chosen. The thermoluminescence during heating was observed with a photomultiplier. Fig. 5 shows glow curves of crystals of types

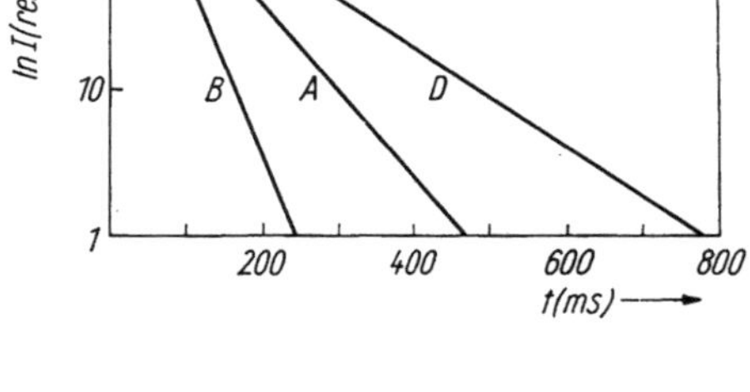


Fig. 4. Decay of delayed fluorescence of anthracene crystals (types A, B, and D) at 4.2 °K

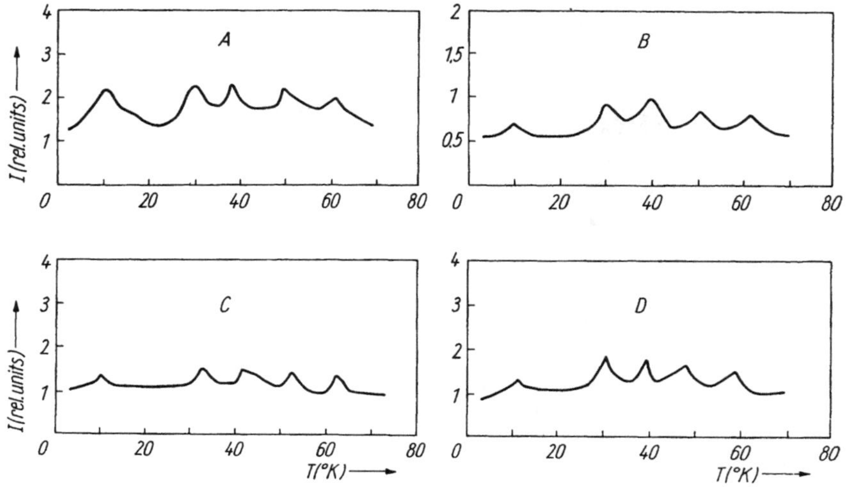


Fig. 5. Optical glow curves of anthracene crystals (types A, B, C, and D) after UV excitation at 4.2 °K

A, B, C, and D obtained with a heating rate of 2 deg/min. The amplitudes of the glow maxima between 2.5 and 77 °K decreased with decreasing concentration of lattice defects. The first glow maximum was observed to dominate only in type A crystals. The amplitudes of the four maxima between 25 and 73 °K decrease continuously with increasing temperature. By studying crystals of same origin but of rather different age it was found that the concentration of centres being responsible for glow peaks is decreasing with increasing age of crystal. In this respect an important dependence on growth conditions and thermal history of crystals was also observed. Excitation of the opposite large crystal surface (the excited surfaces were always *a b*-planes) not illuminated before resulted in identical glow curves. Due to its low intensity the spectrum of the thermoluminescence could not be analyzed with the monochromator, but by means of interference filters it was found that the spectral maximum of the glow emission is located at 400 nm $\equiv \bar{\nu}_{0-0} = 25000\ \text{cm}^{-1}$. At wavelengths $> 400$ nm ($\bar{\nu} < 25000\ \text{cm}^{-1}$) no detectable emission was observed. After excitation with $\lambda_{\text{exc}} > 400\ \text{nm} \equiv \bar{\nu}_{\text{exc}} < 25000\ \text{cm}^{-1}$ no thermoluminescence was observable. Therefore, the glow emission is located within the normal singlet emission range, a result being important.

In order to test whether the crystal defects acting as traps could be annealed thermally a type D crystal was mounted in a closed vessel filled with anthracene vapour and tempered at 80 °C für 5 h. Immediately following this treatment the crystal was cooled to 4.2 °K and irradiated. The thermoluminescence measurement after the usual 5 min dark pause gave no observable glow maxima.

The glow curves of one type B crystal were measured three times with excitation by polarized UV light, each time at different orientations of the polarizer. The results show a preferred direction of the light vector for optimal filling of the traps. It has to be pointed out that this anisotropy of thermoluminescence was observed only with respect to the angle between the direction of polarization and the *a b*-axes of the crystal, not with respect to direction of incidence. This finding refers at equal heating rates for all crystals only to the amplitudes of the glow peaks and not to their temperature. Unexpectedly, however, we observed anisotropy of the distance between the second and the third glow maximum depending on the angle between the incident light and the direction of an applied dc field if this field was applied during the heating process. The distance between these peaks (always at equal heating rates) was increased with increasing field strength when the excited crystal surface was at positive potential and reduced when the surface was negative. The amplitudes of the glow maxima remained unchanged. Fig. 6 shows these results for type A and D crystals. Further measurements indicated that the displacement of the two glow peaks has a maximum at an angle 0° between incident direction (which was always perpendicular to the *a b*-plane) and field direction and that a field parallel to the *a b*-plane produced no displacement. This displacement of the glow maxima by an electric field seems to be not a continuous but a stepwise process. The two possible displacements are shown in Fig. 6. No further displacement is possible by an arbitrary strong field. This seems to prove that at least two of our trap centres have a well revealed polar character and that they are coupled. Their binding energy or depth seems to be enlarged or reduced in two discrete steps.

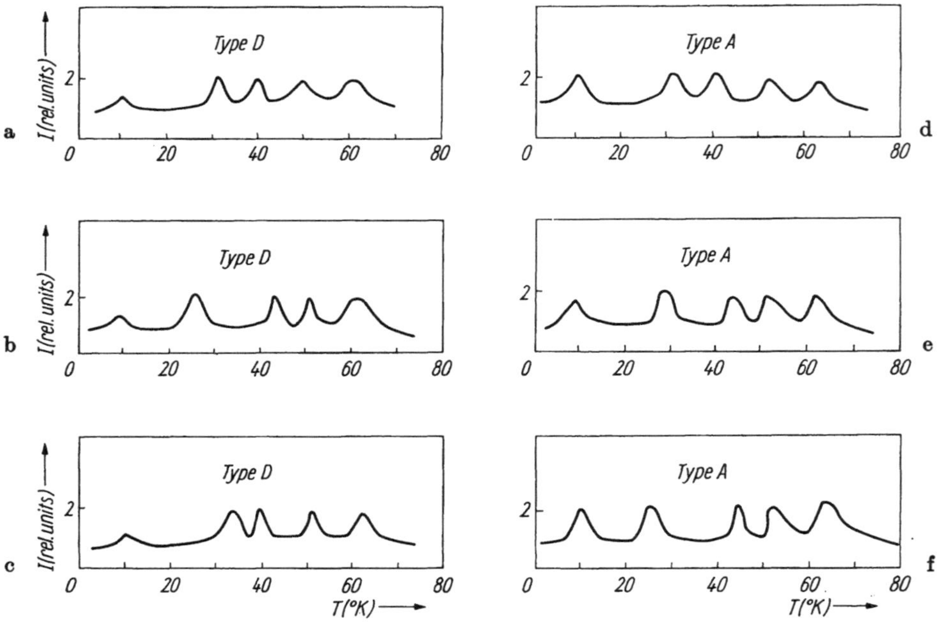


**Fig. 6. Optical glow curves of anthracene crystals (types A and D) after UV excitation at 4.2 °K under the influence of various dc fields. The field is oriented either parallel or antiparallel to incident direction of irradiation before heating (at 5.8 deg/min).**

a) without field,
b) with field 0.56 kV/mm (excited surface is positive),
c) with field 0.56 kV/mm (excited surface is negative),
d) without field,
e) with field 0.63 kV/mm (excited surface is positive),
f) with field 1.4 kV/mm (excited surface is positive)

## 4. Discussion

### *4.1 Decay measurements at 77 °K*

The measured decay times at 77 °K agreed well with the results of SJSSS. For the purpose of evaluation of our data (also at 4.2 °K) we use the equations of Kepler and Switendick [12], and SJSSS:

$$\mathrm{d}n_\mathrm{S}/\mathrm{d}t = - a\, n_\mathrm{S} - k\, n_\mathrm{S} + c\, \gamma\, n_\mathrm{T}^2\,, \tag{1}$$

$$\mathrm{d}n_\mathrm{T}/\mathrm{d}t = k\, n_\mathrm{S} - \beta\, n_\mathrm{T} - c\, \gamma\, b\, n_\mathrm{T}^2\,. \tag{2}$$

At $t = 0$ there is $n_\mathrm{S} = n_{\mathrm{S}_0}$ and $n_\mathrm{T} = n_{\mathrm{T}_0}$ where the time origin $t = 0$ corresponds to the end of the excitation period. For the fast decaying fluorescence the relation

$$n_\mathrm{S} = n_{\mathrm{S}_0}\, \mathrm{e}^{-a t} \tag{3}$$

is valid where $n_\mathrm{S}$ and $n_\mathrm{T}$ are the singlet and triplet concentrations.

From the data of SJSSS we take $a^{-1} = 30$ ns = singlet lifetime. $k$ is the transition rate for the non-radiating transition from the lowest triplet level to the ground level. $\gamma$ is the singlet generation rate by triplet-triplet annihilation. The additional factor $c$ equals 1 if the higher excited triplet states can be neglected, as shown by SJSSS. The triplet decay rate $\beta$ seems to be determined at 77 °K only by non-radiating transition to ground level not by radiative transitions. For $t \gg a^{-1}$

$$\mathrm{d}n_\mathrm{S}/\mathrm{d}t = - a\, n_\mathrm{S} - k\, n_\mathrm{S} + \gamma\, n_\mathrm{T}^2 = 0 \tag{4}$$

is a useful approximation. The solution without the second term

$$n_\mathrm{S} = (\gamma/2\, a)\, n_\mathrm{T}^2 \tag{5}$$

is substituted into (2) giving

$$n_\mathrm{T} = \frac{n_{\mathrm{T}_0}}{\frac{\gamma}{\beta}\, n_{\mathrm{T}_0}\left(1 - \frac{k}{2\, a}\right)(\mathrm{e}^{\beta t} - 1) + \mathrm{e}^{\beta t}}\,. \tag{6}$$

From (5) and (6) follows

$$n_\mathrm{S} = \frac{\beta^2\, n_{\mathrm{T}_0}^2}{2\, a} \cdot \frac{\mathrm{e}^{-2\beta t}}{\left[\gamma\left(1 - \frac{k}{2\, a}\right) + \beta\right]^2} \tag{7}$$

if the term $-\frac{\gamma}{\beta}\, n_{\mathrm{T}_0}\left(1 - \frac{k}{2\, a}\right)$ resulting from multiplication in the denominator of (6) is neglected for times of the order of the total decay time. The equation (7) is used to calculate the triplet lifetimes as done by SJSSS. This results in values for $\beta^{-1}$ at 77 °K between 5 and 15 ms. In agreement with the results of Kepler and Switendick we found appreciably lower values of $\beta^{-1}$ for thin crystals than for thick crystals.

### *4.2 Decay measurements at 4.2 °K*

The decay curves at 4.2 °K show much longer decay times than those at 77 °K. Using the analysis discussed above, at 4.2 °K $\beta^{-1}$ was found to have values between 100 and 500 ms. In this calculation the phosphorescence was neglected. A more exact treatment would require a generalization of equation (1) to

$$dn_T/dt = k\, n_S - \beta\, n_T - \beta_1\, n_T - c\,\gamma\, n_T^2 \tag{8}$$

with $\beta_1^{-1}$ radiative lifetime which would make the system quite unmanageable. The dependence of decay time on crystal thickness was evident also at 4.2 °K. If it is desired to attribute the increase of the calculated triplet lifetimes to triplet diffusion and generation of nearest neighbour pairs near the surface then the diffusion term has to be added to equation (2):

$$dn_T/dt = k\, n_S - \beta\, n_T - c\,\gamma\, n_T^2 + D\, \partial^2 n_T/\partial x^2\,. \tag{9}$$

Here $x$ is the distance measured perpendicular to the surface. The approximate solution of (9) for times $t > 5\times 10^{-2}\, l^2/D$ is given by the equation

$$n_T = \frac{4}{\pi}\, n_{T_0}\, e^{-(p+\pi^2 D/l^2)t} \sin \pi\, x/l\,, \tag{10}$$

where $l$ is the crystal thickness.

The bracketed part of the exponent is the measurable rate $\beta'$. With data available on crystals with thickness $l$ varying between 10 μm and 2 mm and assuming to first approximation similar energy transport mechanisms in all our crystal types the relation $\beta' = \beta + \pi^2\, D/l^2$ can be evaluated. Setting a value of 0.05 $s^{-1}$ for $\beta$ (calculated from the results for a 10 μm thick crystal) and of 1.2 $s^{-1}$ for $\beta'$ (calculated for a 1 mm thick crystal) we find $D_{4.2\,°K} = 10^{-3}$ cm²/s. Kepler and Switendick [12] have calculated $D$-values from their room temperature data, which are larger than our $D$-values at liquid helium temperature. This means that excitons diffuse more slowly with decreasing temperature because of the existence of numerous traps capable of immobilizing the diffusing excited states for varying time intervals. For an ideal crystal, on the other hand, an increase of the diffusion coefficient with decreasing temperature due to the decreasing lattice agitation would have to be expected. Apparently there are many traps which are able to immobilize the diffusing excited states for appreciable lengths of time. Two other approximate relations are available to check the value of $D$:

$$D = a^2/2\, t_h \tag{11}$$

and

$$D = \lambda\, \langle v \rangle\,. \tag{12}$$

(11) is valid for the hopping model of exciton migration and (12) for the band model. The following estimates are useful at a temperature of 4.2 °K [7, 12]: hopping distance $a = 6\times 10^{-8}$ cm (the hopping distance should be independent on temperature), hopping time $t_h = 10^{-11}$ to $10^{-12}$ s, mean free path for exciton $\lambda = 10^{-6}$ to $10^{-5}$ cm (it should be less dependent on temperature than in the case of particle migration), mean value of exciton velocity $\langle v \rangle = 6.6\times 10^2$ cm/s (calculated from the room temperature value of $5\times 10^4$ cm/s given by Kepler). Therefore, from (11) $D_{4.2\,°K} = 1.8\times 10^{-4}$ cm²/s and from (12) $D_{4.2\,°K} = 6.6\times 10^{-3}$ to $7\times 10^{-4}$ cm²/s. The value calculated from the band model expression is in slightly better agreement with the experimental value than that obtained from the hopping model. It should be noted that these approximations are based on relatively raw estimates.

### *4.3 Thermoluminescence measurements*

By the detection of glow peaks in the course of thermoluminescence measurements between 2.5 and 77 °K we were able to show the existence of discrete traps in anthracene. Analyzing the very similar glow curves of our four crystal types with theoretical methods well known in the field of anorganic luminophors it is possible to calculate the depth of the trap terms under the lowest conducting state assuming to have charge carrier traps only. The method of variation of heating rates given by Schön [13] can be used for this purpose. This method is applicable because the results at different heating rates show that glow peaks appear at somewhat higher temperatures if the heating rates are increased. The corresponding equation is

$$E_h = \frac{T_1 T_2}{T_1 - T_2} \ln \frac{q_1 T_2^2}{q_2 T_1^2}, \tag{13}$$

where $T_1$, $T_2$ are the glow-peak temperatures at the heating rates $q_1$, $q_2$. Due to the zero point energy it is not valid at temperatures around 4.2 °K and lower. The results of (13) for the four crystal types possessing glow peaks with equal positions on the temperature axis at equal heating rates are given in the following table:

| Glow peak at 3.5 deg/min | 10 °K | 30 °K | 40 °K | 51 °K | 63 °K |
|---|---|---|---|---|---|
| Trap depth (eV) | (0.007) | 0.01 | 0.03 | 0.05 | 0.059 |
| Trap depth $E_h = 17.5\ kT$ (eV) | 0.009 | 0.026 | 0.035 | 0.045 | 0.057 |

Apparently the glow peaks are not caused by triplet-triplet interaction. Otherwise, anomalies in the delayed fluorescence (higher light sums) near the glow-peak temperatures should appear. To investigate this point the delayed fluorescence near the 30 °K glow peak was measured. We found a continuous decrease of decay times with increasing temperature between 25 and 30 °K. These findings suggest that the trapped states include charge carriers. The trapping of holes by these centres, appears to be excluded by the following observations: i) When a crystal excited by UV light was heated from 4.2 to 77 °K, a voltage was generated such that the irradiated surface became positive. ii) During the heating process this voltage showed none of steps or discontinuities which would be expected, if the charges had been released from hole traps at *discrete* energy levels. The extremely weak minima in this curve reported in [6] could not be confirmed. These discrete trap levels determined by thermoluminescence and conduction glow curve measurements [5, 6, 11] might well be part of a quasi-continuous electron trap distribution located near the lowest conducting state. The existence of such electron trap distribution analogous to the observed quasicontinuous hole trap distribution might also be expected based on considerations of symmetry. The electrical conductivity of crystals was shown in these papers to have maxima near the temperatures of the optical glow peaks (at equal heating rates). From evaluation of optical and electrical glow curve areas density values between $10^{10}$ and $10^{13}$ cm$^{-3}$ are found for the discrete traps in the different crystal types. The highest densities are observed in the crystals grown from the melt (types D and C). A quasicontinuous electron trap distribution, as just discussed, would have to possess much lower concentrations than the quasicontinuous hole trap distribution. This would easily explain that no electron traps are observable at temperatures $> 80$ °K. They might become observable with more sensitive apparatus.

Two important findings contradict the conclusion that these discrete traps are merely charge carrier traps:

i) A crystal supplied with transparent Ag electrodes was cooled to 4.2 °K. Then a weak electric field was applied and the crystal was excited with UV light at constant temperature, until constant fluorescence intensity $J_F$ and constant photocurrent $I_r$ were observed. Then the excitation was turned off and the crystal was heated slowly. This resulted in an optical glow curve of area $\int J dt$ and a conduction glow curve of area $\int I dt$. The ratio of these areas was found to be proportional to the ratio of the constant intensities during excitation:

$$\frac{\int J\,\mathrm{d}t}{\int I\,\mathrm{d}t} = \text{const}\,\frac{J_F}{I_r}\,. \tag{14}$$

With our measurements reported in this and other papers [6, 11] we found this proportionality always to be satisfied but with different proportionality factors for the different crystal types. In all our photocurrent experiments $I_r$ was within the ohmic part of the current-voltage characteristic. This experiment indicates that excited states having been localized at lattice defects are released and react now in the same manner as singlet excitons do in the case of their direct generation by UV light, because the 0—0 transition at 25000 cm$^{-1}$ was observable in glow emission.

ii) The second result of importance is the fact that the lattice defects acting as traps can be annealed thermally at 80 °C or above. Thus our work agrees with the results of Helfrich and Lipsett [14] in an important aspect. They investigated the fluorescence emission of melt-grown anthracene crystals at 4.2 °K at wave numbers below the 0—0 transition ($\bar{\nu} < 25000$ cm$^{-1}$ and $\lambda > 400$ nm). Exciting the crystals at $\bar{\nu}_{exc} > 25000$ cm$^{-1}$ and in the region $19100 < \bar{\nu}_{exc} < 25000$ cm$^{-1}$ they found emission peaks near 23900, 23000, 21500, and 19100 cm$^{-1}$ being superimposed on a structureless background emission. The background decreased with decreasing wave numbers of the excitation.

Designating this structured emission as defect fluorescence, Helfrich and Lipsett have shown that these emission maxima are caused by the recombination radiation of excited states having been localized at lattice point defects. Another of their findings, most essential to the present discussion, is the fact that they were able to anneal these point defects by thermal treatment at temperatures >100 °C in a manner and with results quite similar to our work. This agreement between Helfrich and Lipsett and our experiment plus the conclusions drawn from equation (14) show that the traps observed by us are closely connected with excited states localized at point defects. In attempt to explain charge carrier generation by release of excited states from trap levels deep below the lowest conducting state meets with great difficulties. One can not, by orders of magnitude, account for the rate of carrier generation derived from conduction glow curves under the assumption of singlet generation via phonon interaction with the excited states and succeeding singlet-singlet interaction. The observation of very long lifetimes of the excited states at point defects might be made plausible by their depth below the first excited singlet state and by their connection with a trapped electron. However, there seems to be also no way to account for the observed rate of charge carrier generation due to extremely deep conduction band minima.

On the other hand, the existence of electron traps is well supported by our measurements of conduction glow curves as well as by the behaviour of the voltage generated by release of trapped holes reported in this paper. With certainty the observed glow emission is not due to electron hole recombination because we know this spectrum from other experiments [10].

We are thus led to conclude that point defects are capable of trapping simultaneously excited states and electrons and further that the complex excited state-electron at the defect site has an extremely large lifetime at very low temperatures. In addition there seems to exist a preferred direction for the generation of such complexes, judging from out experiments with polarized light excitation. From the fact that the displacement of two adjacent glow maxima by an applied electric field became zero when the field vector was in the *a b*-plane, we conclude that lattice deformations caused by point defects are strongly oriented with respect to lattice space and that, therefore, strong dipole-dipole interaction is to be expected. The conclusion that the trap levels are consisting of a complex is supported by the polar character of at least two of the trap centres and by the fact that each emitted quantum in glow emission at $\bar{\nu}_{0-0} = 25000\ \text{cm}^{-1}$ is accompanied by the release of at least one electron.

The trap centre consisting of an excited molecule-state and an electron may dissociate by interaction with a phonon being generated by heating. The dissociation process consists of lifting the excited state into the lowest excited singlet-state with spontaneous recombination to the ground level (observation of the 0—0 transition) and of freeing the electron which is transferred into the conducting state (no electron-hole recombination is observable). Therefore, we can conclude that the point defects in anthracene are capable of immobilizing or trapping a) short-lived excited states as observed by Helfrich and Lipsett in defect fluorescence, b) charge carriers (preferentially electrons), c) excited states + electrons.

The ability of such point defects to trap charges and (or) excited states is given by local lattice deformations of different and oriented structure in the lattice.

### *Summarizing these results*

1. The centres found by means of optical and electrical glow curves are complexes consisting of an excited state of a molecule and an electron. By phonon interactions these complexes dissociate into a singlet exciton state which recombines fast and a free electron.

2. These complexes are localized at point defects. These defects may otherwise include excited states alone. In this case one observes defect fluorescence as found by Helfrich and Lipsett. The concentrations of the complexes at defect sites are lower than those of single excited states at defect sites. The point defects may trap charge carriers.

3. The local lattice deformations at such point defects being capable of immobilizing complexes have a preferred orientation in lattice space ($\perp$ *a b*-plane). They are characterized by strong dipole-dipole interaction.

4. The concentration of the point defects is strongly dependent on growth conditions, ambient temperature, and age of crystal.

## References

[1] W. Helfrich, Dissertation TH München, 1961.
[2] P. Mark and W. Helfrich, J. appl. Phys. **33**, 205 (1962).
[3] H. Bässler, G. Becker, and N. Riehl, phys. stat. sol., in the press.
[4] N. Riehl, G. Becker, and H. Bässler, phys. stat. sol., in the press.

[5] W. Helfrich, N. Riehl, and P. Thoma, Phys. Letters (Netherlands) **10**, 31 (1964).
[6] P. Thoma and N. Riehl, Organic Crystal Symposium, Chicago 1965 (p. **39**).
[7] Singh, S., W. J. Jones, W. Siebrand, B. P. Stoicheff, and W. G. Schneider, J. chem. Phys. **42**, **330** (1965).
[8] N. Riehl, P. Thoma, and G. Vaubel, to be published.
[9] W. Helfrich and P. Mark, Z. Phys. **166**, 370 (1962).
[10] H. Miggenrieder, Diplomarbeit TH München, 1966.
[11] P. Thoma and G. Vaubel, to be published.
[12] R. G. Kepler and A. C. Switendick, Phys. Rev. Letters **15**, 56 (1965).
[13] M. Schön, Techn.-wiss. Abh. Osram-Ges. **7**, 175 (1958).
[14] W. Helfrich and F. R. Lipsett, J. chem. Phys., in the press.

*(Received April 12, 1966)*

phys. stat. sol. **16**, 171 (1966)

*Materials Research Laboratory, Martin Company, Orlando, Florida*

# Pumping Induced Imperfections in Glass:$Nd^{3+}$ Lasers

By

L. I. van Torne

A barium crown glass: $Nd^{3+}$ laser rod which developed permanent internal imperfections during laser pumping was investigated by optical metallography, transmission electron microscopy, and electron diffraction techniques. The imperfections were found to be microcracks resulting from differential thermal expansion between crystallites (resulting from devitrification) and the surrounding noncrystalline medium. An analysis of the kinetics of devitrification, the radial temperature distribution during pumping and stresses resulting from differential thermal expansion are in agreement with a proposed mechanism for the development of permanent internal imperfections.

Ein Barium-Kronglas:$Nd^{3+}$-Laserstab, der während des Pumpens permanente innere Defekte entwickelt, wird mit optischer Metallographie, Elektronenmikroskopie und Elektronenbeugung untersucht. Die Defekte erweisen sich als Mikrorisse, die durch unterschiedliche thermische Ausdehnung der Kristallite (die von der Entglasung herrühren) und der nichtkristallinen Umgebung entstehen. Eine Analyse der Kinetik der Entglasung, die radiale Temperaturverteilung während des Pumpens und die Spannungen, die von der unterschiedlichen thermischen Ausdehnung herrühren, befinden sich in Übereinstimmung mit einem vorgeschlagenen Mechanismus für die Bildung von permanenten inneren Defekten.

## 1. Introduction

The most common hosts for laser active ions are crystals. However, noncrystalline (glass) hosts can also be employed. Although the absence of a crystal structure, and consequently a well defined environment for each active ion, in glass hosts generally produces broadening of the emission lines and comparatively higher threshold values, the use of glass hosts is still attractive because of the following advantages [1] over crystalline hosts:

1. The production technology is simpler.
2. Better optical quality is more easily attained.
3. Higher active ion concentrations are possible in addition to the feasibility of optimizing the properties by varying both the glass composition and the proportions of active ion.

Glass hosts for giant pulse lasers however, give rise to material stability problems peculiar principally to noncrystalline hosts.

A brief description of the geometry of a "Q-switch" cavity is presented in order to indicate the conditions imposed on the glass laser rod. A schematic drawing of the principal components of a typical "Q-switch" configuration is shown in Fig. 1. The laser rod A and flash tube B are positioned near the foci of the elliptical cylinder C. A rotating prism mirror D and a fixed mirror E are located, exterior to the cavity, on the axis of the laser rod. The rotation of the "Q-switch" prism is phased with the firing of the flash tube such that radiant energy is periodically stored in and released from the laser rod. A liquid coolant is circulated through the cavity to maintain surface temperatures within ac-

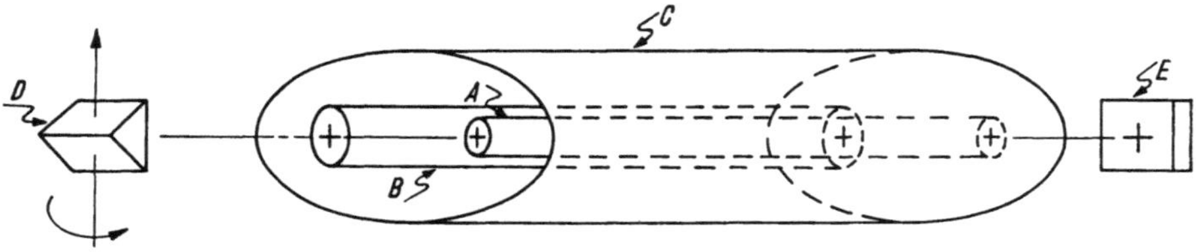


Fig. 1. Schematic representation of a laser "Q-switch" configuration, where A is the laser rod, B the flash tube, C the cavity, D the rotating "Q-switch" prism, and E a fixed mirror

ceptable limits. The pulse rates of the laser output are in the range of 1 to 20 Hz.

Two distinct types of imperfections are observed in glass: $Nd^{3+}$ lasers pumped at power densities of less than $10^7$ W/cm². One type is distinguishable as striae which can develop in the rods with time when pumping levels are moderate. This effect is reversible in the sense that the striae can be removed by annealing the rod at about 200 °C for a few hours. When the pumping levels are near $10^7$ W/cm², internal imperfections, visible to the unaided eye, develop near the laser rod axis. These imperfections cannot be removed by thermal treatments short of melting the rod. A plausible mechanism for the development of permanent imperfections in giant pulse glass: $Nd^{3+}$ lasers is presented. It is based on observations made by optical and transmission electron microscopy.

## 2. Experimental Procedure

The $Nd^{3+}$ doped barium crown glass investigated here had the following nominal chemical composition (in wt%):

| $SiO_2$ | $B_2O_3$ | $Al_2O_3$ | $As_2O_3$ | BaO | $Nd_2O_3$ |
|---|---|---|---|---|---|
| 31 | 13 | 8 | 1 | 42 | 5 |

Active ion ($Nd^{3+}$) concentration $\approx 4.6 \times 10^{20}/\text{cm}^3$.

The pertinent physical properties of this glass were:

Indices of refraction: Sodium $D$: 1.519
1.06 μm: 1.509,

linear coefficient of thermal expansion (50 to 300 °C) $= 10.3 \times 10^{-6}/°\text{C}$,

thermal conductivity $\approx 2.8 \times 10^{-3}$ cal (s cm °C)$^{-1}$,

thermal diffusivity $\approx 5.8 \times 10^{-3}$ cm²/s,

emission wavelength 1.06 μm with 250 Å half width at 300 °K.

Permanent internal imperfections could be developed in a 4.6 mm diameter by 70 mm long rod of this glass by pumping in a focused cavity (such as shown in Fig. 1) at energies approaching 65 J.

Specimens from the "damaged" laser rod were prepared for investigation by optical and transmission electron microscopy in the following manner. An epoxy resin was cast around the periphery of the rod to avoid (mechanical) damage and for ease of handling during sectioning. Discs (1.5 mm thick) were cut perpendicular to the rod axis using a liquid cooled diamond saw. Both surfaces of the transverse sections were polished using 1 μm diamond paste fol-

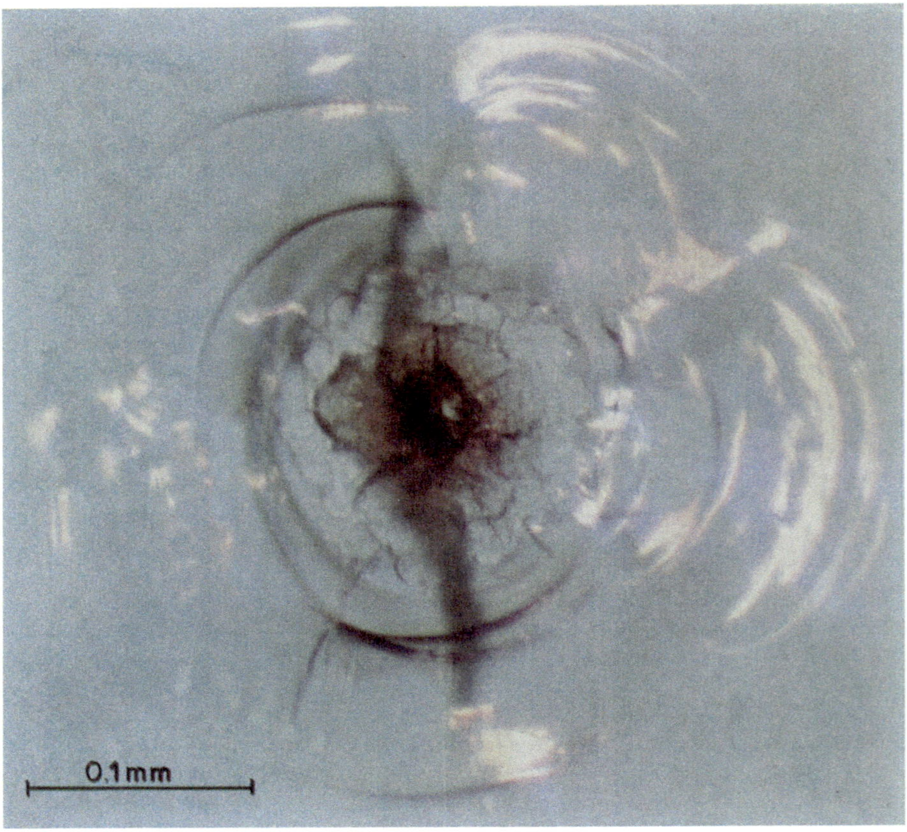


Fig. 2. Optical micrograph of a permanent pumping induced imperfection in a barium crown glass: $Nd^{3+}$ laser rod. The central textured region of diameter $\approx 0.1$ mm has the characteristic appearance of a crystalline phase

lowed by 0.4 μm alumina on a lapping wheel. This procedure produced suitable specimens for optical metallography. Subsequent to the optical metallography examination, the central region of the specimens were thinned (to < 2000 Å) by chemical dissolution with a flowing stream of aqueous hydrofluoric acid solution. A flowing stream was employed to dissipate the heat of chemical solution and thereby avoid specimen heating. Those regions of the thinned specimen which were less than 2000 Å thick were examined by transmission electron microscopy and diffraction in a Siemens Elmiskop operated at 100 kV and at beam currents less than 10 μA.

## 3. Experimental Results

An optical micrograph of an imperfection visible to the unaided eye is shown in Fig. 2. The significant features are a) the nearly circular microcrack lying

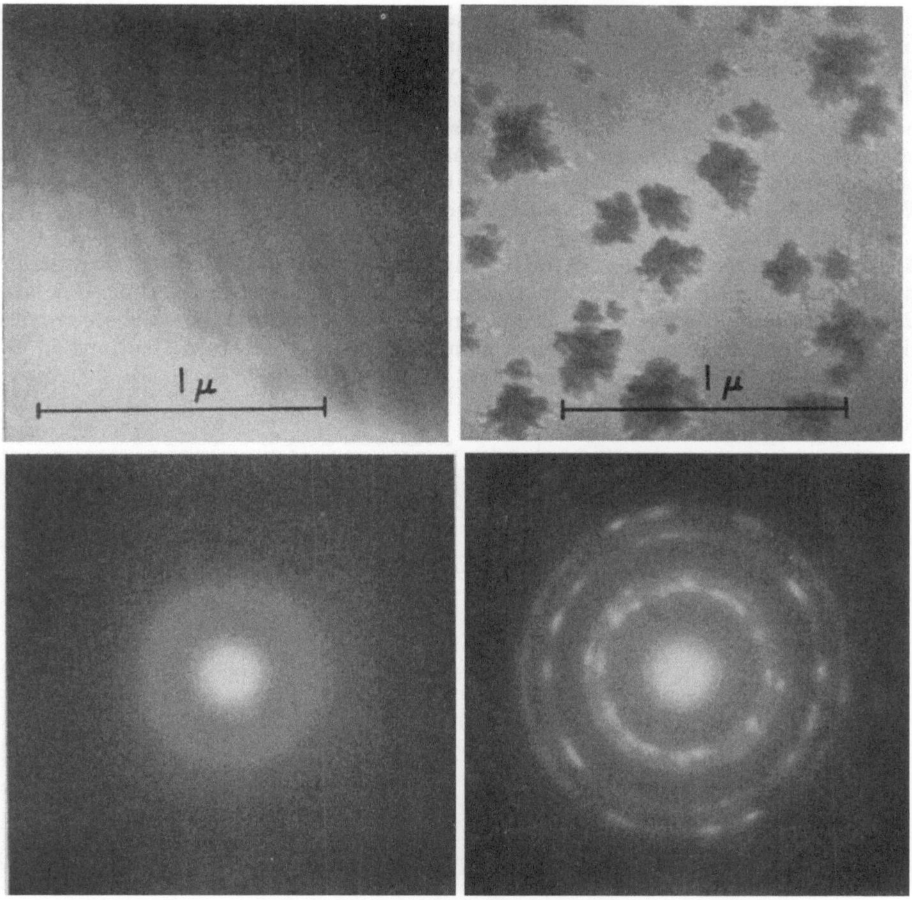


Fig. 3. a) Transmission electron micrograph typical of noncrystalline glass: $Nd^{3+}$ laser rod,
b) Transmission electron diffraction pattern of a) which is characteristic of a noncrystalline solid,
c) Transmission electron micrograph from a region near a) showing the presence of crystallites,
d) Transmission electron diffraction pattern of c) which verifies the presence of crystalline phases

nearly parallel to the plane of the figure, b) the shadow of a microcrack, which is out of focus, lying nearly perpendicular to the figure plane, i.e. parallel to the rod axis, and c) the texture of the central region. The textured region near the center of the imperfection bears a striking resemblance to that of a crystalline phase in a noncrystalline solid [2]. The optical micrograph of Fig. 2 gives conclusive evidence that the pumping induced imperfections are microcracks and suggests that the central region of the imperfection is a crystalline phase.

Transmission electron micrographs and associated diffraction patterns of two different regions of the specimen (Figure 2) are shown in Fig. 3. The bright field image shown in Fig. 3a and the corresponding diffraction pattern (Fig. 3b) are those which are most frequently observed and are typical of a noncrystalline solid. The bright field image shown in Fig. 3c and the corresponding diffraction pattern in Fig. 3d were infrequently observed in isolated regions of the specimen. The diffraction pattern of Fig. 3d is typical of crystallites.

Microdensitometer traces (Fig. 4 and 5) were obtained from the plates of Fig. 3b and d in order to make a more accurate analysis. The radius of the diffuse diffraction ring (Fig. 4) corresponds to a "most frequent" interatomic spacing of 3.52 Å. This is approximately the spacing between two silicon atoms [3] in an uncoordinated aggregate of tetrahedra having a silicon atom at the center of each tetrahedron and oxygen atoms at the apices [4]. The interplanar spacings determined from Fig. 5 are consistent [5] with the interpretation that the crystallites shown in Fig. 3c are a mixture of the three crystalline phases $BaO \cdot 2\,SiO_2$, $SiO_2$ (tridymite), and $3\,BaO \cdot 3\,B_2O_3 \cdot 2\,SiO_2$.

The transmission electron microscopy and electron diffraction results of Fig. 3, 4, and 5 provide conclusive evidence that small (200 to 2000 Å) crystallites are present in isolated regions of the laser rod. Crystallites of dimensions less than 200 Å were also observed in unpumped rods. The possible origin of these crystallites and the mechanism of development of macroscopic imperfections in the laser rod during pumping are discussed below.

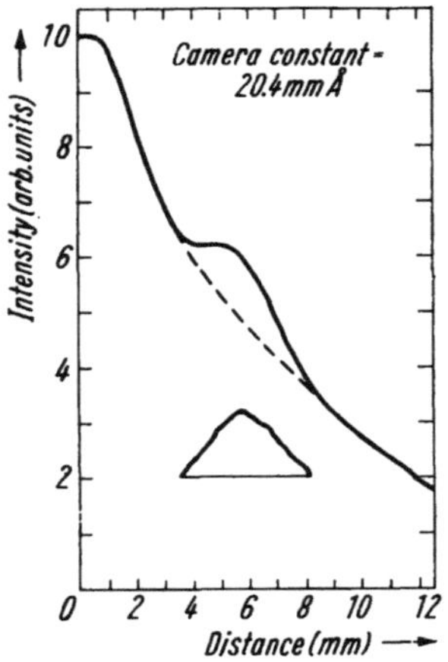


Fig. 4. Microdensitometer trace of Fig. 3b; inset is of peak profile with background subtracted

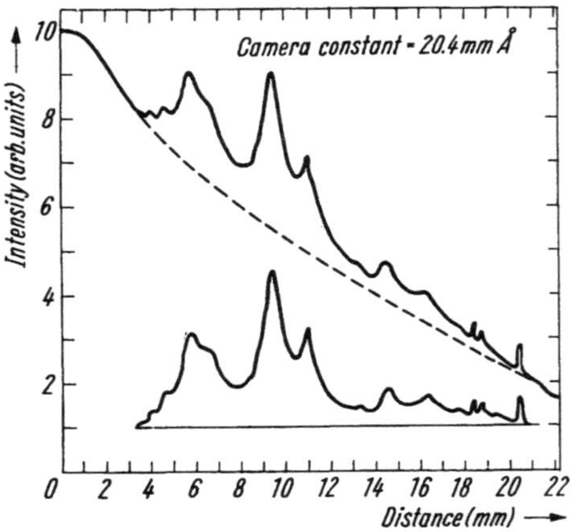


Fig. 5. Microdensitometer trace of Fig. 3d; inset is of peak profile with background subtracted

## 4. Discussion

The discussion can be conveniently divided into five topics which are pertinent to an understanding of material stability of glass: $Nd^{3+}$ lasers:

1. The crystalline structures present when phase equilibrium is achieved;
2. The kinetics of devitrification with respect to time and temperature;
3. Application of the theory of heat transfer in solids to estimate the radial temperature distribution in glass: $Nd^{3+}$ rods subjected to internal heat generation;
4. An evaluation of the possible internal stresses resulting from differential thermal expansion;
5. The mechanism for pumping induced imperfections which is consistent with 1 through 4.

### *4.1 Equilibrium phases in barium crown glass: $Nd^{3+}$ lasers*

The true representation of the equilibrium phases present in the glass: $Nd^{3+}$ laser investigated here requires a knowledge of the phase diagram having the six components: $SiO_2$, $B_2O_3$, $Al_2O_3$, $As_2O_3$, BaO, and $Nd_2O_3$. An experimental determination of this phase diagram is unavailable. Therefore, a phase diagram which is an applicable approximation must be sought. One such approximation is to be found in the ternary phase diagram $SiO_2 \cdot B_2O_3 \cdot BaO$. The following arguments may be invoked to support this choice. Since $Al_2O_3$ and $As_2O_3$ are strong networks formers [6] like $SiO_2$, the $SiO_2$ content of the glass can be taken as the sum of $SiO_2$, $Al_2O_3$, and $As_2O_3$. $Nd_2O_3$ is expected to behave as a network modifier in the same way as BaO [6]. Hence the $Nd_2O_3$ content of the glass can be added to that of the BaO. From this point of view, then, the composition of the glass can be considered to lie within the dashed triangle shown in Fig. 6. The exact location within the dashed triangle is not critical with respect to the equilibrium phases which form as described below.

For compositions within the tridymite field of the dashed triangle, the tridymite crystalline phase will appear first when the liquid is cooled at a rate conducive to the establishment of equilibrium. Tridymite will continue to crystallize out until the composition of the liquid corresponds to that lying on the $BaO \cdot 2\,SiO_2$ — tridymite two phase boundary. Upon further cooling, the two phases $BaO \cdot 2\,SiO_2$ and tridymite will crystallize from the liquid simultaneously while the composition of the liquid will change continuously along the two phase boundary line ($BaO \cdot 2\,SiO_2$-tridymite) until the composition of the liquid corresponds to that of the ternary eutectic $BaO \cdot 2\,SiO_2$, tridymite, and $3\,BaO \cdot 3\,B_2O_3 \cdot 2\,SiO_2$ which is at point "E" in Fig. 6. At point "E" the three phases $BaO \cdot 2\,SiO_2$, tridymite, and $3\,BaO \cdot 3\,B_2O_3 \cdot 2\,SiO_2$ crystallize simultaneously while the temperature of the liquid remains at $950 \pm 20$ °C. On the other hand, if the glass composition within the dashed triangle lies in the $BaO \cdot 2\,SiO_2$ primary field, then $BaO \cdot 2\,SiO_2$ will be the first phase to crystallize from the liquid instead of tridymite as in the above cases. The subsequent crystallizing phases will be tridymite and finally $3\,BaO \cdot 3\,B_2O_3 \cdot 2\,SiO_2$. In summary, the three crystalline phases which will be present at equilibrium for any glass composition within the dashed triangle of Fig. 6 are tridymite, $BaO \cdot 2\,SiO_2$, and $3\,BaO \cdot 3\,B_2O_3 \cdot 2\,SiO_2$.

The interplanar spacings corresponding to the electron diffraction peaks of Fig. 5 are consistent with the interpretation that the crystallites seen in Fig. 3c

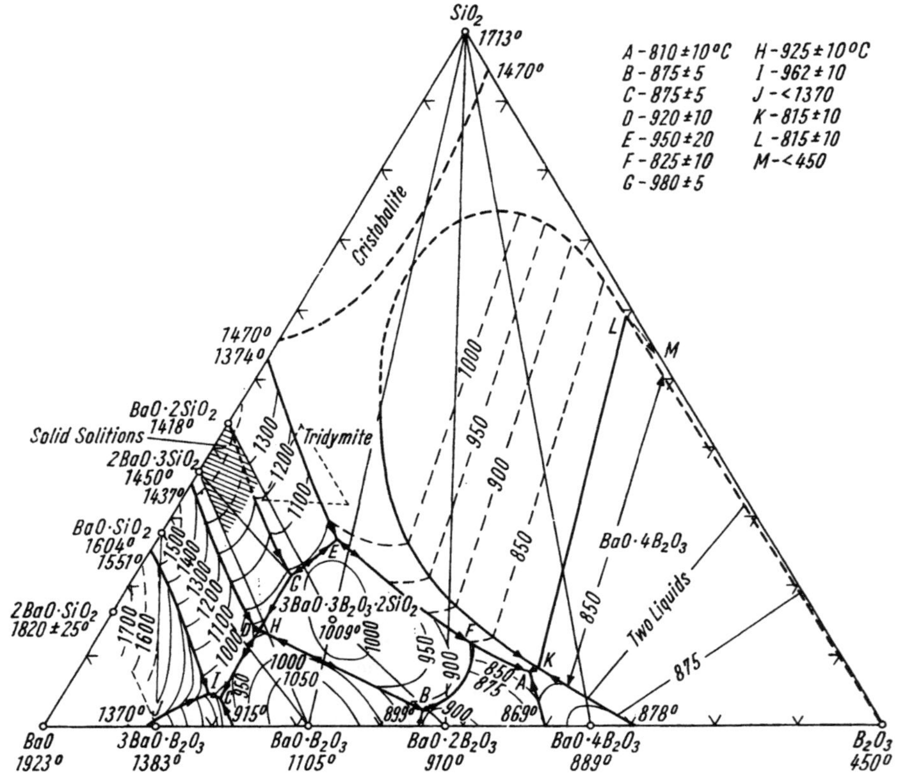


1068/211/6

Fig. 6. The $SiO_2 \cdot B_2O_3 \cdot BaO$ ternary phase diagram taken as being representative of the glass: $Nd^{3+}$ laser composition. The dashed triangle bounds the glass: $Nd^{3+}$ laser composition. From E. M. Levin and G. Ugrinic, Phase Diagrams for Ceramists, The Amer. Ceram. Soc., Columbus, Ohio, 1964 (p. 196)

are a mixture of the three equilibrium phases tridymite, $BaO \cdot 2\,SiO_2$, and $3\,BaO \cdot 3\,B_2O_3 \cdot 2\,SiO_2$. The identification of these phases is considered to provide strong evidence that devitrification occurs in glass: $Nd^{3+}$ lasers, since the phases are the equilibrium ones predicted from the phase diagram for glass of the stated composition.

### *4.2 Kinetics of devitrification*

Elementary considerations of the theory of devitrification [7] can lead to a rough estimate of the phase stability of noncrystalline materials subjected to above ambient temperatures for varying time intervals. For brevity, the presence of a stable crystallite nucleus in a noncrystalline solid is assumed. This assumption is reasonable, since the relatively low activation energy process of heterogeneous nucleation will prevail because of the unavoidable presence of impurities in commercial glass preparations. Further, as the following analysis will show, this assumption has an insignificant effect on the results.

The growth of a crystallite in a noncrystalline solid will occur by an interfacial process, i.e. constituents of the crystalline phase will be required to diffuse to the interface between the crystalline and noncrystalline phases. The linear growth rate of the crystallite is given by

$$\begin{aligned} G(T) &= (\lambda) \times (\text{frequency of species striking the interface}) \times \\ &\quad \times (\text{fraction of species incorporated into the crystallite}) \times \\ &\quad \times (\text{number of successful tries minus the number of unsuccessful tries}) \\ &= \lambda\, f\, \alpha\, (n_s - n_u), \end{aligned} \tag{1}$$

where

$T$ is the temperature in °K (the growth rate being temperature dependent as shown below),

$\lambda$ is the species diameter,

$f$ the frequency of collision of a species with area $\lambda^2$ which is given by

$$f = \frac{3\, D_0 \exp(-Q/k\, T)}{\lambda^2}, \tag{2}$$

where

$D_0$ is the (constant) diffusion coefficient of species having diameter $\lambda$,

$Q$ is the activation energy for diffusion of that species, and

$k$ is Boltzmann's constant.

$\alpha$, the accommodation coefficient, is the probability that the species is incorporated into the crystal upon striking the interface. Here, we set $\alpha$ equal to unity.

The thermodynamic driving force is directly proportional to the net number of successful tries, i.e.

$$\frac{\Delta F}{R\, T} = \frac{\Delta H\, (T_0 - T)}{R\, T_0\, T}, \tag{3}$$

where

$\Delta F = \Delta H\, (T_0 - T)/T_0$ is the free energy change between the noncrystalline and crystalline state,

$\Delta H$ is the heat associated with such a change,

$R$ is the gas constant, and

$T_0$ is an initial temperature taken as being the liquidus temperature corresponding to a given glass composition of the applicable phase diagram.

Upon substitution of equations (2) and (3) into (1), one obtains

$$G(T) = \frac{3\, D_0}{\lambda}\, \frac{\Delta H\, (T_0 - T)}{R\, T_0 T} \exp(-Q/k\, T), \tag{4}$$

where $\alpha$ is set equal to unity. Note that the $(T_0 - T)$ term dominates for $T$ near $T_0$ and the term $\exp(-Q/k\, T)$ dominates at lower temperatures.

In order to obtain an evaluation of equation (4) for devitrification in the glass composition corresponding to the glass: $Nd^{3+}$ laser investigated here, we require the values of $\Delta H$ for the transition of the noncrystalline to the crystalline phase, and the values of $D_0$ and $Q$ for the diffusion of the rate determining species. Unfortunately, these values are presently unavailable. However, a reasonable estimate of $G(T)$ can be obtained if one makes the following approximations. The $\Delta H$ of the transition $SiO_2$ (tridymite) to $SiO_2$ (glass) is 2.3 kcal/mol [8]. If one now assumes that the $\Delta H$ of the transition $BaO \cdot 2\,SiO_2$ (glass) to $BaO \cdot 2\,SiO_2$ (crystalline) is greater than the similar transition for $SiO_2$ by the ratio of the heats of formation of $BaO \cdot 2\,SiO_2$ to $SiO_2$, then one obtains a $\Delta H$ for devitrification of $BaO \cdot 2\,SiO_2$ of about 4 kcal/mol. Further, one may assume that the

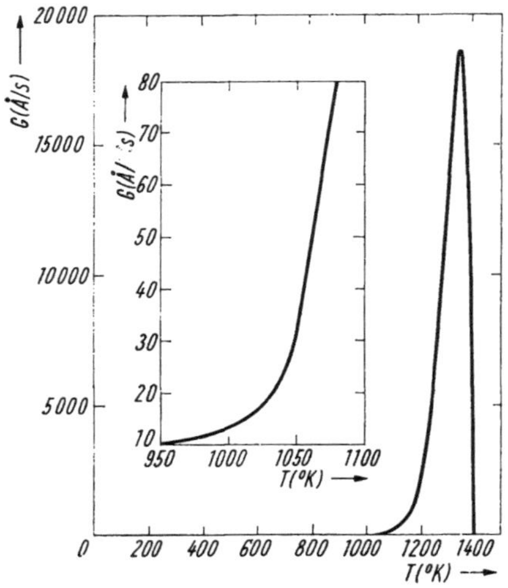


Fig. 7. The approximate linear growth rate (Å/s) of a crystallite in barium crown glass laser rod as a function of temperature (°K)

diffusivity of the largest ion, $Ba^{2+}$, is rate controlling. A review of the published diffusivity data [9] suggests that the diffusivity of $Ba^{2+}$ through $BaO \cdot SiO_2$ (crystalline) will provide the closest approximation to the expected diffusivity behavior of $Ba^{2+}$ in the laser glass composition. Using a value of 4 kcal/mol for $\Delta H$, $\lambda$ corresponding to the diameter of $Ba^{2+}$ and the diffusivity data for the diffusion of $Ba^{2+}$ through $BaO \cdot SiO_2$ (4), one obtains the growth rate shown in Fig. 7. The values of $G(T)$ shown in Fig. 7 are probably only good to within an order of magnitude due to the uncertainty of the approximations.

The low temperature region of $G(T)$ is of importance to devitrification since it gives an indication of the relative phase stability of a noncrystalline solid. The rate of transformation to the crystalline form of a material which is relatively stable in the noncrystalline state will become vanishingly small within a temperature interval of about 150 °C below the liquidus temperature [10]. Consider for example the crystallite growth rate in noncrystalline quartz ($SiO_2$) at 150 °C below the liquidus temperature. One can make the approximation that the diffusion behavior of $Si^{4+}$ through $SiO_2$ (glass) is similar to the experimentally determined [11] diffusivity of $Si^{4+}$ through the noncrystalline solid $40\,CaO \cdot 20\,Al_2O_3 \cdot \cdot 40\,SiO_2$. The pertinent values used in equation (4) are: $D_0 \approx 2 \times 10^{-6}$, $Q \approx 4.1$ eV, $\Delta H \approx 2.3$ kcal/mol, and $\lambda \approx 8.2 \times 10^{-9}$ cm. Upon inserting these values into (4), one obtains a growth rate for a crystallite at 150 °C below the liquidus temperature of about 0.1 Å/s, which is clearly insignificant. Hence, $SiO_2$ is far more stable in the noncrystalline state than the present glass:$Nd^{3+}$ since the crystallite growth rate in the latter at 380 °C below the liquidus temperature is still about 10 Å/s (Fig. 7).

It is instructive to examine whether one can obtain a noncrystalline solid free from any internal crystallites when the glass composition is such as to exhibit a growth rate behavior represented in Fig. 7. A realizable cooling rate between 1400 and 1000 °K by quenching could be about 500 °K/s. If one takes 9000 Å/s as an average growth rate for the above temperature interval, and a cooling rate of 500 °K/s, then one would expect crystallites having linear dimensions of about (300 °K/500 °K $s^{-1}$) (900 Å $s^{-1}$) = 5400 Å to be present in the glass if the time for crystallite nucleation is short compared to the time required to cool through the temperature interval. Crystallite nucleation times are likely to be of the order of seconds, about the same length of time as is probably required to cool through the high growth rate region during commercial glass preparation. Therefore, the presence of crystallites having dimensions of those observed in Fig. 3c could be expected. On this basis, therefore, the following discussion deals with the possibility of inducing temperatures — by pumping the glass:$Nd^{3+}$ laser rod — which are sufficient to cause growth of the already present crystallites.

### *4.3 Internal temperatures produced by pumping*

Internal heat generation can occur within the laser rod during pumping by

1. Hypersonic wave generation through electrooptical coupling with acoustic vibrational quanta, i.e. photon-phonon coupling [12];
2. Nonradiative transitions within the active ions [13].

Regardless of the specific mechanism of internal heat generation, it is an undisputed experimental observation that the internal temperature of the rod markedly increases during laser operation [14]. Further, the internal heat generation is undoubtedly nonuniform throughout the rod because of focusing of radiant energy by the cavity and because mechanism 1 above is far more efficient at local structural imperfections in the rod. However, in order that the calculations of the radial temperature distribution in the rod may be more tractable, the approximations are made that the internal heat generation is uniform throughout the rod and, in addition, that the time variation of heat generation resulting from Q-switching can be averaged. The first approximation will yield minimum temperatures, whereas the second approximation becomes progressively better with time, since the time required for internal temperature changes is large compared to the duration of the radiant energy pump pulse, i.e. steady state can be achieved.

The problem of radial temperature distribution in a right circular cylinder with internal heat generation has been solved [15]. Equation (5) gives the expression for the temperature $T$ (°C) under the following boundary conditions: zero initial temperature, heat generation $A_0$, constant, for time $t > 0$ and radiation at radius $r = a$ into a medium at zero.

$$T = \frac{A_0 a^2}{4K}\left(1 - \frac{r^2}{a^2} + \frac{2}{a h}\right) - \frac{2 h A_0}{a K} \sum_{n=1}^{\infty} \frac{\exp(-\gamma \alpha_n^2 t) J_0(r \alpha_n)}{\alpha_n^2 (h^2 + \alpha_n^2) J_0(a \alpha_n)}, \tag{5}$$

where

$A_0$ internal heat generation in $\frac{\text{cal}}{\text{cm}^3\,\text{s}}$, $a$ rod radius in cm, $K$ rod thermal conductivity in $\frac{\text{cal}}{\text{cm s °C}}$, $r$ rod radial distance in cm, $h = \frac{H}{K}$ in $\frac{1}{\text{cm}}$, $H$ surface conductance in $\frac{\text{cal}}{\text{cm}^2\,\text{s °C}}$, $\gamma$ thermal diffusivity in $\frac{\text{cm}^2}{\text{s}}$, $t$ time in s.

$\alpha_n$ are related to $\beta_n$ (the roots [16] of $\beta J_1(\beta) - n J_0(\beta) = 0$) by $a \alpha_n = \beta_n$, and $J_0$, $J_1$ are Bessel functions of order zero and one, respectively. In the situation of interest here, the laser rod was pumped at an electrical energy input of 65 J with a pulse width $\approx 100$ µs and at a repetition rate of 10 Hz. The flash tube output was a resonable facsimile of a sine wave and a 30% efficiency of conversion to radiant energy could be assumed. Hence, using these values and an averaged sine wave, one obtains a value of 26 cal/cm³ s for $A_0$. The value of $H$ [17] can be computed from

$$H \approx 8 \times 10^{-3} \left(\frac{u}{a}\right)^{1/2}, \tag{6}$$

where $u$ is the coolant velocity normal to the rod axis. The other pertinent values for the terms in equation (5) were given previously. The value of $T$ corresponding to $r = 0$ and $t = 10$ s, for the conditions to which the investigated rod was subjected, is found from (5) to be $\approx 1130$ °K (860 °C). The first three terms of

(5) correspond to the steady state. When $t = 10$ s, these three terms amount to $\approx 850$ °C while the last term (time dependent) amounts to only $\approx 7$ °C. Therefore, the steady state is achieved in a relatively short time for the geometry and conditions employed.

The above analysis suggests that temperatures near the axis of the laser rod can reach $\approx 1100$ °K. The crystallite growth rate corresponding to such a temperature is, from Fig. 7, $\approx 100$ Å/s, which certainly would be significant. Therefore, crystallites which are present from cooling through the maximum growth rate range of Fig. 7 — during glass preparation — could continue to grow during laser pumping.

#### *4.4 Differential thermal expansion stresses*

Crystallites embedded in a glass can exert local tensile stresses when the linear coefficient of thermal expansion is greater than that of the enclosing (noncrystalline) glass. The following analysis shows that for crystallites with thermal expansion coefficients about twice those of the glass, and for a temperature rise of about 200 °C or more over the ambient, internal tensile stresses in excess of the tensile strength of the glass can be produced at the crystallites. The stress $\sigma$ resulting from differential thermal expansion is given by

$$\sigma = \varepsilon E , \tag{7}$$

where

strain $\varepsilon = (\alpha_1 - \alpha_2) T$ ,

$\alpha_1, \alpha_2$ linear coefficients of expansion of phases 1 and 2, respectively,

$T$ temperature, and

$E$ elastic modulus.

As a numerical example, consider the presence of a crystallite of $SiO_2$ within the glass:$Nd^{3+}$ rod. The pertinent values are

$\alpha \approx 18 \times 10^{-6}$/°C for $SiO_2 \perp$ to the $c$-axis [18] and $T$ in the range 0 to 200 °C,

$\alpha \approx 10 \times 10^{-6}$/°C for glass:$Nd^{3+}$ [19] and $T$ in the range 0 to 200 °C,

$E \approx 7 \times 10^3$ kg/mm² for glass:$Nd^{3+}$ [20],

$\sigma_0 \approx 7$ kg/mm², the tensile strength for glass:$Nd^{3+}$ [20].

Upon insertion of these values into equation (1), one obtains a tensile stress of $\approx 11$ kg/mm² which clearly exceeds the tensile strength of the glass. The $SiO_2$ crystallite is, of course, under a compression stress of 11 kg/mm², but the compression strength of $SiO_2$ far exceeds the tensile strength of the glass. Therefore, a microcrack will develop in the glass adjacent to the crystallite similar to that observed in Fig. 2.

The thermal expansion behavior of $SiO_2$ (tridymite), $BaO \cdot 2\,SiO_2$, and $3\,BaO \cdot 3\,B_2O_3 \cdot 2\,SiO_2$ are presently unknown. In fact, the crystal structures of the latter two phases have yet to be determined. However, one can presume that the thermal coefficients of expansion of the crystalline phases are greater than those of the corresponding noncrystalline phases since $\alpha$ of crystalline $SiO_2$ is about 30 times that of noncrystalline $SiO_2$ [18]. Furthermore, internal stresses due to differential thermal expansion vanish only when the thermal expansion coefficients of the two phases in physical contact are identical over the temperature range of interest. The latter condition is rarely realized and is most unlikely to occur when crystallites of $SiO_2$ (tridymite), $BaO \cdot 2\,SiO_2$, and $3\,BaO \cdot 3\,B_2O_3 \cdot 2\,SiO_2$ are present in glass:$Nd^{3+}$ laser rods.

#### *4.5 Mechanism of pumping induced imperfections*

When the molten glass is cooled through the region of maximum crystal growth rate (Fig. 7) at cooling rates commensurate with those of commercial glass preparation methods, a mixture of crystallites of $SiO_2$ (tridymite), $BaO \cdot 2\,SiO_2$, and $3\,BaO \cdot 3\,B_2O_3 \cdot 2\,SiO_2$ nucleate and grow to dimensions of about 200 Å. Crystallites of these dimensions and smaller are observed in glass:$Nd^{3+}$ laser rods which have *not* been pumped, which indicates that these crystallites have nucleated and grown during cooling of the molten glass. The analysis of radial thermal conduction resulting from internal heat generation during laser pumping indicates that temperatures can be achieved near the rod center which correspond to a significant crystallite growth rate. The crystallites which are already present increase their dimensions and possibly new crystallites nucleate and grow during laser pumping. The difference between the thermal coefficients of expansion between the crystallites and the glass produces tensile stresses in the surrounding glass which exceed the glass tensile strength and microcracks result. When the linear dimensions of these microcracks are comparable to or larger than the wavelength of the $Nd^{3+}$ emission line (1.06 μm), they cause internal reflections and thereby seriously limit laser efficiency. Clearly, when the linear dimensions of the crack are comparable to the rod radius, then the rod fractures, since glass is a brittle solid and extremely sensitive to structural imperfections.

## 5. Conclusions

Permanent internal imperfections in pumped barium crown glass:$Nd^{3+}$ lasers are microcracks resulting from differential thermal expansion between crystallites — representing the equilibrium state of the glass — and the surrounding noncrystalline material. These crystallites are present in the glass rod prior to laser pumping and grow as a result of temperatures achieved by internal heat generation during laser pumping. The use of a glass composition which is more stable in the noncrystalline state than barium crown glass for an active ion host would most likely result in the achievement of higher laser pumping levels and consequently higher laser output without producing permanent internal imperfections.

#### *Acknowledgement*

Mr. I. D. Blucher offered valuable assistance in the electron and optical microscopy and Dr. H. M. Otte contributed helpful comments on the manuscript.

## References

[1] P. Görlich et al., phys. stat. sol. **6**, 277 (1964).
[2] W. D. Kingery, Introduction to Ceramics, Wiley, New York 1960 (p. 313).
[3] R. W. G. Wyckoff, Crystal Structures, Vol. I, Interscience Publishers, New York 1951.
[4] W. D. Kingery, l. c. (p. 120).
[5] Powder Diffraction File, Amer. Soc. for Testing and Materials, Philadelphia, Pa., 1964, cards 14—260, 10—45, 46, and 6—0351.
[6] L. H. van Vlack, Physical Ceramics for Engineers, Addison-Wesley, Reading, Mass., 1964 (p. 62).
[7] W. D. Kingery, l. c. (p. 302).
[8] Handbook Chemistry and Physics, 38th ed., Chem. Rubber Publ., Cleveland, Ohio, 1956 (p. 1759).
[9] Amer. Inst. Physics Handbook, 2nd ed., McGraw-Hill Publ. Comp., New York 1963 (p. 9—82).

[10] H. R. Swift, J. Amer. Ceram. Soc. **30**, 165 (1947).
[11] H. Towers and J. Chipman, Trans. AIME **209**, 709 (1957).
[12] N. M. Kroll, J. appl. Phys. **36, 34** (1965).
[13] O. S. Heavens, Optical Masers, Methuen & Co., London 1964 (p. 57).
[14] L. J. Aplet et al., Appl. Phys. Letters 8, 71 (1966), and private communication with Mr. W. M. Hammond and Dr. R. C. Weidler, Martin Co., Orlando, Fla.
[15] H. S. Carslaw and J. C. Jaeger, Conduction of Heat in Solids, Oxford University Press, London 1959 (p. 205).
[16] Ibidem, Appendix IV.
[17] Ibidem (p. 20).
[18] Amer. Inst. Physics Handbook, loc. cit. (p. 4 — 74).
[19] AOLux Bull., Amer. Optical Co., Southbridge, Mass., 1965.
[20] H. R. Lillie, Handbook of Physics, Ed. E. U. Condon and H. Odishaw, McGraw Hill Publ. Comp., New York 1958 (p. 8 — 99, 100).

*(Received April 13, 1966)*

phys. stat. sol. **16**, 183 (1966)

*Physics Department, University of Athens, Athens, Greece (a) and*
*Physics Department, Syracuse University, Syracuse, New York (b)*

# Effect of Neutron and γ-Ray Irradiation on the Carrier Lifetime and Detectivity of InSb[1])

By

P. C. EUTHYMIOU[2]) (a) and A. L. TANENBAUM (b)

Single crystals of p-type InSb were irradiated by 4.5 MeV neutrons or γ-rays at liquid nitrogen temperature. The damage produced by neutron irradiation resulted in a small decrease of the carrier lifetime and a large decrease of the detectivity of the crystal. The damage produced by γ-rays resulted in an increase or decrease of lifetime, the change depending on how strongly p-type the crystal was. The detectivity was decreased upon γ-irradiation. Annealing experiments show that neutron damage is permanent while γ-ray defects can be completely annihilated near or above room temperature, giving a similar result as electron irradiation. Analysis of the effects on lifetime and detectivity shows that neutron damage effects, although severe and permanent, produce no important decrease of lifetime.

Des cristaux de l'antimoniure d'indium du type n et p ont été irradiés par des neutrons d'une énergie de 4.5 MeV ou par des rayons gamma à la temperature de l'azote liquide. Le dommage causé par l'irradiation avec neutrons avait pour suite une petite diminuation de la durée de vie des porteurs et un grand accroissement de la détectivité. Le dommage causé par les rayons gamma avait pour suite un accroissement ou une diminuation de la durée de vie des porteurs, la variation dépendant du cristal. La détectivité des cristaux était diminuée après l'irradiation par les rayons gamma. Des expériments de recuisson indiquent que le dommage causé par les neutrons est permanent tandis que les défauts causés par les rayons gamma disparaissent complètement à des températures voisines ou plus élevées que la température ordinaire comme ceux produits par des électrons. Une analyse de l'influence sur la durée de vie des porteurs et la détectivité indique que les effets du dommage causé par les neutrons, malgré qu'ils sont sévères et permanents, ne produisent guère une diminuation de la durée de vie des porteurs.

## 1. Introduction

When the equilibrium concentration of electrons and holes in a semiconductor is disturbed by some agency, e.g. addition of excess electron-hole pairs by a pulse of light, a finite time is required for the restoration of equilibrium. The recombination of excess electrons and holes usually takes place at localized electronic states (recombination centers) in the crystal. Imperfections and radiation induced defects make excellent recombination centers. Therefore, the lifetime of excess electron-hole pairs or, more usually, the lifetime of minority carriers $\tau$ is an excellent probe for the investigation of radiation damage.

Measurements of carrier lifetime in Ge and Si showed a decrease of lifetime upon neutron irradiation [1 to 4], γ-irradiation [2, 3, 5] or electron irradiation [6]. The same effect must be expected upon neutron irradiation in InSb. The effect

---

[1]) This work was done in the Physics Department, Syracuse University, Syracuse, N.Y

[2]) Postdoctoral Research Associate at Physics Department, Syracuse University (September 1964 to March 1965) under the auspices of the Conference Board of Associated Research Councils (Fulbright Program).

of γ-irradiation on carrier lifetime in InSb depends on how strongly p-type the crystal is [7, 8]. For sufficiently strongly p-type crystal, carrier lifetime $\tau$ is equal to electron lifetime and according to experiments of Laff and Fan [9] it decreases upon electron irradiation. For less strongly p-type crystal the hole lifetime is predominant and this increases upon electron irradiation.

Another physical quantity of practical interest in infrared detectors is the detectivity. It is defined as $D = \frac{S}{N H A}$, where $S$ represents the ac photoconductive signal voltage coming from the detector when incident radiation flux is modulated, $N$ the noise voltage, $H$ the incident r.m.s. power density of the crystal and $A$ the sensitive crystal area.

In the present work the effects of neutron and γ-ray radiation on the carrier lifetime $\tau$ and on the detectivity $D$ of p-type InSb single crystals were measured.

We have also studied the recovery of neutron and γ-ray damage by observing the effects of annealing on carrier lifetime.

## 2. Experimental Part

Photoconductive measurements of the lifetimes of excited states and of detectivity were made with infrared detectors containing p-type InSb single crystals.

As detector envelope we utilized a permanently sealed dewar of about 9 cm length and 2.5 cm diameter. The essential features of the detector assembly are the following: a) a sapphire window which is sealed to the structure by means of a sapphire to glass seal, b) a liquid nitrogen chamber, c) a copper block which reduces thermal variations and acts as a radiation shield and a support for the cell, d) a sapphire plate which supplies an insulated base for the sensitive element, e) the crystal which is cemented with an adhesive and insulating varnish to the sapphire plate, f) an aperture plate which defines the desired angular field of view and g) a getter which assures a permanent vacuum.

### *2.1 Lifetime equipment*

An indium arsenide diode was used as a luminous source. InAs with a spectral emission peak at about 3 μm is ideally suited since its emission spectrum lies within the sensitivity range of an InSb photodetector. The time constant equipment is composed of four parts: a) the exciter, b) the sample box containing the luminous source, the InSb photodetector and its bias supply, c) a preamplifier and d) an oscilloscope-camera arrangement as a final readout.

The exciter was a square wave generator (Tektronix type 105) which was connected by coaxial cable to the InAs diode. The InSb detector was connected in series with a bias supply and a load resistor $R_L$. The $R_L$ terminals (across which we measured the signal) were connected to a low capacity cathode follower. The signal was then applied to a Tektronix type 121 wide band preamplifier connected to an Tektronix type 545 oscilloscope with a camera.

Pictures were taken in a single sweep. On these we measured the time constant $\tau$, i.e. the time required by the signal to rise to 0.63 of its final height or to decay to 0.37 of its initial height. Using a circuitry of time constant $T \ll \tau$ we found a fairly faithful reproduction of the waveform. We tried to keep $T$ small by using a low capacitance probe and short leads and by reducing the load resistance as far as consistent with an acceptable signal to noise level.

### *2.2 Detectivity equipment*

Sinusoidally modulated infrared radiation from a black body at 500 °K was incident on the detector. The voltage variations across $R_L$ were fed through a Model 122 Tektronix preamplifier to a Hewlett Packard Model 300 Harmonic Analyzer. The modulation frequency was 900 Hz and the band width of the harmonic analyzer 5 Hz. For reasons of comparison with other measurements the detectivity $D$ was normalized to a detector area $A$ of 1 cm² and to an amplifier bandwidth $\Delta f$ of 1 Hz. Under these conditions we obtain the normalized detectivity

$$D^* = \frac{S\sqrt{\Delta f}}{N\,H\sqrt{A}}\,.$$

Modulation of the energy flux was obtained by a chopper assembly mounted directly in front of the blackbody. It consisted of an 1800 r.p.m. synchronous motor attached to a wheel containing 30 radial slits. This produced a modulation frequency of 900 Hz. Sinusoidal modulation was obtained by making the chopper opening 1.15 times the diameter of the blackbody aperture. The r.m.s. power density at the detector, when it was 100 cm away from the blackbody at 500 °K, was $10^{-6}$ W/cm². The gain of the arrangement was frequently checked by connecting an oscillator and a microvoltmeter to the input of the preamplifier.

### *2.3 Radiation and annealing experiments*

The irradiation with neutrons and γ-rays and all measurements were performed with the detector at liquid nitrogen temperature. An iron-constantan thermocouple measured the detector temperature.

Two InSb detectors (SIS-30 and BK-259) were irradiated with neutrons and two (BK-252 and BK-35) with γ-rays. The neutrons were generated from a cylindrical Pu-Be source having a nominal activity of 1 C. The average energy of the neutrons is about 4.5 MeV and the integrated flux over space and energy was $1.95 \times 10^6$ neutrons/s. The detector was placed 1 cm above the base of the cylinder. The slow neutrons were intercepted with Cd and In foils. This was necessary in order to prevent the transmutation of In in the sample, as its capture cross section for thermal and resonance neutrons is extremely large and the decay product Sn would act as a donor impurity.

Because of the small dimensions of the liquid nitrogen chamber of the detector we utilized a percolator system for the transport of liquid nitrogen from a big dewar placed at an appropriate distance from the source. Lifetime measurements during the irradiation period were made by removing the detector from the source and placing it in the time constant equipment. Measurements were taken also after annealing of the detector SIS-30 up to 378 °K and during isochronal annealing of the detector BK-259 up to 348 °K.

The $Co^{60}$ γ-ray source had a nominal activity of 1000 C. One detector (BK-252) was irradiated for 5.6 h placed at the nearest possible position, for which the dose rate of γ-rays was 24780 R/h. The detector was kept at liquid nitrogen temperature with the same percolator system as that used during the neutron irradiation. After recovery by isochronal annealing steps up to 348 °K, it was irradiated again for 41.5 h at a position where the dose rate was 16800 R/h. During this irradiation the detector was kept filled with liquid nitrogen by connecting it to the bottom of a 2 l dewar with a small glass tube and a silicon rubber seal. The 2 l dewar had to be replenished every 5 h. We used the same apparatus for the irradiation of the second detector (BK-35) for 62.3 h. This detector was placed at a position where the γ-ray dose rate was 21900 R/h. The annealing of lifetime $\tau$ was followed up to 353 °K.

The annealing time chosen for each isochronal annealing step was half an hour for the detectors irradiated with neutrons and one hour for the detectors irradiated with γ-rays.

Measurements of detectivity were taken each time before irradiation, after irradiation, and after annealing to the highest temperature.

## 3. Results and Discussions

### 3.1 Neutron irradiation

#### 3.1.1 Lifetime

No experiments have been reported on the effect of neutron irradiation on carrier lifetime for InSb. All the existing data on Ge and Si show a decrease of the carrier lifetime in both n- and p-type (the originally n-type went through intrinsic and became p-type) according to the relation [4] $\frac{1}{\tau} = \frac{1}{\tau_0} + K\,\Phi$ where $\tau_0$ initial lifetime, $K$ constant depending on type and energy of the bombarding particle, nature of semiconductors, and temperature and density of excess carriers, and $\Phi$ time integrated particle flux.

In the present measurements on the SIS-30 detector we found also a decrease in lifetime (about 17%) after neutron irradiation which reached a saturation value after the first hours of irradiation (Fig. 1).

Annealing of the detector to 328 °K had no effect whatever on the lifetime.

Irradiation of the BK-259 detector gave erratic results (Fig. 2), mainly in the values obtained in rise-time measurements, perhaps due to the fact that the initial lifetime was much smaller in this detector than in the SIS-30 detector. If we restrict ourselves to lifetimes from decay measurements (indicated by triangles), we observe a small decrease of the lifetime (about 6%) in the first 10 h of irradiation, with a later increase to the initial value.

Isochronal annealing experiments of this detector were equally confusing (Fig. 3).

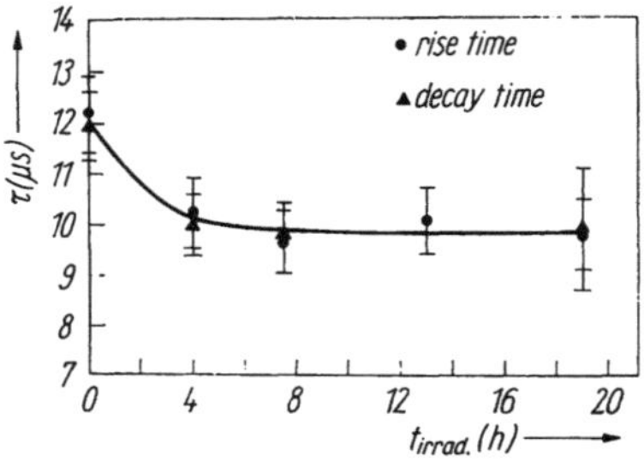


Fig. 1. Effect of neutron irradiation on the carrier lifetime. Detector SIS-30, $T$ = 78 °K, Dose rate: $1.55 \times 10^5$ n/cm² s

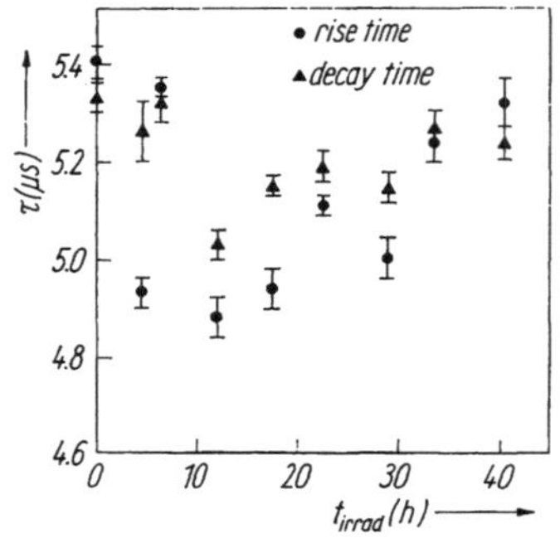


Fig. 2. Effect of neutron irradiation on the carrier lifetime. Detector BK-259, $T$ = 78 °K, Dose rate: $1.55 \times 10^5$ n/cm² s

Fig. 3. Isochronal annealing of the detector BK-259 immediately after neutron irradiation. Dose: $2.23 \times 10^{10}$ n/cm²

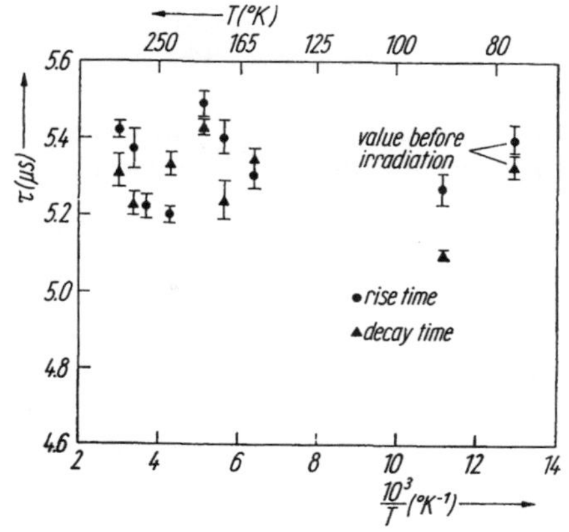


### 3.1.2 *Detectivity*

Kutzsher [10] performed measurements on photoconductive properties of similar IR detectors after neutron irradiation and found no change, using doses of $2.58 \times 10^9$ n/cm². In the present work detectivity experiments were only carried out on BK-259 as the other detector was destroyed. The results are given in Table 1. We notice a very significant decrease of $D^*$ after irradiation and no important change after subsequent annealing. This effect can be attributed to the much stronger dose ($2.23 \times 10^{10}$ n/cm²) which was used in the present experiments.

Comparing the results of lifetimes and detectivities we conclude that *neutron damage* measured on conductivity, Hall effect [11], and detectivity of p-type

Table 1
$D^*$ Measurement

| Measurement | Neutron irradiation | γ-irradiation | | |
|---|---|---|---|---|
| | Detector BK-259<br><br>Dose rate: $1.55 \times 10^5$ n/cm²s<br>Dose: $2.23 \times 10^{10}$ n/cm² | Detector BK-252 (First irradiation)<br>Dose rate: $2.48 \times 10^4$ R/h<br>Dose: $1.39 \times 10^5$ R | Detector BK-252 (Second irradiation)<br>Dose rate: $1.68 \times 10^4$ R/h<br>Dose: $6.97 \times 10^5$ R | Detector BK-35<br><br>Dose rate: $2.19 \times 10^4$ R/h<br>Dose: $1.36 \times 10^6$ R |
| | $D^* \left(\frac{\text{cm c}}{\text{W s}}\right)$ | $D^* \left(\frac{\text{cm c}}{\text{W s}}\right)$ | $D^* \left(\frac{\text{cm c}}{\text{W s}}\right)$ | $D^* \left(\frac{\text{cm c}}{\text{W s}}\right)$ |
| Before irradiation | $8.7 \times 10^9$ | $5.6 \times 10^9$ $\pm 0.5 \times 10^9$ | $5.0 \times 10^9$ $\pm 0.5 \times 10^9$ | $3.0 \times 10^9$ $\pm 0.3 \times 10^9$ |
| After irradiation | $1.5 \times 10^9$ | $5.3 \times 10^9$ $\pm 0.5 \times 10^9$ | $4.3 \times 10^9$ $\pm 0.4 \times 10^9$ | $2.7 \times 10^9$ $\pm 0.3 \times 10^9$ |
| After annealing | $1.3 \times 10^9$ | $5.8 \times 10^9$ $\pm 0.6 \times 10^9$ | $4.5 \times 10^9$ $\pm 0.4 \times 10^9$ | $3.0 \times 10^9$ $\pm 0.3 \times 10^9$ |

InSb, although severe and permanent, *is not accompanied to any important degree by decrease of lifetime.* The detectivity and perhaps some of the other quantities mentioned are known [4] to depend to a high degree on the surface properties, the alteration of which constitutes a permanent damage. It might be that *lifetimes are less dependent on the surface properties.*

### 3.2 γ-irradiation

#### 3.2.1 Lifetime

All the existing data on Ge and Si show that electron and γ-irradiation decreases carrier lifetime in p- or n-type Ge and Si. Similar data for InSb are reported only by Laff and Fan. They found that bombardment of p-type InSb by electrons of 4.5 MeV reduces more than an order of magnitude the electron lifetime and increases less drastically the hole lifetime.

In the present work we find that irradiation of the detector BK-252 with a small dose of γ-rays (dose: $1.39 \times 10^5$ R) produces a small *decrease* of the carrier lifetime (about 5%) reaching a saturation value (Fig. 4), which however does not seem to be the same for the rise time and for the decay time. The scatter in the measured values does not allow us to determine if saturation was reached. The final values can also be seen on Fig. 5 on the right hand side. Saturation for the resistivity, magnetoresistance, and carrier concentration for both n- and p-type of InSb has been found during electron irradiation [12].

The same detector after its complete recovery by annealing up to 348 °K was exposed to a stronger dose ($6.97 \times 10^5$ R) of γ-rays. The decrease of lifetime was larger (about 9%) than the first one since the dose was much higher. The final values can be seen on the right hand side of Fig. 6. Comparison with the small dose results of Fig. 5 shows that the effect of damage on the rise time (points) has not increased thus showing that it had reached saturation during the first irradiation. On the other hand the effect of damage on decay time was found larger.

Irradiation of high dose on the detector BK-35 produced an *increase* in lifetime of about 9.4% both on the rise time and the decay time as can be seen on the right hand side of Fig. 7.

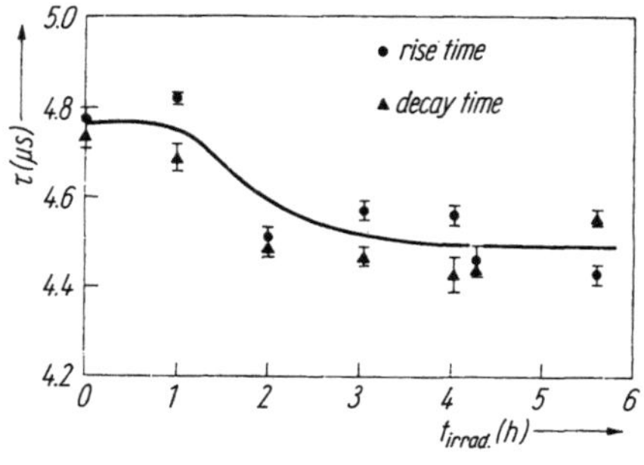


Fig. 4. Effect of γ-ray irradiation on the carrier lifetime. Detector BK-252, $T = 78$ °K, Dose rate: $2.48 \times 10^4$ R/h

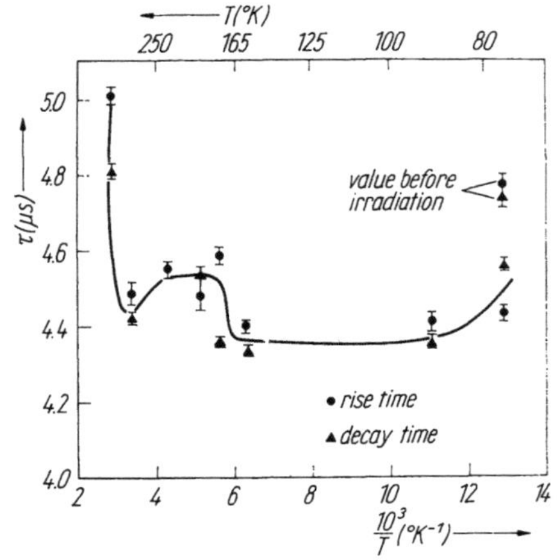


Fig. 5. Isochronal annealing of the detector BK-252 immediately after its first γ-ray irradiation. Dose: $1.39 \times 10^5$ R

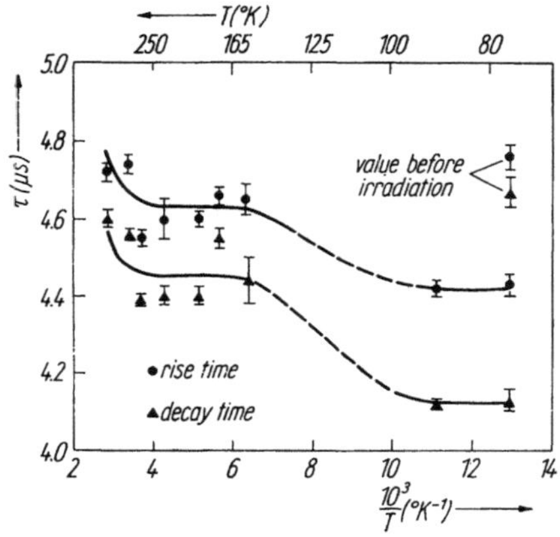


Fig. 6. Isochronal annealing of the detector BK-252 immediately after its second $\gamma$-ray irradiation. Dose: $6.97 \times 10^5$ R

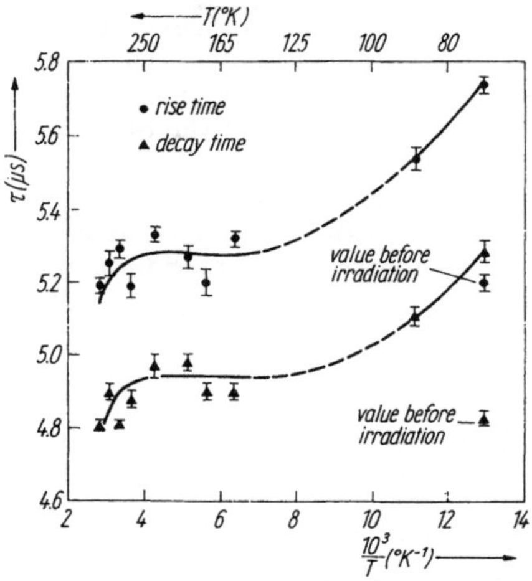


Fig. 7. Isochronal annealing of the detector BK-35 immediately after $\gamma$-ray irradiation. Dose: $1.36 \times 10^6$ R

The fact that carrier lifetime increases in one sample and decreases in the other might be explained if we consider separately the lifetime of electrons and of holes. According to the Hall, Shockley-Read recombination theory the carrier lifetime in a sufficiently strongly p-type material is the electron lifetime which decreases upon irradiation as Laff and Fan found in their experiments. This is probably the case of the detector BK-252, which may be considered as strongly p-type. The detector BK-35 may be less strongly p-type and hence its carrier lifetime is limited by the hole lifetime, which increases upon irradiation.

We will now describe the annealing experiments.

The first isochronal annealing of the detector BK-252 does not show well defined stages (Fig. 5). Between 78 °K and 90 °K there is a decrease (reverse annealing) of lifetime and two other stages of normal annealing one between 150 °K and 190 °K and the other between 296 °K and 348 °K. At 348 °K the lifetime recovers its initial, i.e. pre-irradiation value. These two stages are consistent with the two last stages found by the other investigators [13, 14] on p-type InSb.

After leaving the detector at room temperature for about two weeks we irradiated for a second time with a much stronger dose with the results described in Fig. 6. The annealing curves showed no reverse annealing at low temperatures. However, the two other stages of the first annealing can be fairly well recognized.

Similar annealing experiments were carried out on the detector BK-35, which had been irradiated with a higher dose. The results are given in Fig. 7. They show a full recovery. We notice that the low temperature stage is different for the two detectors. For the detector BK-252 this occurred somewhere between 90 °K and 140 °K and the damage was 60% recovered, while for the detector BK-35 this had started below 90 °K and the damage was 80% recovered. The existence of a low temperature stage below 90 °K has been noticed on annealing experiments on the conductivity, Hall effect [13, 14], and magnetoresistance [12] of p-type InSb. A low temperature stage in the same region has also been found on n-type InSb [15, 16].

The first stage of annealing of the detector BK-252 starts above 90 °K, hence does not coincide with any published data. The present measurements do not allow the detection of an eventual stage around 150 °K as found by all the investigators mentioned above. An indication of it might be assumed in Fig. 5. The stage near room temperature or above it found in all present measurements has also been found by all the other investigators.

It has been usually accepted that γ-rays produce simple defects of vacancy-interstitial type. The low temperature annealing stage includes annihilation of these genetically related vacancy-interstitial pairs. The high temperature stage has been attributed predominantly to the annihilation of interstitials with genetically unrelated vacancies. Both these effects are found to influence carrier lifetime.

### 3.2.2 *Detectivity*

Kutzsher using a total dose of 15450 R from a $Co^{60}$ source found no change of photoconductivity properties of infrared detectors after irradiation. Our results obtained from much stronger doses show for both detectors a decrease of the detectivity. After annealing, the detectivity recovers its initial (pre-irradiated) value, exactly as happened for the carrier lifetime. The exact values are given in Table 1.

***Acknowledgements***

We would like to express our thanks to Professor H. Levinstein for the help he offered us, Professor H. Harth of Syracuse University for the use of neutron source, Professor J. Meyer of State University College of Forestry at Syracuse University for the use of γ-ray source, and the two companies, Texas Instruments and Santa Barbara Research Center, who offered us the detectors. One of us (P. E.) would like to thank Professor W. Fredrickson, Chairman of the Department of Physics, Syracuse University, and others at the same University for the hospitality extended during the visit. She would also like to thank the Conference Board of Associated Research Councils (Fulbright Program) for arranging and supporting her visit.

## References

[1] G. K. WERTHEIM, Phys. Rev. **105**, 1730 (1957).
[2] O. L. CURTIS, J. W. CLELAND, and J. H. CRAWFORD, J. appl. Phys. **29**, 1722 (1958).
[3] G. K. WERTHEIM, Nucleonics **20**, 47 (1962).
[4] J. W. EASLEY, Nucleonics **20**, 51 (1962).
[5] T. NAKANO and Y. INUISHI, J. Phys. Soc. Japan **19**, 851 (1964).
[6] J. J. LOFERSKI and P. RAPPAPORT, Phys. Rev. **111**, 432 (1958).
[7] R. N. HALL, Phys. Rev. **87**, 387 (1952).
[8] W. SHOCKLEY and W. T. READ, Phys. Rev. **87**, 835 (1952).
[9] R. A. LAFF and H. Y. FAN, Phys. Rev. **121**, 53 (1961).
[10] E. W. KUTZSHER, Proc. IRE **47**, 1520 (1959).
[11] J. W. CLELAND and J. H. CRAWFORD JR., Phys. Rev. **93**, 894 (1954).
[12] P. C. EUTHYMIOU, phys. stat. sol. **8**, 131 (1965).
[13] L. W. AUKERMAN, Phys. Rev. **115**, 1125 (1959).
[14] K. D. ALEXOPOULOS and R. B. OSWALD JR., Radiation Damage in Solids, Vol. III, International Atomic Energy Agency, Vienna 1963.
[15] F. H. EISEN, Phys. Rev. **123**, 736 (1961).
[16] C. GEORGOPOULOS, M. VERDONE and D. DAUTREPPE, Academie des Sciences, Seance du 28 Octobre 1963 (p. 2640).

*(Received April 15, 1966)*

phys. stat. sol. 16, 191 (1966)

*A. F. Ioffe Physico-Technical Institute, Academy of Sciences of the USSR, Leningrad*

# Development of Glide Bands During Plastic Deformation of LiF Crystals

By

B. I. Smirnov and B. A. Efimov

The development of the dislocation structure and shear strain in glide bands of LiF crystals under an applied stress is studied. The following parameters are determined: the rate of lateral band growth $\omega$, the change of shear strain $\gamma$ in bands during this growth, and the average slip distance of screw dislocations $\lambda$ during band formation. It is found that the widening of the bands occurs asymmetrically, an increase in the band width resulting in a decrease in $\omega$ and in an increase in $\gamma$ ($\gamma$ eventually reaches a certain limiting value). On the average, the rate of lateral band growth is about an order of magnitude less than the velocity of the individual dislocations. It is found that $\lambda$ has a value of 1.5 mm at 20 °C which decreases with an increase of the magnesium content of the crystal and also decreases with a decrease of the deformation temperature.

Изучалось развитие дислокационной структуры и сдвига в полосах скольжения кристаллов LiF под действием приложенного напряжения. При этом определялись следующие параметры: скорость бокового роста полос $\omega$, изменение относительного сдвига $\gamma$ в полосах при их росте, а также средняя длина пробега винтовых дислокаций $\lambda$ при формировании полос. Оказалось, что расширение полос происходит асимметрично, причём с увеличением ширины полосы скорость $\omega$ уменьшается, а значение $\gamma$ возрастает, достигая некоторого предельного. В среднем скорость бокового роста полос примерно на порядок меньше скорости движения отдельных дислокаций. Установлено, что при 20 °C величина $\lambda$ составляет около 1,5 мм и уменьшается с увеличением содержания магния в кристалле и с понижением температуры деформации.

## 1. Introduction

Variation of the dislocation structure of crystals as a result of their deformation was apparently for the first time observed by Gilman and Johnston [1] in LiF crystals. They showed that the glide process starts, as a rule, with a displacement of dislocation loops already present in the crystal or being formed at points of stress concentration. In their motion these loops exhibit cross-glide and multiplication thus producing new dislocation loops which lie near the original glide plane of the parent dislocation. As a result, a narrow dislocation band appears which will continue to widen under an applied stress. It is assumed that the motion of dislocations takes place predominantly only at the edges of the glide bands. Similar observations were later carried out on other crystals as well [2], however they are all mainly of a qualitative character.

The purpose of this work has been to obtain quantitative information on such parameters related to the process of development of glide bands in LiF crystals as the rate of lateral band growth under stress and the variation of shear strain in the bands during their growth, as well as to estimate the average slip distance of dislocations in the course of formation of glide bands.

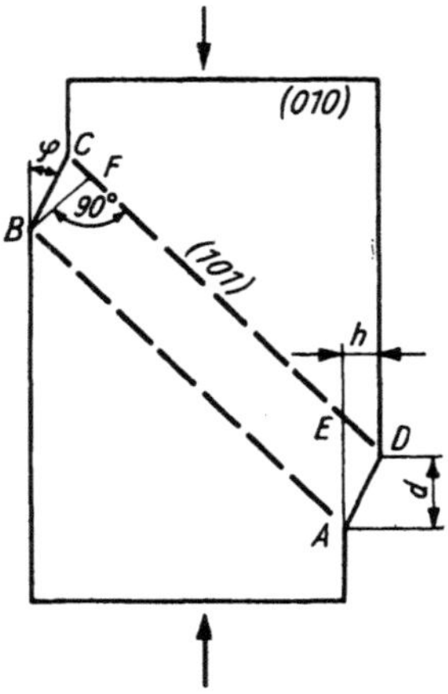


Fig. 1. Shape of crystal after formation of a glide band at a compressive stress

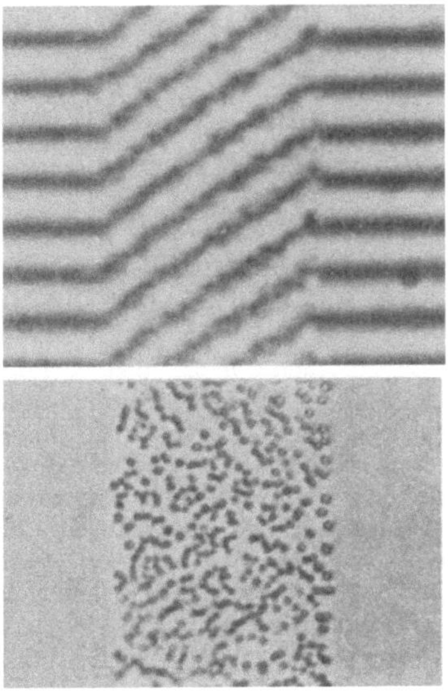

a

b

Fig. 2. Interference pattern of a glide band (a) and its dislocation structure (b). (970 ×)

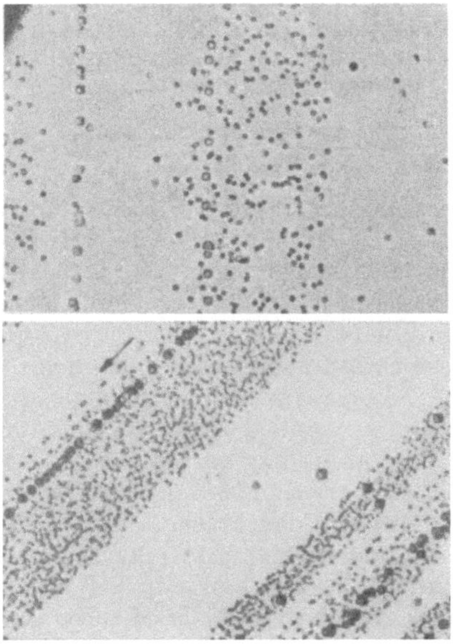

a

b

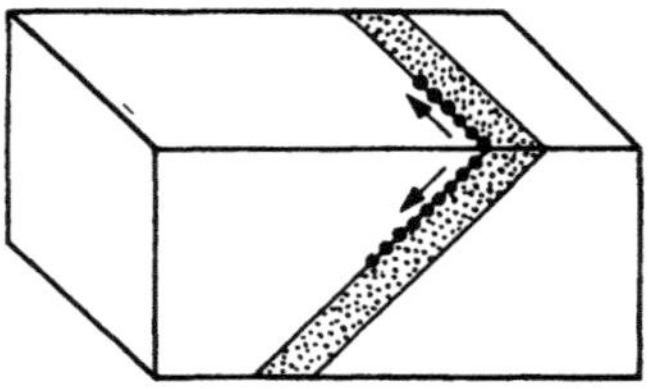

c

Fig. 3. Glide band widening. a) Screw dislocations, b) edge dislocations, c) diagram. Large pits result from etching after first loading, small pits from etching after the second one. (400 ×)

## 2. Experimental Techniques

We have used LiF crystals (containing 0.002% Mg) grown by the Kyropoulos method and subjected to a heat treatment consisting of an annealing at 750 °C for 48 hours with a subsequent cooling at a rate of 10° per hour. The specimens studied measured about $5 \times 5 \times 15\ mm^3$. The density of dislocations in these crystals was $\approx 10^4\ cm^{-2}$ exclusive of grain boundaries, the size of subgrains was 2 to 5 mm. Dislocations were revealed with a $1.5 \times 10^{-2}$% solution of $FeCl_3$ in distilled water.

Specimens were subjected either to compression at a constant strain rate of $10^{-4}\ s^{-1}$ or to a constant compressive stress corresponding to the yield point determined in the previous test.

Some data were obtained also on LiF crystals containing 0.01% Mg and subjected to a similar heat treatment.

Relative shear strain in an individual glide band was determined from band width $d$ and height $h$ of a jog arising on the crystal face with screw glide bands. It is seen from Fig. 1 that $\gamma = ED/BF$ in a glide band, and for the case of small shear strains (when $\varphi$ is small) simple transformations yield

$$\gamma = 2\frac{h}{d}. \tag{1}$$

The values of $d$ and $h$ were determined by means of the pattern observed with a МИИ-4 interference microscope (Fig. 2a), $d$ being found directly, and $h$ calculated from the shift of interference fringes. Besides, the band width can be determined from the dislocation pattern (Fig. 2b).

## 3. Experimental Results

Qualitatively the formation and development of glide bands proceeded in the same way as described by Gilman and Johnston [1]. However, we revealed one more peculiarity of this process consisting in that the widening of glide bands in the majority of cases takes place predominantly in one direction. This effect is seen in Fig. 3 showing "threads" of screw (a) and edge (b) dislocations obtained after the first loading of a crystal and representing a seed of a glide band as well as the band properly formed after a repeated application of stress.

A more detailed study of this phenomenon permitted to establish an unambiguous relation between the direction of the predominant widening of bands and the geometry of their formation. As a rule, the formation of glide bands, that is, the motion of parent dislocations, starts at the edges of a crystal (Fig. 3c). Therefore for each band there is a fixed direction in which the motion of dislocations and the development of the band proper in length (shown by arrows) take place. It has been found that the bands widen predominantly towards the obtuse angle formed by the glide band and the edge acting as the band source (Fig. 3c), that is, in the direction of compressive stresses at the band as revealed by birefringence patterns. The same relationship persisted in asymmetric widening of glide bands on the side of compression in crystals with applied bending stress.

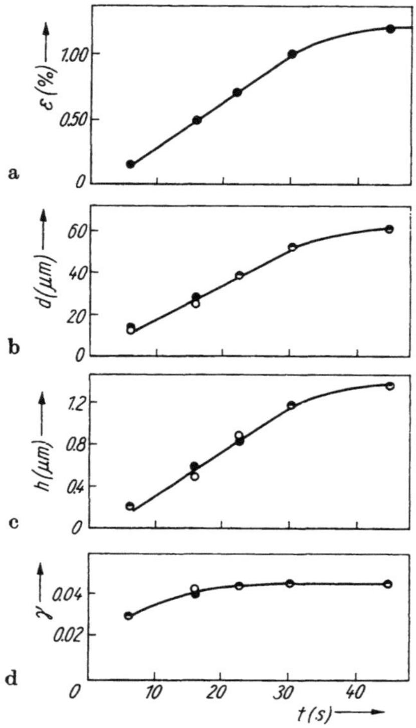


Fig. 4. Dependence of crystal deformation (a), band width (b), jog height (c), and shear strain in a band (d) on duration of stress. Data are given for two bands. $\tau = 300$ g/mm²

The band width, the jog height, and the shear strain in a band as functions of time at a constant load corresponding to a stress of $\tau = 300$ g/mm² are presented in Fig. 4 which shows also a strain vs. time relationship. It is seen that all the above characteristics exhibit at first a fairly rapid variation, after which their growth somewhat slows down, the moment of transition being about the same for all the quantities mentioned. The time interval covered is limited on one side by the time required for the formation of a band revealed clearly in an interference microscope, and on the other side — by the overlapping of bands occurring at large times. It should be noted that the data presented relate to the face on which the glide band is created. On the opposite side, as indicated by preliminary experiments, the pattern changes qualitatively in the same way, however the magnitude of the shear strain is here somewhat lower.

Knowing the change of the width of a band and of the shear strain in it in time, one can determine the dependence of $\gamma$ on $d$. At the same time the relation between these characteristics can be found also through measurements of the magnitude of shear strain in bands of different width. This has been done and the results obtained are presented in Fig. 5. It is seen that in all the cases studied any significant increase of the shear strain in a band occurs only up to a value of 30 μm after which it remains practically constant. This shear strain $\gamma_l$ which may be called a "limiting shear strain" depends on the condition of the crystal and the actual process of its deformation which is indicated by the data listed in Table 1. It is seen that crystal hardening achieved by an introduction of magnesium results in an increase of the relative shear strain whereas a decrease of temperature does not produce a noticeable effect.

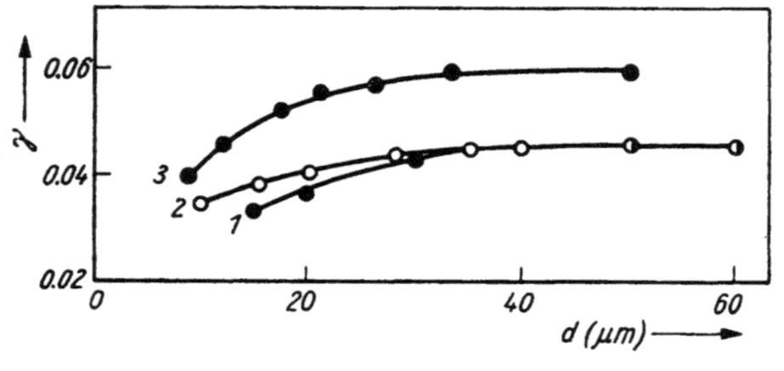


Fig. 5. Dependence of shear strain on band width. Crystal with 0.002% Mg: 1 — from data for one band widening in time, 2 — from data for different bands, 3 — crystal with 0.01% Mg, different bands

Table 1

| | Magnesium content (%) | Yield stress (g/mm²) | $T$ (°C) | $\gamma_l$ | $\varrho_{100}$ ($10^{-7}$ cm$^{-2}$) | $\lambda$ (mm) |
|---|---|---|---|---|---|---|
| 1 | 0.002 | 300 | 20 | 0.044 | 0.75 | 1.5 |
| 2 | 0.002 | 1500 | −196 | 0.040 | 3.5 | 0.3 |
| 3 | 0.01 | 900 | 20 | 0.056 | 2.1 | 0.7 |

## 4. Discussion

The data given in Fig. 4b can be used to evaluate the rate of widening of bands in the course of their growth. During the first stages it constitutes about $1.5 \times 10^{-4}$ cm/s, but after this decreases down to $5 \times 10^{-5}$ cm/s. In absolute magnitude it is about an order of magnitude less than the velocity of individual dislocations which, according to [3], is equal to $\approx 10^{-3}$ cm/s at the yield point of LiF crystals subjected to a strain at the same rate of $10^{-4}$ s$^{-1}$ as used in the present work. The slowing down of lateral growth observed at an increase in band width may be due either to an interaction between the bands (although they do not yet overlap at this stage) or to a change of the influence of the "backing", i.e. the band structure, on the freedom of dislocation motion at the edge of the bands. As an argument in favour of the second supposition may serve the fact that the decrease in band growth rate and the reaching of the limiting shear strain occur simultaneously.

Besides, the existence of an asymmetry in band widening leads one to suggest that the influence of the bands proper on the mobility of dislocations on different edges of the bands is not the same. According to Johnston and Gilman [4], the multiplication of dislocations and the lateral growth of glide bands are realized through multiple cross glide of screw dislocations, the cross glide near the surface occurring, as shown experimentally, predominantly towards the acute angle between screw dislocations and the surface. Now the band widening, as mentioned before, takes place predominantly towards the obtuse angle, that is, it does not correlate with the expected effect of surface. Therefore it may be supposed that the determining role in the process of lateral band growth is played by the stress fields arising near a band during its formation and connected unambiguously with the sign of the first dislocation responsible for the creation of the band. However, experimental data available at present are not sufficient to draw a final conclusion.

As is well known [2], the relative shear strain, the number of dislocations moving in the process of deformation in a unity volume $\varrho$, and their average slip distance $\lambda$ are connected by the relation

$$\gamma = b\,\lambda\,\varrho\,, \tag{2}$$

where $b$ is the Burgers vector. Therefore the knowledge of $\gamma$ and $\varrho$ permits to determine the magnitude of $\lambda$. This method was used in some works with copper [2], the magnitude of $\gamma$ being determined not directly but through calculations from the relative strain of the specimen as a whole. However, this may be done only if the number of the glide systems and the contribution of each of them to the strain is known accurately; besides, one is bound to assume the strain in the specimen to take place uniformly. Otherwise such calculations may lead to considerable errors.

In the present work we were not confronted with these difficulties since the magnitude of $\gamma$ was determined locally and for one definite system of glide planes. One had only to take into account that the measurement of the dislocation density was carried out for the {100} face, whereas the expression (2) contains the value of $\varrho$ on the corresponding glide plane. For LiF crystals $\varrho_{110} = \sqrt{2}\,\varrho_{100}$. Data on dislocation density and average slip distance are given in Table 1. It is seen that a decrease of temperature as well as an addition of magnesium into the crystal reduce the average slip distance of screw dislocations which is quite large in "pure" crystals at 20 °C.

Knowing the average slip distance of dislocations at a shear strain $\gamma_l$ and the time $t_l$ required for this strain to occur, one can determine the average velocity of dislocations:

$$v_{av} = \frac{\lambda_l}{t_l}\,. \tag{3}$$

Assuming for a pure crystal $t_l = 30$ s (Fig. 4), we obtain $v_{av} \approx 5 \times 10^{-3}$ cm/s which is equal by the order of magnitude to the velocity of free dislocations. It may apparently be considered as one more indication that at the yield point the dislocations move mainly on the edges of glide bands.

One may also attempt to evaluate the average slip distance of edge dislocations $\lambda_e$. If we assume that the ratio of velocities of edge and screw dislocations, $v_e$ and $v_s$, respectively, at the formation of glide bands remains the same as for individual loops, where $v_e \approx 50\, v_s$ [1], then we will have at 20 °C $\lambda_e = 50\,\lambda_s \approx 7.5$ cm which is considerably larger than the size of the crystals used ordinarily. In this case the major number of edge dislocations during the formation of a glide band should move out of the crystal (except those, of course, which provide for the difference of shear strains on opposite faces), and etch pits on the faces of the edge dislocation components will represent mainly dislocation dipoles. However, it is possible that the ratio between the magnitudes of $v_e$ and $v_s$ during the motion of dislocations as a whole changes, and the average slip distance of edge dislocations will be different. Since at present this problem is completely unstudied, a more detailed discussion would hardly be justified.

## References

[1] J. J. GILMAN and W. G. JOHNSTON, J. appl. Phys. **27**, 1018 (1956); Solid State Phys. **13**, 147 (1962).
[2] F. R. N. NABARRO, Z. S. BASINSKY, and D. B. HOLT, Adv. Phys. **13**, 193 (1964).
[3] W. G. JOHNSTON, J. appl. Phys. **33**, 2050 (1962).
[4] W. G. JOHNSTON and J. J. GILMAN, J. appl. Phys. **31**, 632 (1960).

*(Received April 18, 1966)*

phys. stat. sol. **16**, 197 (1966)

*Institute of Radio Engineering and Electronics, Academy of Sciences of the USSR, Moscow*

# On the Velocity of Space Charge Waves (Electrical Domains) in Semiconductors

By

S. G. Kalashnikov and V. L. Bonch-Bruevich

The non-linear problem of the velocity of space charge waves (electrical domains) which occur when the volume differential resistance becomes negative is studied. Particular attention is paid to the recombination (concentration controlled) type of instability. The domain velocity is derived for the cases in which the Maxwellian relaxation time is much greater or much smaller than the recombination time. The results obtained are in order of magnitude agreement with experimental data on domains in Au- and Cu-doped Ge [13, 15].

Рассмотрен вопрос о скорости движения волн объёмного заряда (электрических доменов), возникающих при отрицательной объёмной дифференциальной проводимости и электрической неустойчивости концентрационного (рекомбинационного) типа, в нелинейном приближении. При достаточно малых максвелловских временах релаксации скорость волн выражается формулой (19). При больших максвелловских временах справедлива формула (21). Полученные результаты находятся в удовлетворительном порядковом согласии с опытными данными о доменах в германии с золотом [13] и медью [15].

## 1. Introduction

An initially homogeneous semiconductor with negative differential resistance may become unstable with respect to small fluctuations, and separate into regions of high and low electric field strength (electrical domains) [1 to 6]. The domains may move in the field, thus forming space charge waves. Their periodic appearance and disappearance lead to current oscillations in an external circuit.

The negative resistance may be due to a field dependence of either mobility or carrier concentration ("drift" and "concentration" or "recombination" instability, respectively). The first case may take place, for instance, when hot carriers are excited to high effective mass states [7, 8], the second case may occur when the probability of electron capture by some traps rises with the field [9].

At present there is quite a number of experimental data clearly showing a motion of domains in various semiconductors (a review of this work is beyond the scope of the present paper). An example of fast waves under conditions of drift instability is given by Gunn's well-known experiments [10] on GaAs; the domains in Au-doped [11 to 14] and Cu-doped [15, 16] Ge give an example of slow waves in the case of concentration instability.

In spite of the large number of experimental results the theory of space charge waves of the type just considered has not as yet been developed in sufficient detail. The motion of small-amplitude domains was considered in [17] by linearizing the basic equations. However, it is clear that such a linear theory cannot describe many peculiarities of the phenomena: it is enough to remark that it cannot, in principle, give the stationary solutions. A nonlinear treatment was first given in [18] and, independently, in [19] and [20].

Since the complete solution of the nonlinear problem is difficult it seems useful to consider some particular problems. For example, it is interesting to study the velocity of the space charge waves, since this is the main factor determining the period of the current oscillations. In this paper an attempt is made to obtain a simple explicit formula for the velocity of stationary space charge waves in the nonlinear theory. This turns out to be possible in some limiting cases defined by some relations between the relevant characteristic times. However, these relations do correspond to real experimental conditions in many cases of interest.

## 2. Basic Equations

The basic equations for an n-type semiconductor have the form

$$\frac{\partial n}{\partial t} = \frac{\partial j}{\partial x}\,; \qquad n = n_c + n_t\,, \tag{1}$$

$$j = n_c\, v + D\,\frac{\partial n_c}{\partial x}\,; \qquad v = \mu\, E\,, \tag{2}$$

$$\frac{\partial E}{\partial x} = \frac{4\,\pi}{\varepsilon}\,\varrho\,; \qquad \varrho = q\,(N_d - n)\,. \tag{3}$$

Here $n_c$ and $n_t$ are the free and trapped electron concentrations respectively, $N_d$ the concentration of compensating donors, $E$ the electric field strength, $q$, $D$, and $\mu$ the charge, the diffusion coefficient, and the mobility of an electron, respectively. Equations (1) to (3) must be supplemented by an equation describing the kinetics of negative differential resistance formation. In the case of concentration instability this is the recombination-kinetic equation:

$$\frac{\partial n_t}{\partial t} = c\, f\,(N - n_t)\, n_c - g\, n_t\,, \tag{4}$$

where $N$ is the trap concentration, $c$ the low-field electron capture coefficient of the traps, $f = c(E)/c$, $c(E)$ the capture coefficient at field $E$. $g$ is the generation rate per trap:

$$g = c\, n_1 + s\, J\,,$$

where $s$ is the photoionization cross section, $J$ the photon flux density, $n_1$ the Shockley-Read-Hall factor known from recombination statistics.

As shown in [18] equations (1) to (3) possess a particular solution of the form $n = n(z)$, $z = x - u\,t$, where $u$ is an integration constant. This solution describes a unidirectional wave propagation with velocity $u$. Substituting it into (1) we obtain

$$-\,u\, n' = j'\,, \tag{5}$$

where the prime means differentiation with respect to $z$. Since $u$ does not depend on $z$ it is convenient to refer all quantities to the neutrality point ($z = 0$) where the field strength reaches its maximum (see Fig. 1). Then

$$z = 0\colon\quad v' = 0\,, \quad f' = D' = \mu' = 0\,; \quad n' \neq 0\,. \tag{6}$$

Using (5), (2), and (6), we get for the wave velocity

$$-\,u = \frac{n_c'(0)}{n'(0)}\, v(0) + D\,\frac{n_c''(0)}{n'(0)}\,. \tag{7}$$

Fig. 1. Schematic distribution of electric field strength and space charge density in the domain

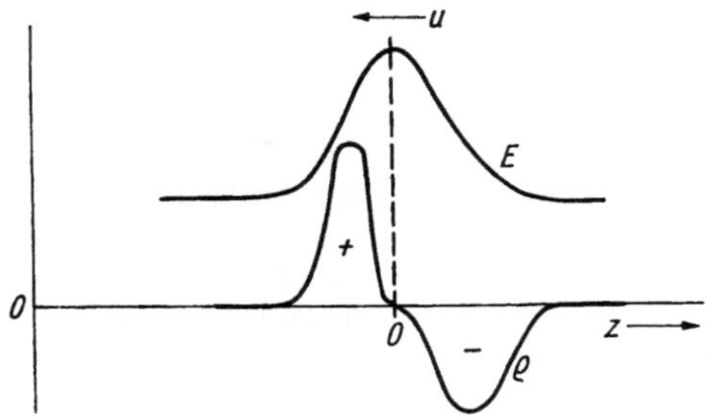


To obtain the ratio of derivatives $n_c'(0)/n'(0)$ we use equation (4). Expressing $n_t$ in terms of $n$ and $n_c$, taking the derivative of (4) with respect to $z$, and using (6), one obtains easily

$$\frac{n_c'(0)}{n'(0)} = \frac{\tau}{\tau_g} + u\,\tau\,\frac{n_c''(0) - n''(0)}{n'(0)}\,. \tag{8}$$

Here the time $\tau$ is defined by

$$\frac{1}{\tau} = \frac{1}{\tau_r} + \frac{1}{\tau_g}\,, \tag{9}$$

where $\tau_r$ and $\tau_g$ are the "capture" and "generation" time, respectively:

$$\frac{1}{\tau_r} = c\,f\,(N - N_d + n_c)\,, \qquad \frac{1}{\tau_g} = g + c\,f\,n_c\,. \tag{10}$$

Evidently $\tau$ is just the recombination time which determines the rate of attaining the band-traps equilibrium in the homogeneous sample.

To find the as yet undetermined ratios $n''(0)/n'(0)$ and $n_c''(0)/n'(0)$ we differentiate (5) with respect to $z$, put $z = 0$, and take account of (6). If $D$ and $\mu$ may be treated as constants we obtain

$$-\frac{1}{v(0)\,\tau_M}\frac{n'(0)}{n''(0)} + \frac{n_c''(0)}{n''(0)} + \frac{D}{v(0)}\frac{n_c'''(0)}{n''(0)} = -\frac{u}{v(0)}\,. \tag{11}$$

Here $\tau_M = \varepsilon/4\,\pi\,q\,\mu\,n_c(0)$ is the Maxwellian relaxation time corresponding to the electron concentration at the neutrality point.[1])

Equations (11) and (8) contain some terms that may be neglected under certain conditions. This makes it possible to find the relevant ratios of derivatives and, thus, obtain the wave velocity.

## 3. Case $\tau_M \ll \tau$

If the domain were symmetrical all the odd derivatives $E^{(2n+1)}(0)$ would be equal to zero. It is seen from the experimental data that the observed domain

---

[1]) To take account of the field dependence of $D$ and $\mu$ one has merely to replace $\tau_M$ by

$$\tau_M^* = \tau_M \left(1 + \frac{d \ln \mu}{d \ln E} + \frac{j_{dif}(0)}{j_{dr}(0)}\frac{d \ln D}{d \ln E}\right)^{-1}$$

in all the subsequent equations. Note that usually $\dfrac{j_{dif}}{j_{dr}} = \dfrac{D}{v}\dfrac{n_c'}{n_c} \ll 1\,.$

asymmetry is rather small. Therefore we may put

$$\frac{E'''(0)\,\Delta}{E''(0)} = \frac{n''(0)\,\Delta}{n'(0)} \lessapprox 1\,, \qquad \frac{n_c'''(0)\,\Delta}{n_c''(0)} \lessapprox 1\,, \tag{12}$$

where $\Delta$ is the domain width whose order of magnitude may be taken from experiment.

Note that if, e.g., the first of the inequalities (12) would not hold there would be just a local maximum at the point $z = 0$, while all the experimental data now available show that this maximum is in fact an absolute one.

Under typical conditions the ratio $D/v(0)$ is usually small compared to $\Delta$ ($\approx 10^{-4}$ to $10^{-5}$ cm, see Section 5). Thus one may neglect the third term on the left-hand side of (11) as compared to the second term. This is valid provided

$$\frac{D}{v(0)\,\Delta} \ll 1\,. \tag{13}$$

The possibility of other simplifications depends upon the relation between $\tau_M$ and the other characteristic times. Consider first the case of small $\tau_M$ (see the exact condition below). Then the right-hand side of (11) may be neglected compared with the first term. Taking account of the first of the inequalities (12) this leads to the condition

$$\frac{|u|\,\tau_M}{\Delta} \ll 1\,. \tag{14}$$

Then (11) gives

$$\frac{n_c''(0)}{n'(0)} = \frac{1}{v(0)\,\tau_M}\,. \tag{15}$$

Finally, to estimate the ratio $n''(0)/n_c''(0)$ entering (8), we obtain by combining (15) with the first of the inequalities (12)

$$\frac{n''(0)}{n_c''(0)} < \frac{v(0)\,\tau_M}{\Delta}\,. \tag{16}$$

Note that this is probably an overestimate due to the use of (12), and actually the left-hand side of (16) is much smaller than the right-hand side. Thus, putting $v(0) \approx 10^6$ cm s$^{-1}$ and $\Delta \approx 10^{-1}$ to $10^{-2}$ cm, the left-hand side of (16) becomes $< 1$ when $\tau_M < 10^{-7}$ to $10^{-8}$ s.

In case of small $\tau_M$ obeying the condition

$$\frac{v(0)\,\tau_M}{\Delta} < 1\,, \tag{17}$$

inequality (14) is satisfied a fortiori since $|u| \ll v$. Further one may neglect $n''(0)$ compared with $n_c''(0)$ in (8). Then we obtain, using (7) and (8),

$$-\,u = \left(\frac{\tau}{\tau_g}\,\mu\,E + \frac{D}{\mu\,E\,\tau_M}\right)\left(1 + \frac{\tau}{\tau_M}\right)^{-1}, \tag{18}$$

where $\mu\,E = v(0)$. The minus sign means that in a n-type sample the space charge wave moves like a negative charge, in accordance with experimental data.

If the recombination time $\tau \gg \tau_M$, (18) gives

$$-\,u = \frac{\tau_M}{\tau_g}\,\mu\,E + \frac{D}{\mu\,E\,\tau}\,. \tag{19}$$

Note that (18) is analogous to the equation obtained by Ridley (equation (19) in [17]) using a linearization of the basic equations. It is seen, however, that in the nonlinear case the quantities entering (18) should be referred to the neutrality point (field strength maximum). Further we see that since $u$ = const the group velocity (18) coincides with the phase velocity.

## 4. Case $\tau_M \gg \tau$

If the recombination time $\tau$ is small compared with $\tau_M$ one may approximately assume the band-traps equilibrium to be established at every point. Then (4) gives

$$n_c \approx \frac{g}{c f} \frac{n_t}{N - n_t} . \tag{20}$$

Hence it follows, in view of (6) and (10),

$$\frac{n'_c(0)}{n'(0)} = \frac{\tau}{\tau_g} . \tag{8a}$$

In contrast to (8) there is no second term on the right-hand side.

As before we assume the validity of (13) and neglect the right-hand side of (11). (Now this is possible if the relation (22) holds; see below.) Then the former equation (15) is rederived, and the wave velocity is found to be

$$- u = \frac{\tau}{\tau_g} \mu E + \frac{D}{\mu E \tau_M} . \tag{21}$$

Substituting this into (14) and taking account of (13), we see that (14) is satisfied if

$$\frac{v(0) \tau_M}{\Delta} \frac{\tau}{\tau_g} \ll 1 . \tag{22}$$

Thus (21) is valid provided the additional conditions (13) and (22) hold. In typical experiments on the concentration instability free electrons are created by external light and $\tau$ is several orders of magnitude smaller than $\tau_g$. Therefore condition (22) is easily met.

Note that when the second term on the right-hand side of (21) is small compared to the first term and also $\tau_r \ll \tau_g$, (21) reduces to the expression which was obtained in [21] by linearizing the equations determining the explicit form of the particular solution that describes the stationary motion of domains. This had to be expected since in [21] the solution was in fact linearized near the recombination equilibrium point at a field strength close to the maximum value.

## 5. Discussion

In the preceding sections we considered only the slow waves under conditions of concentration instability.

In case of the fast waves observed by Gunn [10] there are no indications on the role of the traps. Therefore, we may put $n_t$ = const and, accordingly, we get $n'_c(0) = n'(0)$ in (7). If one further assumes the diffusion current to be negligible [20], (7) gives immediately

$$- u = v(0) = \mu E . \tag{23}$$

Thus the domain moves approximately with the drift velocity at the neutrality point [18]. This is in order-of-magnitude agreement with the experimental data [10].

Above we made an attempt to obtain a simple expression for the domain velocity without solving the full non-linear problem. This turned out to be possible provided some reasonable approximations were used, based on the smallness of the two parameters, $D/v(0)\,\Delta$ and $|u|/v(0)$. However, in doing so the wave velocity was expressed in terms of the maximum field at the neutrality point, and not in terms of the total voltage applied to the sample. If the high-field region is not too sharp the maximum field may be approximately replaced by the average one. In the general case the maximum field must be determined independently (cf., for example, [19] and [20]).

We will now compare our results with some experimental data. A domain motion was recently observed in Ge containing partly compensated Cu (upper level) [15]. The conductivity (before domain formation) was $\approx 10^{-5}$ to $10^{-3}\ \Omega^{-1}\ \text{cm}^{-1}$, corresponding to $\tau_M \approx 10^{-9}$ to $10^{-7}$ s. The recombination time in the samples varied from several milliseconds (low field) to $\approx 10^{-5}$ s (high field). Thus this is the case $\tau_M \ll \tau$. The drift velocity was $v \approx 10^6\ \text{cm s}^{-1}$ and the domain width $\Delta \approx 0.1$ cm. Thus, putting $D \approx 10\ \text{cm}^2\ \text{s}^{-1}$ we have $D/v\,\Delta \approx$ $\approx 10^{-4}$, in good agreement with the approximation used. As to the second inequality (17), we obtain $v\,\tau_M/\Delta \approx 1$ to $10^{-2}$. Since, as we pointed out, this relation is probably an overestimate, the approximation used seems justified in this case as well. It is difficult to estimate the values of $\tau_g$ in these experiments (due to possible excitation of two copper levels). However the $\tau_g$ values needed for the equation (19) to be compatible with the experimental data seem quite acceptable.

The domain motion was studied in detail in the case of Ge with partly compensated upper gold level [13]. It may be deduced from the data [13] that the Maxwellian relaxation time under various illumination conditions was $\tau_M \approx$ $\approx 10^{-6}$ to $10^{-10}$ s, while the recombination time varied from $10^{-1}$ to $10^{-6}$ s (weak illumination) depending on the field strength. Approximately we have $v \approx 10^5\ \text{cm s}^{-1}$ and $D \approx 10\ \text{cm}^2\ \text{s}^{-1}$ (neglecting the heating of the electron gas). The order of magnitude of $\tau_g$ may be estimated from the low-field data. Using (10) and (20) we obtain $\tau_g = n_t/N\,c\,n_c$, where $n_t/N \approx 1$. Since $c = c_n^= \approx$ $\approx 10^{-12}\ \text{cm}^3\ \text{s}^{-1}$ for gold at hydrogen temperature [22, 23] we obtain $\tau_g \approx 10^3$ to $10^{-1}$ s at $n_c \approx 10^9$ to $10^{13}\ \text{cm}^{-3}$. Therefore, (19) is expected to be valid in this case as well, the second term on the right-hand side being greater than the first. It was pointed out in [17] that this expression is in satisfactory agreement with experimental data.

We should like to thank M. S. Kagan for informing us of his experimental results concerning the domains in Cu-doped Ge and for discussing the manuscript. We should also like to thank Yu. F. Sokolov for discussing the manuscript.

## References

[1] A. L. Zakharov, Zh. eksper. teor. Fiz. **38**, 665 (1960).
[2] B. K. Ridley, Proc. Phys. Soc. **82**, 9541 (1963).
[3] K. W. Böer and W. E. Wilhelm, phys. stat. sol. **3**, 1704 (1963).
K. W. Böer, Proc. Internat. Conf. Semiconductors Phys. 1964, Dunod, Paris 1964 (p. 987); Phys. Rev. **139**, A 1949 (1965).

[4] H. KIESS and F. STÖCKMANN, phys. stat. sol. **4**, 117 (1964).
[5] V. L. BONCH-BRUEVICH and SH. M. KOGAN, Fiz. tverd. Tela **7**, 23 (1965).
[6] V. L. BONCH-BRUEVICH, Fiz. tverd. Tela **8**, 356 (1966).
[7] B. K. RIDLEY and T. B. WATKINS, Proc. Phys. Soc. **78**, 293 (1961).
[8] C. HILSUM, Proc. IRE **50**, 185 (1962).
[9] B. K. RIDLEY and T. B. WATKINS, J. Phys. Chem. Solids **22**, 155 (1961); see also V. L. BONCH-BRUEVICH, Fiz. tverd. Tela **6**, 2047 (1964); V. L. BONCH-BRUEVICH and S. G. KALASHNIKOV, Fiz. tverd. Tela **7**, 750 (1965).
[10] I. B. GUNN, Solid State Commun. **1**, 88 (1963); IBM J. Res. Developm. **8**, 141 (1964). I. B. GUNN, Plasma Effects in Solids, Dunod, Paris 1964 (p. 199).
[11] B. K. RIDLEY and R. G. PRATT, J. Phys. Chem. Solids **4**, 300 (1963).
[12] I. A. KUROVA and S. G. KALASHNIKOV, Fiz. tverd. Tela **5**, 3221 (1963).
[13] B. K. RIDLEY and R. G. PRATT, J. Phys. Chem. Solids **26**, 21 (1965); Proc. Phys. Soc. **85**, 293 (1965); Proc. Internat. Conf. Semiconductors Phys. 1964, Dunod, Paris 1964 (p. 487).
[14] V. I. STAFEEV, Fiz. tverd. Tela **5**, 3095 (1963). V. P. SONDAEVSKII, V. I. STAFEEV, and E. I. KARAKUSHAN, Proc. Internat. Conf. Semiconductors Phys. 1964, Dunod, Paris 1964 (p. 481).
[15] M. S. KAGAN, S. G. KALASHNIKOV, and N. G. ZHDANOVA, phys. stat. sol. **11**, 415 (1965); Fiz. tverd. Tela **8**, 774 (1966); **8**, 788 (1966).
[16] B. V. KORNILOV, Fiz. tverd. Tela **7**, 329 (1965).
[17] B. K. RIDLEY, Proc. Phys. Soc. **86**, 637 (1965).
[18] V. L. BONCH-BRUEVICH, Phys. Letters (Netherlands) **20**, 249 (1966); Fiz. tverd. Tela **8**, in press (1966).
[19] A. F. VOLKOV, Phys. Letters (Netherlands) **20**, 598 (1966).
[20] P. N. BUTCHER, Phys. Letters (Netherlands) **19**, 547 (1965).
[21] V. L. BONCH-BRUEVICH, preprint IRE Moscow, T−01008 (1966).
[22] I. A. KUROVA, S. G. KALASHNIKOV, and N. D. TYAPKINA, Fiz. tverd. Tela **4**, 1503 (1962).
[23] R. G. PRATT and B. K. RIDLEY, J. Phys. Chem. Solids **26**, 11 (1965).

*(Received April 18, 1966)*

phys. stat. sol. **16**, 205 (1966)

*Institute of Physics, Academy of Sciences of the Azerbaidzhan SSR, Baku*

# Electrical Conductivity of p-GaSe Single Crystals in Strong Electric Fields

By

G. B. Abdullaev, E. S. Guseinova, and B. G. Tagiev

The electrical conductivity of p-GaSe single crystals is measured in strong electric fields up to $3 \times 10^4$ V/cm. Pulse duration is used with 4 μs and a repetition rate of 50 Hz. The critical field for p-GaSe is about 1600 V/cm, above which Ohm's law fails and the conductivity increases with field according to Frenkel's law: $\sigma = \sigma_0 \, e^{\beta \sqrt{E}}$. From the deduced temperature dependence of $\beta$ ($\beta = \sqrt{e^3}/k\,T\sqrt{\varepsilon}$) the dielectric constant $\varepsilon$ of p-GaSe single crystals is found to be approximately five. The hole activation energy determined from the temperature dependence of the conductivity, decreases with increasing field, which agrees with Frenkel's theory. By extrapolating the $\lg \sigma = f(1/T)$ straight lines a thermal breakdown temperature of about 25000 °K is found, which can be referred to as the metallization temperature ($T_m$). It is found that $k\,T_m \approx 2$ eV is the thermal energy gap at absolute zero. By extrapolating the $\lg \sigma = f(\sqrt{E})$ straight lines a breakdown intensity, $E_b$, of about $10^7$ V/cm is found, which characterizes the electrical strength. The width of the energy gap $\Delta E_0$, obtained from the expression: $\Delta E_0 = 2\,e\sqrt{e\,E_b/\varepsilon}$ appears to be about 1.9 eV.

Измерена электропроводность монокристаллов p-GaSe в сильных электрических полях до $3 \cdot 10^4$ в/см в импульсном режиме с длительностью импульса 4 мк.сек, частотой следования 50 гц. Показано, что критическое поле для p-GaSe равно примерно 1600 в/см, начиная с которого закон Ома нарушается и электропроводность возрастает с ростом напряженности электрического поля. Рост электропроводности с напряжением следует закону Френкеля: $\sigma = \sigma_0 \, e^{\beta \sqrt{E}}$. Из температурной зависимости наклона $\beta = \sqrt{e^3}/k\,T\sqrt{\varepsilon}$ определена диэлектрическая проницаемость монокристаллов p-GaSe и получена $\varepsilon \approx 5$. Энергия активации дырок, определенная из температурной зависимости электропроводности, уменьшается с ростом электрического поля, что также согласуется с выводами теории Френкеля. Путем экстраполяции прямых $\lg \sigma = f(1/T)$ найдена температура теплового пробоя около 25000 °K, что характеризуется как температура металлизации $T_m$, при которой ширина запрещенной зоны $\Delta E_0$ становится нулем. Найдено, что $k\,T_m \approx 2$ эв — есть энергетическая ширина запрещенной зоны, что соответствует ширине запрещенной зоны p-GaSe при абсолютном нуле. Экстраполяцией прямых $\lg \sigma = f(\sqrt{E})$ найдена напряженность пробоя $E_{\text{нр.}}$ около $10^7$ в/см, которое харатеризует электрическую прочность. Из выражения $\Delta E_0 = 2\,e\sqrt{e\,E_{\text{нр.}}/\varepsilon}$ также найдена $\Delta E_0$ — ширина запрещенной зоны, которая оказалась примерно 1,9 эв.

p-GaSe single crystals possess a high photosensitivity with respect to light in the visible region [1], and in the infrared region up to 3 μm [2], as well as to X- and γ-radiation [3]. Induced emission under the action of accelerated electrons [4] and glow effects during the passage of an electric current [5] have been observed. The lifetime of carriers in p-GaSe single crystals is of the order of milliseconds [6].

In a number of papers the electrical conductivity of p-GaSe single crystals in weak electric fields has been studied [1, 7, 8, 9].

Studies of the electrical conductivity of semiconductors in strong electric fields can give some information concerning the mechanism of the increase of conductivity, electroluminescence, the existence and position of impurity levels, the width of the forbidden gap, the dielectric constant, the metallization temperature, the value of the critical field above which deviations from Ohm's law occur, and the breakdown field. This is why the authors were interested in studying the conductivity of a p-GaSe single crystal in strong electric fields.

The p-GaSe single crystals were prepared according to a method described in [10]. The specimens had a length of 4 to 5 mm, a width of 2 to 4 mm, and a thickness of 0.1 to 0.4 mm, and were supplied with indium contacts. The area of the electrodes was 0.3 to 0.8 mm$^2$.

The conductivity of p-GaSe in strong electric fields has been measured in an arrangement described in [11].

To avoid heating, square pulses with a duration of 4 μs and a repetition rate of 50 Hz were applied to the specimens. An analysis of the experimental conditions showed that at this pulse duration and the repetition rate thermal effects for specimens with a conductivity of $10^{-3}\ \Omega^{-1}\ \text{cm}^{-1}$ may be neglected.

The conductivity of six p-GaSe single-crystal specimens has been measured along the $c$-axis in strong electric fields up to $3 \times 10^4$ V/cm within the temperature range 77 to 293 °K. At room temperature the specimens showed a conductivity of about $10^{-4}\ \Omega^{-1}\ \text{cm}^{-1}$ in the ohmic range.

It has been found that at room temperature for the specimens studied Ohm's law is valid in fields up to about 1600 V/cm. Above 1600 V/cm the conductivity of p-GaSe increases with growing electrical field strength.

Results for those field strengths for which Ohm's law is violated are presented here.

In Fig. 1 experimental data are given for specimen 4. As shown by this figure, a plot of lg $\sigma$ vs. $\sqrt{E}$ yields straight lines. This is an indication of the validity

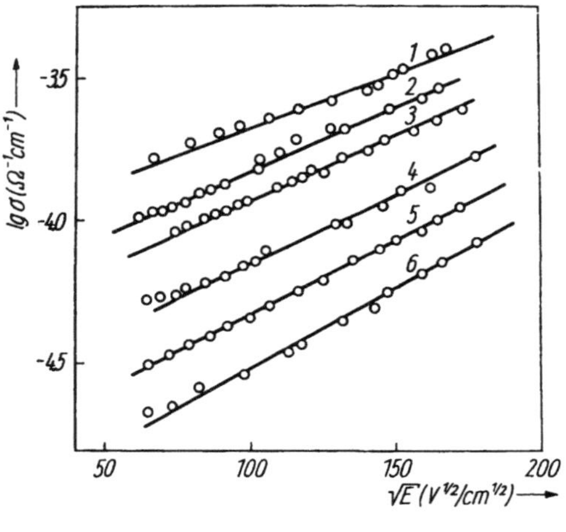


Fig. 1. Dependence of the conductivity $\sigma(\Omega^{-1}\ \text{cm}^{-1})$ of p-GaSe on electric field strength $E$ (V/cm) at different temperatures: 1 — 216 °K, 2 — 198 °K, 3 — 185 °K, 4 — 169 °K, 5 — 158 °K, 6 — 144 °K

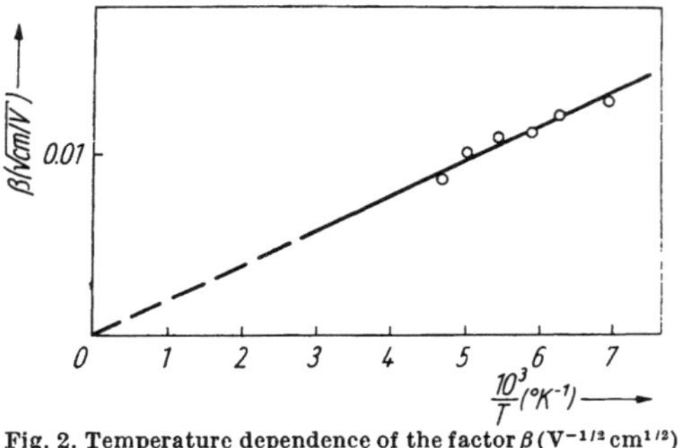


Fig. 2. Temperature dependence of the factor $\beta$ ($\text{V}^{-1/2}\ \text{cm}^{1/2}$) in Frenkel's formula $\left(\beta = \sqrt{e^3}/k\,T\sqrt{\varepsilon}\right)$

of Frenkel's law [12]

$$\sigma = \sigma_0 \, e^{\beta \sqrt{E}} \, . \tag{1}$$

The same relationship has been found for a number of specimens.

Fig. 1 shows that the slope of the straight lines increases with decreasing temperature. This change is in agreement with Frenkel's theory. According to this theory the slope of the $\lg \sigma = f(\sqrt{E})$ straight lines in a $\lg \sigma$ vs. $\sqrt{E}$ plot may be determined from the following formula

$$\beta = \frac{\sqrt{e^3}}{k\,T\,\sqrt{\varepsilon}} \tag{2}$$

($e$ electronic charge, $k$ Boltzmann constant, $T$ absolute temperature, $\varepsilon$ dielectric constant, pertaining to the electronic part of polarization).

The dependence of the slope $\beta$ on temperature is presented in Fig. 2. It should be noted that not only a growth of $\beta$ with decreasing temperature is observed, but also a rather good proportionality $\beta \sim 1/T$ in the temperature range 77 to 293 °K, which is in agreement with formula (2).

As the value of $\varepsilon$ is unknown, it is impossible to compare quantitatively our data with Frenkel's theory.

By using the values of $\beta$ determined from the straight lines in Fig. 1 at different temperatures, the dielectric constant of p-GaSe along the $c$-axis was calculated according to formula (2). It was found that the dielectric constant of p-GaSe crystals parallel to the $c$-axis is about 5 and almost independent of temperature.

The conductivity–temperature relationship in strong electric fields is presented in Fig. 3. It is shown here that $\lg \sigma$ depends linearly on $1/T$, which is also observed in weak fields [1, 7, 8, 9].

The slope of the straight lines in the $\lg \sigma$ vs. $1/T$ plot decreases with growing electric field strength and according to Frenkel's theory it is determined by the expression $\Delta E = \Delta E_0 - 2\,e\,\sqrt{e\,E/\varepsilon}$ where $\Delta E_0$ and $\Delta E$ are the activation energies of the majority carriers in weak and strong fields, respectively. The activation energy of the majority carriers was determined from the temperature dependence of conductivity in strong electric fields. For example, this activation energy is about 0.16 and 0.12 eV at 4900 and 28900 V/cm, respectively.

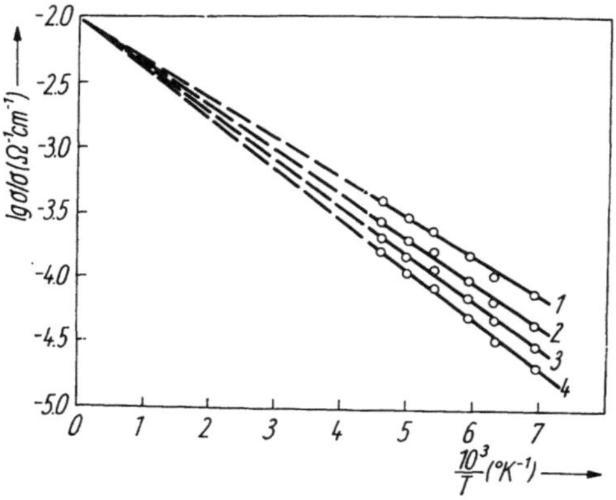


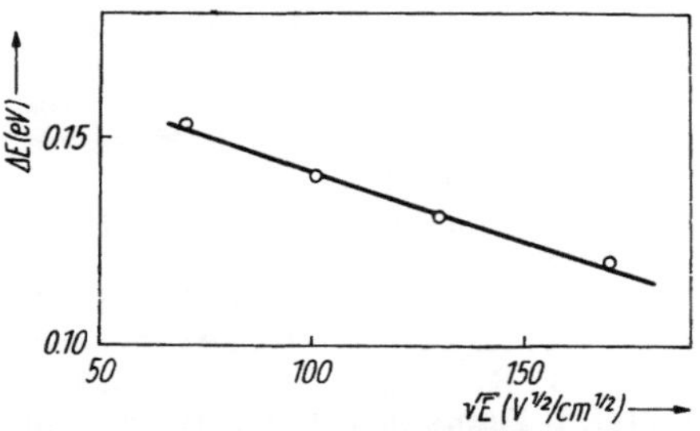


Fig. 4. Activation energy of majority carriers vs. electric field strength

◀ Fig. 3. Temperature dependence of the conductivity of p-GaSe. Curves 1, 2, 3, and 4 have been obtained at 28900, 16900, 10000, and 4900 V/cm, respectively

The dependence of the activation energy of carriers on the electric field strength is given in Fig. 4. As shown by this figure, the value of $\Delta E$ decreases linearly with growing $\sqrt{E}$.

The foregoing results can be explained in the following way:

It is known that the Frenkel mechanism involves a field-induced increase of the thermal release of electrons.

The decreasing slope of the $\lg \sigma$ vs. $1/T$ curves indicates that the concentration of current carriers increases with the electric field strength. By extrapolating straight lines to high $T$ we get a conductivity which is independent of the electric field strength. This gives a metallization temperature, $T_m$, equal to about 25000 °K. A thermal breakdown of p-GaSe apparently takes place at this temperature. Therefore, the value of $kT_m$ which is equal to nearly 2 eV corresponds to the width of the forbidden gap of p-GaSe single crystals, which is in agreement with optical data [1].

By extrapolating the straight lines in the $\lg \sigma$ vs. $\sqrt{E}$ plot (Fig. 1) a breakdown field of about $10^7$ V/cm was obtained, which is the field strength at which the width of the forbidden gap becomes zero. Under the effect of this critical field "metallization" of p-GaSe, i.e. a fusion of the valence and conduction band, takes place.

Using the values of the electric breakdown field $E_b$ and the dielectric constant determined from the temperature dependence of the slopes of the $\lg \sigma$ vs. $\sqrt{E}$ curves a value of the band gap of 1.9 eV was found from the formula $\Delta E_0 = 2\, e \sqrt{e\, E_b/\varepsilon}$; this value agrees with optical data. This is an additional support of the validity of Frenkel's law, according to which the increase of conductivity in strong electric fields is due to an increase of the concentration of current carriers.

In our opinion, one can determine the band gap of semiconductors by studying the electrical conductivity in strong electric fields. Such a possibility may be especially important for non-transparent semiconductors and for semiconductors having a small width of the forbidden gap.

## References

[1] R. H. BUBE and E. L. LIND, Phys. Rev. **115**, 1159 (1959).
[2] R. F. MEKHTIEV, L. G. PARITSKII, and S. M. RYVKIN, Fiz. tverd. Tela **5**, 1649 (1963).
[3] E. S. GUSEINOVA and R. F. MEKHTIEV, Izv. Akad. Nauk Azerb. SSR, **1**, 63 (1965).
[4] N. G. BASOV, O. B. BOGDANOVICH, A. N. PECHENOV, G. B. ABDULLAEV, G. A. AKHUNDOV, and E. Y. SALAEV, Dokl. Akad. Nauk SSSR **161**, 5 (1965).
[5] G. A. AKHUNDOV, Optika i Spektroskopiya, **19**, 302 (1965).
[6] G. B. ABDULLAEV, R. F. MEKHTIEV, A. F. MAMEDOVA, and E. S. GUSEINOVA, phys. stat. sol. **14**, K 127 (1966).
[7] R. F. FILDING, G. FISHER, and E. MOOSER, J. Phys. Chem. Solids **8**, 434 (1959).
[8] G. FISCHER and J. L. BREBNER, J. Phys. Chem. Solids **23**, 1363 (1962).
[9] G. FISCHER, Helv. phys. Acta **36**, 317 (1963).
[10] R. F. MEKHTIEV, O. OSMANOV, and Y. V. RUD, Prib. Tekh. Eksper. **2**, 179 (1964).
[11] A. P. REGEL and B. G. TAGIEV, Fiz. tverd. Tela **5**, 2914 (1963).
[12] J. I. FRENKEL, Selected Works, Vol. 2, published by the Academy of Sciences of the USSR, Moscow and Leningrad 1958 (p. 217).

*(Received March 7, 1966)*

phys. stat. sol. 16, 209 (1966)

*Institute of Physics, Academy of Sciences of the Azerbaidzhan SSR, Baku*

# Rectifying and Photoconductive Properties of p-GaSe Single Crystals

By

G. B. ABDULLAEV, M. R. AKHUNDOV, and G. A. AKHUNDOV

The current-voltage characteristics of a Cd–GaSe–Bi diode structure are given for different values of temperature and illumination. The photovoltage and photocurrent characteristics and the photocurrent spectral distribution are also presented. The mean rectification coefficient is $10^4$ at $U = 2$ V. The activation energies obtained from the temperature dependence of the exponential part of the forward current characteristic for different samples are grouped around the values 0.60 and 0.34 eV. The values of the open circuit voltage ($V_{oc}$) and of the short circuit current ($I_{sc}$) lie between 250 to 450 mV and 300 to 900 $\mu$A/cm$^2$ respectively. It is found that photocells based on GaSe have a better time-response than GaSe photoresistors.

Приводятся вольтамперные характеристики диодной сртуктуры Cd-GaSe-Bi при различных температурах и освещенностях, а также люксвольтные; люксамперные характеристики и спектральное распределение фототока. В среднем коэффициент выпрямления составляет $10^4$ при $U = 2$ в. Энергии активации, найденные из температурной зависимости экспоненциального участка прямой ветви для различных образцов группируются вокруг типичных значения 0,60 и 0,34 эв. Значения напряжения холостого хода ($V_{oc}$) и тока короткого замывания ($I_{к.з.}$) находились в интервалах 250 до 450 мв и 300 до 900 мка/см$^2$ соответственно. Выяснилось, что фотоэлементы на основе GaSe менее инерционные, чем фотосопротивления из GaSe.

Investigations of GaSe single crystals [1 to 4] have shown that this semiconductor has a number of interesting properties which are valuable for the construction of energy converters. Due to the appearance of induced emission (at $\lambda = 5925$ Å) under fast electron irradiation ($\varepsilon \approx 200$ keV) [5] and electroluminescence [6], GaSe is an interesting object for a study of rectifying and photoelectrical phenomena. Such investigations are necessary also for the explanation of the mechanism of electroluminescence.

In the present paper the rectifying and photoelectrical properties of p-GaSe single crystals are investigated for the first time. The samples were prepared as plates, cleaved from a large p-GaSe single crystal perpendicular to the $c$-axis. The thickness of the initial crystals was 50 to 200 $\mu$m, the resistivity ($\varrho$) was $\approx 10^2$ $\Omega$cm, with a hole concentration ($p$) of $10^{15}$ cm$^{-3}$. The contacts were made out of cadmium (evaporated in vacuum) and a bismuth alloy (also synthetised in vacuum). The obtained system (diode structure) is shown schematically in Fig. 1a.

It will be noted that the mentioned values of $\varrho$ and $p$ were determined in the direction perpendicular to the $c$-axis, and a rectifying contact was made along this axis.

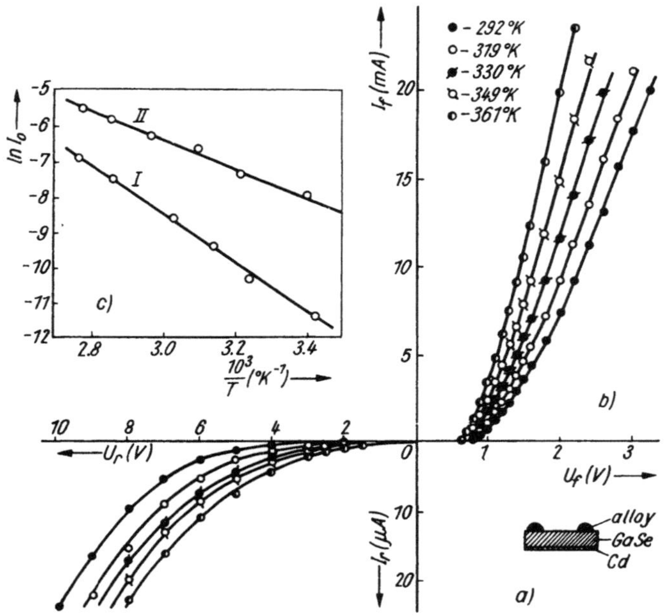


Fig. 1. a) Scheme of Cd-GaSe-Bi diode structure. b) Dark current-voltage characteristics for a typical sample at various temperatures. c) Dependence of the pre-exponential factor on temperature

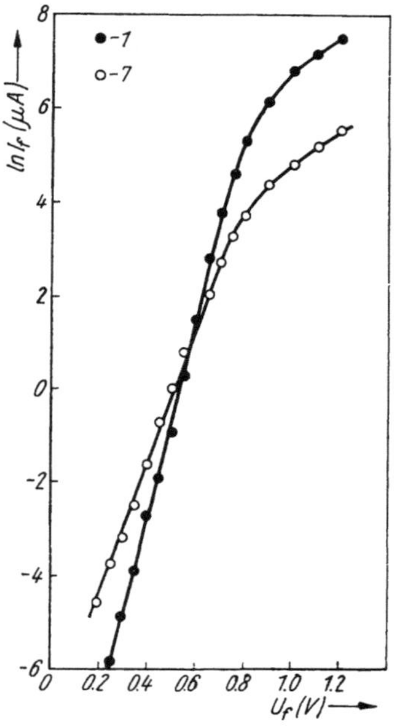


a

Fig. 2. Current-voltage characteristics for two samples.
a) Forward-current branch.
b) Reverse-current branch

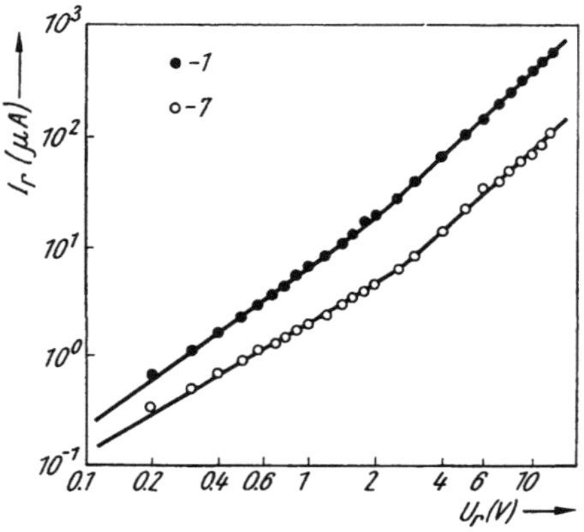


b

## 1. Rectifying Properties

The dark current-voltage characteristics under dc conditions is shown in Fig. 1b for one typical sample at various temperatures. As will be seen from the figure, in the temperature range 292 to 361 °K the density of the forward current at $U = 2$ V increases from 0.37 to 1.0 A/cm², and that of the reverse current at $U = 10$ V increases from 1.25 to 2 mA/cm².

The rectification coefficient for the given sample is $10^4$ for $U = 2$ V. For different samples this value varies within the range $10^3$ to $10^4$.

The rectifying barrier in this system is formed between Cd and GaSe. This seems to be confirmed by the fact that the reverse current direction always corresponds to positive sign of Cd.

A semi-logarithmic plot of forward current ($I_f$) versus applied voltage is presented in Fig. 2a for the samples 1 and 7. In the voltage range 0.2 to 0.8 V there is a linear dependence which can be described by the formula

$$I_f = I_0 \exp\left(\frac{q\,U_f}{\beta\,k\,T}\right),$$

where $\beta$ is a dimensionless coefficient. For all samples investigated the values of $I_0$ and $\beta$ at room temperature were found to lie in the ranges $I_0 = (1 \text{ to } 40) \times \times 10^{-9}$ A/cm² and $\beta = 1.9$ to 2.8.

As the temperature increases from 293 to 363 °K, the region having an exponential current-voltage dependence is narrowed and shifted towards low voltages. Then the pre-exponential factor ($I_0$) as a function of temperature can be represented in the usual form: $I_0 = A \exp(-\Delta E/k\,T)$, where $\Delta E$ means the activation energy. All values of $\Delta E$ obtained from the slopes of plots $I_0\,(1/T)$ for different samples, are grouped near the values 0.60 and 0.34 eV (Fig. 1c, curves I and II). It may be noted that for the samples in which $\beta$ is larger (2.8), the values of $\Delta E$ obtained from the slopes of $I_0\,(1/T)$ are smaller (0.34 eV). The nature of these centres has not yet been investigated.

In the higher voltage region (1 to 5 V) the forward current approximately obeys a square law: $I_f \sim U_f^2$. This behaviour is probably due to space charge limited current [7].

The reverse current as a function of applied voltage for samples 1 and 7 is given in Fig. 2b on a double logarithmic scale. For all samples investigated the reverse current obeys the law $I_r \sim U_r^n$, where $n = 1$ to 1.5 at low voltage (0.1 to 3 V), and $n = 1.7$ to 2 at higher voltage (3 to 12 V).

The current-voltage characteristics were obtained under photodiode conditions too, for several illumination intensities, at room temperature. The results of the measurements for one typical sample are given in Fig. 3a, where curve 1 is the dark characteristics, and curves 2, 3, 4, and 5 correspond to increasing intensity of illumination. The same characteristics are obtained on the oscillograph screen, at 50 Hz (Fig. 3b).

It is known [8] that germanium and silicon photocells and photodiodes have a smaller response time than photoresistors made from the same semiconductors. In the case of GaSe, photocells are also appreciably faster than photoresistors.

## 2. Photoelectrical Properties

The photoelectrical properties of the diode structure Cd–GaSe–Bi were studied under valve conditions. The measurements of current vs. light intensity and

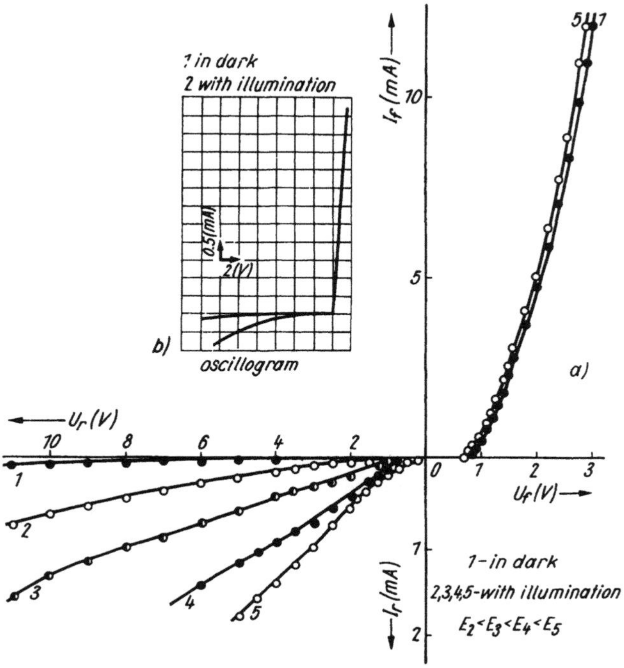


Fig. 3.
a) Current-voltage characteristics at various illumination intensities ($E$).
b) Oscillograms of current-voltage characteristics in the dark and under illumination

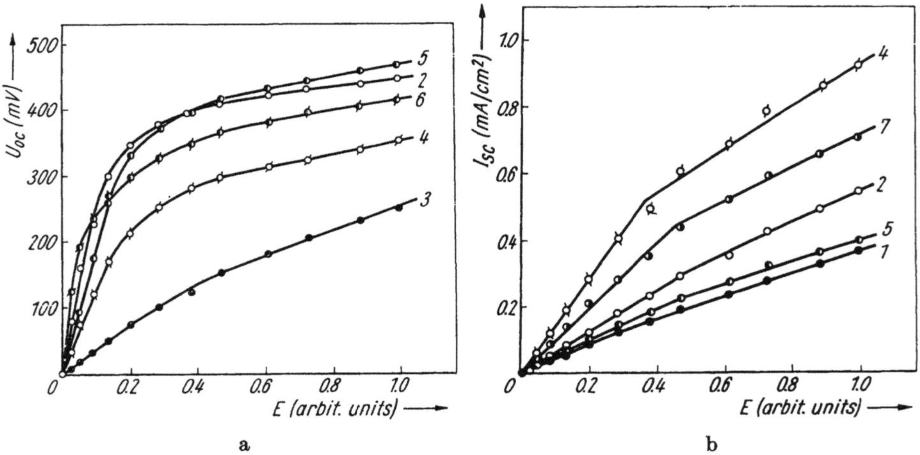


Fig. 4 a) Open-circuit voltage as a function of illumination intensity. b) Short-circuit current as a function of illumination intensity

voltage vs. light intensity characteristics and of the spectral distribution of photocurrent were made at room temperature.

We used a 170 W tungsten filament lamp as light source, and a water filter of 2 cm thickness which cut off the inactive and infrared spectral regions to reduce heating of the sample. The lamp was placed at a distance of 0.6 m from the sample to be illuminated. A mirror monochromator was used to obtain spec-

Fig. 5. Spectral characteristics of two samples

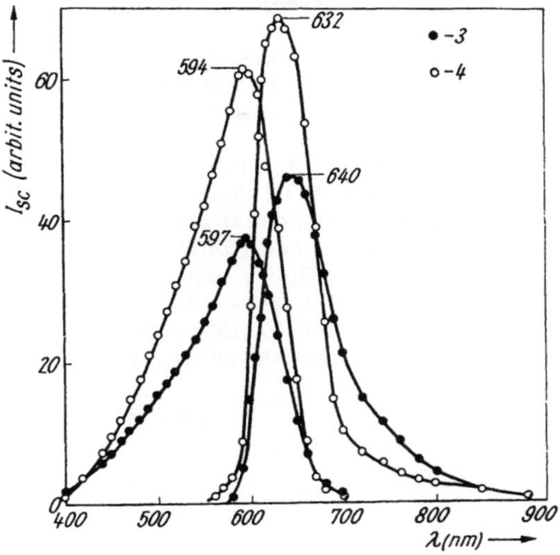


tral characteristics. The samples were illuminated on the Cd side for obtaining current-intensity and voltage-intensity characteristics, and on both sides for obtaining spectral characteristics. In both cases the Cd was found to be charged negatively, and this indicates again that the potential barrier is formed between Cd and GaSe.

The open-circuit voltage ($V_{oc}$) as a function of illumination intensity ($E$) for different samples is shown in Fig. 4a. As can be seen from the figure, the photo-e.m.f. sharply increases at low intensities and then saturates, the increase and saturation both having a linear character but with different slopes. For some samples (for example, samples 2, 5, and 6 in Fig. 4a) a logarithmic law is observed: $V_{oc} \sim \lg E$. The short-circuit current density ($I_{sc}$) as a function of illumination intensity is given in Fig. 4b. It can be seen that the short-circuit current density for all samples obeys the law $I_{sc} \sim E$, and obviously two intensity regions can be distinguished: For weak illumination the current increase is larger than for strong illumination. It may be noted that for different samples at maximum intensity ($E \approx 10^4$ lx) $V_{oc}$ is 250 to 450 mV and $I_{sc}$ is 300 to 900 μA/cm$^2$.

The dependence of the short-circuit current on the wavelength of the incident light is shown for two samples in Fig. 5. For illumination of the Cd side the photocurrent maxima for all samples are about 600 nm and cover the region from 400 to 700 nm. If the GaSe side is illuminated photosensitivity is observed between 560 and 900 nm, and the photocurrent maxima are shifted to the long-wavelength side (630 to 640 nm) and coincide with the photoconductivity maximum of GaSe. Another peculiarity of the spectral distribution of the photocurrent in GaSe diode structures lies in the fact that, for illumination of the GaSe side, the short-wavelength part of the spectrum sharply increases with increasing wavelength, and the long-wavelength part is less extended than for a photoconductor.

If the Cd side is illuminated there is a sharp decrease of photocurrent behind the maximum.

## References

[1] G. Fischer and I. Brebner, J. Phys. Chem. Solids **23**, 1363 (1962).
[2] G. A. Akhundov, G. B. Abdullaev, G. D. Guseinov, R. F. Mekhtiev, and M. Kh. Alieva, Congrès International sur la Physique des Semiconducteurs, Paris 1964 (p. 1227).
[3] R. H. Bube and E. L. Lind, Phys. Rev. **115**, 5 (1959); **119**, 1535 (1960).
[4] R. F. Mekhtiev, L. G. Paritskii, and S. M. Ryvkin, Fiz. tverd. Tela **5**, 1649 (1963).
[5] N. G. Basov, O. V. Bogdankevich, A. N. Pechenov, G. B. Abdullaev, G. A. Akhundov, and E. Yu. Salaev, Dokl. Akad. Nauk SSSR **161**, 5 (1965).
[6] G. A. Akhundov, Optika i Spektroskopiya **18**, 743 (1965); **19**, 302 (1965).
[7] R. W. Smith and A. Rose, Phys. Rev. **97**, 1531 (1955).
A. Rose, Phys. Rev. **97**, 1538 (1955).
[8] S. M. Ryvkin, Fotoelektricheskie yavleniya v poluprovodnikakh, Fizmatgiz 1963.

*(Received April 25, 1966)*

phys. stat. sol. 16, 215 (1966)

*Physikalisches Institut der Universität Frankfurt am Main*

# EPR Measurements on Diatomic Halogen Centers in Alkali Halides

By

W. DREYBRODT and D. SILBER

EPR spectra of diatomic halogen centers in alkali halides are investigated. Centers of the type $FBr^-$, $FJ^-$, and $ClJ^-$ are identified. The constants of the spin Hamiltonian are derived from the $FBr^-$ spectra by a perturbation calculation up to fourth order. By studying the photochemical properties of centers of the type $X_2^-$ and $XY^-$ information regarding the formation mechanismus of the $FCl^-$ and $FBr^-$ centers are obtained.

EPR-Spektren zweiatomiger Halogenzentren in Alkalihalogeniden werden untersucht. Spektren von Zentren des Typs $FBr^-$, $FJ^-$ und $ClJ^-$ werden identifiziert. Aus dem Spektrum des $FBr^-$-Zentrums werden mit Hilfe von Störungsrechnung bis zur 4. Ordnung die Konstanten des Spin-Hamilton-Operators berechnet. Untersuchungen photochemischer Eigenschaften von Zentren des Typs $X_2^-$ und $XY^-$ geben Aufschluß über die Bildungsmechanismen der Zentren $FCl^-$ und $FBr^-$.

## 1. Introduction

Diatomic halogen molecule ions of the type $XY^-$ in alkali halides are produced by X-irradiation at liquid nitrogen temperature. The homonuclear molecule ions of the type $X_2^-$, the $V_K$-centers, have been investigated by Känzig et al. [1, 2, 3] using EPR technique. Optical properties have been studied by Delbecq et al. [4, 5].

Heteronuclear centers have been produced in alkali halide crystals doped with extraneous halide ions. Wilkins and Gabriel found an $FCl^-$ center in KCl:$F^-$ by EPR measurements [6]. Schoemaker et al. [7] investigated a $ClBr^-$ center by EPR technique and optical measurements. Recently measurements at a $ClJ^-$ center have been published by Goldberg and Mahr [8]. In this paper we report measurements on the $FBr^-$ center in KCl and other alkali halide crystals. Furthermore we have found two new centers of this type, an $FJ^-$ center and a center which we believe to be a $ClJ^-$ associated with a fluorine ion.

As all these centers can be produced in KCl, it is now possible to compare the properties of at least six different centers in one alkali halide lattice. Besides some new thermochemical and photochemical relations between these centers have been found.

## 2. Experimental Techniques

The crystals were grown from the melt by Kyropoulos technique. For measurements on $FBr^-$, the melt of reagent grade KCl was doped with 0.2 mol% $Br^-$, 0.5 mol% $F^-$, and 0.1 mol% $Pb^{++}$.

For investigations of centers containing $J^-$, ultrapure KCl was used (ELH of E. Merck AG, Darmstadt). It was doped with $F^-$, $Pb^{++}$, and 1 mol% $J^-$.

The crystals were sawn into bars of $3 \times 3 \times 20$ mm³ parallel to a [111] axis of the crystal. They were X-rayed at 100 kV, 10 mA for several hours at 80 °K. The resonance measurements were performed with a Varian EPR spectrometer V 4500 with 100 kHz modulation and Fieldial V–FR 2503. The frequency was measured with a Sperry "microline" cavity wavemeter 12 X 1. For calibration a DPPH signal was used.

For measurements of the angular dependence of the spectra a cryostat was used with pivoted crystal holder. EPR measurements of the $Br^-$ doped crystals at 80 °K yield signals of the $Cl_2^-$, $FCl^-$, $ClBr^-$, and $FBr^-$ centers. UV irradiation between 300 and 400 nm destroys all centers except $FCl^-$, which increases considerably. After warming the crystal up to 150 °K and cooling it down again to 80 °K, $FCl^-$ has disappeared, and new strong lines appear. Signals of this kind have already been observed by Wilkins and Gabriel and tentatively interpreted as an $FBr^-$ center [6].

## 3. Results

### *3.1 EPR spectra of the $FBr^-$ center*

Fig. 1 shows three EPR spectra of the $FBr^-$ center in KCl, the magnetic field being directed along the [100], [111], and [110] axes of the crystal.

In the case of ***B*** along [100] only one spectrum is seen. It consists of 16 lines with nearly equal intensities. 8 lines should arise as F has a nuclear spin of 1/2

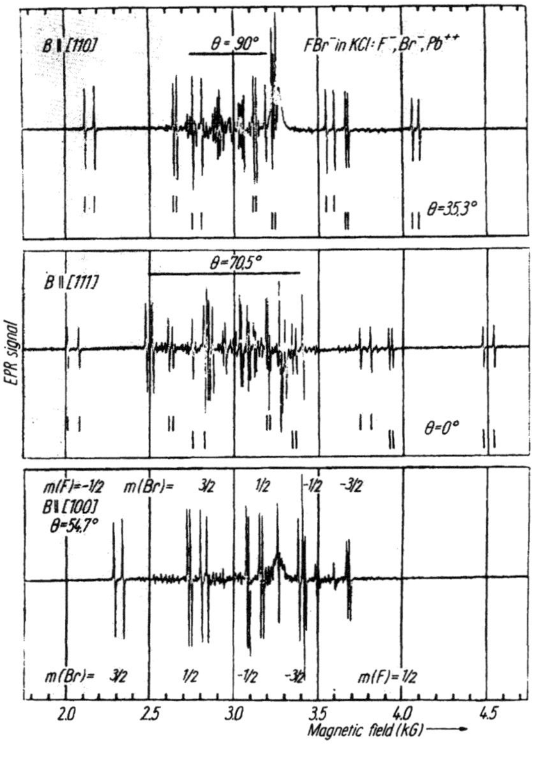


Fig. 1. EPR spectra of the $FBr^-$ center in KCl with field along [100], [110], and [111] axes, respectively. The narrow line at $B = 3170$ G is due to DPPH

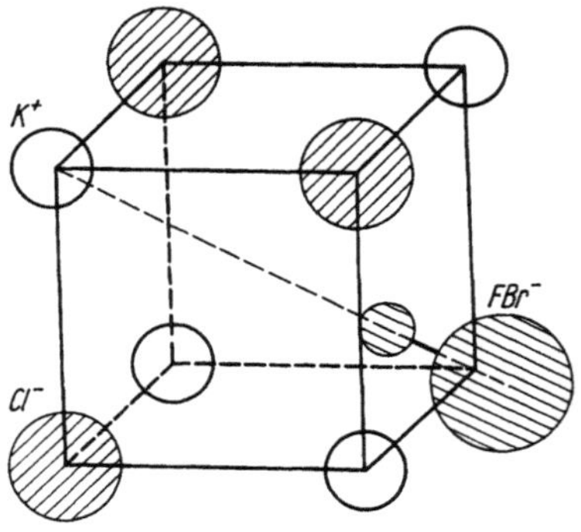


Fig. 2. Schematic model of the $FBr^-$ center

and Br one of 3/2, but since Br has two isotopes, $Br^{79}$ and $Br^{81}$, with natural abundance 50.57 and 49.43 respectively, this number is doubled.

Due to the large hfs splitting of Br and F the spectra are spread over a range of about 2 kG. Most of the spectra are strongly asymmetrical. From the fact that in the case of $\boldsymbol{B} \| [100]$ there are no spectra of different center orientations to the magnetic field, we conclude that the center has an axis parallel to the [111] direction, the symmetry being $C_{3v}$. The center axis and the magnetic field include an angle of 54.7°.

Consequently in $\boldsymbol{B} \| [111]$ orientation, a quarter of the centers have their axes parallel to the magnetic field. The others are equivalent at an angle of 70.5° to the magnetic field. The two spectra can be observed in Fig. 1.

In $\boldsymbol{B} \| [110]$ we also have two different spectra. Two of the four [111] directions are perpendicular to the $\langle 110 \rangle$ direction, and the other two include an angle of 35.3°.

### *3.2 Interpretation of the spectra*

Fig. 2 shows a schematic model of the $FBr^-$ center. Similar to the $FCl^-$ center the F should be located interstitially along the [111] direction, whereas the Br should relax from lattice site in the opposite direction. EPR measurements cannot determine the exact positions of the nuclei.

The hfs splittings of the centers parallel to the magnetic field are similar to those of $F_2^-$ and $Br_2^-$ respectively. The increase of the Br splitting and the decrease of the F splitting show the increasing polarity of the molecule due to the difference of the electronegativities.

To describe the line positions, we use a spin Hamiltonian of the form

$$\mathcal{H} = \beta_0 \boldsymbol{B}\,\mathbf{g}\,\mathbf{S} + g_0 \beta_0 \left\{ \sum_{j=1}^{2} \mathbf{S}\,\mathbf{A}_j\,\boldsymbol{I}_j + \boldsymbol{I}_2\,\mathbf{Q}\,\boldsymbol{I}_2 \right\};$$

$$\mathbf{g} = \begin{pmatrix} g_\perp & 0 & 0 \\ 0 & g_\perp & 0 \\ 0 & 0 & g_\| \end{pmatrix}, \quad \mathbf{A} = \begin{pmatrix} A_\perp & 0 & 0 \\ 0 & A_\perp & 0 \\ 0 & 0 & A_\| \end{pmatrix}, \quad \mathbf{Q} = \frac{1}{3} Q \begin{pmatrix} -1 & 0 & 0 \\ 0 & -1 & 0 \\ 0 & 0 & 2 \end{pmatrix},$$

where $j = 1$ refers to the F, $j = 2$ to the Br nucleus. The quadrupole term expressed by the tensor $\mathbf{Q}$ must be included because of the large quadrupole moment of Br.

To fit the constants, the operator must be transformed into appropriate sets of axes [9, 10]

$$\begin{aligned} \mathcal{H} = g\,\beta_0\,B\,S_{z'} + \sum_{j=1}^{2} g_0\,\beta_0 \Big\{ & S_{z'}\,K_j\,I_{zj} + S_{x'}\,I_{xj} \cdot A_{\perp j}\,A_{\| j}\,\frac{1}{K_j} + \\ & + S_{x'}\,I_{zj}\,(A_{\perp j}^2 - A_{\| j}^2) \sin\theta \cos\theta\,\frac{g_\|\,g_\perp}{g^2\,K_j} + A_{\perp j}\,I_{yj}\,S_{y'} \Big\} + \\ & + Q\,\frac{g_0\,\beta_0}{K_2^2\,g^2}\,(I_{z2}\,A_{\| 2}\,g_\| \cos\theta - I_{x2}\,A_{\perp 2}\,g_\perp \sin\theta)^2 \,; \end{aligned}$$

$$g^2 = g_\|^2 \cos^2\theta + g_\perp^2 \sin^2\theta\,, \qquad K^2\,g^2 = A_\|^2\,g_\|^2 \cos^2\theta + A_\perp^2\,g_\perp^2 \sin^2\theta\,.$$

$\theta$ is the angle between the center axis and the magnetic field. It should be noted that electron spin and the two nuclear spins must be described in three different coordinate systems. The hfs splittings due to the Br and F nuclei are rather large and cannot be regarded as small perturbations of the Zeeman splitting. Therefore, a Schrödinger perturbation calculation at which the energy denominator is an averaged Zeeman splitting, is not adequate, as it converges rather slowly.

Consequently we have added the diagonal term of the perturbation to the undisturbed Zeeman energy. So, in spite of different Zeeman splittings at the positions of the single lines, the energy denominators become approximately equal. They are given by the energy $h\nu$, $\nu$ being the frequency of the klystron radiation. Thus all odd orders of the perturbation series disappear and faster convergence is achieved.

As energy denominator we always use the Zeeman energy corresponding to the $g$-value. A closer examination shows that we have expanded — apart from small negligible terms — into a new kind of perturbation series in which the energy denominator is the difference of the optimal energies [11]. We have calculated all terms to second order and the important ones of fourth order. To second order, the positions of the lines are given by

$$\frac{\Delta E}{g_0 \beta_0} = \frac{g}{g_0} B + \sum_{j=1}^{2} K_j M_j + \frac{g_0 \beta_0 g_{\|}^2 g_{\perp}^2 \cos^2\theta \sin^2\theta}{2 g^4 h\nu} \left( \sum_{j=1}^{2} \frac{M_j}{K_j} [A_{\perp j}^2 - A_{\| j}^2] \right)^2 +$$
$$+ \frac{g_0 \beta_0}{4} \sum_{j=1}^{2} \left\{ \frac{J_j^2 + J_j - M_j^2 - M_j}{2h\nu + K_j} + \frac{J_j^2 + J_j - M_j^2 + M_j}{2h\nu - K_j} \right\} \left\{ \frac{A_{\| j}^2}{K_j^2} + 1 \right\} A_{\perp j}^2 .$$

Here the advantage of equal energy denominators for all lines becomes apparent. For all lines the perturbation terms are in a simple way dependent on the quantum numbers only. In the calculation of the line distances many of them cancel out. Thus the number of terms for the calculation of the spin Hamiltonian constants decreases considerably.

Especially, the terms of second order shift the terms of opposite quantum numbers equally, whereas the shift from the corrections of fourth order is of nearly equal amount, but in opposite direction. Fig. 3 shows this schematically. These properties allow a simple iteration method to be used in calculating the constants. From the parallel spectrum, the constants $A_{\|}^{F}$, $A_{\|}^{Br}$, and $g_{\|}$ can easily be determined by perturbation calculation to second order, as the contributions of higher orders are small in this case. A crude estimation of $A_{\perp}^{Br}$ can also be obtained from that spectrum.

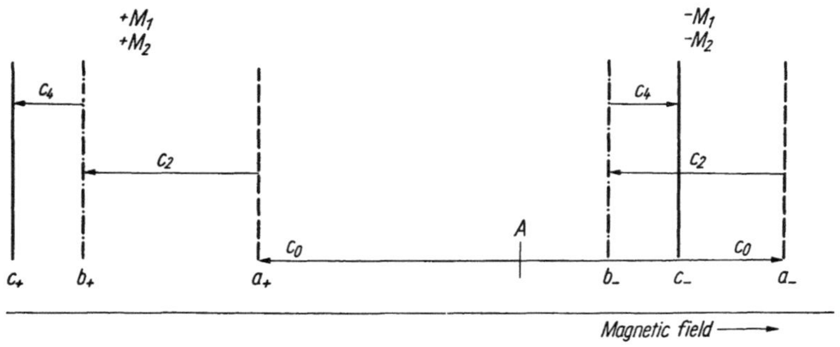


Fig. 3. Schematic representation of the corrections arising from second and fourth order perturbation terms respectively. $a_+$ and $a_-$ are the line positions calculated from the diagonal terms. Corrections from second order shift the lines by $c_2$ to positions $b_+$ and $b_-$. Those corrections from fourth order shift by $c_4$ to $c_+$ and $c_-$. The positions $c_+$ and $c_-$ are approximately the true line positions. From the point A the magnetic field refering to the $g$-value is determined

The other constants are calculated from the spectrum with $\theta = 54.7°$ as the spectrum with $\theta = 90°$ is too complex for identification of the lines. For this calculation we collect respectively two lines with opposite nuclear quantum numbers. According to Fig. 3 the center of two such lines is shifted from the position corresponding to the true $g$-value by the correction energies of second order. The symmetric corrections from fourth order may be neglected for the moment. With the aid of reasonable estimated constants $A_\perp$ and $g_\perp$ we calculate the second order corrections, and from these an effective $g$-value for each pair of lines. From the average of these values $g$ is calculated in a better approximation. This procedure may be iterated. With this improved $g$-value, the $A_\perp$ values are corrected by calculating the line distances from fourth order.

From the corrected $A_\perp$ a better $g$-value can be calculated in turn from the second order and then the whole procedure may be repeated as often as necessary. In doing this iteration, it becomes evident, however, that an exact fitting is not possible in principle.

An additional small correction is necessary, which is given by the quadrupole term [9, 10]

$$-\frac{2\,Q^2 \cos^2\theta \sin^2\theta}{K_2}\left(\frac{A_{\parallel 2}\,A_{\perp 2}\,g_\parallel\,g_\perp}{K_2^2\,g^2}\right)^2 M_2\,[4\,J_2\,(J_2+1) - 8\,M_2^2 - 1] +$$
$$+\frac{Q^2 \sin^4\theta}{2\,K_2}\left(\frac{A_{\perp 2}\,g_\perp}{K_2\,g}\right)^4 \cdot M_2\,[2\,J_2\,(J_2+1) - 2\,M_2^2 - 1]\,.$$

This correction has no significant influence on the $g$-value. Proceeding thus, the spectrum was fitted to within 5 G of the line positions. With the calculated constants, the 35.7° spectrum can be fitted satisfactorily.

The parameters of the spin Hamiltonian are:

$$A_\parallel^{\mathrm{F}} = 724.5 \pm 2\ \mathrm{G}\,,\quad A_\parallel^{\mathrm{Br^{81}}} = 588.5 \pm 2\ \mathrm{G}\,,\quad g_\parallel = 1.9893 \pm 0.0005\,,$$
$$A_\perp^{\mathrm{F}} = 89 \pm 10\ \mathrm{G}\,,\quad A_\perp^{\mathrm{Br^{81}}} = 125 \pm 15\ \mathrm{G}\,,\quad g_\perp = 2.128 \pm 0.003\,,$$
$$Q = 54 \pm 8\ \mathrm{G}\,.$$

### *3.3 The $FBr^-$ center in other crystals*

In NaCl the $FBr^-$ center is produced in the same way as in KCl. In KBr the $FBr^-$ center appears immediately after X-raying and bleaching out $V_K$. In NaBr, after X-raying and bleaching out $V_K$, a weak spectrum is observed, which seems to be due to $FBr^-$. Fig. 4 and 5 show the spectra in NaCl and KBr.

The parallel values of the hfs tensor have been calculated and turn out to be both smaller than in KCl. They are given by

$$A_\parallel^{\mathrm{F}} = 670\ \mathrm{G}\,,\quad A_\parallel^{\mathrm{Br^{81}}} = 576\ \mathrm{G}\quad \text{in NaCl},$$
$$A_\parallel^{\mathrm{F}} = 695\ \mathrm{G}\,,\quad A_\parallel^{\mathrm{Br^{81}}} = 579\ \mathrm{G}\quad \text{in KBr}.$$

The lines are considerably wider than in KCl. The line widths in the parallel spectra are

$$\Delta B = 5.2\ \mathrm{G}\ \text{in KCl}\,,\qquad \Delta B = 22\ \mathrm{G}\ \text{in NaCl}\,,\qquad \Delta B = 15\ \mathrm{G}\ \text{in KBr}.$$

The change of line widths is larger than expected from the ratio of magnetic moments and from the lattice constants.

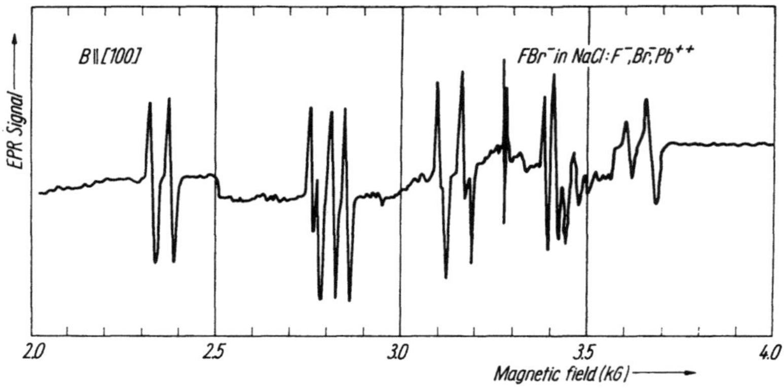


Fig. 4. EPR spectrum of the $FBr^-$ center in NaCl with magnetic field along [100] axis

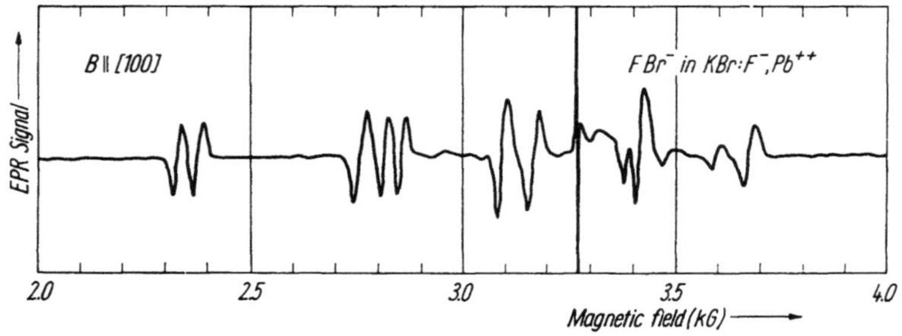


Fig. 5. EPR spectrum of the $FBr^-$ center in KBr with magnetic field along [100] axis

### *3.4 Mechanism of $FBr^-$ center formation*

By irradiation in the near uv the $FBr^-$ center is destroyed and other centers appear. In KCl the $FCl^-$ center is formed. In KBr the $Br_2^-$ center reappears, if irradiation into the absorption band of $Br_2^-$ is avoided. Irradiation with a Bausch & Lomb grating monochromator shows the wavelengths of about 310 nm to be most effective in destroying $FBr^-$. Irradiation into the $Br_2^-$ band reproduces $FBr^-$. Absorption measurements with a Perkin Elmer spectrophotometer 350 yield a weak absorption band with maximum at about 310 nm, behaving like the EPR signals of $FBr^-$ in optical experiments. The $FCl^-$ center in KCl cannot be destroyed by near — uv — irradiation ($\lambda \geqq 300$ nm). In KBr : $F^-$, $Cl^-$, $Pb^{++}$ the $FCl^-$ center cannot be produced by bleaching out $V_K$ and $FBr^-$ simultaneously.

Wilkins and Gabriel [6] explain the formation of $FCl^-$ from $V_K$ and of $FBr^-$ from $FCl^-$ by assuming that $F^\circ$ atoms migrate insterstitially. Hatcher and Dienes [12] have calculated the excitation energy of interstitial $Cl^\circ$ to be 0.5 eV. If the excitation energy of $F^\circ$ is of the same order, such a migration would not be likely by thermal excitations at 150 °K, the annealing temperature of $FCl^-$. Therefore, we should like to propose a different mechanism. During X-irradiation, $F^-$ ions are produced insterstitially, and we assume that there is relatively high probability for finding these interstitials as $F^-Br^-$ in lattice cells containing a $Br^-$ ion which widens the lattice. Additionally $V_K$ and $FCl^-$ centers and — to a certain extent — $FBr^-$ and $ClBr^-$ centers are found. If $V_K$ is bleached out holes are produced, and captured by an interstitial $F^-$. As there are by far more interstitials in cells containing no $Br^-$ ion and as the $FBr^-$ absorption wavelength is not far from that of the radiation used, new $FCl^-$ centers are formed. If the crystal is warmed up, the $FCl^-$ dissociates releasing a hole. These holes migrate through the lattice.

They might be trapped by two different mechanisms: formation of a $V_K$ center, or capture by an $F^-Br^-$ interstitial to form an $FBr^-$. We assume the cross section for the former process to be much smaller than that for the latter.

This is reasonable since the $F^-Br^-$ has an additional negative charge. Thus no $V_K$ centers but $FBr^-$ centers are produced. Experimentally this assumption is supported by the observation that after extended X-irradiation KCl crystals doped with approximately equal amounts of $F^-$ and $Br^-$ (0.3 mol%) exhibit strong signals of $FCl^-$ being of the same magnitude as the $V_K$ signals; but only weak $ClBr^-$ signals appear. This shows that the cross section for a hole to form either $V_K$ or $ClBr^-$ is much smaller than that for forming $FCl^-$. This model also explains the formation of $FCl^-$ centers by bleaching out $FBr^-$ in KCl.

In KBr, $F^-$ interstitials in cells containing $Cl^-$ are by far less likely, as the lattice is contracted. Thus no $FCl^-$ centers are produced in KBr.

We have made measurements on KCl crystals doped with $J^-$ and $Br^-$ in equal amounts, and additionally $F^-$ and $Pb^{++}$. In these crystals we have produced $FJ^-$ centers by X-raying, bleaching out $V_K$ and warming up, but we have not observed $FBr^-$. This means that $F^-$ interstitials are found more readily in cells containing $J^-$ because of the large lattice distortion.

### 3.5 Hole centers in KCl:$F^-$, $J^-$, $Pb^{++}$

In KCl crystals containing $F^-$, $J^-$, $Pb^{++}$ signals of $V_K$, $FCl^-$ and a new center are observed after X-irradiation (Fig. 6). After irradiation in the near uv only $FCl^-$ remains. Warming the crystals up to 150 °K destroys $FCl^-$ and a new strong spectrum is obtained. This spectrum has its largest extension in $\boldsymbol{B} \parallel [111]$

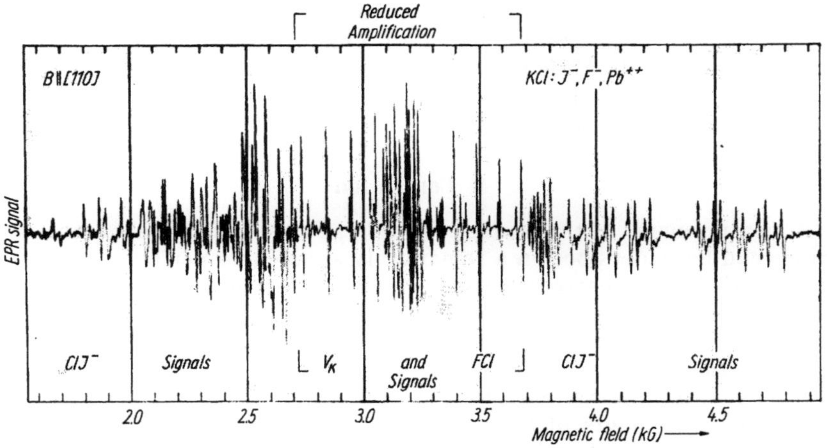


Fig. 6. EPR spectrum in KCl observed immediately after X-raying. The $V_K$ and $FCl^-$ signals are recorded with reduced amplification

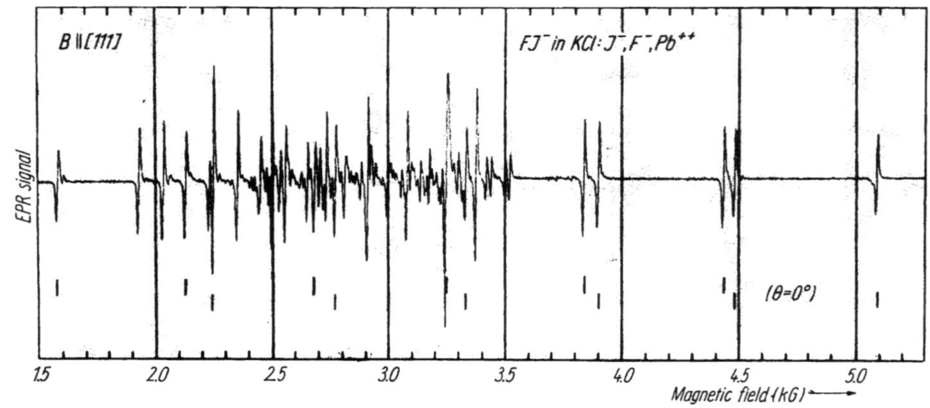


Fig. 7. EPR spectrum of $FJ^-$ with magnetic field along [111] axis

direction. Investigations of the angular dependence show that a spectrum of twelve lines can be separated out. In Fig. 7 twelve lines are marked. They are interpreted as the parallel spectrum of an $FJ^-$ center which has an axis of symmetry in the [111] direction of the crystal. The constants which can be derived from this spectrum are

$$g_{\|} = 1.93\,, \quad A_{\|}^{F} = 633\ \text{G}, \quad A_{\|}^{J} = 552\ \text{G}\,.$$

The interpretation as an $FJ^-$ center is supported by the fact that no isotopic lines are seen and the parallel splittings of J and F are reasonable from the magnetic moment of the J nucleus and the electronegativity difference of F and J. The other lines of the spectrum $\boldsymbol{B}\|$ [111] must be due to centers with an angle of 70.5° between center axis and the magnetic field. Irradiation in the near uv destroys this center, and $FCl^-$ appears.

The spectrum of the new center observed immediately after X-raying of this crystal has its largest extension in $\boldsymbol{B}\|$ [110]. The hfs splittings and the isotopic ratio suggest a $ClJ^-$ center. An additional splitting is seen, however. This splitting may be due to a hfs interaction with a nucleus with nuclear spin 1/2. As this center is observed only in F-doped crystals, we believe this center to be an aggregate of $ClJ^-$ with an $F^-$.

## 4. Conclusion

EPR measurements of the centers $F_2^-$, $FCl^-$, $FBr^-$, and $FJ^-$ show that the hfs splittings can be understood in a remarkably simple manner by considering the differences of electronegativity between the halogen atoms. The F hfs splittings decrease with increasing differences in electronegativity. The analogous rows show a similar behaviour.

The orientations of the centers containing $F^-$ show that $F^-$ tends to go to an interstitial site after X-raying rather than to remain in a lattice site.

Further measurements of mixed centers, especially in other alkali halides, should lead to a comprehensive understanding of their properties and their formation mechanisms.

*Acknowledgements*

We wish to thank Prof. Dr. W. Martienssen and Prof. Dr. H. A. Müser for many stimulating discussions. We also wish to thank Dr. R. Wehner for valuable discussions concerning the perturbation treatment.

## References

[1] T. G. Castner and W. Känzig, J. Phys. Chem. Solids **3**, 178 (1957).
[2] T. O. Woodruff and W. Känzig, J. Phys. Chem. Solids **5**, 268 (1958).
[3] W. Känzig and T. O. Woodruff, J. Phys. Chem. Solids **9**, 70 (1958).
[4] C. J. Delbecq, B. Smaller, and P. H. Yuster, Phys. Rev. **111**, 1235 (1958).
[5] C. J. Delbecq, W. Hayes, and P. H. Yuster, Phys. Rev. **121**, 1043 (1961).
[6] J. W. Wilkins and J. R. Gabriel, Phys. Rev. **132**, 1950 (1963).
[7] D. Schoemaker, C. J. Delbecq, and P. H. Yuster, Bull. Amer. Phys. Soc. **9**, 629 (1964).
[8] L. S. Goldberg and H. Mahr, Bull. Amer. Phys. Soc. **11**, 246 (1966).
[9] B. Bleaney, Phil. Mag. **42**, 441 (1951).
[10] W. Low, Paramagnetic Resonance in Solids, Solid State Phys. Suppl. **2**, 60 (1960).
[11] R. Wehner, private communication.
[12] R. D. Hatcher and G. J. Dienes, Phys. Rev. **134**, A 214 (1964).

*(Received April 27, 1966)*

phys. stat. sol. **16**, 225 (1966)

*Institut für Magnetische Werkstoffe der Deutschen Akademie der Wissenschaften zu Berlin, Jena*

# Zur Theorie des Ferromagnetismus magnetisch anisotroper Schichten mit kubischer Kristallstruktur

## I. Berechnung der Magnetisierung und der Curietemperatur mit der Methode der Greenschen Funktionen

Von

W. Brodkorb

Es wird eine ferromagnetische dünne Schicht betrachtet, deren Atomspins $S > 1/2$ in einem kubischen Gitter mit (100)-Oberflächenorientierung angeordnet sind. Der Hamiltonoperator des Systems enthält den Zeeman-, den Austausch- und magnetische Anisotropieterme. Es wird ein expliziter Ausdruck für die Magnetisierung angegeben, und dieser Ausdruck wird dazu benutzt, um die ferromagnetische und paramagnetische Curietemperatur zu berechnen. Dabei wird die Methode der Greenschen Funktionen angewendet.

A ferromagnetic thin film is considered with atomic spins $S > 1/2$ arranged in a cubic lattice with (100) surface orientation. The Hamiltonian of the system contains the Zeeman, exchange, and magnetic anisotropy terms. An explicit expression is given for the magnetization and this expression is used to obtain the ferromagnetic and paramagnetic Curie temperature. The Green function method is used.

### 1. Der Hamiltonoperator des Systems

In einer früheren Arbeit [1] wurden mit Hilfe der Methode der Greenschen Funktionen die Magnetisierung und die Curietemperatur für eine einkristalline ferromagnetische dünne Schicht nach dem Heisenbergmodell berechnet. Die in dieser Arbeit angegebenen Formeln für die Magnetisierung und die Curietemperatur enthalten die vom Flächengitter her bekannten Divergenzen [2], die im Felde $H = 0$ durch den Null-Spinwellenmodus verursacht werden. Am einfachsten lassen sich die Divergenzen nach Döring [2] dadurch beseitigen, daß man im Hamiltonoperator einen in jeder realen Schicht vorhandenen einachsigen Anisotropieterm berücksichtigt. In der vorliegenden Arbeit wollen wir deshalb den Einfluß einer einachsigen Volumenanisotropie in der Schichtebene und zugleich den Einfluß der Formanisotropie auf die magnetischen Eigenschaften einer dünnen ferromagnetischen Schicht untersuchen. Wir betrachten eine ferromagnetische dünne Schicht mit kubischer Gitterstruktur, deren Oberfläche parallel zur (100)-Ebene liegt und in der Schichtebene translationsinvariant ist. Für eine aus $n$ Netzebenen parallel zur Oberfläche bestehende Schicht hat der effektive Spin-Hamiltonoperator für unser System folgende Gestalt:

$$\mathcal{H} = -g\,\mu_B\,H \sum_{\nu} \sum_{j\in\nu} S^z_{\nu_j} - \frac{1}{2} K_{\|} \sum_{\nu} \sum_{j\in\nu} \left(S^z_{\nu_j}\right)^2 -$$

$$- \frac{1}{2} I \sum_{\langle\nu,\nu'\rangle} \sum_{\langle j\in\nu, j'\in\nu'\rangle} \left\{S^+_{\nu_j} S^-_{\nu'_{j'}} + S^z_{\nu_j} S^z_{\nu'_{j'}}\right\} +$$

$$+ \frac{1}{2} \sum_{\nu,\nu'} \sum_{j\in\nu, j'\in\nu'} D_{\nu_j \nu'_{j'}} \left\{\left[S^z_{\nu_j} S^z_{\nu'_{j'}} - \frac{1}{2} S^-_{\nu_j} S^+_{\nu'_{j'}}\right] \times\right.$$

$$\times \left[1 - 3\left(\beta^z_{\nu_j \nu'_{j'}}\right)^2\right] - \frac{3}{4} S^+_{\nu_j} S^+_{\nu'_{j'}} \left(\beta^-_{\nu_j \nu'_{j'}}\right)^2 - \frac{3}{4} S^-_{\nu_j} S^-_{\nu'_{j'}} \left(\beta^+_{\nu_j \nu'_{j'}}\right)^2 -$$

$$\left. - 3\, S^+_{\nu_j} S^z_{\nu'_{j'}}\, \beta^-_{\nu_j \nu'_{j'}} \beta^z_{\nu_j \nu'_{j'}} - 3\, S^-_{\nu_j} S^z_{\nu'_{j'}}\, \beta^+_{\nu_j \nu'_{j'}} \beta^z_{\nu_j \nu'_{j'}}\right\}, \tag{1}$$

wobei wir folgende Abkürzungen verwendet haben:

$$S^{\pm}_{\nu_j} = S^x_{\nu_j} \pm i\, S^y_{\nu_j}, \qquad \beta^{\pm}_{\nu_j \nu'_{j'}} = \beta^x_{\nu_j \nu'_{j'}} \pm i\, \beta^y_{\nu_j \nu'_{j'}}.$$

Der Hamiltonoperator (1) setzt sich aus dem Zeemanterm, einem Term für die einachsige Anisotropie, dem Diracschen Austauschoperator für die Wechselwirkung zwischen den nächsten Nachbarn und dem dipolaren Spin-Spin-Wechselwirkungsoperator (er liefert die Formanisotropie) zusammen. Der von uns gewählte einachsige Anisotropieterm ist der in der Spinwellentheorie oft benutzte Ausdruck für die einachsige Volumenanisotropie und besitzt dieselben Symmetrieeigenschaften wie die im Mikromagnetismus verwendete Energie für die einachsige Anisotropie. Die Berücksichtigung des Einflusses der magnetostatischen Wechselwirkungsenergie auf die magnetischen Eigenschaften der Ferromagnetika im Nichtkontinuumsbild geht auf van Vleck [3] zurück, wobei man sich auf die Dipol-Dipol-Wechselwirkung beschränkt und sich die Dipole im Atomkern konzentriert denkt. Der Hamiltonoperator (1) bezieht sich auf ein rechtwinkliges $\alpha = (x, y, z)$-Kristallkoordinatensystem, dessen Achsen $x, y, z$ mit den Richtungen [100], [010], [001] des Kristalls zusammenfallen sollen. $S^\alpha_{\nu_j}$ bedeutet die $\alpha$-Komponente des Spinoperators $S_{\nu_j}$ am Gitterpunkt $\boldsymbol{R}_{\nu_j}$ bei beliebiger Spinquantenzahl $S > 1/2$, wobei $\nu$ die Netzebene ($\nu = 1, 2, \ldots n$) und $j$ einen Gitterpunkt in der $\nu$-ten Netzebene charakterisiert. $I$ ist das positive Austauschintegral zwischen den benachbarten Gitterpunkten mit den Vektoren $\boldsymbol{R}_{\nu_j}$ und $\boldsymbol{R}_{\nu'_{j'}}$, $K_{\|}$ die Konstante der einachsigen Volumenanisotropie, $D_{\nu_j \nu'_{j'}} = g^2 \mu_B^2 |\boldsymbol{R}_{\nu_j} - \boldsymbol{R}_{\nu'_{j'}}|^{-3}$ die dipolare Spin-Spin-Kopplungskonstante ($D_{\nu_j \nu_j} \equiv 0$), $H$ der Betrag des in $z$-Richtung angelegten äußeren Magnetfeldes, $g$ der Landé-Faktor und $\mu_B$ das Bohrsche Magneton. Die $\beta^\alpha_{\nu_j \nu'_{j'}}$ sind die Richtungskosinus des Vektors $(\boldsymbol{R}_{\nu_j} - \boldsymbol{R}_{\nu'_{j'}})$ bezüglich des Kristallkoordinatensystems. Das Summationszeichen $\langle\cdots\rangle$ in (1) bedeutet die Summation über sämtliche benachbarte Gitterpunktpaare des Kristalls, in allen anderen Summen laufen die Indizes $\nu_j$, $\nu'_{j'}$ unabhängig voneinander über sämtliche Gitterpunkte des Kristalls.

## 2. Die Methode der Greenschen Funktionen

Bei den folgenden Rechnungen wollen wir die zweizeitigen temperaturabhängigen retardierten Greenschen Funktionen mit der in der Arbeit [4] angegebenen Definition benutzen. Die Fouriertransformierten dieser Greenschen Funktionen (im weiteren wollen wir die Fouriertransformierten selbst als Greensche Funktionen bezeichnen) genügen der Bewegungsgleichung [4]

$$E \langle\langle A|B\rangle\rangle = \frac{1}{2\pi} \langle [A, B]\rangle + \langle\langle [A, \mathcal{H}]|B\rangle\rangle . \tag{2}$$

Das Symbol $[A, B] = A\,B - B\,A$ bezeichnet den Kommutator und das Zeichen $\langle\cdots\rangle$ die statistische Mittelung über eine kanonische Gesamtheit. $A$ und $B$ sind Operatoren in Heisenbergdarstellung. Zur Berechnung des Energiespektrums der Spinmoden für den Hamiltonoperator (1) benötigen wir die folgenden zwei Greenschen Funktionen:

$$\langle\langle S^{\pm}_{\varrho l} | B_{\varphi m}\rangle\rangle . \tag{3}$$

Mit dem Hamiltonoperator (1) lauten die Bewegungsgleichungen der Greenschen Funktionen (3)

$$\begin{aligned}
&(E \mp g\,\mu_B\,H) \langle\langle S^{\pm}_{\varrho l}|B_{\varphi m}\rangle\rangle \mp K_{||} \langle\langle S^z_{\varrho_l} S^{\pm}_{\varrho_l}|B_{\varphi m}\rangle\rangle \pm \\
&\quad \pm I \sum_{\langle\underline{\varrho}, \nu\rangle} \sum_{\langle \underline{l \in \varrho}, j \in \nu\rangle} \left\{ \langle\langle S^z_{\varrho_l} S^{\pm}_{\nu_j}|B_{\varphi m}\rangle\rangle - \langle\langle S^{\pm}_{\varrho_l} S^z_{\nu_j}|B_{\varphi m}\rangle\rangle \right\} \pm \\
&\quad \pm \frac{1}{2} \sum_{\nu} \sum_{j\in\nu} D_{\varrho_l \nu_j} \left\{ \left[1 - 3\left(\beta^z_{\varrho_l \nu_j}\right)^2\right] \left[ \langle\langle S^z_{\varrho_l} S^{\pm}_{\nu_j}|B_{\varphi m}\rangle\rangle + 2 \langle\langle S^{\pm}_{\varrho_l} S^z_{\nu_j}|B_{\varphi m}\rangle\rangle \right] + \right. \\
&\quad + 3 \left(\beta^{\pm}_{\varrho_l \nu_j}\right)^2 \langle\langle S^z_{\varrho_l} S^{\mp}_{\nu_j}|B_{\varphi m}\rangle\rangle + 6\, \beta^{\pm}_{\varrho_l \nu_j} \beta^z_{\varrho_l \nu_j} \langle\langle S^z_{\varrho_l} S^z_{\nu_j}|B_{\varphi m}\rangle\rangle - \\
&\quad \left. - 3\, \beta^z_{\varrho_l \nu_j} \beta^{\mp}_{\varrho_l \nu_j} \langle\langle S^{\pm}_{\varrho_l} S^{\pm}_{\nu_j}|B_{\varphi m}\rangle\rangle - 3\, \beta^z_{\varrho_l \nu_j} \beta^{\pm}_{\varrho_l \nu_j} \langle\langle S^{\pm}_{\varrho_l} S^{\mp}_{\nu_j}|B_{\varphi m}\rangle\rangle \right\} = \frac{1}{2\pi} \delta_{\varrho\varphi}\, \delta_{lm}\, \omega^{\pm}_{\varrho}
\end{aligned} \tag{4}$$

mit $\omega^{\pm}_{\varrho} = \left\langle \left[S^{\pm}_{\varrho_l}, B_{\varrho_l}\right] \right\rangle$. (Wegen der Translationssymmetrie des Gitters in der Schichtebene hängt $\omega^{\pm}_{\varrho}$ nicht vom Index $l$ ab.)

Die unterstrichenen Indizes an den Summen sind als fixiert zu betrachten. Für die auf der linken Seite der Bewegungsgleichungen (4) auftretenden Greenschen Funktionen höherer Ordnung des Typs $\langle\langle S^z_{\sigma_p} S^{\pm}_{\nu_j}|B_{\varphi m}\rangle\rangle$ definieren wir eine modifizierte Tyablikovsche Entkopplungsvorschrift [5]

$$\langle\langle S^z_{\sigma_p} S^{\pm}_{\nu_j}|B_{\varphi m}\rangle\rangle \approx \langle S^z\rangle \langle\langle S^{\pm}_{\nu_j}|B_{\varphi m}\rangle\rangle , \tag{5}$$

wobei wir unter $\langle S^z\rangle$ den arithmetischen Mittelwert sämtlicher Netzebenenmagnetisierungen $\langle S^z_{\varrho}\rangle$ verstehen wollen,

$$\langle S^z\rangle = \frac{1}{n} \sum_{\varrho=1}^{n} \langle S^z_{\varrho}\rangle . \tag{6}$$

Die verbleibenden Greenschen Funktionen höherer Ordnung können in erster Näherung gegenüber den Greenschen Funktionen $\langle\langle S^z_{\sigma_p} S^{\pm}_{\nu_j}|B_{\varphi m}\rangle\rangle$ vernachlässigt werden. Die Bewegungsgleichungen (4) spezialisieren sich jetzt zu

$$\Big\{E \mp \Big[g\,\mu_B H + K_{||}\langle S^z\rangle + \langle S^z\rangle \sum_{\langle\underline{\varrho},\nu\rangle}\ \sum_{\langle l\in\underline{\varrho}, j\in\nu\rangle} I - \langle S^z\rangle \sum_{\nu}\sum_{j\in\nu} D_{\varrho_l\nu_j}\Big(1 - 3\left(\beta^z_{\varrho_l\nu_j}\right)^2\Big)\Big]\Big\} \times$$
$$\times \langle\langle S^{\pm}_{\varrho_l} | B_{\varphi_m}\rangle\rangle \pm$$
$$\pm \langle S^z\rangle\, I \sum_{\langle\underline{\varrho},\nu\rangle}\ \sum_{\langle l\in\underline{\varrho}, j\in\nu\rangle} \langle\langle S^{\pm}_{\nu_j} | B_{\varphi_m}\rangle\rangle \pm$$
$$\pm \frac{1}{2}\langle S^z\rangle \sum_{\nu}\sum_{j\in\nu} D_{\varrho_l\nu_j}\Big(1 - 3\left(\beta^z_{\varrho_l\nu_j}\right)^2\Big)\langle\langle S^{\pm}_{\nu_j} | B_{\varphi_m}\rangle\rangle \pm$$
$$\pm \frac{3}{2}\langle S^z\rangle \sum_{\nu}\sum_{j\in\nu} D_{\varrho_l\nu_j}\left(\beta^{\pm}_{\varrho_l\nu_j}\right)^2 \langle\langle S^{\mp}_{\nu_j} | B_{\varphi_m}\rangle\rangle = \frac{1}{2\pi}\,\delta_{\varrho\varphi}\,\delta_{lm}\,\omega^{\pm}_{\varrho}\,. \tag{7}$$

Die Translationssymmetrie des Gitters in der Schichtebene bedingt für die Greenschen Funktionen folgenden Ansatz

$$\langle\langle S^{\pm}_{\nu_j} | B_{\varphi_m}\rangle\rangle = \frac{1}{N}\sum_{h} e^{ih(j-m)}\, Q^{h\pm}_{\nu\varphi}\,. \tag{8}$$

$N$ ist die Anzahl der Gitterpunkte in einer Netzebene. $h$ bezeichnet den Wellenvektor in der Schichtebene, die Summation ist über die erste Brillouinzone zu erstrecken. Mit dem Ansatz (8) erhalten wir aus (7) $2n$ Gleichungen zur Bestimmung der $Q^{h\pm}_{\nu\varphi}$,

$$\Big\{E \mp \Big[g\,\mu_B H + \langle S^z\rangle K_{||} + \langle S^z\rangle \sum_{\langle\underline{\varrho},\nu\rangle} I_{\varrho\nu}(0) - \langle S^z\rangle \sum_{\nu} m^z_{\varrho\nu}(0)\Big]\Big\}\, Q^{h\pm}_{\varrho\varphi} \pm$$
$$\pm \langle S^z\rangle \sum_{\langle\underline{\varrho},\nu\rangle} I_{\varrho\nu}(h)\, Q^{h\pm}_{\nu\varphi} \pm \frac{1}{2}\langle S^z\rangle \sum_{\nu} m^z_{\varrho\nu}(h)\, Q^{h\pm}_{\nu\varphi} \pm$$
$$\pm \frac{3}{2}\langle S^z\rangle \sum_{\nu} m^{\pm}_{\varrho\nu}(h)\, Q^{h\mp}_{\nu\varphi} = \frac{1}{2\pi}\,\delta_{\varrho\varphi}\,\omega^{\pm}_{\varrho}\,, \tag{9}$$

wobei wir die Größen

$$I_{\varrho\nu}(h) = I \sum_{\langle l\in\underline{\varrho}, j\in\underline{\nu}\rangle} e^{ih(j-l)}\,,\quad I_{\varrho\varrho}(h) \equiv I_1(h)\,,\quad I_{\varrho\varrho\pm 1}(h) \equiv I_2(h)\,, \tag{10a}$$

$$m^z_{\varrho\nu}(h) = \sum_{j\in\underline{\nu}} D_{\varrho_l\nu_j}\Big[1 - 3\left(\beta^z_{\varrho_l\nu_j}\right)^2\Big] e^{ih(j-l)}\,,\quad m^{\pm}_{\varrho\nu}(h) = \sum_{j\in\underline{\nu}} D_{\varrho_l\nu_j}\left(\beta^{\pm}_{\varrho_l\nu_j}\right)^2 e^{ih(j-l)} \tag{10b}$$

eingeführt haben. Die Dipolsummen $m^{\alpha}_{\varrho\nu}(h)$ $(\alpha = z, \pm)$ lassen sich für $h \neq 0$ leicht berechnen, wenn man sich mit der Langwellennäherung begnügt. Wir können in diesem Falle von der Summation zur Integration über die zweidimensionalen Ortskoordinaten übergehen und die Integration ähnlich wie in der Arbeit [6] durchführen. Das Ergebnis dieser Integration lautet

$$\left.\begin{aligned}
m^z_{\varrho\nu}(h) &= m^z_{\varrho\nu}(0) + 2\pi N_2 g^2 \mu_B^2 |h| \cos^2\Theta_h \exp[-|h|\,a\,\delta\,|\nu-\varrho|] \\
&\hspace{16em}\text{für alle } \nu, \\
m^{\pm}_{\varrho\varrho}(h) &= m^z_{\varrho\varrho}(0) + \frac{2\pi}{3} N_2 g^2 \mu_B^2 |h| (1 + \sin^2\Theta_h)\,, \\
m^{\pm}_{\varrho\nu}(h) &= m^z_{\varrho\nu}(0) + \frac{2\pi}{3} N_2 g^2 \mu_B^2 |h| \left[1 \mp \frac{|\nu-\varrho|}{(\nu-\varrho)} \sin\Theta_h\right]^2 \times \\
&\qquad \times \exp[-|h|\,a\,\delta\,|\nu-\varrho|] \qquad \text{für } \nu \neq \varrho\,.
\end{aligned}\right\} \tag{11}$$

$N_2$ ist die Anzahl der Gitterpunkte pro Flächeneinheit, $\delta$ ein Zahlenfaktor, der für das einfach kubische Gitter den Wert 1, für kubisch raumzentriertes und kubisch flächenzentriertes Gitter den Wert 1/2 annimmt, $a$ die Gitterkonstante und $\Theta_h$ der Winkel zwischen dem Wellenvektor $h$ und der $z$-Richtung. Da die Dipolsummen für $h \neq 0$ im wesentlichen dem Betrag des Wellenvektors $h$ direkt proportional sind, können wir im Rahmen der Langwellennäherung die $h$-abhängigen Terme gegenüber $m^z_{\varrho\nu}(0)$ vernachlässigen,

$$m^{\pm}_{\varrho\nu}(h) \approx m^z_{\varrho\nu}(0) , \qquad m^z_{\varrho\nu}(h) \approx m^z_{\varrho\nu}(0) . \tag{12a}$$

Zur weiteren Vereinfachung der Rechnungen beschränken wir uns in den Dipolsummen für $h = 0$, die sich auf verschiedene Netzebenen beziehen, auf die Dipolsummen benachbarter Schichtebenen,

$$m^z_{\varrho\varrho}(0) = n^z_1 = -\frac{1}{2} n^x_1 , \quad m^z_{\varrho\varrho\pm 1}(0) = n^z_2 = -\frac{1}{2} n^x_2 , \quad m^z_{\varrho\nu}(0) = 0 \text{ sonst} . \tag{12b}$$

(Eine numerische Abschätzung zu (12b) erfolgt im Teil II dieser Arbeit. Für kubisch raumzentrierte Gitter erhielten wir z.B. $|m^z_{\varrho\varrho\pm 1}(0)| \approx 40\%$ von $|m^z_{\varrho\varrho}(0)|$, $|m^z_{\varrho\varrho\pm 2}(0)| \approx 3\%$ von $|m^z_{\varrho\varrho}(0)|$.)
Unter Beachtung von (10a, b) und (12a, b) vereinfachen sich die Bewegungsgleichungen (9) zu

$$\left.\begin{aligned}
&(E \mp E_h) Q^{h\pm}_{1\varphi} \pm \langle S^z\rangle \left[I_2(0) + \frac{1}{2} n^x_2\right] Q^{h\pm}_{1\varphi} \pm \langle S^z\rangle \left[I_2(h) - \frac{1}{4} n^x_2\right] Q^{h\pm}_{2\varphi} \mp \\
&\quad \mp \frac{3}{4} \langle S^z\rangle n^x_1 Q^{h\mp}_{1\varphi} \mp \frac{3}{4} \langle S^z\rangle n^x_2 Q^{h\mp}_{2\varphi} = \frac{1}{2\pi} \delta_{1\varphi}\, \omega^{\pm}_1 , \\
&(E \mp E_h) Q^{h\pm}_{\varrho\varphi} \pm \langle S^z\rangle \left[I_2(h) - \frac{1}{4} n^x_2\right] (Q^{h\pm}_{\varrho+1\varphi} + Q^{h\pm}_{\varrho-1\varphi}) \mp \\
&\quad \mp \frac{3}{4} \langle S^z\rangle n^x_1 Q^{h\mp}_{\varrho\varphi} \mp \frac{3}{4} \langle S^z\rangle n^x_2 (Q^{h\mp}_{\varrho+1\varphi} + Q^{h\mp}_{\varrho-1\varphi}) = \frac{1}{2\pi} \delta_{\varrho\varphi}\, \omega^{\pm}_\varrho , \\
&(E \mp E_h) Q^{h\pm}_{n\varphi} \pm \langle S^z\rangle \left[I_2(0) + \frac{1}{2} n^x_2\right] Q^{h\pm}_{n\varphi} \pm \langle S^z\rangle \left[I_2(h) - \frac{1}{4} n^x_2\right] Q^{h\pm}_{n-1\varphi} \mp \\
&\quad \mp \frac{3}{4} \langle S^z\rangle n^x_1 Q^{h\mp}_{n\varphi} \mp \frac{3}{4} \langle S^z\rangle n^x_2 Q^{h\mp}_{n-1\varphi} = \frac{1}{2\pi} \delta_{n\varphi}\, \omega^{\pm}_n ; \\
&\qquad\qquad \varrho = 2, 3, \ldots (n-1)
\end{aligned}\right\} \tag{13}$$

mit

$$E_h = g\,\mu_B\, H + \langle S^z\rangle \left\{K_{||} + [I_1(0) - I_1(h)] + \frac{1}{2}(n^x_1 + 2\,n^x_2) + 2\,I_2(0) + \frac{1}{4} n^x_1\right\}.$$

Wir führen in (13) die unitäre Transformation [7]

$$Q^{h\pm}_{\varrho\varphi} = \sum_{\tau=1}^{n} T^h_{\varrho\tau} L^{h\pm}_{\tau\varphi} \tag{14}$$

durch und stellen an die $Q_{\varrho\varphi}^{h\pm}$ folgende Transformationsforderungen:

$$\left.\begin{aligned} \left[I_2(0)+\frac{1}{2}n_2^x\right]Q_{1\varphi}^{h\pm}+\left[I_2(h)-\frac{1}{4}n_2^x\right]Q_{2\varphi}^{h\pm} &= \sum_{\tau=1}^{n} T_{1\tau}^h x_\tau^h L_{\tau\varphi}^{h\pm}\,,\\ \left[I_2(h)-\frac{1}{4}n_2^x\right]\left(Q_{\varrho+1\varphi}^{h\pm}+Q_{\varrho-1\varphi}^{h\pm}\right) &= \sum_{\tau=1}^{n} T_{\varrho\tau}^h x_\tau^h L_{\tau\varphi}^{h\pm}\,,\\ \left[I_2(0)+\frac{1}{2}n_2^x\right]Q_{n\varphi}^{h\pm}+\left[I_2(h)-\frac{1}{4}n_2^x\right]Q_{n-1\varphi}^{h\pm} &= \sum_{\tau=1}^{n} T_{n\tau}^h x_\tau^h L_{\tau\varphi}^{h\pm}\,;\\ \varrho &= 2,3,\ldots(n-1)\,,\\ \sum_{\varrho=1}^{n}\left(Q_{\varrho\varphi}^{h\pm}\right)^2 &= \sum_{\tau=1}^{n}\left(L_{\tau\varphi}^{h\pm}\right)^2. \end{aligned}\right\}\tag{15}$$

Die Forderungen (15) transformieren sich in die Gleichungen

$$\left.\begin{aligned} \left(\beta_h-\bar{x}_\tau^h\right)T_{1\tau}^h+T_{2\tau}^h &= 0\,,\\ T_{\varrho-1\tau}^h-\bar{x}_\tau^h T_{\varrho\tau}^h+T_{\varrho+1\tau}^h &= 0\,,\\ T_{n-1\tau}^h+\left(\beta_h-\bar{x}_\tau^h\right)T_{n\tau}^h &= 0\,;\\ \varrho &= 2,3,\ldots(n-1)\,,\\ \sum_{\varrho=1}^{n}T_{\varrho\tau}^h T_{\varrho\tau'}^h &= \delta_{\tau\tau'}\,, \end{aligned}\right\}\tag{16}$$

mit

$$x_\tau^h=\bar{x}_\tau^h\left[I_2(h)-\frac{1}{4}n_2^x\right]\quad\text{und}\quad\beta_h=\frac{\left[I_2(0)+\frac{1}{2}n_2^x\right]}{\left[I_2(h)-\frac{1}{4}n_2^x\right]}.$$

Aus (13) erhalten wir mit (14) und (15)

$$\left.\begin{aligned} &\left(E\mp E_{h\tau}^{(1)}\right)L_{\tau\varphi}^{h\pm}\pm E_{h\tau}^{(2)}L_{\tau\varphi}^{h\mp}\pm\sum_{\tau'(\neq\tau)}\tilde{\gamma}_{\tau\tau'}^h L_{\tau'\varphi}^{h\mp}=\frac{1}{2\pi}\omega_\varphi^{\pm}T_{\varphi\tau}^h\,,\\ &E_{h\tau}^{(1)}=E_h-\langle S^z\rangle x_\tau^h\,,\quad E_{h\tau}^{(2)}=-\frac{3}{4}\langle S^z\rangle\left[n_1^x+n_2^x\bar{x}_\tau^h\right]+\tilde{\gamma}_{\tau\tau}^h\,,\\ &\tilde{\gamma}_{\tau\tau'}^h=\langle S^z\rangle\gamma_{\tau\tau'}^h=\langle S^z\rangle\cdot\frac{3}{4}n_2^x\beta_h\left(T_{1\tau}^h T_{1\tau'}^h+T_{n\tau}^h T_{n\tau'}^h\right);\\ &\tau=1,2,\ldots n\,. \end{aligned}\right\}\tag{17}$$

Die Gleichungen (17) lassen sich praktisch nur näherungsweise lösen. Um eine analytische Lösung zu erhalten, vernachlässigen wir die Terme $\sum_{\tau'(\neq\tau)}\cdots$ (für raumzentrierte Gitter gilt z. B. $\gamma_{\tau\tau'}^h<10\%$ von $n_1^x$) und bekommen dadurch aus (17)

$$\begin{aligned} L_{\tau\varphi}^{h\pm}=\frac{1}{4\pi}\omega_\varphi^{\pm}T_{\varphi\tau}^h&\left\{\frac{E_{h\tau}\mp E_{h\tau}^{(1)}}{E_{h\tau}(E+E_{h\tau})}+\frac{E_{h\tau}\pm E_{h\tau}^{(1)}}{E_{h\tau}(E-E_{h\tau})}\right\}\pm\\ &\pm\frac{1}{4\pi}\omega_\varphi^{\mp}T_{\varphi\tau}^h\left\{\frac{-E_{h\tau}^{(2)}}{E_{h\tau}(E-E_{h\tau})}+\frac{E_{h\tau}^{(2)}}{E_{h\tau}(E+E_{h\tau})}\right\} \end{aligned}\tag{18}$$

mit

$$\left.\begin{aligned} E_{h\tau} &= \{(E_{h\tau}^{(1)})^2 + (E_{h\tau}^{(2)})^2\}^{1/2} = \{[g\,\mu_B H + \langle S^z\rangle E_{h\tau}^{(1)}]\,[g\,\mu_B H + \langle S^z\rangle E_{h\tau}^{(2)}]\}^{1/2}\,,\\ E_{h\tau}^{(1)} &= K_{||} + [I_1(0) - I_1(h)] + \frac{3}{2}(n_1^x + 2\,n_2^x) + 2\,I_2(0)\,\times\\ &\quad\times\left[1 - \frac{1}{2}\frac{I_2(h)}{I_2(0)}\bar{x}_\tau^h\right] - 2\,n_2^x\left(1 - \frac{1}{2}\bar{x}_\tau^h\right) - \gamma_{\tau\tau}^h\,,\\ E_{h\tau}^{(2)} &= K_{||} + [I_1(0) - I_1(h)] + 2\,I_2(0)\left[1 - \frac{1}{2}\frac{I_2(h)}{I_2(0)}\bar{x}_\tau^h\right] +\\ &\quad + n_2^x\left(1 - \frac{1}{2}\bar{x}_\tau^h\right) + \gamma_{\tau\tau}^h\,. \end{aligned}\right\} \tag{18a}$$

Die gefundenen Lösungen $L_{\tau\varphi}^{h\pm}$ gestatten es, die gesuchten Greenschen Funktionen $\langle\langle S_{\varrho_l}^{\pm} | B_{\varphi_m}\rangle\rangle$ sofort aufzuschreiben (unter Beachtung von (14) und (8)):

$$\begin{aligned} \langle\langle S_{\varrho_l}^{\pm} | B_{\varphi_m}\rangle\rangle &= \frac{1}{4\pi N}\sum_h e^{ih(l-m)} \sum_{\tau=1}^{n} T_{\varrho\tau}^h\, T_{\varphi\tau}^h\,\times\\ &\times\left\{\omega_\varphi^{\pm}\left[\frac{E_{h\tau} \mp E_{h\tau}^{(1)}}{E_{h\tau}(E + E_{h\tau})} + \frac{E_{h\tau} \pm E_{h\tau}^{(1)}}{E_{h\tau}(E - E_{h\tau})}\right] \pm\right.\\ &\left.\pm\,\omega_\varphi^{\mp}\left[\frac{-E_{h\tau}^{(2)}}{E_{h\tau}(E - E_{h\tau})} + \frac{E_{h\tau}^{(2)}}{E_{h\tau}(E + E_{h\tau})}\right]\right\}, \end{aligned} \tag{19}$$

wobei die Größen $E_{h\tau}$, die Pole der Greenschen Funktionen (18), die Energien der Spinmoden sind.

### 3. Die Berechnung der Magnetisierung

Nach dem Spektraltheorem gilt folgender Zusammenhang zwischen den Greenschen Funktionen $\langle\langle A | B\rangle\rangle$ und den Korrelationsfunktionen $\langle B(t')\,A(t)\rangle$ [4]:

$$\langle B(t')\,A(t)\rangle = i\lim_{\varepsilon\to 0}\int_{-\infty}^{+\infty}\frac{\langle\langle A|B\rangle\rangle_{E+i\varepsilon} - \langle\langle A|B\rangle\rangle_{E-i\varepsilon}}{e^{\beta E} - 1}\, e^{-iE(t-t')}\,dE \tag{20}$$

mit $\beta = \frac{1}{K_B\,T}$; $K_B$ Boltzmannkonstante, $T$ absolute Temperatur.

Aus den Korrelationsfunktionen kann man die statistischen Mittelwerte makroskopischer Größen berechnen.

Wir wenden das Spektraltheorem (20) auf die erste Greensche Funktion (19) an. Dabei setzen wir $t = t'$, $B_{\varphi_m} = e^{a S_{\varphi_m}^z}\, S_{\varphi_m}^-$ und $\varrho_l = \varphi_m$ (es gilt dann $\omega_\varrho^- = 0$):

$$\langle e^{a S_\varrho^z}\, S_\varrho^-\, S_\varrho^+\rangle = \omega_\varrho^+\, A_\varrho^{(S)}\,,$$

$$A_\varrho^{(S)} = \frac{1}{2N}\sum_{\tau=1}^{n}\sum_h (T_{\varrho\tau}^h)^2\,\frac{E_{h\tau}^{(1)}}{E_{h\tau}}\coth\frac{\beta\,E_{h\tau}}{2} - \frac{1}{2}\,. \tag{21}$$

Nach Callen [8] können wir aus (21) eine Differentialgleichung für den statistischen Mittelwert $\langle e^{a S^z_\varrho} \rangle$ gewinnen, deren Lösung für $a = 0$ lautet

$$\langle S^z_\varrho \rangle = \frac{[S - A^{(S)}_\varrho]\,[1 + A^{(S)}_\varrho]^{2S+1} + [S + 1 + A^{(S)}_\varrho]\,[A^{(S)}_\varrho]^{2S+1}}{[1 + A^{(S)}_\varrho]^{2S+1} - [A^{(S)}_\varrho]^{2S+1}} \tag{22}$$

### 3.1 Der Tieftemperaturbereich

Um die Magnetisierungsformel (22) besser mit den Ergebnissen der Spinwellentheorie vergleichen zu können, wollen wir sie nach Potenzen von $A^{(S)}_\varrho$ entwickeln (für $T \to 0$ wird $A^{(S)}_\varrho \approx 0$) [9],

$$\langle S^z_\varrho \rangle = S - A^{(S)}_\varrho + 0[(A^{(S)}_\varrho)^3]\,. \tag{23}$$

Die Entwicklung (23) führt (bei Benutzung von (6)) zu folgender Formel für die Magnetisierung:

$$\langle S^z \rangle = S - \frac{1}{2\,n\,N} \sum_{\tau=1}^{n} \sum_{h} \left( \frac{E^{(1)}_{h\tau}}{E_{h\tau}} - 1 \right) - \frac{1}{n\,N} \sum_{\tau=1}^{n} \sum_{h} \frac{E^{(1)}_{h\tau}}{E_{h\tau}} \frac{1}{e^{\beta E_{h\tau}} - 1}\,. \tag{24}$$

Die Gleichung (24) ist die bekannte (bei Schichten entsprechend modifizierte) Holstein-Primakoff-Formel [10]. Der zweite Term der rechten Seite von (24) gibt die schichtdickenabhängige Abweichung der Größe $\langle S^z \rangle_{T=0}$ von $S$ an, ist aber so klein, daß er gewöhnlich vernachlässigt werden kann. Bei Spezialisierung der Formel (24) auf den isotropen Fall für $S = 1/2$ erhalten wir exakt die Resultate der Blochschen Spinwellentheorie [11].

### 3.2 Paramagnetische und ferromagnetische Curietemperatur

In der Nähe der Curietemperatur ist $A^{(S)}_\varrho$ groß gegen Eins. Wir können deshalb (22) nach inversen Potenzen von $A^{(S)}_\varrho$ entwickeln [9],

$$\langle S^z_\varrho \rangle = \frac{1}{3} S\,(S+1)\,\frac{1}{A^{(S)}_\varrho} + 0\,[(A^{(S)}_\varrho)^{-2}]\,. \tag{25}$$

Wenn wir noch in $A^{(S)}_\varrho$ die Funktion coth ... durch das erste Glied ihrer Potenzreihenentwicklung bei hohen Temperaturen ersetzen und mit diesem Ausdruck in (25) eingehen, vereinfacht sich die Magnetisierungsformel (25) auf die Form

$$\langle S^z_\varrho \rangle = \frac{S\,(S+1)}{3\,K_B\,T} \left\{ \frac{1}{2\,N} \sum_{\tau=1}^{n} \sum_{h} \frac{(T^h_{\varrho\tau})^2}{g\,\mu_B\,H + \langle S^z \rangle\, E^{(1)}_{h\tau}} + \frac{1}{2\,N} \sum_{\tau=1}^{n} \sum_{h} \frac{(T^h_{\varrho\tau})^2}{g\,\mu_B\,H + \langle S^z \rangle\, E^{(2)}_{h\tau}} \right\}^{-1} \tag{26}$$

Die Gleichung (26) dient uns als Ausgangspunkt für die Definition der ferromagnetischen und paramagnetischen Curietemperatur.

Als ferromagnetische Curietemperatur definieren wir diejenige Curietemperatur, die sich aus (26) und (6) bei abwesendem äußerem Feld ($H = 0$) für $\langle S^z \rangle \to 0$ ergibt,

$$K_B\,T^f_c(n) = \frac{2}{3} S\,(S+1)\,\frac{1}{n} \sum_{\varrho=1}^{n} \left\{ \frac{1}{N} \sum_{\tau=1}^{n} \sum_{h} (T^h_{\varrho\tau})^2 \left[ \frac{1}{E^{(1)}_{h\tau}} + \frac{1}{E^{(2)}_{h\tau}} \right] \right\}^{-1}. \tag{27}$$

Bei einer Unterdrückung der anisotropen Glieder in (27) erhalten wir für das einfach kubische Gitter den in der Arbeit [1] angegebenen Ausdruck für die ferromagnetische Curietemperatur. Durch die Einführung der einachsigen Volumenanisotropie ist in (27) die durch den Null-Spinwellenmodus ($h = 0$, $\tau = 1$) bedingte Divergenz beseitigt, so daß sich für jede endliche Schichtdicke eine von Null verschiedene Curietemperatur ergibt.

Wir wollen jetzt einen Ausdruck für die paramagnetische Suszeptibilität im Felde $H = 0$ ableiten. Zu diesem Zweck führen wir durch die Gleichungen

$$g\,\mu_{\rm B}\,\langle S_\varrho^z\rangle = \chi_\varrho\,H\,, \qquad \chi = \frac{1}{n}\sum_{\varrho=1}^{n}\chi_\varrho \tag{28}$$

die Suszeptibilität $\chi_\varrho$ bzw. $\chi$ in (26) ein und gehen dann zum Limes $H \to 0$ über:

$$\frac{1}{3}\,S\,(S+1)\,g^2\,\mu_{\rm B}^2 = \chi_\varrho\,K_{\rm B}\,T\,\frac{1}{2\,N}\sum_{\tau=1}^{n}\sum_{h}(T_{\varrho\tau}^{h})^2 \times$$
$$\times\left\{\frac{1}{1+\dfrac{\chi}{g^2\,\mu_{\rm B}^2}\,\boldsymbol{E}_{h\tau}^{(1)}}+\frac{1}{1+\dfrac{\chi}{g^2\,\mu_{\rm B}^2}\,\boldsymbol{E}_{h\tau}^{(2)}}\right\}. \tag{29}$$

Wir entwickeln (29) in eine Reihe nach Potenzen von $\chi$,

$$\chi_\varrho - \frac{g^2\,\mu_{\rm B}^2\,S\,(S+1)}{3\,K_{\rm B}\,T} =$$
$$= -\,\chi_\varrho\,\frac{1}{2\,N}\sum_{\tau=1}^{n}\sum_{h}\left(T_{\varrho\tau}^{h}\right)^2\sum_{r=1}^{\infty}(-1)^r\left(\frac{\chi}{g^2\,\mu_{\rm B}^2}\right)^r\left[\left(\boldsymbol{E}_{h\tau}^{(1)}\right)^r+\left(\boldsymbol{E}_{h\tau}^{(2)}\right)^r\right]. \tag{30}$$

Die Gleichung (30) lösen wir durch Iteration (zunächst für $\chi_\varrho$, womit wir sodann mit (28) die gesuchte Größe $\chi$ gewinnen). Bis zu Termen der Ordnung $T^{-2}$ finden wir

$$\chi = \frac{g^2\,\mu_{\rm B}^2\,S\,(S+1)}{3\,K_{\rm B}\,T}\left\{1+\frac{T_{\rm c}^{\rm p}(n)}{T}\right\}. \tag{31}$$

Der Ausdruck (31) stimmt bis auf Entwicklungsglieder höherer Ordnung mit dem Curie-Weissschen Gesetz überein, weshalb $T_{\rm c}^{\rm p}(n)$ als paramagnetische Curietemperatur angesehen werden kann,

$$K_{\rm B}\,T_{\rm c}^{\rm p}(n) = \frac{1}{3}\,S\,(S+1)\,\frac{1}{2\,n\,N}\sum_{\tau=1}^{n}\sum_{h}\left\{\boldsymbol{E}_{h\tau}^{(1)}+\boldsymbol{E}_{h\tau}^{(2)}\right\}. \tag{32}$$

Die Gleichung (26) kann auch als Grundlage für die Definition einer ferromagnetischen und paramagnetischen Curietemperatur für die Netzebene $\varrho$ dienen. Auf Grund der Struktur der Gleichung (22) bzw. (26) bedingt, im Gegensatz zum Molekularfeldmodell für dünne Schichten (Valenta [12]), das Verschwinden der spontanen Magnetisierung bzw. der reziproken Suszeptibilität in einer Netzebene nicht das Verschwinden der spontanen Magnetisierung bzw. der reziproken Suszeptibilität in den restlichen Netzebenen der Schicht, so daß nicht alle Netzebenen den gleichen Curiepunkt haben.

Es ist interessant, die Formeln (27) und (33) für die ferromagnetische und paramagnetische Curietemperatur im Bereich großer $n$ zu betrachten. Wir wollen in (27) und (32) die anisotropen Terme vernachlässigen und uns bei der Lösung des Eigenwertproblems (16) näherungsweise auf die sogenannten Raumlösungen [6, 11] beschränken $\left(\text{aus } \beta_h \approx 1 \text{ folgt } \bar{x}_\tau^h \approx 2 \cos \frac{\pi}{n} (\tau - 1)\right)$. Für die Ausdrücke (18a) erhalten wir dann

$$E_{h\tau}^{(1)} = E_{h\tau}^{(2)} = I_1(0) - I_1(h) + 2\, I_2(0) - 2\, I_2(h) \cos \frac{\pi}{n} (\tau - 1)\,. \tag{33}$$

Wir gehen mit (33) in (27) und (32) ein und ersetzen die Summation durch die Integration über den Wellenvektor $h$ in der ersten Brillouinzone. Ferner können wir für große $n$ von der Summation über $\tau$ zur $\tau$-Integration übergehen, wobei wir gleichzeitig die Translationssymmetrie der Schicht in Richtung der Schichtnormalen fordern $\left((\text{d. h. } (T_{\varrho\tau}^0)^2 \to \frac{1}{n}\right)$. Wir finden somit schließlich

$$K_B\, T_c^f(\infty) = \frac{1}{3} S\,(S + 1)\, z\, I\, \{F(-1)\}^{-1} \tag{34}$$

für die ferromagnetische und

$$K_B\, T_c^p(\infty) = \frac{1}{3} S\,(S + 1)\, z\, I \tag{35}$$

für die paramagnetische Curietemperatur des massiven Materials. Hierbei bedeuten $z$ die Koordinationszahl und $F(-1)$ die in der Arbeit [9] angegebenen Summen. Gleichung (34) stimmt exakt mit der nach dem sphärischen Modell [13] und (35) mit der nach der Molekularfeldtheorie berechneten Curietemperatur überein.

## 4. Schlußbetrachtung

Bei der Berechnung der Magnetisierung für den gesamten Temperaturbereich und der ferromagnetischen und paramagnetischen Curietemperatur unter Berücksichtigung von magnetisch anisotropen Termen beschränkten wir uns auf Schichten mit kubischer Kristallstruktur und (100)-Oberflächenorientierung. Jedoch lassen sich unsere Betrachtungen ohne prinzipielle Schwierigkeiten auch auf andere Oberflächenorientierungen ausdehnen. Zur Entkopplung der Greenschen Funktionen höherer Ordnung benutzten wir eine modifizierte Tyablikovsche Entkopplungsvorschrift. In den magnetostatischen Energietermen beschränkten wir uns bei der Auswertung der Dipolsummen auf die Langwellennäherung und die Wechselwirkung benachbarter Schichtebenen, wodurch wir etwas modifizierte makroskopische Entmagnetisierungsfaktoren erhielten. Die vom Flächengitter her bekannte Divergenz, die im Felde $H = 0$ durch den Null-Spinwellenmodus verursacht wird, wurde nach Döring durch die Einführung der einachsigen Volumenanisotropie beseitigt. Im Bereich tiefer Temperaturen erhielten wir die bekannte (für Schichten entsprechend modifizierte) Holstein-Primakoff-Formel. Bei dem Grenzübergang $n \to \infty$ und der Forderung der Translationssymmetrie der Schicht in $x$-Richtung sowie für den magnetisch isotropen Fall geht die ferromagnetische Curietemperatur einer Schicht exakt in die nach dem sphärischen Modell für massives Material berechnete Curietemperatur über. Analog erhält man für die paramagnetische Curietemperatur einer Schicht die aus der Molekularfeldtheorie in erster Näherung (Curie-Weiss-Gesetz) resultierende Curietemperatur für massives Material.

Im Teil II der Arbeit erfolgt eine ausführliche Diskussion der Formeln (27) und (32) für eine Fe-(100)-Schicht. Teilergebnisse sind in [14] enthalten.

Herrn Dr. Haubenreißer danke ich für die Anregung dieser Untersuchungen und sein ständiges Interesse an ihrem Fortgang, Herrn Dr. Andrä für wertvolle Diskussionsbemerkungen.

## Anhang

Wenn man den Energieausdruck (18a) mit der Kittelschen Resonanzformel [15] vergleicht, sieht man, daß an die Stelle des Entmagnetisierungsfeldes $N^x M$ der Ausdruck $\frac{3}{2} \frac{n_1^x + 2 n_2^x}{g \mu_B} \langle S^z \rangle$ und somit an die Stelle des makroskopischen Entmagnetisierungsfaktors $N^x = 4\pi$ ein etwas modifizierter Entmagnetisierungsfaktor tritt. Diese Modifikation liegt in unserem Modell begründet:

Die Dipolsummen $m^\alpha_{\varrho\nu}(0)$ $(\alpha = x, y, z)$ sind mit den makroskopischen Entmagnetisierungsfaktoren durch die Relation

$$\frac{\sum_\nu m^\alpha_{\varrho\nu}(0)}{N_1 N_2 g^2 \mu_B^2} + \frac{4\pi}{3} = N^\alpha \tag{A1}$$

verbunden [16] ($N_1$ bedeutet die Anzahl der Netzebenen pro Längeneinheit). Bei jedem endlichen Körper sind die in (A1) definierten Größen $N^\alpha$ von dem Ort $\varrho_l$ des Gitterpunktes abhängig. In der Theorie für massive Materialien umgeht man diesen Umstand, indem man nach dem Kontinuumsmodell die Entmagnetisierungsfaktoren $N^\alpha$ für das Zentralatom berechnet und die so gewonnenen $N_\alpha$ als gültig für sämtliche anderen Atome annimmt, wobei die besonders bei den Oberflächenatomen auftretenden spezifischen Beiträge zu den $N^\alpha$ vernachlässigt werden. Gleichzeitig geht durch die Einführung des Kontinuummodells die Strukturabhängigkeit der $N^\alpha$ verloren. (Hierzu siehe auch [17].)

Bei der Berechnung der Entmagnetisierungsfaktoren $N^\alpha$ für eine Schicht hat man zu beachten, daß nach (A1) die $N^\alpha$ vom Netzindex $\varrho$ abhängig werden. Dadurch wird die praktische Lösung der Bewegungsgleichungen (9) sehr kompliziert. Um eine analytische Lösung für die Funktionen $Q^{h\pm}_{\varrho\varphi}$ zu erhalten, beschränkten wir uns u. a. in den Dipolsummen (für $h = 0$) auf die Wechselwirkung benachbarter Schichtebenen, wodurch sich (A1) auf die Form

$$\frac{(n_1^\alpha + 2 n_2^\alpha)}{N_1 N_2 g^2 \mu_B^2} + \frac{4\pi}{3} = \tilde{N}^\alpha \tag{A2}$$

reduziert. Die Entmagnetisierungsfaktoren (A2) sind auf Grund unseres Modells strukturabhängig. Mit den im Teil II der Arbeit angegebenen numerischen Werten erhält man z. B. für kubisch raumzentrierte Gitter

$$\tilde{N}^x = 11{,}51\,, \qquad \tilde{N}^y = \tilde{N}^z = 0{,}528$$

anstatt der makroskopischen Entmagnetisierungsfaktoren

$$N^x = 4\pi\,, \qquad N^y = N^z = 0\,.$$

## Literatur

[1] W. Brodkorb und W. Haubenreisser, phys. stat. sol. **8**, K 21 (1965).
[2] W. Döring, Z. Naturf. **16a**, 1008 (1961).
[3] J. H. van Vleck, Phys. Rev. **78**, 266 (1950).
[4] D. N. Zubarev, Fortschr. Phys. **9**, 275 (1961).
[5] S. V. Tyablikov, Metody kvantovoi teorii magnetizma, Nauka, Moskva 1965.
[6] T. Holstein und H. Primakoff, Phys. Rev. **58**, 1098 (1940).
[7] A. Corciovei, Acta phys. Polon. **24**, 91 (1963).
[8] H. B. Callen, Phys. Rev. **130**, 890 (1963).
[9] R. A. Tahir-Kheli und D. ter Haar, Phys. Rev. **127**, 88 (1962).
[10] C. W. Haas, Phys. Rev. **132**, 228 (1963).
[11] W. Döring, Z. Naturf. **16a**, 1146 (1961).
R. Abbel, Z. Naturf. **18a**, 371 (1963).
R. Jelitto, Z. Naturf. **19a**, 1567 (1964); **19a**, 1580 (1964).
[12] L. Valenta, phys. stat. sol. **2**, 112 (1962).
[13] T. H. Berlin und M. Kac, Phys. Rev. **86**, 821 (1952).
M. Lax, Phys. Rev. **97**, 1419 (1955).
[14] W. Brodkorb und W. Haubenreisser, Preprint 1.1 to the Internat. Colloquium on Magnetic Thin Films, Jena, April 1966.
[15] C. Kittel, Phys. Rev. **71**, 270 (1947); **73**, 155 (1948).
[16] D. Fraitova, Czech. J. Phys. **B 11**, 500 (1961).
[17] M. H. Cohen und F. Keffer, Phys. Rev. **99**, 1128 (1955).

*(Received April 28, 1966)*

phys. stat. sol. 16, 237 (1966)

*Institut für Maschinelle Rechentechnik der Deutschen Akademie der Wissenschaften zu Berlin, Dresden, Außenstelle für Magnetische Signalspeicher, Berlin (a), und Theoretisch-Physikalisches Institut der Karl-Marx-Universität, Leipzig (b)*

# Zur Berechnung der Wechselwirkung einer Ansammlung von ausgerichteten magnetischen Einbereichsteilchen

Von

R. STRAUBEL (a)[1]) und D. ROBASCHIK (b)[2])

Es wird ein Verfahren zur Berechnung der magnetischen Kennwerte, insbesondere des Verlaufs der Magnetisierung einer großen Ansammlung von länglichen Einbereichsteilchen bei frei wählbaren Magnetisierungszyklen angegeben. Bei beliebiger Anordnung der Teilchen sollen jedoch deren Längsachsen vollständig parallel ausgerichtet sein, so daß jedem Teilchen eine exakt rechteckige individuelle Magnetisierungskurve zugeordnet werden kann. Dabei wirkt jedes Teilchen mit seinem äußeren Streufeld magnetisierungsabhängig auf jedes andere Teilchen. Das Verfahren wird zunächst an dem Néel-Wohlfarthschen Pärchenmodell demonstriert.

A method of calculating magnetic criteria is proposed with particular reference to the course of magnetizing a large number of oblong single-domain particles. In any arrangement the longitudinal axes of the particles are taken to be adjusted to be exactly parallel, so that a rectangular hysteresis-loop can be associated with each particle. Each particle acts with its external dispersion magnetization dependent on each of the other particles. The Néel and Wohlfarth interaction of particle pairs is treated as an example to demonstrate the present method.

## 1. Problemstellung

Hartmagnetische heterogene Werkstoffe bestehen aus einem unmagnetischen Bindemittel, in das kugel- oder nadelförmige ferro- bzw. ferrimagnetische Einbereichsteilchen eingebettet sind. Dabei steht jedes Teilchen unter dem Einfluß der äußeren Streufelder aller seiner Nachbarn. Eine erste Veranschaulichung des feldstärkeabhängigen Vielkörperproblems gestattet das von Preisach [1] für den Rayleigh-Bereich vorgeschlagene Modell, dessen Gültigkeit in den Arbeiten [6] und [7] bis in das Sättigungsgebiet erweitert wurde.

Hierbei wird jedem Einbereichsteilchen eine exakt rechteckförmige Hystereseschleife zugeordnet, die wegen der Wechselwirkung mit den Nachbarn positiv oder negativ „vorgespannt" ist. Gleichzeitig gilt die Annahme, daß diese „Vorspannfelder" unabhängig vom Magnetisierungszustand der Probe, zumindest aber „statistisch stabil" [5] sind. Gegen diese Annahmen wurden verschiedentlich (siehe z. B. [4, 5 und 6] Bedenken erhoben.

Es liegt nahe, die Teilchenansammlung als ein Kollektiv von Teilchenpaaren aufzufassen (siehe z. B. [6 bis 10]), bei denen jeweils die Wechselwirkung zwischen den Partnern eines Paares berücksichtigt wird. Obwohl damit die Wechselwirkungsfelder von dem Magnetisierungszustand der Teilchenansammlung abhängen, gestattet dieses Pärchenmodell nur einen sehr groben Einblick in die tatsächlichen Magnetisierungsvorgänge der Probe.

---

[1]) Anschrift: 1107 Berlin, Am Nesselweg 15.

[2]) Jetzige Anschrift: Москва-Дубна, ул. Ленинградская 2-22.

Deshalb soll in der vorliegenden Arbeit in Erweiterung dieses Modells die Hystereseschleife für das folgende Modell berechnet werden:

Es ist eine Ansammlung beliebig angeordneter, formanisotroper, vollständig ausgerichteter Einbereichsteilchen gegeben. Infolge der vorausgesetzten totalen Ausrichtung der Teilchen besitzt jedes Teilchen eine individuelle rechteckförmige Hystereseschleife. Mit seinem äußeren magnetischen Streufeld wirkt jedes Teilchen auf jedes andere Teilchen des Kollektivs. Die individuellen Wechselwirkungsfelder wechseln ihre Orientierung bei Ummagnetisierung der die Felder verursachenden Teilchen.

## 2. Modell für ein Einbereichsteilchen

Das zu betrachtende Einbereichsteilchen soll die Form eines verlängerten Rotationsellipsoides mit dem Teilchenvolumen $v_T$ besitzen. Seine spontane Magnetisierung sei $\hat{\boldsymbol{M}}$, deren Betrag $\hat{M}$ konstant, deren Richtung aber von der Lage des angelegten Erregerfeldes $\boldsymbol{H}$ abhängig ist. Liegt die Längsachse des Teilchens in Richtung des magnetischen Erregerfeldes, so besitzt das Teilchen in dieser Richtung eine individuelle rechteckförmige Hystereseschleife $M(H)$ mit der Koerzitivfeldstärke $H_c$ (Fig. 1a). Der Betrag der Koerzitivfeldstärke hängt wesentlich von der Differenz der Entmagnetisierungsfaktoren $N_a$ und $N_b$ in Richtung der Hauptachsen $a$ und $b$ des Ellipsoides ab ($a > b$). Es gilt im Falle kohärenter Ummagnetisierungsprozesse (siehe z. B. [12])

$$H_c = (N_b - N_a)\,\hat{M}\,. \tag{1}$$

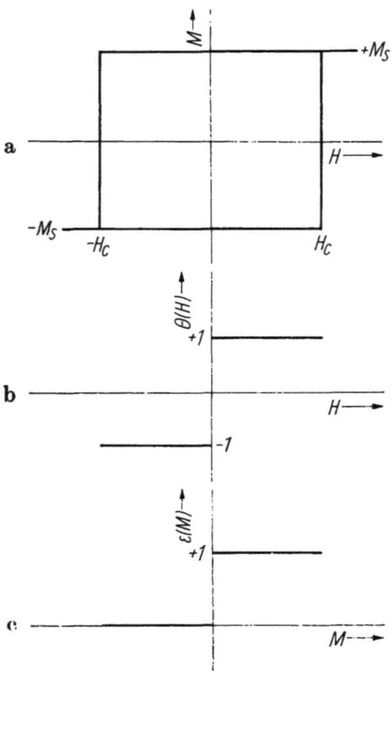


Fig. 1. Zur Definition einer rechteckigen Hystereseschleife (a) und der sie darstellenden Funktionen $\Theta(H)$ und $\varepsilon(M)$ (b, c)

Eine für Rechenautomaten zweckmäßige Darstellung der rechteckigen Hystereseschleife läßt sich durch geeignete Zusammenstellung von Sprungfunktionen $\Theta(H)$ und $\varepsilon(M)$ nach Fig. 1b und c angeben. Der linke und der rechte Ast der Hystereseschleife des Einbereichsteilchens ist dabei durch die um $\pm H_c$ verschobenen $\Theta$-Funktionen $\Theta\,(H + H_c)$ bzw. $\Theta\,(H - H_c)$ beschreibbar. Die Eindeutigkeit des Kurvenverlaufes kann durch Multiplikation der $\Theta$-Funktionen mit $\varepsilon(-M)$ bzw. $\varepsilon(+M)$ erreicht werden, indem diese je nach dem Vorzeichen der Magnetisierung $M$ entweder $\Theta\,(H + H_c)$ oder $\Theta\,(H - H_c)$ zum Verschwinden bringen müssen. Dabei ist — wenn $H_\nu = \{H_1, H_2 \ldots H_{n-1}, H_n\}$ eine Folge der zu durchlaufenden Erregerfeldwerte bedeutet — zur Bestimmung der $\varepsilon$-Funktionen der jeweils vorhergehende $H$-Wert aus der Folge heranzuziehen. Damit werden Mehrdeutigkeiten an den Sprungstellen der $\Theta$-Funktionen vermieden. Die Differenz $\Delta H$ zweier Erregerfeldwerte der Folge $\{H_\nu\}$ ist kleiner als die doppelte Koerzitivfeldstärke zu wählen,

$$|\Delta H| = |H_{\nu-1} - H_\nu| < 2\, H_c\,. \tag{2}$$

Daraus ergibt sich als eindeutige Darstellung der rechteckigen Hystereseschleife eines Einbereichsteilchens

$$M(H_\nu) = \hat{M}\,\{\Theta\,(H_\nu - H_c)\,\varepsilon\,[-\,M(H_{\nu-1})] + \Theta\,(H_\nu + H_c)\,\varepsilon\,[M(H_{\nu-1})]\}\,. \tag{3}$$

Das äußere Feld eines solchen Teilchens kann in erster Näherung als das Feld eines Dipols mit dem Dipolmoment

$$\boldsymbol{M}\, v_\mathrm{T} = \pm\,\hat{\boldsymbol{M}}\, v_\mathrm{T} \tag{4}$$

aufgefaßt werden. In einem kartesischen Koordinatensystem $(x, y, z)$ erzeugt nach [13] ein parallel der $x$-Achse orientierter Dipol, dessen Mittelpunkt die Koordinaten $(x_0, y_0, z_0)$ besitzt, am Aufpunkt $(x, y, z)$ im Abstand $r$ ein äußeres Feld mit einer $x$-Komponente

$$H_x = \frac{M(H_\nu)\, v_\mathrm{T}}{4\,\pi\,\mu_0}\left[\frac{3\,(x - x_0)^2}{r^5} - \frac{1}{r^3}\right], \tag{5}$$

wobei $r^2 = (x - x_0)^2 + (y - y_0)^2 + (z - z_0)^2$ gilt.

Gleichung (5) nimmt mit Gleichung (3) folgende Form an:

$$H_x = \frac{\hat{M}\, v_\mathrm{T}}{4\,\pi\,\mu_0}\left[\frac{3\,(x - x_0)^2}{r^5} - \frac{1}{r^3}\right]\{\Theta\,(H_\nu - H_c)\,\varepsilon\,[-\,M(H_{\nu-1})] + \\ + \Theta\,(H_\nu + H_c)\,\varepsilon\,[M(H_{\nu-1})]\}\,. \tag{6}$$

## 3. Modell für eine Ansammlung orientierter Einbereichsteilchen

### *3.1 Wechselwirkung zwischen $\lambda$ isolierten Teilchen*

In einer Ansammlung von $\lambda$ im Raum beliebig verteilten Einbereichsteilchen beeinflußt jedes einzelne Teilchen jedes andere durch sein äußeres Feld. Nach Voraussetzung sind alle Dipole in der $x$-Richtung ausgerichtet. So kommt bei der Wechselwirkung jeweils nur die oben angegebene $x$-Komponente der äußeren Teilchenfelder zur Geltung, und es entfällt künftig der Index $x$. Für die weitere Rechnung möge jedes Teilchen durch drei Indizes $(i, j, k)$ bzw. $(l, m, n)$ charakterisiert werden. Diese Indizes sollen den Laufzahlen in den drei Dimensionen eines karthesischen Raumes entsprechen, die vom Ursprung (0, 0, 0) aus positiv und negativ gezählt werden. Mit der Definition

$$g_{ijk} = \frac{\hat{M}_{ijk}\, v_{\mathrm{T}\,ijk}}{4\,\pi\,\mu_0} \tag{7}$$

ruft ein Teilchen $(i\,j\,k)$ am Ort eines anderen Teilchens $(l\,m\,n)$ nach (6) die Feldstärke

$$H_{ijk;\,lmn} = g_{ijk}\left[\frac{3\,(x_{lmn} - x_{ijk})^2}{r^5_{ijk;\,lmn}} - \frac{1}{r^3_{ijk;\,lmn}}\right] \times \\ \times\,\{\Theta\,(H_\nu - H_{c\,ijk})\,\varepsilon\,[-\,M_{ijk}(H_{\nu-1})] + \Theta\,(H_\nu + H_{c\,ijk})\,\varepsilon\,[M_{ijk}(H_{\nu-1})]\} \tag{8}$$

hervor, wobei gilt

$$r^2_{ijk;\,lmn} = (x_{lmn} - x_{ijk})^2 + (y_{lmn} - y_{ijk})^2 + (z_{lmn} - z_{ijk})^2\,.$$

Umgekehrt wirken alle anderen $\lambda - 1$ Teilchen aus der Gesamtheit gleichfalls mit einer entsprechenden Feldkomponente auf das betrachtete Teilchen $(i\,j\,k)$ zurück.

Die Summe dieser Feldkomponenten

$$\sum_l \sum_m \sum_n H_{l\,m\,n;\,i\,j\,k}$$

ist in Gleichung (8) dem äußeren Erregerfeld $H_\nu$ hinzuzufügen, so daß schließlich für die Feldkomponente des Teilchens $(i\,j\,k)$ am Orte des Teilchens $(l\,m\,n)$ geschrieben werden kann

$$\begin{aligned} H_{i\,j\,k;\,l\,m\,n} = g_{i\,j\,k} \left[\frac{3\,(x_{l\,m\,n} - x_{i\,j\,k})^2}{r^5_{i\,j\,k;\,l\,m\,n}} - \frac{1}{r^3_{i\,j\,k;\,l\,m\,n}}\right] \times \\ \times \left\{\Theta\left(H_\nu + \sum_l \sum_m \sum_n H_{l\,m\,n;\,i\,j\,k} - H_{\mathrm{c}\,i\,j\,k}\right) \varepsilon\,[-\,M_{i\,j\,k}(H_{\nu-1})] + \right. \\ \left. + \,\Theta\left(H_\nu + \sum_l \sum_m \sum_n H_{l\,m\,n;\,i\,j\,k} + H_{\mathrm{c}\,i\,j\,k}\right) \varepsilon\,[M_{i\,j\,k}(H_{\nu-1})]\right\}. \end{aligned} \tag{9}$$

Der Ausdruck (9) stellt ein implizites Gleichungssystem für die Wechselwirkungsfelder $H_{i\,j\,k;\,l\,m\,n}$ als Funktion von $H_\nu$ dar. Daraus kann mit $\bar{v}_{\mathrm{T}}$ als mittlerem Teilchenvolumen die resultierende Magnetisierung der Teilchenansammlung

$$M_p(H_\nu) = \frac{1}{\lambda\,\bar{v}_{\mathrm{T}}} \sum_i \sum_j \sum_k v_{\mathrm{T}\,i\,j\,k}\, M_{i\,j\,k}(H_{\mathrm{c}\,i\,j\,k}, M_{i\,j\,k}, H_\nu, \text{Anordnung}) \tag{10}$$

für eine beliebig vorzugebende Teilchenanordnung berechnet werden. Es ergeben sich je nach Wahl der Parameter $H_{\mathrm{c}\,i\,j\,k}$, $\hat{M}_{i\,j\,k}$ und der Teilchenanordnung für diesen Ausdruck bezüglich der $H_\nu$ stückweise konstante Funktionen.

### 3.2 $\lambda$ Teilchen in „verschmierter" Umgebung

Es sei eine große Zahl magnetischer Einbereichsteilchen mit der Form von verlängerten Rotationsellipsoiden gegeben. Die Teilchen sollen mit ihren Längsachsen einander parallel, sonst aber mit einer Packungsdichte $p$ innerhalb eines großen Volumens statistisch verteilt sein. Zur Vereinfachung ist es möglich, dieses Teilchenkollektiv als „verschmiertes" Kontinuum gleichförmig polarisierter Dipole aufzufassen. Innerhalb des Kontinuums möge ein von einem außen an das Volumen angelegten Magnetfeld herrührendes homogenes Feld $H_{\mathrm{a}}$ in Richtung der Längsachsen aller Teilchen wirken.

In diesem Kontinuum kann innerhalb einer kleinen Kugel mit dem Volumen $V_{\mathrm{K}}$ und dem Radius $R$ das magnetische Material an endlich vielen „Gitterpunkten" konzentriert werden (Fig. 2). Dabei soll innerhalb der Kugel weder zusätzliche Magnetisierung entstehen noch irgendwelche verlorengehen. Somit liegen an diesen Gitterpunkten wiederum als Dipole aufzufassende verlängerte Rotationsellipsoide, deren Längsachsen einander parallel auszurichten sind. Dann kann das Kollektiv von $\lambda$ Teilchen innerhalb der Kugel mit den in Abschnitt 3.1 behandelten $\lambda$ Teilchen identifiziert werden. Auf die Teilchen wirkt jetzt zusätzlich ein Feld, das von dem Kontinuum verursacht wird. Es ist das Feld in einem kugelförmigen Hohlraum, der aus einem bis an den inneren Rand gleichförmig polarisierten Kontinuum ausgespart ist. Nach [13] herrscht, wenn

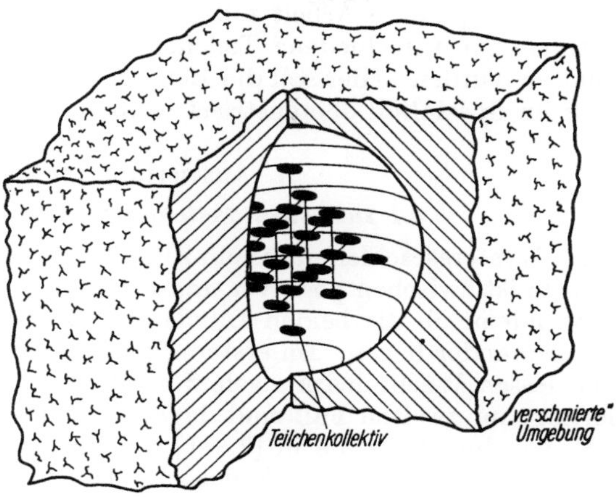


Fig. 2. Modellvorstellung von an Gitterpunkten angeordneten Einbereichsteilchen innerhalb eines Hohlraumes, der aus einer „verschmierten Umgebung" ausgespart ist

$1/3\,\mu_0 = \beta$ gesetzt wird, in dem Hohlraum ein homogenes Feld mit der Stärke

$$H_\mathrm{i} = H_\mathrm{a} + \beta\, M_\mathrm{Kontinuum}\,. \tag{11}$$

Es ist in weiten Grenzen unabhängig von der Größe des Hohlraumes.

Die Magnetisierungsbeiträge innerhalb und außerhalb des Hohlraumes unterscheiden sich in der Art ihrer Berechnung nur modellmäßig. Prinzipiell sind sie gleich, so daß zunächst mit Gleichung (10) gilt

$$M_\mathrm{Kontinuum}(H_\nu) = M_p(H_\nu)\,. \tag{12}$$

So soll zur Berechnung der gesamten Magnetisierung des Teilchenkollektivs, das sich innerhalb und außerhalb der Hohlkugel erstreckt, ein iteratives Verfahren angewendet werden. Zunächst ist die Magnetisierung $M_p^{(\mathrm{I})}$ der Teilchen innerhalb der Hohlkugel nach Gleichung (10) zu berechnen. Diese Magnetisierung wird nach Gleichung (12) dem Kontinuum zugeschrieben. Nun ist erneut die Magnetisierung $M_p^{(\mathrm{II})}$ der Teilchen innerhalb der Hohlkugel unter dem Einfluß des Kontinuums zu berechnen.

Für den $(\varkappa + 1)$-Schritt des Verfahrens lautet die Gleichung

$$\begin{aligned} M_p^{(\varkappa+1)}(H_\nu) = {} & \frac{1}{\lambda\,\bar{v}_\mathrm{T}} \sum_i \sum_j \sum_k v_{\mathrm{T}\,ijk}\, \hat{M}_{ijk} \times \\ & \times \Big\{ \Theta \Big( H_\nu + \sum_l \sum_m \sum_n H_{lmn;\,ijk} \big(H_\nu + \beta\, M_p^{(\varkappa)}(H_\nu)\big) + \\ & + \beta\, M_p^{(\varkappa)}(H_\nu) - H_{\mathrm{c}\,ijk} \Big)\, \varepsilon \left[ - M_{ijk} \big(H_{\nu-1} + \beta\, M_p^{(\varkappa)}(H_{\nu-1})\big)\right] + \\ & + \Theta \big( H_\nu + \sum \sum \sum H_{lmn;ijk} \left(H_\nu + \beta\, M_p^{(\varkappa)}(H)\right) + \\ & + \beta\, M_p^{(\varkappa)}\,(H_\nu) + H_{\mathrm{c}\,ijk} \big)\, \varepsilon \left[ M_{ijk} \big(H_{\nu-1} + \beta\, M_p^{(\varkappa)}(H_{\nu-1})\big)\right] \Big\}\,. \end{aligned} \tag{13}$$

Dieses Iterationsverfahren zur Berechnung der $M_p^{(\varkappa)}$ beginnt bei $M_p^{(0)} = 0$ und ist bis zur Konvergenz fortzusetzen, d. h., die Hystereseschleife soll sich bei weiteren Iterationsschritten nicht mehr ändern.

Es dürfte schwierig sein, ein allgemeines Konvergenzkriterium anzugeben, so daß die Konvergenz in jedem Spezialfall geprüft werden muß.

In einer späteren Arbeit sollen Ergebnisse mitgeteilt werden, die nach dieser Methode für verschiedene Teilchenanordnungen mit einem Rechenautomaten erhalten wurden. Hier soll die Methode an einem einfachen, in der Literatur oft zitierten Spezialfall demonstriert werden.

## 4. Auswertung für Spezialfälle

### *4.1 Das Zwei-Teilchen-Kollektiv (Pärchenmodell)*

Die Hystereseschleifen für ein Zwei-Teilchen-Modell sind in [7, 11] angegeben. Danach ist zwischen positiver und negativer Wechselwirkung zu unterscheiden, je nachdem, ob die beiden Dipole mit ihren Längsachsen hintereinander oder nebeneinander liegen. Im ersten Fall haben die Magnetisierung eines Teilchens und dessen Wechselwirkungsfeld am Ort des anderen Teilchens gleiche Orientierung; im zweiten Fall sind dagegen beide Vektoren stets entgegengerichtet.

Es soll hier kurz die in [7, 11] fehlende Ableitung der Ergebnisse mit Gleichung (9) für *hintereinander*liegende Dipole (Fig. 3a) angegeben werden. Mit

$$c_i = \frac{v_{\mathrm{T}i}}{16\,\mu_0\,\pi\,x_0^3} \tag{14}$$

hat Gleichung (9) für diesen Fall die Form

$$\left.\begin{aligned} H_{2,1} &= c_2\,\hat{M}_2\,\{\Theta\,(H_\nu + H_{1,2} - H_{c\,2})\,\times \\ &\quad\times\,\varepsilon\,[-\,M_2(H_{\nu-1})] + \Theta\,(H_\nu + H_{1,2} + H_{c\,2})\,\varepsilon\,[M(H_{\nu-1})]\}\;, \\ H_{1,2} &= c_1\,\hat{M}_1\,\{\Theta\,(H_\nu + H_{2,1} - H_{c\,1})\,\times \\ &\quad\times\,\varepsilon\,[-\,M_1(H_{\nu-1})] + \Theta\,(H_\nu + H_{2,1} + H_{c\,1})\,\varepsilon\,[M(H_{\nu-1})]\}\;. \end{aligned}\right\} \tag{15}$$

Gesucht wird die Hystereseschleife des Kollektivs nach Gleichung (10):

$$\begin{aligned} M_p^{(1)}(H_\nu) = \frac{\hat{M}_1\,v_{\mathrm{T}\,1}}{2\,\bar{v}_{\mathrm{T}}}\,&\{[\Theta\,(H_\nu + c_2\,M_2(H_\nu) - H_{c\,1})\,\varepsilon\,[-\,M_1(H_{\nu-1})]] + \\ &+ [\Theta\,(H_\nu + c_2\,M_2(H_\nu) + H_{c\,1})\,\varepsilon\,[M_1(H_{\nu-1})]]\} + \\ &+ \frac{\hat{M}_2\,v_{\mathrm{T}\,2}}{2\,\bar{v}_{\mathrm{T}}}\,\{[\Theta\,(H_\nu + c_1\,M_1(H_\nu) - H_{c\,2})\,\varepsilon\,[-\,M_2(H_{\nu-1})]] + \\ &+ [\Theta\,(H_\nu + c_1\,M_1(H_\nu) + H_{c\,2})\,\varepsilon\,[M_2(H_{\nu-1})]]\}\;. \end{aligned} \tag{16}$$

Für das Teilchenkollektiv kommen lediglich vier mögliche Magnetisierungszustände in Frage. Diese mögen in der folgenden Weise bezeichnet werden:

| | | |
|---|---|---|
| Zustand (A): | $M_1 > 0$ ; | $M_2 > 0$ , |
| Zustand (B): | $M_1 < 0$ ; | $M_2 < 0$ , |
| Zustand (C): | $M_1 < 0$ ; | $M_2 > 0$ , |
| Zustand (D): | $M_1 > 0$ ; | $M_2 < 0$ . |

Bei beliebiger Änderung der Erregerfeldstärke $\Delta H = H_{\nu-1} - H_\nu$ kann jeder Magnetisierungszustand (A) bis (D) des Kollektivs — abhängig von den Parametern — in einen der vier möglichen stationären Zustände übergehen.

Die Bedingungen für die Übergänge zwischen den Ausgangszuständen $H = H_{\nu-1}$ und den stationären Zuständen $H = H_\nu$ können unmittelbar aus den Gleichungen (15) entnommen werden. Sie sind in Tabelle 1 zusammengefaßt.

**Tabelle 1**

**Zustandsänderungen hintereinanderliegender Dipole**

Zustand (A): $M_1(H_\nu) > 0;\ M_2(H_\nu) > 0$

Ausgangszustand (A): $M_1(H_{\nu-1}) > 0;\ M_2(H_{\nu-1}) > 0;\ H_\nu > \begin{cases} -c_2\hat{M}_2 - H_{c\,1} \\ -c_1\hat{M}_1 - H_{c\,2} \end{cases}$

Ausgangszustand (B): $M_1(H_{\nu-1}) < 0;\ M_2(H_{\nu-1}) < 0;\ H_\nu > \begin{cases} -c_2\hat{M}_2 + H_{c\,1} \\ -c_1\hat{M}_1 + H_{c\,2} \end{cases}$

Ausgangszustand (C): $M_1(H_{\nu-1}) < 0;\ M_2(H_{\nu-1}) > 0;\ H_\nu > \begin{cases} -c_2\hat{M}_2 + H_{c\,1} \\ -c_1\hat{M}_1 - H_{c\,2} \end{cases}$

Ausgangszustand (D): $M_1(H_{\nu-1}) > 0;\ M_2(H_{\nu-1}) < 0;\ H_\nu > \begin{cases} -c_2\hat{M}_2 - H_{c\,1} \\ -c_1\hat{M}_1 + H_{c\,2} \end{cases}$

Zustand (B): $M_1(H_\nu) < 0;\ M_2(H_\nu) < 0$

Ausgangszustand (A): $M_1(H_{\nu-1}) > 0;\ M_2(H_{\nu-1}) > 0;\ H_\nu < \begin{cases} c_2\hat{M}_2 - H_{c\,1} \\ c_1\hat{M}_1 - H_{c\,2} \end{cases}$

Ausgangszustand (B): $M_1(H_{\nu-1}) < 0;\ M_2(H_{\nu-1}) < 0;\ H_\nu < \begin{cases} c_2\hat{M}_2 + H_{c\,1} \\ c_1\hat{M}_1 + H_{c\,2} \end{cases}$

Ausgangszustand (C): $M_1(H_{\nu-1}) < 0;\ M_2(H_{\nu-1}) > 0;\ H_\nu < \begin{cases} c_2\hat{M}_2 + H_{c\,1} \\ c_1\hat{M}_1 - H_{c\,2} \end{cases}$

Ausgangszustand (D): $M_1(H_{\nu-1}) > 0;\ M_2(H_{\nu-1}) < 0;\ H_\nu < \begin{cases} c_2\hat{M}_2 - H_{c\,1} \\ c_1\hat{M}_1 + H_{c\,2} \end{cases}$

Zustand (C): $M_1(H_\nu) < 0;\ M_2(H_\nu) > 0$

Ausgangszustand (A): $M_1(H_{\nu-1}) > 0;\ M_2(H_{\nu-1}) > 0;\quad c_1\hat{M}_1 - H_{c\,2} < H_\nu < -c_2\hat{M}_2 - H_{c\,1}$

Ausgangszustand (B): $M_1(H_{\nu-1}) < 0;\ M_2(H_{\nu-1}) < 0;\quad c_1\hat{M}_1 + H_{c\,2} < H_\nu < -c_2\hat{M}_2 + H_{c\,1}$

Ausgangszustand (C): $M_1(H_{\nu-1}) < 0;\ M_2(H_{\nu-1}) > 0;\quad c_1\hat{M}_1 - H_{c\,2} < H_\nu < -c_2\hat{M}_2 + H_{c\,1}$

Ausgangszustand (D): $M_1(H_{\nu-1}) > 0;\ M_2(H_{\nu-1}) < 0;\quad c_1\hat{M}_1 + H_{c\,2} < H_\nu < -c_2\hat{M}_2 - H_{c\,1}$

Zustand (D): $M_1(H_\nu) > 0;\ M_2(H_\nu) < 0$

Ausgangszustand (A): $M_1(H_{\nu-1}) > 0;\ M_2(H_{\nu-1}) > 0;\quad c_2\hat{M}_2 - H_{c\,1} < H_\nu < -c_1\hat{M}_1 - H_{c\,2}$

Ausgangszustand (B): $M_1(H_{\nu-1}) < 0;\ M_2(H_{\nu-1}) < 0;\quad c_2\hat{M}_2 + H_{c\,1} < H_\nu < -c_1\hat{M}_1 + H_{c\,2}$

Ausgangszustand (C): $M_1(H_{\nu-1}) < 0;\ M_2(H_{\nu-1}) > 0;\quad c_2\hat{M}_2 + H_{c\,1} < H_\nu < -c_1\hat{M}_1 - H_{c\,2}$

Ausgangszustand (D): $M_1(H_{\nu-1}) > 0;\ M_2(H_{\nu-1}) < 0;\quad c_2\hat{M}_2 - H_{c\,1} < H_\nu < -c_1\hat{M}_1 + H_{c\,2}$

Ohne Einschränkung der Allgemeinheit lassen sich bei Annahme der Bedingung $v_{T1} \hat{M}_1 > v_{T2} \hat{M}_2$ damit nach Fig. 3b und c zunächst zwei mögliche Grenzhystereseschleifen für das Kollektiv angeben. Die Schleifen unterscheiden sich lediglich nach dem Verhältnis der Koerzitivfeldstärken der beiden Teilchen $H_{c1}/H_{c2}$.

Das Teilchen I schaltet bei zunehmender Erregerfeldstärke für den Fall

$$\left.\begin{array}{ll} H_{c1} \leqq H_{c2} \quad \text{bei einer Erregerfeldstärke} & H_{s1}^{(1)} = H_{c1} + c_2 \hat{M}_2 \\ \text{und für den Fall} & \\ H_{c1} > H_{c2} \quad \text{bei einer Erregerfeldstärke} & H_{s1}^{(1)} = H_{c1} - c_2 \hat{M}_2 . \end{array}\right\} \quad (17)$$

Das Teilchen II schaltet für den Fall

$$H_{c1} \leqq H_{c2} \quad \text{bei einer Feldstärke} \quad H_{s2}^{(1)} = H_{c2} - c_1 \hat{M}_1$$

und für den Fall

$$H_{c1} > H_{c2} \quad \text{bei einer Feldstärke} \quad H_{s2}^{(1)} = H_{c2} + c_1 \hat{M}_1 .$$

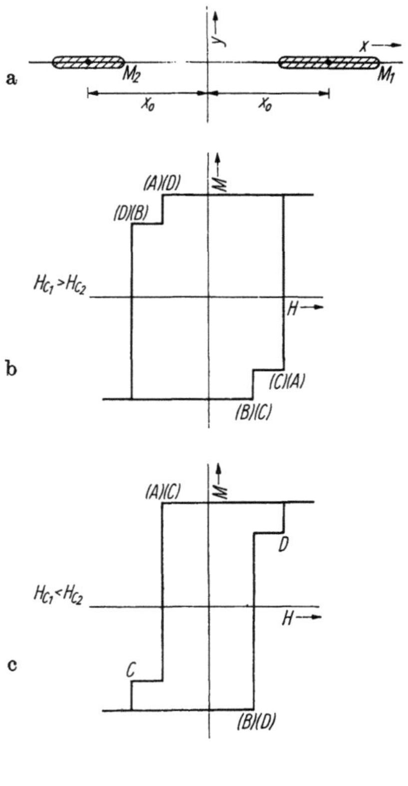


Fig. 3. Das Zwei-Teilchen-Kollektiv (a) und zwei typische Hystereseschleifen des Zwei-Teilchen-Kollektivs (b, c)

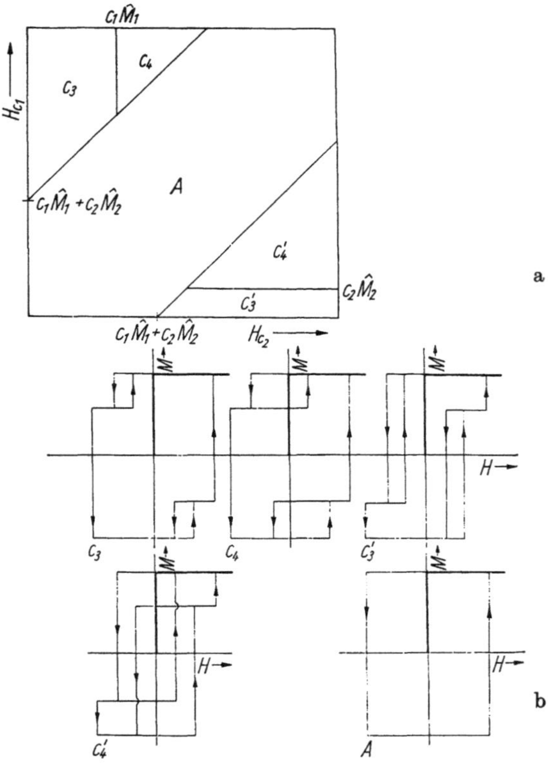


Fig. 4. Die $H_{c1}$, $H_{c2}$-Ebene (a) und die dazugehörigen charakteristischen Hystereseschleifentypen (b)

Aus den angegebenen Bedingungen der Tabelle 1 folgt weiter, daß die Zustandsänderungen (A) nach (C) oder (B) nach (D) bzw. (A) nach (D) oder (B) nach (C) nicht möglich sind, wenn gilt

$$\left.\begin{aligned} - c_1 \hat{M}_1 + H_{c\,2} &= c_2 \hat{M}_2 + H_{c\,1} \quad \text{für} \quad (A) \to (C) \quad \text{oder} \quad (B) \to (D)\,, \\ - c_2 \hat{M}_2 + H_{c\,1} &= c_1 \hat{M}_1 + H_{c\,2} \quad \text{für} \quad (A) \to (D) \quad \text{oder} \quad (B) \to (C)\,. \end{aligned}\right\} \tag{18}$$

Dann besitzt die Hystereseschleife des Kollektivs eine rechteckige Form (Fig. 1a). Zur Veranschaulichung ist es nach [3, 4] möglich, eine $H_{c\,1}$, $H_{c\,2}$-Ebene (Fig. 4a) aufzuspannen und in dieser die Bereiche unterschiedlicher Schleifentypen (Fig. 4b) abzugrenzen. In dieser Ebene beschreiben die Bedingungen (18) einen um 45° gegen das Koordinatensystem gedrehten Parallelstreifen. Die weiteren Unterteilungen der Fig. 4a ergeben sich aus Tabelle 1 für die Gestalt von inneren Schleifen.

#### *4.2 Das Zwei-Teilchen-Kollektiv innerhalb einer „verschmierten Umgebung"*

Für das Zwei-Teilchen-Modell läßt sich das in Abschnitt 3.2 beschriebene Iterationsverfahren in verhältnismäßig übersichtlicher Form angeben und die Konvergenz des Vorgehens zeigen. Danach wird der „verschmierten Umgebung" im ersten Iterationsschritt eine Hystereseschleife nach Gleichung (16) zugeschrieben. Das innere Feld $H_i$ innerhalb der Hohlkugel in der „verschmierten Umgebung" hat nach Gleichung (11) die Form

$$H_{i\,\nu} = H_\nu + \beta\, M_p^{(1)}(H_\nu)\,. \tag{19}$$

Für die damit zu verbessernde Hystereseschleife $M_p^{(2)}$ des Zwei-Teilchen-Kollektivs gilt

$$\begin{aligned} M_p^{(2)}(H_\nu) = \frac{1}{2\,\bar{v}_T} \sum_i v_{T\,i}\, \hat{M}_i \times \\ &\times \{[\Theta\,(H_\nu + c_j\, M_j\,(H_\nu + \beta\, M_p^{(1)}(H_\nu)) + \beta\, M_p^{(1)}(H_\nu) - H_{c\,i}) \times \\ &\times \varepsilon\,[- M_i\,(H_{\nu-1} + \beta\, M_p^{(1)}(H_{\nu-1}))]] + \\ &+ [\Theta\,(H_\nu + c_j\, M_j\,(H_\nu + \beta\, M_p^{(1)}(H_\nu)) + \beta\, M_p^{(1)}(H_\nu) + H_{c\,i}) \times \\ &\times \varepsilon\,[M_i\,(H_{\nu-1} + \beta\, M_p^{(1)}(H_{\nu-1}))]]\}\;; \quad i, j = 1, 2\,. \end{aligned} \tag{20}$$

Das Verfahren kann in der gleichen Weise fortgesetzt werden, bis sich die Hystereseschleife bei weiteren Iterationsschritten nicht mehr ändert. In dem hier behandelten Beispiel ist dies bereits nach zwei Iterationsschritten der Fall. Das zeigt sich bei Betrachtung der Änderung der Sprungfeldstärken $H_s$ von Teilchen I und II nach den einzelnen Iterationsschritten:

Für das Teilchen I liefert eine Diskussion der Gleichung (20) unabhängig von dem Verhältnis der Koerzitivfeldstärken stets die gleichen Sprungfeldstärken $H_{s\,1}^{(1)}$. Dagegen ändert sich die Sprungfeldstärke des Teilchens II beim ersten Iterationsschritt,

$$\left.\begin{aligned} H_{s\,2}^{(2)} &= H_{s\,2}^{(1)} - \beta\,(\hat{M}_1 - \hat{M}_2) \quad \text{für} \quad H_{c\,1} \leqq H_{c\,2} \quad \text{bzw.} \\ H_{s\,2}^{(2)} &= H_{s\,2}^{(1)} + \beta\,(\hat{M}_1 - \hat{M}_2) \quad \text{für} \quad H_{c\,1} > H_{c\,2}\,, \end{aligned}\right\} \tag{21}$$

um nach weiteren Iterationsschritten konstant zu bleiben, da dann „die Teilchen II" der Umgebung und das Teilchen II des Kollektivs bei der gleichen Erregerfeldstärke die Vorzeichen ihrer Magnetisierung synchron wechseln.

Damit konvergiert das Verfahren. Es liefert Hystereseschleifen, die nach der idealen Rechteckform tendieren. Insbesondere wird durch den Einfluß der Umgebung der durch die Gleichungen (18) gegebene Parallelstreifen in der $H_{c1}$, $H_{c2}$-Ebene für rechteckige Hystereseschleifentypen verbreitert. Durch Gleichsetzen der entsprechenden Ausdrücke aus den Gleichungen (17) und (21) ergibt sich für die Begrenzung des Streifens

$$\left.\begin{aligned} H_{c2} &= H_{c1} + c_1 \hat{M}_1 + c_2 \hat{M}_2 + \beta (\hat{M}_1 - \hat{M}_2)\,, \\ H_{c2} &= H_{c1} - c_1 \hat{M}_1 - c_2 \hat{M}_2 - \beta (\hat{M}_1 - \hat{M}_2)\,, \end{aligned}\right\} \qquad (22)$$

im Gegensatz zu den Gleichungen (18), die ohne den Einfluß der „verschmierten Umgebung" abgeleitet waren.

## Literatur

[1] F. PREISACH, Z. Phys. **94**, 277 (1935).
[2] R. FELDTKELLER und H. WILDE, Elektrotech. Z. (Germany) **A 77**, 449 (1956).
[3] H. WILDE, NTZ **10**, 497 (1957).
[4] D. HAROSKE und G. VOGLER, Z. angew. Phys. **15**, 150 (1963).
[5] G. BATE, J. appl. Phys. **33**, 2263 (1962).
[6] W. P. BROWN, J. appl. Phys. Suppl. **33**, 1307 (1962).
[7] E. P. WOHLFARTH, J. appl. Phys. **35**, 783 (1964).
[8] O. HENKEL, phys. stat. sol. **2**, 1393 (1962).
[9] O. HENKEL, phys. stat. sol. **7**, 81 (1964).
[10] O. HENKEL, phys. stat. sol. **7**, 919 (1964).
[11] L. NÉEL, C. R. Acad. Sci. (France) **246**, 2313 (1958).
[12] E. P. WOHLFARTH, Phil. Mag. Suppl. 8, 87 (1959).
[13] F. HUND, Einführung in die theoretische Physik, Bd. IV, VEB Bibliographisches Institut, Leipzig 1950.

*(Received April 29, 1966)*

phys. stat. sol. **16**, 247 (1966)

*Institut für Physik am Max-Planck-Institut für Metallforschung, Stuttgart*

# A Model for the Change in Elastic Modulus of a Solid Caused by Point Defects

By

J. Melngailis[1])

Point defects introduced into copper at low temperatures by irradiation are known to produce bulk expansion. We suggest here that an external stress might cause the defect (represented as oversized spherical inclusions) to relax. This would change the expansion produced by the defects so that the deformation due to the stress would be greater and the elastic modulus would thus appear smaller. This relaxation is represented by a small deformation of the spherical inclusions, and elastic energy considerations are used to calculate the change in the elastic modulus of Cu due to such defects. If the radius of the inclusions is chosen to be two lattice constants, the experimentally observed value of 18% for the decrease in the elastic modulus at saturation density of defects is obtained.

Die bei tiefen Temperaturen durch Bestrahlung in Kupfer erzeugten Punktfehler rufen bekannterweise eine Volumenausdehnung hervor. Hier wird die Vorstellung entwickelt, daß die Defekte — dargestellt durch eingezwängte Kugeln — bei Anwendung einer äußeren Spannung relaxieren. Dadurch ändert sich die ursprünglich erzeugte Ausdehnung, und die durch die Spannung hervorgerufene Deformation erscheint größer, der Elastizitätsmodul also kleiner. Diese Relaxation wird durch eine kleine Deformation der sphärischen Einschlüsse dargestellt. Mit Hilfe von Betrachtungen über die elastische Energie wird die Änderung des Elastizitätsmoduls von Kupfer infolge solcher Defekte berechnet. Wählt man den Radius der Einschlüsse gleich zwei Gitterkonstanten, so erhält man den experimentell gefundenen Wert von 18% für die Erniedrigung des Elastizitätsmoduls bei der Sättigungsdichte der Defekte.

## 1. Introduction

The irradiation of Cu at temperatures below 40 °K and the subsequent annealing have shown that much of the damage introduced anneals out rather abruptly at about 40 to 50 °K [1]. This part of the annealing process has been named stage I and has been observed mainly by measuring the changes in electrical resistivity and in mechanical properties. Among the mechanical properties that have been studied after low temperature radiation damage of Cu are: change in length [2], change in lattice constant [3], internal friction [4], and change in elastic modulus [5]. The latter measurements by König, Völkl, and Schilling are of particular interest here. These measurements showed that the Young's modulus of polycrystalline, high purity, low dislocation density Cu was decreased 18% by saturation $\alpha$-particle irradiation at 32 °K.

The purpose of the present note is to describe a simple model for the softening of a crystal by point defects and to show that this model might be used to account for this surprisingly large observed decrease of Young's modulus of Cu. The model may be understood in simple terms from the following considerations:

[1]) National Science Foundation Postdoctoral Fellow. Present address: C.N.R.S. 92, Bellevue, France.

Copper is known also to suffer a relative expansion of about $3 \times 10^{-4}$ by saturation irradiation at low temperatures [2, 3]. Like the observed change in elastic modulus this expansion anneals out to a large extent in stage I. The expansion can be accounted for by visualizing the point defects as spheres forced into slightly smaller holes in an isotropic continuum [6]. According to more detailed calculations the misfit should be about 1 atomic volume for an interstitial [7, 8] and $-0.2$ for a vacancy [8]. We will consider for the moment only the interstitials and represent them by the misfit spheres. Now, one can imagine that, if a stress is applied to the crystal, the spheres might become slightly ellipsoidal. This would produce a corresponding change in the expansion originally produced by the defects. Thus, the response to the stress would be effectively greater and the material would appear softer. Since the expansion of the solid due to the defects is large ($\Delta l/l = 3 \times 10^{-4}$) compared to the strains used in the measuring process [5] ($< 10^{-5}$), the distortion could be small and still yield a comparatively large response. Just how large these effects are expected to be, will be estimated with some simple calculations.

## 2. The Elastic Energy of the Defect in an External Strain Field

The displacement field $\boldsymbol{u}(\boldsymbol{x})$ in an isotropic medium due to a force density $\boldsymbol{f}$ is the solution of the equation

$$(C_{11} - C_{44}) \nabla (\nabla \cdot \boldsymbol{u}) + C_{44} \nabla^2 \boldsymbol{u} + \boldsymbol{f} = 0 . \tag{1}$$

(Although the material is assumed to be isotropic, the elastic stiffness constants $C_{11}$, $C_{12}$, and $C_{44}$ are used with the isotropy condition that $2\,C_{44} = C_{11} - C_{12}$. They are related to the Lamé constants by $C_{12} = \lambda$, $C_{44} = \mu$.) The displacement field due to a non-spherical inclusion can be easily calculated if it is assumed to arise from unequal crossed double forces [6] at $\boldsymbol{x} = 0$ (equal crossed double forces represent a spherical inclusion):

$$f_j = -\,\alpha_j \frac{\partial}{\partial x_j} \delta(\boldsymbol{x}) . \tag{2}$$

$\delta(\boldsymbol{x})$ is the three-dimensional Dirac delta function; $\alpha_1 \neq \alpha_2 = \alpha_3$. Assuming an infinite medium, the solution to equation (1) with this force density is [12]

$$\boldsymbol{u}(\boldsymbol{x}) = \frac{C}{4\pi} \frac{1}{r^3} \{\boldsymbol{x} + \beta\,[\hat{\boldsymbol{a}}_1\, x_1 - \gamma\, \boldsymbol{x}\, (1 - 3\, x_1^2/r^2)]\} , \tag{3}$$

where $C$ is the volume displaced before distortion (for $\beta = 0$), $\beta = (\alpha_1 - \alpha_2)/\alpha_2$, $r^2 = \boldsymbol{x} \cdot \boldsymbol{x}$, $\gamma = 1/2\,(C_{11} - C_{44})/C_{44}$, and $\hat{\boldsymbol{a}}_1$ is a unit vector in the $x_1$ direction. This solution is assumed to represent the displacement outside of the inclusion.

The interaction energy of a defect, whose displacement field is given by (3), and a constant strain field can be estimated as follows: Consider a cube of side $2b$ centered around the isotropic defect ($\beta = 0$) in an infinite medium. Due to the presence of a constant strain field $\eta_{11}$ and the surrounding material, the two opposite sides of the cube (perpendicular to $x_1$) are constrained to a certain displacement resulting in a strain energy $1/2\, C_{11}\, \eta_{11}^2 (8\, b^3)$ being stored in the cube. Suppose now the defect "relaxes" becoming non-spherical and producing an extra displacement of the surface. This will replace some of the original displacement and reduce the strain by an amount $\Delta\eta_{11}$ and the stored strain energy by $C_{11}\, \eta_{11}\, \Delta\eta_{11} (8\, b^3)$ (to first order in $\Delta\eta_{11}$). This interaction energy can be approximated by

$$E^{\text{int}} = -2\, C_{11}\, \eta_{11} \int_{-b}^{b} \mathrm{d}x_2 \int_{-b}^{b} \mathrm{d}x_3 \left[u_1(b, x_2, x_3) - u_1(b, x_2, x_3)|_{\beta=0}\right] . \tag{4}$$

This is seen to be merely the work done in displacing the two surfaces of the cube against a constant stress $C_{11}\, \eta_{11}$ by an average amount as the defect relaxes.

When the integrals in (4) are evaluated, the interaction energy to first order in $\beta$ is

$$E^{\text{int}} = -\frac{C}{2\pi}\, C_{11}\, \eta_{11}\, \beta \left[\frac{2\pi}{3} + \frac{2\sqrt{3}}{3}\left(\frac{C_{11} - C_{44}}{C_{44}}\right)\right] \tag{5}$$

and is independent of the size of the cube $b$. For an applied strain of $10^{-5}$ and for $C = a^3/4$ (one atomic volume) $E^{\text{int}} = -2.2 \times 10^{-16}\, \beta$ erg.

For the elastic constants we have used the effective isotropic constants related to the cubic lattice constants $C_{ij}^0$ by [8, 9]

$$\left.\begin{aligned} C_{11} &= \frac{1}{5}\,(3\, C_{11}^0 + 2\, C_{12}^0 + 4\, C_{44}^0)\,, \\ C_{12} &= \frac{1}{5}\,(C_{11}^0 + 4\, C_{12}^0 - 2\, C_{44}^0)\,, \\ C_{44} &= \frac{1}{5}\,(C_{11}^0 - C_{12}^0 + 3\, C_{44}^0)\,. \end{aligned}\right\} \tag{6}$$

If for the $C_{ij}^0$ the values measured at 30 °K [10] are used, then $C_{11} = 22.1 \times 10^{11}$, $C_{12} = 10.2 \times 10^{11}$, and $C_{44} = 5.9 \times 10^{11}$ dyn/cm².

Thus the relaxation of the defect represented by a mismatched inclusion will reduce the strain energy stored by an amount given by equation (5). Working against this relaxation is the fact that the energy stored in the strain field produced by the defect as well as the energy stored in the inclusion will become greater as the displacement fields becomes anisotropic.

The energy in the strain field in a volume $V$ is given by

$$E = \frac{1}{2} \int_V \mathrm{d}V\, \frac{\partial u_i}{\partial x_j}\, C_{ijkl}\, \frac{\partial u_k}{\partial x_l}\,, \tag{7}$$

which can also be written as a surface integral over the surface $S$ enclosing $V$,

$$E = \frac{1}{2} \int_S \mathrm{d}S\, u_i\, \hat{n}_j\, C_{ijkl}\, \frac{\partial u_k}{\partial x_l}\,, \tag{8}$$

where $\hat{\boldsymbol{n}}$ is the outward pointing unit vector normal to the surface at the element $\mathrm{d}S$. The energy $E^{\text{out}}$ in the field outside of the inclusion of radius $\varrho$ can be calculated by using either equation (7) or (8) and by taking $\boldsymbol{u}(\boldsymbol{x})$ as given by (3) and the volume of integration $V$ as a spherical shell of inner radius $\varrho$ and an infinite outer radius

$$E^{\text{out}} = \frac{C^2\, C_{44}}{4\pi\varrho^3}\left[2 + \frac{4}{3}\beta + \beta^2\left(\frac{1}{3} + \frac{8}{15}\gamma + \frac{12}{5}\gamma^2\right)\right], \tag{9}$$

where $\gamma$ is defined above (3).

The energy inside the inclusion can similarly be calculated if the displacement field inside is known. Like the displacement field outside (3), this field will contain a term independent of $\beta$ and one linear in $\beta$. However, the inside solution may not contain negative powers of $r$. For the term independent of $\beta$ we notice that $\boldsymbol{u} = \text{const.} \times \boldsymbol{x}$ satisfies equation (1) (with $\boldsymbol{f} = 0$). The constant is determined by requiring that, when the inclusion is spherical, the stress should be continuous at the surface of the inclusion, $r = \varrho$. The terms linear in $\beta$ are obtained by assuming that during deformation the inclusion remains in contact with the outside material. The solution to equation (1) (with $\boldsymbol{f} = 0$) inside a sphere with known surface displacement is given by Love [12]. Thus the displacement field inside is

$$\boldsymbol{u}' = \frac{C}{4\pi}\left\{-\frac{4\,C_{44}}{C_{11} + 2\,C_{12}} \times \frac{\boldsymbol{x}}{\varrho^3} + \frac{\beta}{\varrho^3}\left[\hat{\boldsymbol{a}}_1\, x_1 - \gamma\, \boldsymbol{x}\,(1 - 3\,x_1^2/\varrho^2) + \right.\right.$$
$$\left.\left. + 3\,\hat{\boldsymbol{a}}_1\,(M + \gamma)\, x_1\,(1 - r^2/\varrho^2) - M\,\boldsymbol{x}\,(1 - r^2/\varrho^2)\right]\right\}, \tag{10}$$

where

$$M = 3\,\gamma\, C_{12}/(2\,C_{12} + 7\,C_{44}).$$

The energy inside the inclusion $E^{\text{in}}$ is obtained, as was $E^{\text{out}}$ above, by substituting this solution in either (7) or (8):

$$E^{\text{in}} = \frac{C^2\,C_{44}}{4\,\pi\,\varrho^3}\left\{\frac{8\,C_{44}}{C_{11} + 2\,C_{12}} - \frac{4}{3}\beta + \frac{\beta^2}{2}\left[\left(\frac{1}{3} + \frac{2}{3}\,M\right)\frac{C_{11}}{C_{44}} + \frac{4\,\gamma\,(C_{11} + 2\,C_{12})}{15\,C_{44}} - \right.\right.$$
$$\left.\left. - 6\,M\left(\frac{2}{5}\gamma + \frac{1}{3}\right) + \frac{8\,\gamma}{15} + \frac{4\,\gamma^2\,(C_{11} + 2\,C_{12})}{5\,C_{44}} - \frac{8\,\gamma\,M\,C_{11}}{5\,C_{44}}\right]\right\}. \tag{11}$$

Thus the total energy per defect in the presence of a constant strain is the sum (appart from constant terms)

$$E^{\text{out}} + E^{\text{in}} + E^{\text{int}} = 1.6 \times 10^{-13}\,\beta^2 - 2.2 \times 10^{-16}\,\beta\,.$$

We have taken $\varrho = 2\,a$ and $C = a^3/4$. (As expected, the term linear in $\beta$ in $E^{\text{out}} + E^{\text{in}}$ cancels.) This total energy is a minimum for $\beta = 0.7 \times 10^{-3}$. This then is the amount by which the inclusion becomes anisotropic in response to an external strain of $10^{-5}$.

## 3. Results and Discussion

At the saturation density of Frenkel defects of 0.14% [11] the volume of crystal per defect is about $(5.7\,a)^3$. Considering a defect centered in such a cube, we calculate the average displacement of the sides in the $x_1$ direction caused by the defects non-spherical deformation. This was already calculated above in estimating the interaction energy (5). The result is $1.14 \times 2.85\,a \times 10^{-3}\,\beta$. The displacement of the same surface relative to the centre of the cube due to the constant strain field of $10^{-5}$ is $2.85\,a \times 10^{-5}$. The ratio of the two is $114\,\beta$ or for the above value of $\beta$, 0.08. Thus the response of the material to an applied stress will be on the average 8% greater due to the relaxation of the defects. The elastic modulus will then appear decreased by about this same amount.

So far we have considered the crystal to be infinite. The presence of surfaces would increase the displacements caused by the inclusions. More precise considerations [6] indicate that to calculate the volume or external dimension change due to spherical inclusions one should multiply the result for the infinite crystal by a factor $\nu = 3(1 - \sigma)/(1 + \sigma)$ where $\sigma$ is Poisson's ratio. In this case $\nu = 1.5$. We will assume that the effect of the non-spherical relaxation on external dimensions is likewise enhanced by the same factor in passing to a finite crystal. The extra displacement of the external surface enters twice in our result, firstly in the interaction energy (4), and secondly above in calculating the response of the material to a constant stress. In both cases the result is multiplied by $\nu$, so that the 8% above is multiplied by $\nu^2$. Thus the elastic modulus would appear decreased by 18%. (The effect of a finite crystal enters in an additional way to increase this result. The energy in the field (9) and in the inclusion (11) is reduced in a finite crystal. The term independent of $\beta$ is reduced by about 10%. The reduction of the $\beta^2$ term is difficult to estimate. Since it is probably also small, we will neglect it.)

There are, of course, many rather obvious assumptions underlying this result. Perhaps the most important is the choice of the size of the inclusion. The solution for the displacement field, (3) as well as the energy in this field (9) diverges rapidly as $x \to 0$. Hence, the delta function force field at the origin is imagined replaced [6] (for $\beta = 0$) by a spherical inclusion forced into a smaller hole. The radius of this inclusion is within limits arbitrary and was treated here as an adjustable parameter. A choice of this radius equal to two lattice constants in calculating the energy $E^{out} + E^{in}$ yielded the final change in elastic modulus in agreement with experiment.

If the defects which produce expansion can be pictured as such inclusions, this calculation suggests that through their deformation they also produce a large change in elastic modulus.

So far we have imagined the defect to be an interstitial. With simply a change of sign of $C$ the model is also applicable to a vacancy. An interstitial in the dumb-bell configuration [7] can be treated by taking the equilibrium defect as being already non-spherical to an extent corresponding to $\beta = 0.14$ [7] and considering additional changes in $\beta$ due to the external strain. The same results would be obtained.

Since much of the observed decrease of Young's modulus anneals out in stage I [5], it must partly be due to close interstitial vacancy pairs. These pairs might be visualized as a single inclusion where the atoms are mispacked. Most of the atoms in this region may have considerably relaxed from their original lattice sites. Hence, this disordered region might have a reduced resistance to shear and might perhaps be well represented by a liquid. To this situation our model is readily applicable. The inclusion would have no resistance to deformation and would contribute no $\beta^2$ term to the energy. On this account we could increase the energy in the field outside the inclusion by choosing its radius smaller: $1.7\,a$ instead of $2\,a$. Again the same change in elastic modulus would be obtained.

Regardless of the detailed nature of the low temperature radiation defects in Cu, that they cause expansion of the sample is a well established fact [2, 3]. We have attempted to show that the observed decrease of Young's modulus [5] is intimately connected with this expansion. The modulus is determined by how much a surface of the sample moves in response to an applied stress. If the crystal contains defects and if, in response to the stress, they change their effective expansion in the direction of the stress by only 0.5%, the observed decrease of the modulus can be accounted for. The elastic energy considerations for a simple, but hopefully sufficiently realistic, deformable spherical inclusion model described above indicate that such a response of the defects might be expected.

***Acknowledgements***

The author is grateful to Professor A. Seeger for suggesting the problem and for his guidance and encouragement, and to Dr. E. Mann and Dr. R. von Jan for helpful discussions.

## References

[1] See, for example, the review by D. K. HOLMES, Proc. Internat. Summer School on Solid State Physics, Mol 1963, Ed. R. STRUMANE et al., North-Holland Publ. Comp. 1964 (p. 147).
[2] R. VOOK and C. WERT, Phys. Rev. **109**, 1529 (1958).
[3] R. O. SIMMONS and R. W. BALLUFFI, Phys. Rev. **109**, 1142 (1958).
[4] D. KÖNIG, J. VÖLKL, and W. SCHILLING, phys. stat. sol. **7**, 577 (1964).
[5] D. KÖNIG, J. VÖLKL, and W. SCHILLING, phys. stat sol. **7**, 591 (1964).
[6] J. D. ESHELBY, Solid State Phys. **3**, 79 (1956).
[7] A. SEEGER, E. MANN, and R. v. JAN, J. Phys. Chem. Solids **23**, 639 (1962).
[8] A. SEEGER and E. MANN, J. Phys. Chem. Solids **12**, 326 (1960).
[9] G. LEIBFRIED, Z. Phys. **135**, 23 (1953).
[10] W. OVERTON and J. GAFFNEY, Phys. Rev. **98**, 969 (1955).
[11] G. BURGER, H. MEISSNER, and W. SCHILLING, phys. stat. sol. **4**, 281 (1964).
[12] A. E. H. LOVE, A Treatise on the Mathematical Theory of Elasticity, 4th Ed., Cambridge 1952 (p. 185 and p. 265).

*(Received May 2, 1966)*

phys. stat. sol. 16, 253 (1966)

*Max-Planck-Institut für Metallforschung, Stuttgart, und*
*Institut für theoretische und angewandte Physik der Technischen Hochschule Stuttgart*

# Magnetostriktiver und magnetokristalliner elektrischer Restwiderstand in Ferromagnetika

Von

H. Kronmüller und A. Seeger

Im Rahmen der Bornschen Näherung wird in einem einfachen Modell die Streuung von s-Elektronen in ferromagnetischen Metallen an räumlich inhomogenen Spinverteilungen der d-Elektronen behandelt. Als maßgebend für die Streuung der s-Elektronen wird die sd-Austauschkopplung betrachtet. Die allgemeine Theorie wird zur Berechnung des elektrischen Restwiderstandes der Versetzungen infolge magnetischer Inhomogenitäten herangezogen. Der magnetostriktive Restwiderstand einer Versetzung verschwindet wie $1/H^2$ mit wachsender äußerer Feldstärke. Bei kleinen Feldstärken nähert sich der Restwiderstand einem konstanten Wert. Bei plausiblen Annahmen über die Stärke der sd-Austauschkopplung ergibt sich der magnetostriktive Restwiderstand als hinreichend groß, um experimentell nachgewiesen werden zu können.

The scattering of s-electrons due to an inhomogeneous arrangement of the spins of the d-electrons in ferromagnetic metals is studied using a simple model based on the Born approximation. The scattering mechanism for the s-electrons is attributed to the sd-exchange coupling forces. The general theory is applied to the calculation of the residual resistance of dislocations due to magnetic inhomogeneities. The magnetostrictive residual resistance of a dislocation is found to decrease as $1/H^2$ with increasing magnetic field. At small fields the magnetostrictive residual resistance is independent of the magnetic field. With reasonable assumptions for the magnitude of the sd-coupling, it's found that the magnetostrictive residual resistance may become large enough to be detected experimentally.

## 1. Einführung

Der bei Annäherung an den absoluten Nullpunkt verbleibende elektrische Widerstand eines Metalls, der sogenannte Restwiderstand, kommt bekanntlich durch die Streuung der Leitfähigkeitselektronen an den durch Gitterfehlstellen verursachten Störungen des Gitterpotentials zustande. Theoretisch untersucht wurde vor allem die Streuung an Versetzungen [1 bis 7], Stapelfehlern [8 bis 12], Korngrenzen [13] sowie Leerstellen und Leerstellenagglomeraten [14 bis 22]. Sie erfolgt an Störungen des periodischen Verlaufs des elektrostatischen Potentials, die entweder von elastischen Verzerrungen oder, wie im Falle der Leerstellen infolge der Herausnahme von Gitterionen, von der Störung der Elektroneutralität herrühren.

Während bei gewöhnlichen Metallen allein die Streuung am elektrostatischen Potential eine Rolle spielt, tritt in ferromagnetischen Metallen eine zusätzliche Streuung an Störungen des periodischen Verlaufs des von der Spinstellung abhängigen Austauschpotentials auf. Derartige Störungen des periodischen Austauschpotentials kommen durch einen inhomogenen Magnetisierungsverlauf zustande. Inhomogene Spinverteilungen können z. B. von der thermischen Anregung von Spinwellen oder von der magnetostriktiven Kopplung zwischen Magnetisierung und inneren Spannungen herrühren. Domänenwände stellen ebenfalls inhomogene Spinverteilungen dar und geben daher Anlaß zur Streuung von Elektronen. Die Magnetisierungsverteilung wird bei inneren Spannungen vor allem durch die Magnetostriktion, bei den Blochwänden jedoch vor allem durch die Kristallenergie bestimmt; wir unterscheiden im folgenden die beiden Fälle als *magnetostriktiver* bzw. als *magnetokristalliner* elektrischer Restwiderstand.

Die unelastische Streuung von Leitfähigkeitselektronen an thermisch angeregten Spinwellen (Magnonen) wurde von Vonsovski [23], de Gennes und Friedel [24], Kasuya [25] und Mannari [26] untersucht. In diesen Arbeiten wird als maßgebendes Streupotential die Austauschkopplung zwischen s- und d-Elektronen betrachtet. Der Widerstand ergibt sich bei tiefen Temperaturen als proportional zu $T^2$ und bestimmt daher bei tiefen Temperaturen den temperaturabhängigen Anteil des elektrischen Widerstandes, da ja der von der Elektron-Phonon-Wechselwirkung herrührende Widerstand wie $T^5$ verschwindet.

Wir müssen also bei ferromagnetischen Metallen demnach zwischen fünf verschiedenen Streuursachen unterscheiden:

1. Elektron-Phonon-Wechselwirkung,
2. Elektron-Magnon-Wechselwirkung,
3. Streuung am elektrostatischen Potential der Gitterfehler,
4. Streuung an inhomogenen Spinverteilungen infolge innerer Spannungen (magnetostriktiver Restwiderstand),
5. Streuung an inhomogenen Spinverteilungen in Domänenwänden u. ä. (magnetokristalliner Restwiderstand).

Die unter 1. und 2. angeführten Prozesse liefern den temperaturabhängigen Anteil des elektrischen Widerstandes. Bei 0 °K verschwinden die von Phononen und Magnonen herrührenden Beiträge zum elektrischen Widerstand. Die Prozesse 3., 4. und 5. liefern den elektrischen Restwiderstand, der, abgesehen von einer indirekten Abhängigkeit über die Stoffkonstanten und eventuellen Abweichungen von der Matthiessenschen Regel, temperaturunabhängig ist.

Da Domänenwände auch in idealen Einkristallen bei 0 °K vorhanden sind, sollte in fehlerfreien, ferromagnetischen Kristallen immer ein endlicher Restwiderstand vorhanden sein. Die Messungen von Semenenko und Sudovtsov [27, 28] an Eisen scheinen in der Tat auf einen derartigen endlichen Restwiderstand der Domänenwände hinzudeuten.

Zahlreiche Diskussionen wurden in den letzten Jahren über den Anteil der unter 1. und 2. aufgeführten Prozesse zum Gesamtwiderstand in ferromagnetischen Metallen geführt [24, 25, 26, 29]. Entscheidend ist hierbei, ob die Elektron-Phonon- oder die Elektron-Magnon-Wechselwirkung überwiegt.

Der relativ große elektrische Widerstand der Übergangsmetalle wurde ursprünglich von Mott [30, 31] auf die Streuung der s-Elektronen in das nur teilweise gefüllte d-Band zurückgeführt. Die Streuwahrscheinlichkeit eines solchen Prozesses sollte auf Grund der großen Zustandsdichte im d-Band sehr groß sein und eine Erklärung für den gegenüber den Edelmetallen größeren elektrischen Widerstand bei den Übergangsmetallen abgeben.

Für das Hinzutreten des Elektron-Magnon-Anteils zum elektrischen Widerstand in Ferromagnetika sprechen vor allem drei experimentelle Tatsachen:

1. Oberhalb der Curietemperatur entspricht die Temperaturabhängigkeit des elektrischen Widerstandes derjenigen eines normalen Metalls, während unterhalb der Curietemperatur der Widerstand meist schneller abnimmt als bei einem normalen Metall.

2. Der elektrische Widerstand der Ferromagnetika verschwindet am absoluten Nullpunkt im Einklang mit der Temperaturabhängigkeit des Elektron-Magnon-Anteils proportional zu $T^2$.
3. Bei dem seltenen Erdmetall Gd [32] wurde dieselbe Temperaturabhängigkeit gefunden wie bei den Übergangsmetallen, obwohl hier die den s–d-Übergängen entsprechenden s–f-Übergänge sicher nicht stattfinden.

Diese drei Befunde werden durch die Streuung von s-Elektronen an thermisch angeregten Spinwellen, wie Kasuya [25] gezeigt hat, befriedigend erklärt. Nach Kasuya [25] und Mannari [26] nimmt der magnetische Anteil des elektrischen Widerstandes bei tiefen Temperaturen wie $T^2$ zu und bleibt oberhalb der Curietemperatur konstant, da in diesem Temperaturbereich die Spinunordnung vollkommen ist und nicht mehr anwachsen kann. Die Zunahme des elektrischen Widerstandes erfolgt dann nur noch durch das Anwachsen der Zahl der thermisch angeregten Phononen.

Die Tatsache, daß auch bei ferromagnetischen seltenen Erden ein ähnlicher Verlauf der Temperaturabhängigkeit des elektrischen Widerstandes wie bei den Übergangsmetallen gefunden wird, obwohl in den seltenen Erden keine s–f-Übergänge stattfinden, weist darauf hin, daß zwischen den s- und f-Elektronen eine Austauschkopplung besteht, die für den magnetischen Beitrag zum elektrischen Widerstand verantwortlich ist. Direkte Hinweise auf eine Austauschkopplung zwischen s- und d-Elektronen wurden von Shull [33] bei Fe gefunden. Mit Hilfe von Neutronenstreuexperimenten konnte Shull zeigen, daß die s-Elektronen in Fe ein antiferromagnetisches Moment von $-0{,}2\,\mu_B$ besitzen. In den nun folgenden Abschnitten soll untersucht werden, welchen Beitrag die s–d-Austauschkopplung zum Restwiderstand der Versetzungen liefert. In einer noch unveröffentlichten Arbeit ist der Restwiderstand von Domänenwänden berechnet worden.

In den folgenden beiden Abschnitten werden wir uns mit dem sd-Austauschoperator sowie dessen Einfluß auf die Bandstruktur befassen.

## 2. Der sd-Austauschoperator

Ursache der Spinwechselwirkung zwischen den s- und den d-Elektronen sind die auf das Antisymmetrieprinzip zurückführenden Austauschkräfte. Diese Austauschkräfte wurden erstmals von Heisenberg [34] zur Erklärung der spontanen Magnetisierung herangezogen. Heisenberg's Überlegungen bezogen sich ursprünglich nur auf die Austauschkräfte zwischen d-Elektronen benachbarter Atome. Wie Vonsovski [23] und Kasuya [25] gezeigt haben, gelten für die sd-Austauschwechselwirkung qualitativ dieselben Beziehungen wie für die dd-Austauschwechselwirkung. Bezeichnen $\sigma_s^i$ und $\sigma_d^n$ die Paulischen Spinmatrizen der s- bzw. der d-Elektronen am $i$-ten bzw. am $n$-ten Gitterpunkt, so lautet der Operator der sd-Austauschkopplung zwischen den s- und den d-Elektronen

$$H_{sd} = -2 \sum_{n,i} J_{sd}\,(|\boldsymbol{r}_i - \boldsymbol{R}_n|)\,\sigma_d^n\,\sigma_s^i\,, \tag{2.1}$$

wobei $J_{sd}\,(|\boldsymbol{r}_i - \boldsymbol{R}_n|)$ das Austauschintegral zwischen dem $i$-ten s-Elektron und $n$-ten d-Elektron bedeutet. Das Austauschintegral $J_{sd}\,(|\boldsymbol{r}_i - \boldsymbol{R}_n|)$ ist sicher zwischen Elektronen, die sich am gleichen Atom befinden, am größten. Da wir ferner voraussetzen wollen, daß sich die d-Elektronen näherungsweise wie gebundene Elektronen verhalten, dürfen wir uns bei der Summation in Gl. (2.1) auf die Elektronen an ein und demselben Atom beschränken. Dann gilt

$$H_{sd} = -2\,J_{sd}(0) \sum_{i,j} \sigma_d^i\,\sigma_s^j\,, \tag{2.2}$$

wobei $\sigma_d^i$ und $\sigma_s^j$ jetzt die Spinmatritzen der s- und d-Elektronen am $n$-ten Atom bedeuten. Das Vorzeichen von $J_{sd}(0)$ entscheidet darüber, ob die s-Elektronen ferromagnetisch oder antiferromagnetisch an die d-Elektronen gekoppelt sind. Nach Shull [33] ist bei Fe diese Kopplung antiferromagnetisch. Deshalb besitzt $J_{sd}(0)$ in diesem Fall einen negativen Wert. Bei Ni und Co wurden bisher keine Experimente durchgeführt, die eine Aussage über Größe und Vorzeichen des Austauschintegrals $J_{sd}(0)$ zu machen gestatten. Für unsere spätere Rechnung ist es erforderlich, anstelle der Spinmatrizen $\sigma_d^i$ die makroskopische Induktion $\boldsymbol{B}$ einzuführen. Dabei wird vorausgesetzt, daß die magnetische Induktion $4\pi \boldsymbol{J}_s(\boldsymbol{r})$ der spontanen Magnetisierung im wesentlichen durch die d-Elektronen zustandekommt. Zwischen $\sigma_d^i$ und $\boldsymbol{J}_s$ besteht der Zusammenhang

$$\sum_i \sigma_d^i = \frac{4\pi \boldsymbol{J}_s(\boldsymbol{r})}{2\pi g \mu_B} \Omega_0 . \tag{2.3}$$

Dabei bedeutet $\Omega_0$ das Atomvolumen, $g$ den gyromagnetischen Faktor, der $\approx 2$ beträgt, und $\mu_B$ das Bohrsche Magneton. Der Operator der sd-Austauschkopplung lautet nun

$$H_{sd} = J_0 \, 4\pi \boldsymbol{J}_s \cdot \sigma_s \tag{2.4}$$

mit

$$J_0 = -\frac{J_{sd}(0)}{\pi g \mu_B} \Omega_0 . \tag{2.5}$$

Der Operator (2.4) beschreibt die Wechselwirkung eines s-Elektrons mit der spontanen Magnetisierung.

Den Verlauf der Magnetisierung $\boldsymbol{J}_s$ als Funktion des Ortes bei Anwesenheit innerer Spannungen oder in Domänenwänden kann mit Hilfe der Theorie des Mikromagnetismus berechnet werden. Diese Theorie wurde mit Erfolg zur Berechnung kleiner Abweichungen der Magnetisierung aus einer durch ein äußeres Feld oder durch die Kristallanisotropie festgelegten Vorzugsrichtung bei Anwesenheit innerer Spannungen angewandt [35 bis 39]. Maßgebend für die Ausdehnung der durch innere Spannungen hervorgerufenen magnetischen Inhomogenität ist die sogenannte Austauschlänge

$$\varkappa_H^{-1} = l_H = \left(\frac{2A}{H J_s}\right)^{1/2} . \tag{2.6}$$

$l_H$ ist diejenige Länge, innerhalb der eine magnetische Störung auf das 1/e-fache abgefallen ist. In Gl. (2.6) bedeutet $A$ die Konstante der dd-Austauschkopplung, $H$ das äußere magnetische Feld und $J_s$ die Sättigungsmagnetisierung. Für $H = 2000$ Oe ergibt sich bei Ni $l_H = 200$ Å. Die magnetische Inhomogenität wurde in früheren Arbeiten [37 bis 39] (zur Berechnung des magnetischen Einmündungsgesetzes) im Bereich hoher Magnetfelder für geradlinige Versetzungen, Versetzungsringe und Zwischengitteratome bestimmt. Wir können somit an diese Arbeiten anknüpfen und ihre Ergebnisse zur Berechnung des magnetostriktiven Restwiderstandes verwenden.

## 3. Wellenfunktion und Energie der s-Elektronen

Den folgenden Überlegungen legen wir das von Mott [30] für die Übergangsmetalle eingeführte Zweibandmodell zugrunde. Diese durch Messungen der spezifischen Elektronenwärme und der optischen Konstanten ausreichend gesicherte Elektronenbandstruktur besteht aus einem breiten s-Band mit s-Elektronen der effektiven Masse $m_s^*/m \approx 1$ und einem schmalen d-Band, dessen Elektronen eine effektive Masse $m_d^*/m \approx 20$ bis 30 besitzen. Beide Bänder besitzen eine gemeinsame Fermigrenze. Auf Grund der effektiven Massen sind Leitfähigkeitseigenschaften im wesentlichen durch das s-Band bestimmt, während die ferromagnetischen Eigenschaften auf das d-Band zurückgeführt werden. In diesem einfachen Modell sollen die s-Elektronen durch ebene Wellen

$$\psi_0^{\uparrow,\downarrow} = \frac{1}{\sqrt{V_0}} e^{i\boldsymbol{k}^{\uparrow,\downarrow}\cdot\boldsymbol{r}} \chi^{\uparrow,\downarrow} \tag{3.1}$$

beschrieben werden. Hierbei bedeutet $V_0$ das Volumen der Probe, $\boldsymbol{r}$ den Ortsvektor, $\boldsymbol{k}^{\uparrow,\downarrow}$ den Wellenvektor eines Elektrons mit (+)- bzw. (—)-Spin und $\chi^{\uparrow,\downarrow}$ die dazugehörige Spinwellenfunktion.

Für Elektronen mit (+)-Spin gilt $\qquad \chi^{\uparrow} = \binom{1}{0}$ (3.2.a)

und für Elektronen mit (—)-Spin $\qquad \chi^{\downarrow} = \binom{0}{1}$ . (3.2.b)

In nicht-ferromagnetischen Metallen lautet die Energie der quasifreien Elektronen

$$E = U_0 + \frac{\hbar^2}{2\,m_s^*}\,k^2\,, \tag{3.3}$$

wo $U_0$ das mittlere Gitterpotential bedeutet und das zweite Glied in Gl. (3.3) der kinetischen Energie des Elektrons entspricht. Sofern nun eine sd-Kopplung vorhanden ist, so wird die potentielle Energie der s-Elektronen von ihrer Spinstellung bezüglich der Richtung der spontanen Magnetisierung abhängig. Es kommt daher zu einer Polarisation des s-Bandes in ein (+)- und ein (—)-Band. Dabei wird je nach Vorzeichen des Austauschintegrals das (+)-Band oder das (—)-Band eine größere Bandbreite besitzen als das andere Teilband. Die Verhältnisse sind in Fig. 1 dargestellt. Bezeichnen wir mit $\zeta^{\uparrow}$ und $\zeta^{\downarrow}$ die energetische Bandbreite der Elektronen mit (+)- und (—)-Spin, so ergibt sich für die kinetische Energie der Elektronen

$$E^{\uparrow} = \frac{\hbar^2}{2\,m_s^*}\,k^2 + E_M\;; \quad E^{\downarrow} = \frac{\hbar^2}{2\,m_s^*}\,k^2\,. \tag{3.4}$$

Dabei wird vorausgesetzt, daß die effektiven Massen für Elektronen mit (+)- und (—)-Spin gleich groß sind, was sicher nur näherungsweise richtig ist. Ferner bedeutet $E_M$ die Differenz der beiden Bandbreiten, für die gilt

$$E_M = \zeta^{\downarrow} - \zeta^{\uparrow} = \frac{\hbar^2}{2\,m_s^*}\,4\,k_M^2\,. \tag{3.5}$$

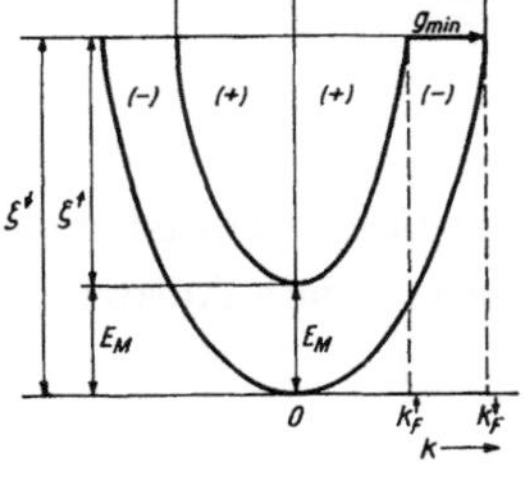


Fig. 1. s-Bandstruktur ferromagnetischer Metalle mit sd-Austauschkopplung

Die Wellenzahlen der Elektronen mit (+)- und (−)-Spin sind an der Fermigrenze verschieden groß und betragen

$$k_F^\uparrow = \left(\frac{2\, m_s^*\, \zeta^\uparrow}{\hbar^2}\right)^{1/2}, \qquad k_F^\downarrow = \left(\frac{2\, m_s^*\, (\zeta^\uparrow + E_M)}{\hbar^2}\right)^{1/2}. \tag{3.6}$$

Zwischen den Elektronendichten $n^\uparrow$ und $n^\downarrow$ der Elektronen beider Spinrichtungen und den Wellenzahlen $k_F^{\uparrow,\downarrow}$ besteht die Beziehung

$$k_F^{\uparrow,\downarrow} = (6\, \pi^2\, n^{\uparrow,\downarrow})^{1/3}. \tag{3.7}$$

Die Geschwindigkeit der Elektronen an der Fermigrenze ist für die Elektronen mit (+)- und (−)-Spin verschieden groß und beträgt

$$v^{\uparrow,\downarrow} = \frac{\hbar\, k_F^{\uparrow,\downarrow}}{m_s^*}. \tag{3.8}$$

## 4. Lösung der Schrödingergleichung für freie Elektronen bei Anwesenheit einer sd-Austauschkopplung und Berechnung der Streuwahrscheinlichkeit

Die Einelektronschrödingergleichung lautet für ein Elektron mit der Gesamtenergie $E$, das sich in einem Potential $V$ bewegt,

$$-\frac{\hbar^2}{2\, m_s^*}\, \Delta\psi + V\, \psi = E\, \psi. \tag{4.1}$$

Entsprechend unseren Ausführungen in Abschnitt 1 können wir $V$ in einen elektrostatischen Anteil $V_e$, herrührend von der Volumdilatation der Gitterfehlstellen, und einen Anteil $V_{sd}$, der auf die sd-Austauschkopplung zurückzuführen ist, zerlegen. Eine Näherungslösung der Schrödingergleichung mit dem Potential $V_e$ einer Stufenversetzung wurde von Seeger und Bross [7] angegeben. Im folgenden werden wir voraussetzen, daß $V_{sd} \gg V_e$ gilt. In diesem Falle dürfen wir den Term $V_e$ vernachlässigen. Dies ist sicher berechtigt bei Gitterfehlstellen, die keine Volumdilatation erzeugen, und in Metallen mit großer Magnetostriktion. Die Schrödingergleichung (4.1) wurde für $V = V_{sd}$ bereits früher in Bornscher Näherung für den Fall der Wechselwirkung des Neutronenspins mit einer inhomogenen Magnetisierung gelöst [40], so daß wir hier auf eine ausführliche Ableitung verzichten können und nur die Ergebnisse unserer früheren Arbeit mitzuteilen brauchen.

Mit dem Ansatz

$$\psi^{\uparrow,\downarrow} = \psi_0^{\uparrow,\downarrow} + \psi_s^{\uparrow,\downarrow} \tag{4.2}$$

für die gestreute Welle folgt aus Gl. (4.1) in der Bornschen Näherung

$$\psi_s^{\uparrow,\downarrow} = \frac{m_s^*(2\,\pi)^{3/2}}{2\,\pi\,\hbar^2\sqrt{V_0}}\, \tilde{V}_{sd}(\boldsymbol{g})\, \frac{e^{i\boldsymbol{k}\cdot\boldsymbol{R}}}{R}\, \chi^{\uparrow,\downarrow}, \tag{4.3}$$

wobei

$$\tilde{V}_{sd} = \frac{1}{(2\,\pi)^{3/2}} \int\limits_{V_0} H_{sd}(\boldsymbol{r})\, e^{-i\boldsymbol{g}\cdot\boldsymbol{r}}\, d\tau_{\boldsymbol{r}} \tag{4.4}$$

die Fouriertransformierte des sd-Austauschpotentials bedeutet. Der in Gl. (4.3) auftretende Ausdruck $\tilde{V}_{sd}$ liefert bei Berücksichtigung von Gl. (2.4)

$$\tilde{V}_{sd} = -\, J_0\, (i\, \tilde{B}_x + \tilde{B}_z), \tag{4.5}$$

und bei Einführung der Richtungscosinus $\gamma_x$, $\gamma_z$ der Magnetisierung gemäß $B_{x,z} = 4\pi J_s \gamma_{x,z}$, lautet Gl. (4.5)

$$\tilde{V}_{sd}\,\chi^{\uparrow,\downarrow} = 4\pi J_s J_0 (i\,\tilde{\gamma}_x + \tilde{\gamma}_z)\,\chi^{\downarrow,\uparrow}\,. \tag{4.6}$$

Wie Gl. (4.6) zu entnehmen ist, erfolgt die Streuung der Elektronen unter Spinumkehr. Für den Streuvektor gilt somit

$$\boldsymbol{g}^{\uparrow} = \boldsymbol{k}^{\downarrow} - \boldsymbol{k}^{\uparrow}$$

für den Übergang vom (+)- zum (−)-s-Band und

$$\boldsymbol{g}^{\downarrow} = \boldsymbol{k}^{\uparrow} - \boldsymbol{k}^{\downarrow}$$

für den Übergang vom (−)- zum (+)-s-Band.

Da bei Streuprozessen an statischen Gitterfehlern die Energie erhalten bleibt, besteht folgender Zusammenhang:

$$E^{\uparrow}(\boldsymbol{k}^{\uparrow}) = E^{\downarrow}(\boldsymbol{k}^{\downarrow}) \tag{4.7a}$$

oder nach Gl. (3.5)

$$k_F^{\downarrow 2} - k_F^{\uparrow 2} = 4\,k_M^2\,. \tag{4.7b}$$

Gl. (4.7b) ist zu entnehmen, daß sich beim Streuprozess die absolute Größe des Wellenvektors ändert und der Streuvektor $\boldsymbol{g}$ den Betrag

$$g_{\min} = k_F^{\downarrow} - k_F^{\uparrow} = \frac{4\,k_M^2}{k_F^{\downarrow} + k_F^{\uparrow}} \tag{4.8}$$

nicht unterschreiten kann.

Zur Berechnung der Leitfähigkeit benötigen wir die Zahl der in der Zeiteinheit in das Volumelement $d\tau_{k\uparrow}$ bzw. $d\tau_{k\downarrow}$ gestreuten Elektronen. Diese beträgt

$$P^{\uparrow\downarrow}(\boldsymbol{k}^{\uparrow}, \boldsymbol{k}^{\downarrow})\,d\tau_{k\uparrow} = (2\pi)^3 R^2 V_0\, v^{\uparrow}\, \psi_s^{\uparrow}\, \psi_s^{*\uparrow}\, d\omega_{k\uparrow}\,. \tag{4.9}$$

Eine entsprechende Beziehung mit vertauschter Reihenfolge der Spinrichtungen gilt für die Übergangswahrscheinlichkeit $P^{\downarrow\uparrow}$. In Gl. (4.9) bedeutet $d\omega_{k\uparrow}$ das Oberflächenelement auf der Einheitskugel im $\boldsymbol{k}^{\uparrow}$-Raum. Für die Größe $\psi_s^{\uparrow}\psi_s^{*\uparrow}$ erhalten wir bei Berücksichtigung von Gl. (4.6) und Gl. (4.3)

$$\psi_s^{\uparrow}\psi_s^{*\uparrow} = (2\pi)^3 \left(\frac{4\pi J_s m_s^* J_0}{2\pi\hbar^2}\right)^2 \left(|\tilde{\gamma}_x|^2 + |\tilde{\gamma}_z|^2 - i\,(\tilde{\gamma}_x^*\,\tilde{\gamma}_z - \tilde{\gamma}_x\,\tilde{\gamma}_z^*)\right). \tag{4.10}$$

Somit lautet Gl. (4.9)

$$P^{\uparrow\downarrow}(\boldsymbol{k}^{\uparrow}, \boldsymbol{k}^{\downarrow})\,d\tau_{k\uparrow} = v^{\uparrow}\cdot T\,\Delta\tilde{\gamma}^2\,d\omega_{k\uparrow} \tag{4.11}$$

mit

$$T = \left(\frac{(2\pi)^3\,4\pi J_s m_s^* J_0}{2\pi\hbar^2}\right)^2 \tag{4.12}$$

und

$$\Delta\tilde{\gamma}^2 = |\tilde{\gamma}_x|^2 + |\tilde{\gamma}_z|^2 - i\,(\tilde{\gamma}_x^*\,\tilde{\gamma}_z - \tilde{\gamma}_x\,\tilde{\gamma}_z^*)\,. \tag{4.13a}$$

Im folgenden werden wir nur Fälle mit rein imaginären Fouriertransformierten betrachten. Dann gilt

$$\Delta\tilde{\gamma}^2 = |\tilde{\gamma}_x|^2 + |\tilde{\gamma}_z|^2\,. \tag{4.13b}$$

Mit Gl. (4.11) und Gl. (4.13b) ist das Streuproblem auf die Berechnung der Fouriertransformierten der Richtungskosinus $\gamma_x$ und $\gamma_z$ der Magnetisierung zurückgeführt.

## 5. Lösung der Boltzmanngleichung

In einem Kristall ohne äußere Felder gilt für die Wahrscheinlichkeit, daß ein Elektronenzustand mit dem Wellenvektor $\boldsymbol{k}^{\uparrow}$ oder $\boldsymbol{k}^{\downarrow}$ besetzt ist,

$$f_0^{\uparrow,\downarrow} = \frac{1}{e^{\frac{E^{\uparrow,\downarrow} - \zeta^{\uparrow,\downarrow}}{kT}} + 1}, \tag{5.1}$$

wobei entsprechend den beiden Bandbreiten in ferromagnetischen Metallen mit einer sd-Austauschkopplung auch zwei Verteilungsfunktionen $f^{\uparrow}$ und $f^{\downarrow}$ der Elektronen mit (+)- und (—)-Spin eingeführt werden müssen. Durch ein elektrisches Feld wird die durch Gl. (5.1) gegebene sog. Fermische Verteilungsfunktion gestört, denn die Elektronen erfahren hierdurch eine zusätzliche Beschleunigung. Für die zeitliche Änderung der Verteilungsfunktion infolge des elektrischen Feldes $\boldsymbol{E}$ gilt

$$\left.\frac{\partial f^{\uparrow,\downarrow}}{\partial t}\right|_{\text{Feld}} = \frac{\mathrm{d}\boldsymbol{k}}{\mathrm{d}t} \cdot \operatorname{grad}_{\boldsymbol{k}} f^{\uparrow,\downarrow} = -\frac{e\hbar}{m_s^*} \frac{\partial f_0^{\uparrow,\downarrow}}{\partial E^{\uparrow,\downarrow}} \boldsymbol{E}\cdot\boldsymbol{k}. \tag{5.2}$$

($e$ ist die Elementarladung und $\hbar$ das durch $2\pi$ dividierte Plancksche Wirkungsquantum). Die unbeschränkte Beschleunigung der Elektronen wird durch die Streuung an Gitterfehlstellen gehemmt. Im stationären Zustand wird die Beschleunigung der Elektronen durch das äußere Feld gerade kompensiert, so daß die Bedingung

$$\left.\frac{\partial f}{\partial t}\right|_{\text{Feld}} + \left.\frac{\partial f}{\partial t}\right|_{\text{Gitter}} = 0 \tag{5.3}$$

erfüllt ist, wo $\left.\frac{\partial f}{\partial t}\right|_{\text{Gitter}}$ die zeitliche Änderung der Verteilungsfunktion durch die Streuung an Gitterfehlstellen darstellt. Bei Berücksichtigung der Übergänge vom (+)- ins (—)-Band und umgekehrt ergibt sich

$$\left.\frac{\partial f^{\uparrow}}{\partial t}\right|_{\text{Gitter}} = \frac{1}{V_0}\int \{P^{\uparrow\downarrow}(\boldsymbol{g})\, f^{\uparrow}(1-f^{\downarrow}) - P^{\downarrow\uparrow}(\boldsymbol{g})\, f^{\downarrow}(1-f^{\uparrow})\}\, \mathrm{d}\tau_{\boldsymbol{k}^{\downarrow}} \tag{5.4}$$

und eine entsprechende Gleichung mit umgekehrten Spinrichtungen für $\left.\frac{\partial f^{\downarrow}}{\partial t}\right|_{\text{Gitter}}$. Gegenüber unmagnetischen Metallen liegen die Verhältnisse bei den Ferromagnetika etwas komplizierter, da zwei Verteilungsfunktionen $f^{\uparrow}$ und $f^{\downarrow}$ zu bestimmen sind und deshalb anstelle *einer* Integralgleichung zwei gekoppelte Integralgleichungen zu lösen sind. Diese beiden Integralgleichungen ergeben sich durch Einsetzen von Gl. (5.4) und Gl. (5.2) in Gl. (5.3) und lauten

$$\frac{e\hbar}{m_s^*}\frac{\partial f_0^{\uparrow}}{\partial E}\boldsymbol{E}\cdot\boldsymbol{k}^{\uparrow} = \frac{1}{V_0}\int \{P^{\uparrow\downarrow}(\boldsymbol{g})\, f^{\uparrow}\,(1-f^{\downarrow}) - P^{\downarrow\uparrow}_{(\boldsymbol{g})}\, f^{\downarrow}\,(1-f^{\uparrow})\}\, \mathrm{d}\tau_{\boldsymbol{k}^{\downarrow}} \tag{5.5a}$$

für Elektronen mit (+)-Spin und

$$\frac{e\hbar}{m_s^*}\frac{\partial f_0^{\downarrow}}{\partial E}\boldsymbol{E}\cdot\boldsymbol{k}^{\downarrow} = \frac{1}{V_0}\int \{P^{\downarrow\uparrow}(\boldsymbol{g})\, f^{\downarrow}\,(1-f^{\uparrow}) - P^{\uparrow\downarrow}(\boldsymbol{g})\, f^{\uparrow}\,(1-f^{\downarrow})\}\, \mathrm{d}\tau_{\boldsymbol{k}^{\uparrow}} \tag{5.5b}$$

für Elektronen mit (—)-Spin. Zur Lösung dieses gekoppelten Gleichungssystems machen wir für die Verteilungsfunktionen $f^{\uparrow}$ und $f^{\downarrow}$ folgenden Ansatz

$$f^{\uparrow,\downarrow} = f_0^{\uparrow,\downarrow} - \frac{e\hbar}{m_s^*}\frac{\partial f_0^{\uparrow\downarrow}}{\partial E}\boldsymbol{E}\cdot\boldsymbol{k}^{\uparrow,\downarrow}\cdot\tau^{\uparrow,\downarrow}. \tag{5.6}$$

In Gl. (5.6) hat $\tau$ die Bedeutung einer Relaxationszeit. Die Einführung der Relaxationszeiten $\tau^{\uparrow}$ und $\tau^{\downarrow}$ ist, wie von Bross und Seeger [41] gezeigt wurde, immer dann erlaubt, falls die Übergangswahrscheinlichkeit nur vom Winkel zwischen $k^{\uparrow}$ und $k^{\downarrow}$ abhängig ist. Wir behandeln im folgenden Fälle, bei denen diese Bedingung erfüllt ist. Nehmen wir das elektrische Feld als parallel zur $x$-Richtung an, so ergibt sich mit Hilfe von Gl. (5.6) aus Gl. (5.5a) und Gl. (5.5b)

$$k_x^{\uparrow,\downarrow} = -\frac{T}{V_0}\int (v^{\uparrow,\downarrow}\,\tau^{\uparrow,\downarrow}\,k_x^{\uparrow,\downarrow} - v^{\downarrow,\uparrow}\,\tau^{\downarrow,\uparrow}\,k_x^{\downarrow,\uparrow})\,\Delta\tilde{\gamma}^2\,\mathrm{d}\omega_{\boldsymbol{k}^{\downarrow,\uparrow}}\,. \tag{5.7}$$

Gl. (5.7) stellt den Ausgangspunkt zur Berechnung des magnetostriktiven Widerstandes von Versetzungen dar. Für geradlinige Versetzungen wurde der in Gl. (5.7) auftretende Ausdruck $\Delta\tilde{\gamma}^2$ der mittleren quadratischen Abweichung der Magnetisierung aus der Magnetisierungsrichtung in einer früheren Arbeit [39] für folgendes Versetzungsmodell berechnet: Die Versetzungen sollen in sogenannten Versetzungsgruppen angeordnet sein und parallel zur $z$-Richtung verlaufen. Die Versetzungsgruppen bestehen aus $n$ Versetzungen, die die Abstände $R_{ij}$ voneinander besitzen und in einer Gleitebene liegen. Der Abstand zwischen den einzelnen Versetzungsgruppen beträgt $R_0$ und die Versetzungen besitzen die Länge $L$. Pro Volumeinheit existieren deshalb $N_0 = 1/(R_0^2\,L)$ Versetzungsgruppen. Wir verzichten im folgenden auf die Angabe des exakten Ausdrucks für $\Delta\tilde{\gamma}^2$ und betrachten nur den maßgebenden radialsymmetrischen Anteil. In dieser Näherung gilt

$$\Delta\tilde{\gamma}^2 = A_0\cdot\frac{n - \sum\limits_{i\neq j} J_0(g\,R_{ij})}{g_\varrho\,(g^2 + \varkappa_H^2)^2}\cdot\frac{\sin^2\frac{1}{2}\,g_z\,L}{g_z^2}\,, \tag{5.8}$$

wobei $A_0 = \left(\frac{b\,G}{2\,A}\right)^2\frac{c_0}{32\,\pi^2}$, $b$ die Versetzungsstärke, $G$ den Schubmodul, $J_0(g\,R_{ij})$ die Besselfunktion nullter Ordnung und $c_0$ eine Konstante von der Größenordnung $10^{-8}$ bis $10^{-9}$ bedeutet, die für Nickel von Kronmüller und Seeger [39] berechnet wurde.

Der Faktor $\dfrac{\sin^2\frac{1}{2}\,g_z\,L}{g_z^2}$ in Gl. (5.8) trägt der Zylindersymmetrie des Problems

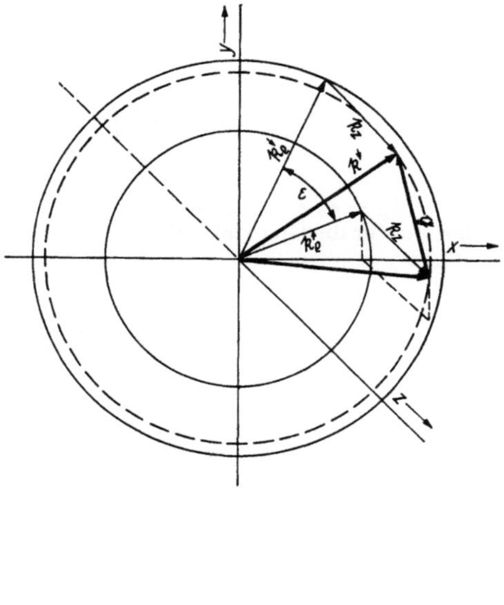


Fig. 2. Zur Streuung an einer unendlich ausgedehnten Versetzung.
Die Versetzung verläuft parallel zur $z$-Achse

Rechnung und gewährleistet, daß die $z$-Komponente des Wellenvektors des gestreuten Elektrons konstant bleibt. Dies hat zur Folge, daß die gestreuten Wellenvektoren einen Kegelmantel um die $z$-Achse bilden. Die Verhältnisse sind in Fig. 2 veranschaulicht. Zur Integration von Gl. (5.7) über den $\boldsymbol{k}^{\uparrow}$- bzw. den $\boldsymbol{k}^{\downarrow}$-Raum führen wir Zylinderkoordinaten ein. Wir setzen also

$$g^2 = g_\varrho^2 + g_z^2\,, \quad k_z = k_{\mathrm{F}} \cos\vartheta\,, \quad d\omega_{\boldsymbol{k}^{\uparrow,\downarrow}} = \sin\vartheta \, d\vartheta \, d\varphi = \frac{d\, g_z^{\uparrow,\downarrow}}{k_{\mathrm{F}}^{\uparrow,\downarrow}} \, d\varphi\,. \tag{5.9}$$

Die Ausführung der Integration über $g_z$ und $\varphi$ [1]) liefert folgende Bestimmungsgleichungen für $\tau^{\uparrow}$ und $\tau^{\downarrow}$:

$$k_\varrho^{\uparrow} k_{\mathrm{F}}^{\downarrow} = \frac{\pi}{2} N_0 L T \left(I_0 v^{\uparrow} \tau^{\uparrow} k_\varrho^{\uparrow} - I_1 v^{\downarrow} \tau^{\downarrow} k_\varrho^{\downarrow}\right), \tag{5.10a}$$

$$k_\varrho^{\downarrow} k_{\mathrm{F}}^{\uparrow} = \frac{\pi}{2} N_0 L T \left(I_0 v^{\downarrow} \tau^{\downarrow} k_\varrho^{\downarrow} - I_1 v^{\uparrow} \tau^{\uparrow} k_\varrho^{\uparrow}\right). \tag{5.10b}$$

$I_0$ und $I_1$ bedeuten die folgenden Integrale:

$$I_n = \int_0^{2\pi} \Delta\tilde{\gamma}^2 \left(g_\varrho(\varepsilon)\right) \cos n\,\varepsilon \, d\varepsilon\,; \quad n = 0{,}1 \tag{5.11}$$

mit

$$g_\varrho^2 = k_\varrho^{\uparrow 2} + k_\varrho^{\downarrow 2} - 2\, k_\varrho^{\uparrow} k_\varrho^{\downarrow} \cos\varepsilon\,, \tag{5.12}$$

wobei $\varepsilon$ den Winkel zwischen $\boldsymbol{k}_\varrho^{\uparrow}$ und $\boldsymbol{k}_\varrho^{\downarrow}$ bedeutet.

Die Auflösung von Gl. (5.10a) und Gl. (5.10b) nach $\tau^{\uparrow}$ und $\tau^{\downarrow}$ ergibt schließlich

$$\left.\begin{aligned} \tau^{\uparrow} &= 2\,\frac{k_\varrho^{\uparrow} k_{\mathrm{F}}^{\downarrow} I_0 + k_\varrho^{\downarrow} k_{\mathrm{F}}^{\downarrow} I_1}{\pi N_0 L T v^{\uparrow} k_\varrho^{\uparrow}(I_0^2 - I_1^2)}\,, \\ \tau^{\downarrow} &= 2\,\frac{k_\varrho^{\downarrow} k_{\mathrm{F}}^{\uparrow} I_0 + k_\varrho^{\uparrow} k_{\mathrm{F}}^{\downarrow} I_1}{\pi N_0 L T v^{\downarrow} k_\varrho^{\downarrow}(I_0^2 - I_1^2)}\,. \end{aligned}\right\} \tag{5.13a}$$

Diesen etwas komplizierten Ausdruck für die Relaxationszeiten können wir unter der plausiblen Voraussetzung $k_{\mathrm{F}}^{\uparrow} \sim k_{\mathrm{F}}^{\downarrow}$ stark vereinfachen und erhalten dann in guter Näherung

$$\tau^{\uparrow,\downarrow} = \frac{2\, k_{\mathrm{F}}^{\downarrow,\uparrow}}{\pi N_0 L T v^{\uparrow,\downarrow} (I_0 - I_1)}\,. \tag{5.13b}$$

Zur Berechnung des Ausdrucks $I_0 - I_1$ führen wir die Integration in Gl. (5.11) mit Hilfe von Gl. (5.12) in eine Integration über $g_\varrho$ über. Dann gilt

$$I_0 - I_1 = A_0 \int_\alpha^\beta \frac{n - \sum_{i \neq j} J_0(g_\varrho R_{ij})}{g_\varrho (g_\varrho^2 + \varkappa_H^2)^2} \sqrt{\frac{\alpha^2 - g_\varrho^2}{g_\varrho^2 - \beta^2}} \, d g_\varrho \tag{5.14}$$

mit $\alpha = k_\varrho^{\downarrow} - k_\varrho^{\uparrow} = g_{\varrho,\min}$ und $\beta = k_\varrho^{\uparrow} + k_\varrho^{\downarrow} = g_{\varrho,\max}$. Die Integration von Gl. (5.14) ist bisher nur für die Grenzfälle $R_{ij} = 0$ und $R_{ij}\, g_{\varrho,\min} \gg 1$ gelungen. In diesen Fällen lautet Gl. (5.14)

$$I_0 - I_1 = n^i A_0 \int_\alpha^\beta \frac{1}{g_\varrho (g_\varrho^2 + \varkappa_H^2)^2} \sqrt{\frac{\alpha^2 - g_\varrho^2}{g_\varrho^2 - \beta^2}} \, d g_\varrho\,, \tag{5.15}$$

wobei $i = 1$ für $R_{ij}\, g_{\varrho,\min} \gg 1$ und $i = 2$ für $R_{ij} = 0$ zu wählen ist.

---

[1]) Die Integration über $\varphi$ wird nach dem von Seeger und Stehle [6] angegebenen Verfahren durchgeführt.

Die Integration von Gl. (5.15) liefert

$$I_0 - I_1 = - n^i \frac{\mathrm{d}}{\mathrm{d}\varkappa_H^2}\left(\frac{\pi A_0}{2\, k_\varrho^\uparrow k_\varrho^\downarrow}\right)\frac{1}{\varkappa_H^2}\left(\sqrt{\frac{\alpha^2+\varkappa_H^2}{\beta^2+\varkappa_H^2}} - \frac{\alpha}{\beta}\right). \qquad (5.16)$$

## 6. Berechnung der elektrischen Leitfähigkeit

Zur Bestimmung des magnetostriktiven elektrischen Widerstandes knüpfen wir an die von Seeger und Stehle [6] für die Schraubenversetzungen durchgeführten Rechnungen an. Wie bereits in Abschnitt 4 erwähnt wurde, besitzt eine Versetzung parallel zur Versetzungslinie keinen elektrischen Widerstand, da der Streuvektor in dieser Richtung immer Null ist. Deshalb spielt nur der Widerstand senkrecht zur Versetzungslinie eine Rolle. Erweitern wir die von Seeger und Stehle [6] gegebene Beziehung für die elektrische Leitfähigkeit einer Schraubenversetzung auf den ferromagnetischen Fall mit zwei Relaxationszeiten, so gilt

$$\sigma = \frac{1}{8\pi^2}\frac{e^2}{m_s^*}\int_{-k_F^\downarrow}^{k_F^\uparrow}\{(k_F^{\uparrow 2} - k_z^2)\,\tau^\uparrow + (k_F^{\downarrow 2} - k_z^2)\,\tau^\downarrow\}\,\mathrm{d}k_z\,. \qquad (6.1)$$

Zusammen mit Gl. (5.13b) und Gl. (5.16) erhält man für Gl. (6.1)

$$\sigma = \frac{e^2}{2\pi^4 \hbar A_0 N_0\, T\, L\, n^i}\int_{-k_F^\downarrow}^{k_F^\uparrow}\left\{(k_F^{\uparrow 2} - k_z^2)\frac{k_F^\downarrow}{k_F^\uparrow} + (k_F^{\downarrow 2} - k_z^2)\frac{k_F^\uparrow}{k_F^\downarrow}\right\} k_\varrho^\uparrow k_\varrho^\downarrow \frac{\mathrm{d}k_z}{F(k_z)}\,, \qquad (6.2)$$

wobei $F(k_z) = -\frac{\mathrm{d}}{\mathrm{d}\varkappa_H^2}\left\{\frac{1}{\varkappa_H^2}\left(\sqrt{\frac{\alpha^2+\varkappa_H^2}{\beta^2+\varkappa_H^2}} - \frac{\alpha}{\beta}\right)\right\}$ gesetzt wurde. Eine exakte Integration von Gl. (6.2) konnte nicht ausgeführt werden. Jedoch können für die Funktion $F(k_z)$ bestimmte Näherungen eingeführt werden, die den magnetostriktiven Widerstand in befriedigender Genauigkeit zu berechnen gestatten. Zunächst betrachten wir den Fall a) eines nur geringfügig polarisierten s-Bandes. Dies bedeutet, daß näherungsweise $k_\varrho^\uparrow \approx k_\varrho^\downarrow$ und $k_F^\uparrow \approx k_F^\downarrow$ gilt. Außerdem soll die Bedingung $\alpha^2, \beta^2 \ll \varkappa_H^2$ erfüllt sein. Die Funktion $F(k_z)$ lautet dann

$$F(k_z) = -\frac{\mathrm{d}}{\mathrm{d}\varkappa_H^2}\frac{1}{\varkappa_H^2}\left(1 - \frac{\alpha}{\beta}\right) = \frac{k_\varrho^\uparrow}{\varkappa_H^4\, k_\varrho^\downarrow}\,. \qquad (6.3)$$

Für die Leitfähigkeit ergibt sich

$$\sigma = \frac{e^2 \varkappa_H^4}{n^i \pi^4 \hbar A_0 N_0 L\, T}\int_{-k_F^\uparrow}^{k_F^\uparrow}(k_F^{\uparrow 2} - k_z^2)^2\,\mathrm{d}k_z = \frac{16\, e^2}{15\, n^i \pi^4 \hbar A_0 N_0 L\, T}\varkappa_H^4\, k_F^{\uparrow 5}\,. \qquad (6.4)$$

Als Fall b) berechnen wir nun den magnetostriktiven Widerstand, falls das s-Band eine nennenswerte Polarisation besitzt. Dann ist sicher die Bedingung $\alpha^2, \beta^2 \gg \varkappa_H^2$ erfüllt. In dieser Näherung erhält man durch Reihenentwicklung von Gl. (6.2), wenn wir uns auf das wichtigste Glied beschränken,

$$F(k_z) = \frac{k_\varrho^{\downarrow 2}}{64\, k_M^6}\,. \qquad (6.5)$$

Für die elektrische Leitfähigkeit ergibt sich nunmehr

$$\sigma = \frac{32\, e^2\, k_M^6}{n^i\, \pi^4\, \hbar\, A_0\, N_0\, L\, T} \int\limits_{-k_F^\uparrow}^{k_F^\uparrow} \left\{ (k_F^{\uparrow 2} - k_z^2) \frac{k_F^\downarrow}{k_F^\uparrow} + (k_F^{\downarrow 2} - k_z^2) \frac{k_F^\uparrow}{k_F^\downarrow} \right\} \frac{k_\varrho^\uparrow}{k_\varrho^\downarrow}\, dk_z =$$

$$= \frac{128\, e^2}{n^i\, \pi^4\, \hbar\, A_0\, N_0\, L\, T}\, k_F^{\uparrow 3}\, k_M^6 \left( 1 - \frac{2\, k_M^2}{k_F^{\downarrow 2}} \ln \frac{2\, k_F^\downarrow}{k_M} - \frac{k_M^2}{k_F^{\downarrow 2}} \right). \qquad (6.6)$$

Die Glieder mit einem Faktor $k_M^8$ dürfen wir vernachlässigen, da im allgemeinen die Bedingung $k_M^2/k_F^2 \ll 1$ erfüllt sein wird.

Es ist üblich, den spezifischen Widerstand pro cm Versetzungslinienlänge anzugeben. Dieser beträgt im Falle a) $(\alpha^2 \ll \varkappa_H^2)$

$$\varrho_m = \frac{15\, \pi^4\, n^{i-1}\, c_0}{128} \left( \frac{\pi\, J_s\, b\, G}{2\, \mu_B\, A} \right)^2 \left( \frac{J_{sd}}{\zeta\uparrow} \right)^2 \frac{\hbar\, a^3}{e^2} \frac{\Omega_0}{\varkappa_H^4\, k_F^\uparrow} \qquad (6.7)$$

und im Falle b) $(\alpha^2 \gg \varkappa_H^2)$

$$\varrho_m = \frac{\pi^4\, n^{i-1}\, c_0}{1024} \left( \frac{\pi\, J_s\, b\, G}{2\, \mu_B\, A} \right)^2 \left( \frac{J_{sd}}{\zeta\uparrow} \right)^2 \frac{\hbar\, a^3}{e^2} \frac{\Omega_0\, k_F^\uparrow}{k_M^6}. \qquad (6.8)$$

## 7. Diskussion

Wie unsere Ausführungen in Abschnitt 6 ergeben haben, ist der magnetostriktive elektrische Widerstand empfindlich von der Bandstruktur des betreffenden Metalls abhängig. Dabei spielt vor allem der Polarisationsgrad des s-Bandes eine entscheidende Rolle. Die Polarisation des s-Bandes hat zur Folge, daß die Wellenvektoren der Elektronen an der Fermifläche für (+)- und (—)-Elektronen verschieden groß sind. Da die Streuprozesse unter Spinumkehr erfolgen, kann daher ein kleinster Streuvektor $g_{\min} = k_F^\downarrow - k_F^\uparrow$ nicht unterschritten werden. Die Größe dieses Streuvektors $g_{\min}$ bestimmt die Abhängigkeit des Restwiderstandes vom äußeren Felde. Unsere in Abschnitt 6 gewonnenen Ergebnisse zeigen, daß bei großen Feldstärken der magnetostriktive Restwiderstand wie $1/H^2$ abnimmt. Dieses Verhalten ist darauf zurückzuführen, daß mit wachsender Feldstärke die Magnetisierungsinhomogenitäten geglättet werden und dadurch das Störpotential kleiner wird. Bei kleinen Feldstärken dagegen strebt der Restwiderstand einem vom äußeren Feld unabhängigen konstanten Wert zu, der nach Gl. (6.8) zu berechnen ist. Dieses Verhalten läßt sich qualitativ an Hand von Gl. (5.8) leicht verstehen. Da die Streuwahrscheinlichkeit proportional zu $\Delta\tilde{\gamma}^2$ ist und deren Feldabhängigkeit durch den Faktor $1/(g^2 + \varkappa_H^2)^2$ gegeben ist wird sofort klar, daß eine ausgeprägte Feldstärkeabhängigkeit nur für $g_{\min} \ll \varkappa_H^2$ zu erwarten ist. Da ferner zwischen $g_{\min}$ und dem Polarisationsgrad der s-Elektronen eine enge Beziehung besteht, wird verständlich, warum die Feldstärkeabhängigkeit indirekt von der Polarisation der s-Elektronen abhängt. Die physikalische Interpretation der Feldstärkeabhängigkeit hat davon auszugehen, daß die Elektronen nur Magnetisierungsinhomogenitäten, deren Wellenlänge $\lambda < 1/g_{\min}$ ist, noch aufzulösen in der Lage sind. Die Ausdehnung der Magnetisierungsinhomogenitäten ist aber von der Größenordnung $\varkappa_H^{-1} = \left(\frac{2\, A}{H J_s}\right)^{1/2}$, so daß bei kleinen Feldstärken die Wellenlänge der Magnetisierungsinhomogenität sehr groß wird und ihre Variation mit der Feldstärke von den Elektronen nicht mehr wahrgenommen werden kann, da die Wellenlänge des Streuvektors den Betrag $\lambda_{\max} = 2\pi/g_{\min}$ nicht überschreiten kann. Inwiefern in den ferromagnetischen Metallen Eisen, Nickel und Kobalt die beiden in Abschnitt 6 behandelten Fälle a) oder b) auftreten, hängt in starkem Maße von deren Elektronenbandstruktur ab. Die reziproke Austauschlänge $\varkappa_H$ beträgt bei Nickel für $H = 2000$ Oe $\approx 0{,}5 \times 10^6$ cm$^{-1}$. Der Wellenvektor an der Fermifläche ist von der Größenordnung einiger $10^8$ cm$^{-1}$. Bei nicht zu kleinen Polarisationsgraden und Feldstärken $H < 10^7$ Oe dürfte daher im allgemeinen die Bedingung $\varkappa_H^2 < g_{\min}^2$ und damit der Fall b) erfüllt sein. Dies bedeutet, daß der magnetostriktive Zusatzwiderstand in erster Näherung als von der Feldstärke unabhängig betrachtet werden kann.

Zur quantitativen Berechnung des Restwiderstandes setzen wir $J_{sd}/\zeta^{\uparrow} \approx 1/10$. Dies entspricht einer schwachen sd-Kopplung mit einer Austauschkonstante von der Größenordnung $J_{sd} \approx 0{,}1$ eV. Die Konstante $c_0$ beträgt im Falle von Nickel nach früheren Rechnungen [39] $c_0 = 4 \times 10^8$. Im Falle von Nickel gilt ferner $\frac{\pi J_s b G}{2 \mu_B A} = 2{,}2 \times 10^{33}$ cm$^{-4}$ und $\frac{\hbar a^3}{e^2} = 19{,}36 \times 10^{-20}$ cm$^3$. Legen wir unserer Rechnung eine Polarisation von $n^-/n^+ = 1{,}05$ zugrunde, so erhält man $k_F = 1{,}16 \times 10^8$ cm$^{-1}$ und $k_M^2 = k_F^2/60$, wobei die Zahl der s-Elektronen pro Atom zu 0,6 angenommen wurde. Mit diesen Zahlenwerten für Nickel ergibt sich im Falle b) für den magnetostriktiven Restwiderstand

$$\varrho_m = 7 \times 10^{-21}\ \Omega\ \mathrm{cm}^3\,.$$

Dieser Zahlenwert ist vergleichbar mit den von Seeger und Bross [7] berechneten Versetzungswiderständen in Kupfer infolge der eingangs erwähnten Dilatationsstreuung. Der magnetostriktive Restwiderstand stellt somit einen nicht zu vernachlässigenden Anteil des gesamten Versetzungsbeitrags zum Restwiderstand dar.

Zum Schluß wollen wir die Frage aufwerfen, wie der hier diskutierte Restwiderstand experimentell ermittelt werden kann. Besonders einfach liegen die Verhältnisse, falls der magnetostriktive Restwiderstand wesentlich größer ist als der auf die Dilatationsstreuung zurückzuführende Anteil. Der magnetostriktive Anteil zum elektrischen Widerstand ergibt sich in diesem Falle aus der Differenz der auf die Fehlstellen zurückzuführenden Widerstände, die oberhalb und weit unterhalb der Curietemperatur gemessen wurden. Den von Versetzungen herrührenden Anteil des elektrischen Widerstandes erhält man nach Ausheilen der punktförmigen Gitterfehlstellen durch Anlassen oberhalb der Rekristallisationstemperatur. Oberhalb der Curietemperatur ist dann der magnetostriktive Restwiderstand Null; unterhalb der Curietemperatur besitzt er einen endlichen Wert. Falls die Dilatationsstreuung eine Rolle spielt, ist zu beachten, daß deren Restwiderstand temperaturabhängig sein kann. Nach Bross [42] spielt diese Temperaturabhängigkeit des Dilatationsrestwiderstandes jedoch nur bei tiefen Temperaturen eine Rolle, so daß das oben beschriebene Verfahren zur Bestimmung von $\varrho_m$ grundsätzlich immer anwendbar sein sollte, falls der Restwiderstand unterhalb der Curietemperatur bei nicht zu tiefen Temperaturen gemessen wird.

## Literatur

[1] J. S. KOEHLER, Phys. Rev. **75**, 106 (1949).
[2] J. K. MACKENZIE und E. H. SONDHEIMER, Phys. Rev. **77**, 264 (1950).
[3] R. LANDAUER, Phys. Rev. **82**, 520 (1951).
[4] D. L. DEXTER, Phys. Rev. **86**, 770 (1952).
[5] S. C. HUNTER und F. R. N. NABARRO, Proc. Roy. Soc. **A 220**, 542 (1953).
[6] A. SEEGER und H. STEHLE, Z. Phys. **146**, 217, 242 (1956).
[7] A. SEEGER und H. BROSS, Z. Naturf. **15a**, 663 (1960).
[8] P. G. KLEMENS, Austral. J. Phys. **6**, 122 (1953); Canad. J. Phys. **34**, 1212 (1956).
[9] A. SEEGER, Canad. J. Phys. **34**, 1219 (1956).
[10] J. U. ZIMAN, Nuovo Cimento Suppl. **7**, 353 (1958).
[11] A. HOWIE, Phil. Mag. **5**, 251 (1960).
[12] A. SEEGER und H. STATZ, phys. stat. sol. **2**, 857 (1962).
[13] A. SEEGER und G. SCHOTTKY, Acta metall. **7**, 455 (1959).
[14] P. JONGENBURGER, Appl. sci. Res. **B 3**, 237 (1953).
[15] F. J. BLATT, Phys. Rev. **99**, 1708 (1955).
[16] A. SEEGER, Z. Phys. **144**, 637 (1956).
[17] A. SEEGER und H. BROSS, Z. Phys. **145**, 161 (1956).
[18] D. L. DEXTER, Phys. Rev. **87**, 768 (1952); **103**, 107 (1956).
[19] F. ABELES, C.R. Acad. Sci. (France) **237**, 796 (1953).
[20] P. JONGENBURGER, Phys. Rev. **90**, 710 (1953).
[21] A. SEEGER und H. BROSS, J. Phys. Chem. Solids **6**, 324 (1958).
[22] M. ASDENTE und J. FRIEDEL, J. Phys. Chem. Solids **11**, 115 (1959).
[23] S. V. VONSOVSKI, Zh. eksper. teor. Fiz. **16**, 981 (1946).
[24] P. G. DE GENNES und J. FRIEDEL, J. Phys. Chem. Solids **4**, 71 (1958).
[25] T. KASUYA, Progr. theor. Phys. **16**, 45, 58 (1956).
[26] I. MANNARI, Progr. theor. Phys. **22**, 335 (1959).
[27] A. U. SUDOVTSOV und E. E. SEMENENKO, Soviet Phys. — J. exp. theor. Phys. **8**, 211 (1959).
[28] E. E. SEMENENKO und A. U. SUDOVTSOV, Soviet Phys. — J. exp. theor. Phys. **15**, 708 (1962).
[29] N. F. MOTT und K. W. H. STEVENS, Phil. Mag. **2**, 1364 (1957).
[30] N. F. MOTT, Proc. Phys. Soc. **47**, 571 (1935).
[31] N. F. MOTT, Proc. Roy. Soc. **153**, 699 (1936).
[32] S. LEGVOLD, F. H. SPEDDING, F. DARSON und J. E. ELLIOT, Rev. mod. Phys. **25**, 129 (1953).
[33] C. G. SHULL und Y. YAMADA, J. Phys. Soc. Japan **17**, 1 (1962).
[34] W. HEISENBERG, Z. Phys. **49**, 619 (1928).
[35] W. F. BROWN, Phys. Rev. **58**, 736 (1940).
[36] W. F. BROWN, Phys. Rev. **60**, 132 (1941).
[37] A. SEEGER und H. KRONMÜLLER, J. Phys. Chem. Solids **12**, 298 (1960).
[38] H. KRONMÜLLER, Z. Phys. **154**, 574 (1959); **168**, 478 (1962).
[39] H. KRONMÜLLER und A. SEEGER, J. Phys. Chem. Solids **18**, 93 (1961).
[40] H. KRONMÜLLER, A. SEEGER und M. WILKENS, Z. Phys. **171**, 299 (1963).
[41] H. BROSS und A. SEEGER, J. Phys. Chem. Solids **4**, 161 (1958).
[42] H. BROSS, Z. Naturf. **14a**, 560 (1963).

*(Received May 4, 1966)*

phys. stat. sol. 16, 267 (1966)

*Physikalisch-Technisches Institut der Deutschen Akademie der Wissenschaften zu Berlin, Abteilung Theoretische Physik*

# Zur Bestimmung der freien Weglänge der Elektronen aus Messungen der Ultraschallabsorption in Metallen

Von

J. MERTSCHING

Es werden verschiedene Methoden zur Bestimmung der freien Weglänge der Elektronen in Metallen eingehend diskutiert, die auf Messungen der longitudinalen Ultraschallabsorption im äußeren Magnetfeld beruhen. Es handelt sich dabei um die Bestimmung der Anzahl geometrischer Resonanzen und der Dämpfung ihrer Amplituden sowie um die Änderung der Absorption beim Einschalten eines hohen Magnetfeldes. Es wird gezeigt, daß die gebräuchliche Auswertung der Messungen der geometrischen Resonanz inkorrekt ist und viel zu große freie Weglängen ergibt. Dadurch werden die bestehenden Diskrepanzen zu den Ergebnissen der Messungen im hohen Magnetfeld erheblich verringert. Die restlichen Diskrepanzen weisen auf Abweichungen realer Metalle vom Modell freier Elektronen hin, auf dessen Grundlage die Messungen im hohen Magnetfeld ausgewertet wurden.

A discussion is given of various methods of determining the mean free path of electrons in metals, based on longitudinal ultrasonic absorption in an external magnetic field. In particular the number of geometric resonances, their attenuation, and the absorption change on application of a high magnetic field are considered. The usual analysis of the geometric resonance measurements is shown to be incorrect and gives to high mean free paths. A correct analysis gives better agreement with the experimental results in a high magnetic field. The remaining discrepancies are probably due to the limitations of the free electron model, previonsly used in the analysis of the high field measurements.

## 1. Einleitung und Formulierung der Ergebnisse

Es gibt verschiedene Methoden, aus Messungen der Ultraschallabsorption von Metallen im äußeren konstanten Magnetfeld bei tiefen Temperaturen die mittlere freie Weglänge der Leitungselektronen zu bestimmen.

Eine dieser Methoden beruht auf der Messung der Oszillationen der Absorption in Abhängigkeit vom (reziproken) Magnetfeld infolge geometrischer Resonanz. Diese Resonanzen treten besonders deutlich auf, wenn das äußere Magnetfeld senkrecht zur Ausbreitungsrichtung des Schalls gerichtet ist und die Elektronenbahnen senkrecht zum Magnetfeld geschlossen sind. In Fig. 1 sind eine solche geschlossene Bahn sowie die Ebenen gleicher Phase der Ultraschall-

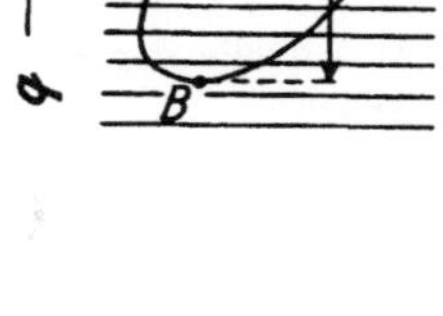


Fig. 1. Projektion einer Elektronenbahn und der Phasenebenen der Ultraschallwelle in eine Ebene senkrecht zum Magnetfeld

welle dargestellt. Für longitudinalen Ultraschall hat die Absorption dann Maxima, wenn der extremale Abstand $D$ der Punkte stationärer Phase A und B auf der sogenannten „Extremalbahn" bis auf eine Phasenkorrektur von $\frac{3}{8}\lambda$ ein ganzzahliges Vielfaches der Wellenlänge $\lambda$ beträgt:

$$D_n = \left(n + \frac{3}{8}\right)\lambda \qquad (n = 1, 2, 3, \ldots)\,. \tag{1.1}$$

In diesem Fall wirkt das von der Ultraschallwelle erzeugte Wechselfeld an den Punkten stationärer Phase annähernd gleichphasig auf die Extremalbahnelektronen ein, so daß diese den Schall maximal absorbieren.

Aus der Periode der geometrischen Resonanzen sind für viele Metalle die extremalen Bahndurchmesser, insbesondere im reziproken Raum auf der Fermifläche bestimmt worden. Uns interessiert jedoch hier besonders die Frage, unter welchen Bedingungen die geometrischen Resonanzen zu beobachten sind und wie sie mit wachsender Ordnung $n$ bzw. abnehmendem Magnetfeld abklingen.

Galkin und Korolyuk [1] sowie Rayne und Chandrasekhar [2] haben angenommen, daß eine geometrische Resonanz nur dann beobachtbar ist, wenn der Bahnumfang mindestens eine mittlere freie Weglänge beträgt, so daß die Elektronen auf ihrer Bahn im Mittel mindestens einmal ungestreut umlaufen können. Sind die Bahnen kreisförmig, so ist ihr Umfang $\pi D$, und die freie Weglänge $l$ kann aus der Anzahl $N$ der beobachteten Resonanzen bestimmt werden. Unter Einführung der Wellenzahl $q = \frac{2\pi}{\lambda}$ ergibt sich aus (1.1)

$$q\,l = q\,\pi\,D_N = 2\,\pi^2\left(N + \frac{3}{8}\right).^{1)} \tag{1.2}$$

Im Gegensatz dazu haben Easterling und Bohm [3] angenommen, daß es für die Beobachtung einer geometrischen Resonanz ausreicht, wenn das Elektron im Mittel nur den halben Bahnumfang von einem Punkt stationärer Phase zum anderen ungestreut durchfliegt, so daß

$$q\,l = \pi^2\left(N + \frac{3}{8}\right) \tag{1.3}$$

gelten würde.

Nach diesen Vorstellungen wären Werte von $q\,l \approx 20$ bzw. 10 nötig, um überhaupt eine geometrische Resonanz zu sehen. Dieses Ergebnis steht jedoch im Widerspruch zu Ultraschallabsorptionskurven, die Kjeldaas und Holstein [4] im Modell freier Elektronen berechnet haben und die zeigen, daß die $q\,l$-Werte nur etwa 4 mal größer als die Anzahl der deutlichen Resonanzen ist. Der Grund hierfür liegt darin, daß nicht alle Elektronen, sondern nur ein gewisser Bruchteil den halben Umfang der Extremalbahn im Mittel ohne Streuung zurückzulegen brauchen, um eine beobachtbare Resonanz hervorzurufen. Wie groß dieser Bruchteil sein muß, hängt natürlich von der Meßgenauigkeit ab.

Ausgehend von einer allgemeinen Theorie der Ultraschallabsorption [5] wird im folgenden Abschnitt der vorliegenden Arbeit der $q\,l$-Wert in Abhängigkeit von der Anzahl $N$ der Resonanzen sowie der relativen Meßgenauigkeit $\varepsilon$ be-

---

1) Anstelle der Phasenkorrektur 3/8 ist in [1] der Wert 1 und in [2] der Wert 1/2 angegeben.

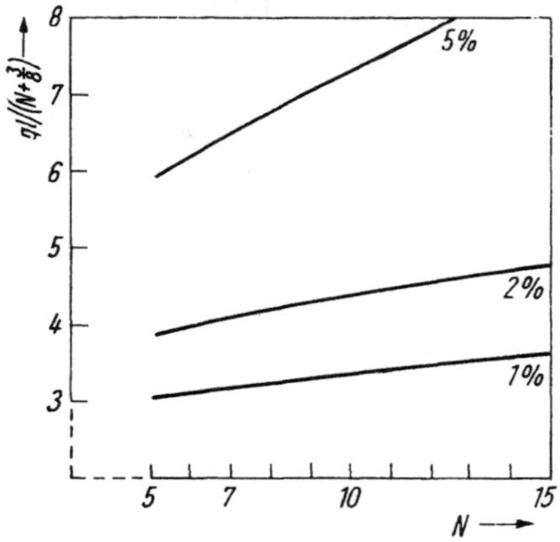


Fig. 2. $\frac{q\,l}{N+\frac{3}{8}}$ in Abhängigkeit von der Anzahl $N$ der Resonanzen für relative Meßgenauigkeiten von $\varepsilon = 1\%$, 3% bzw. 5%

rechnet (Gleichung (2.7)). Das Ergebnis ist in Fig. 2 graphisch dargestellt:

Bei vorgegebenen Meßgenauigkeiten von 1 bis 5% hängt $\frac{q\,l}{N+\frac{3}{8}}$ noch schwach von $N$ ab und nimmt Werte zwischen 3 und 8 an, in Übereinstimmung mit [4].

Mit Hilfe einer anderen Methode haben Bohm und Easterling [6] die freie Weglänge bestimmt, indem sie die Abnahme der Oszillationsamplituden mit abnehmendem Magnetfeld gemessen haben. Hierbei wird angenommen, daß die Amplitude $a_n$ der $n$-ten Resonanz der Wahrscheinlichkeit proportional ist, daß ein Elektron die Extremalbahn einmal ungestreut durchläuft:

$$a_n \sim \mathrm{e}^{-\frac{\pi D_n}{l}} = \mathrm{e}^{-\frac{2\pi^2\left(n+\frac{3}{8}\right)}{q\,l}}. \tag{1.4}$$

Daraus resultiert ein konstantes Verhältnis $\mathrm{e}^{-\frac{2\pi^2}{q\,l}}$ zweier aufeinanderfolgender Amplituden, aus dem $q\,l$ zu bestimmen ist.

Dieses heuristische Verfahren führt jedoch ebenso wie (1.2) oder (1.3) nicht zu korrekten Zahlenwerten von $q\,l$. Hierfür gibt es verschiedene Gründe:

1. Führen die Elektronen viele Umläufe hintereinander aus, ist also $l \gg \pi D_n$ bzw. $n \ll N$, so sind die Amplituden näherungsweise der Anzahl $l/\pi D_n$ der Umläufe proportional, so daß sie mit zunehmender Ordnung wegen (1.1) wie $a_n \sim \left(n+\frac{3}{8}\right)^{-1}$ und nicht exponentiell wie in (1.4) abfallen. Diese Amplitudenabnahme ist jedoch nicht spezifisch für die geometrische Resonanz, da der nichtoszillatorische Anteil zur (elektronischen) Absorption in gleicher Weise abnimmt.[2])

2. Durchfliegen die Elektronen den halben Bahnumfang zwischen den Punkten stationärer Phase nur selten ohne Streuung, ist also $l \ll \pi D_n$ bzw. $n \gtrsim N$,

[2]) Nach Gl. (2.3) ist diese nichtoszillatorische Absorption $\approx \alpha_0 \frac{\omega_c \tau}{\pi} = \alpha_0 \frac{q\,l}{\pi^2\left(n+\frac{3}{8}\right)}$ ($\alpha_0$ elektronische Absorption für $B = 0$), so daß im Bereich $\omega_c \tau \gg \pi$ bzw. $q\,l \gg \pi^2 n$ bereits aus dem linearen Anstieg der Untergrundabsorption mit dem Magnetfeld (ohne Messung der Oszillationsamplituden) $\tau$ bzw. die freie Weglänge $l$ bestimmt werden könnte.

so wären die Amplituden tatsächlich einer Wahrscheinlichkeit (1.4) proportional, allerdings mit dem halben Exponenten. Dieser Bereich beginnt jedoch erst etwa an der Beobachtungsgrenze $n \approx N$ der Oszillationen.

3. Ferner muß noch berücksichtigt werden, daß mit zunehmender Ordnung die Anzahl der effektiv zur geometrischen Resonanz beitragenden Elektronenbahnen aus der Umgebung der Extremalbahn und damit die Oszillationsamplitude $\sim \left(n + \frac{3}{8}\right)^{-1/2}$ abnimmt.

Die relative Oszillationsamplitude $A_n$, d. h. das Verhältnis der Oszillationsamplitude $a_n$ zum nichtoszillatorischen (elektronischen) Untergrund der Absorption verhält sich also wie

$$A_n \sim \begin{cases} \left(n + \frac{3}{8}\right)^{-1/2} & (n \ll N) \\ \left(n + \frac{3}{8}\right)^{-1/2} e^{-\frac{\pi^2\left(n+\frac{3}{8}\right)}{q\,l}} & (n \gtrsim N)\,. \end{cases} \tag{1.5}$$

In (2.4) ist eine Formel für $A_n$ angegeben, die im gesamten Bereich gültig ist. Bemerkenswert ist, daß sich $A_n$ für $n < N$ annähernd exponentiell verhält, allerdings mit einem anderen Exponenten als in (1.4) oder (1.5). Um diesen aufzufinden, haben wir die logarithmischen Dekremente

$$\log \frac{A_n}{A_{n+1}} = d_n \frac{\pi^2}{q\,l} \tag{1.6}$$

bzw. den Koeffizienten $d_n$ in Abhängigkeit von $q\,l$ und $n$ berechnet (Gleichung (2.8)). Das Ergebnis zeigt Fig. 3: Für $q\,l$-Werte zwischen 14 und 50 nimmt $d_n$ Werte zwischen 1,1 und 1,3 an und variiert nur schwach mit der Ordnung $n$ (nach Gleichung (1.4) wäre $d_n = 2$).

Eine weitere Methode zur Bestimmung der freien Weglänge bzw. der Relaxationszeit der Elektronen wurde von Deaton und Gavenda [7, 8] angewandt. Sie beruht darauf, daß beim Übergang zu hohen Magnetfeldern $\boldsymbol{B} \perp \boldsymbol{q}$ die longitudinale Ultraschallabsorption einem Sättigungswert zustrebt, der proportional zur Relaxationszeit $\tau$ ist. Im Modell freier Elektronen ist dieser Sättigungswert $\alpha_\infty$ genau so groß wie die Absorption $\alpha_0$ ohne Magnetfeld, wenn $q\,l \approx 6{,}8$ ist [7]. Die Frequenz $\omega$, bei der $\alpha_\infty = \alpha_0$ ist, liefert also direkt die freie Weglänge. Diese Methode ergibt für Cu etwa 20 mal kleinere freie Weglängen, als man gemäß (1.1) aus der Anzahl der geometrischen Resonanzen erhält [7]. Auch nach der neuen Auswertung gemäß Fig. 2 verbleibt eine Diskrepanz um einen Faktor $\approx 3$.

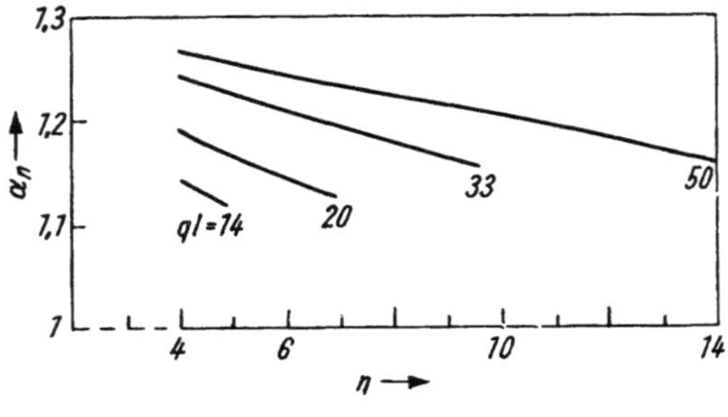


Fig. 3. Der Koeffizient $d_n$ des logarithmischen Dekrements der relativen Oszillationsamplitude in Abhängigkeit von der Ordnung $n$ der Resonanz für $q\,l$-Werte von 14, 20, 33 bzw. 50

Deshalb haben wir $\alpha_0$ und $\alpha_\infty$ für allgemeine Fermiflächen mit geschlossenen Bahnen näherungsweise berechnet. Während zu $\alpha_\infty$ im allgemeinen alle Elektronen auf der Fermifläche mehr oder weniger beitragen, geben zu $\alpha_0$ nur die Elektronen an der sogenannten effektiven Zone $\boldsymbol{q} \cdot \boldsymbol{v} = \omega$ einen Beitrag. Daraus folgt, daß für $\alpha_\infty = \alpha_0$ (Gleichung (2.11)) ein mittlerer $q\,l$-Wert auch größerals 6,8 sein kann, wenn der Beitrag der effektiven Zone relativ größer und der anderen Bereiche der Fermifläche relativ kleiner als im Modell freier Elektronen ist.

In sämtlichen Rechnungen wird das (selbstkonsistente) elektromagnetische Feld, das durch die Ultraschallwelle im Metall erzeugt wird, vernachlässigt und nur die sogenannte „Deformations-Absorption" berücksichtigt. Außerdem wird meist der Ultraschall als stationäre Welle betrachtet ($\omega \approx 0$). Beide Näherungen werden im letzten Abschnitt diskutiert und gerechtfertigt.

## 2. Ableitung der Ergebnisse

Nach [5] ist die elektronische longitudinale Ultraschallabsorption

$$\alpha = \frac{\omega^2}{8\pi^3 \hbar^2 \varrho\, v_s^3} \operatorname{Re} \int \mathrm{d}k_z \frac{m_c}{\omega_c} \int_0^{2\pi} \mathrm{d}\Theta\, \Lambda(\Theta) \int_0^{2\pi} \mathrm{d}\Theta'\, \Lambda(\Theta')\, \mathrm{e}^{-\frac{i}{\omega_c}\int_{\Theta'}^{\Theta} \boldsymbol{q}\cdot\boldsymbol{v}(\Theta'')\,\mathrm{d}\Theta''} \times$$

$$\times\, \mathrm{e}^{-\frac{\frac{1}{\tau} - i\omega}{\omega_c}(\Theta - \Theta')} \left[\operatorname{sgn}(\Theta - \Theta') + \coth \frac{\pi}{\omega_c}\left(\frac{1}{\tau} - i\,\omega + i\,\boldsymbol{q}\cdot\tilde{\boldsymbol{v}}\right)\right]. \tag{2.1}$$

Darin haben die bisher noch nicht erklärten Symbole folgende Bedeutung:

$\varrho$ Dichte des Metalls,

$v_s$ Schallgeschwindigkeit,

$k_z$ Komponente des Wellenvektors der Elektronen in Richtung des Magnetfeldes,

$\Theta$ Winkelvariable auf den Zyklotronbahnen der Elektronen ($\dot{\Theta} = \omega_c$),

$m_c$, $\omega_c$ Zyklotronmasse bzw. Zyklotronfrequenz der Elektronen,

$\boldsymbol{v}$ Geschwindigkeit der Elektronen; $\tilde{\boldsymbol{v}} = \frac{1}{2\pi}\int_0^{2\pi} \boldsymbol{v}(\Theta)\,\mathrm{d}\Theta$,

$\Lambda$ longitudinale Deformationspotentialkonstante auf der Fermifläche.

(Im Modell freier Elektronen ist $\Lambda = m\left(\frac{v^2}{3} - v_x^2\right)$, wenn $x$ die Schallrichtung bezeichnet.)

In (2.1) ist der Beitrag des (selbstkonsistenten) elektrischen Feldes der Ultraschallwelle vernachlässigt, was im nächsten Abschnitt begründet wird. Außerdem ist der Einfachheit halber die Relaxationszeit als konstant angenommen.

Im Bereich der geometrischen Resonanzen ($q\,D \ll 1$) berechnen wir die Absorption aus (2.1) unter folgenden vereinfachenden Voraussetzungen und Approximationen:

1. Das Magnetfeld sei senkrecht zum Ultraschall angelegt, und die Zyklotronbahnen der Elektronen seien geschlossen, so daß $\boldsymbol{q} \cdot \tilde{\boldsymbol{v}} = 0$ gilt.

2. Es sei $\omega\,\tau < 1$ und $\omega \ll \omega_c$, so daß $\omega$ im Exponenten sowie im coth von (2.1) vernachlässigt werden kann (die Ultraschallwelle ist annähernd stationär).

3. Die Winkel-Integrationen werden in der Sattelpunktsnäherung ausgeführt und ergeben nur Beiträge an den Bahnpunkten stationärer Phase (Fig. 1).

4. Für den oszillatorischen Anteil zur Absorption wird auch die $k_z$-Integration in der Sattelpunktsnäherung ausgeführt und ergibt nur Beiträge für die Extremalbahnen, wo der Bahndurchmesser $D$ extremal ist.

5. Wir nehmen an, daß es nur eine einzige Extremalbahn gibt, wo $D$ maximal ist.

6. Ferner habe die Extremalbahn nur zwei Punkte stationärer Phase A und B, die wie in Fig. 1 symmetrisch zueinander liegen, so daß

$$\Theta_A - \Theta_B = \pi\,, \qquad \Lambda(\Theta_A) = \Lambda(\Theta_B)\,, \qquad \left|\boldsymbol{q}\cdot\frac{\partial \boldsymbol{v}}{\partial\Theta}\right|_A = \left|\boldsymbol{q}\cdot\frac{\partial \boldsymbol{v}}{\partial\Theta}\right|_B \tag{2.2}$$

gilt.

Unter Beachtung dieser sechs Punkte ergibt sich aus (2.1) die Absorption

$$\alpha = \frac{\omega^2}{2\pi^2\hbar^2\varrho\,v_s^3}\left[\int \mathrm{d}k_z\,\frac{m_c\,\Lambda^2}{\left|\boldsymbol{q}\cdot\frac{\partial \boldsymbol{v}}{\partial\Theta}\right|}\coth\frac{\pi}{\omega_c\,\tau} + \sqrt{\frac{2\pi}{\left|\frac{\partial^2}{\partial k_z^2}q\,D\right|}}\;\frac{m_c\,\Lambda^2}{\left|\boldsymbol{q}\cdot\frac{\partial \boldsymbol{v}}{\partial\Theta}\right|}\;\frac{\cos\left(q\,D-\frac{3\pi}{4}\right)}{\sinh\frac{\pi}{\omega_c\,\tau}}\right]. \tag{2.3}$$

Im ersten Integral ist über die Punkte stationärer Phase aller Bahnen zu integrieren. Der zweite Summand ist der oszillatorische Beitrag der Punkte stationärer Phase der Extremalbahn. Er beschreibt die geometrischen Resonanzen und ist unter der Bedingung (1.1) maximal.

Der Quotient beider Summanden an den Maxima $q\,D_n = 2\pi\left(n+\frac{3}{8}\right)$ ist die relative Resonanzamplitude $A_n$. Um diese übersichtlich diskutieren zu können, soll ferner angenommen werden, daß alle Elektronenbahnen annähernd dieselbe Zyklotronfrequenz haben. Dann gilt

$$A_n = \frac{C}{\pi\sqrt{n+\frac{3}{8}}\,\cosh\frac{\pi}{\omega_c\,\tau}} = \frac{C}{\pi\sqrt{n+\frac{3}{8}}\,\cosh\frac{\pi^2\left(n+\frac{3}{8}\right)}{q\,l}}\,, \tag{2.4}$$

wobei

$$C = \pi\left[\left|\frac{\partial^2}{\partial k_z^2}\,\frac{q\,D_n}{2\pi\left(n+\frac{3}{8}\right)}\right|^{-1/2}\frac{\Lambda^2}{\left|\boldsymbol{q}\cdot\frac{\partial \boldsymbol{v}}{\partial\Theta}\right|}\right]_{\mathrm{Extr}}\cdot\left[\int \mathrm{d}k_z\,\frac{\Lambda^2}{\left|\boldsymbol{q}\cdot\frac{\partial \boldsymbol{v}}{\partial\Theta}\right|}\right]^{-1} \tag{2.5}$$

ein Zahlenfaktor ist, der kaum vom Magnetfeld abhängt und im Modell freier Elektronen den Wert 1 hat, und die Länge $l$ durch

$$l = \pi\,D\cdot\frac{\omega_c\,\tau}{2\pi} = \frac{\hbar\,K\,\tau}{2\,m_c} \tag{2.6}$$

($K$ Durchmesser der extremalen Zyklotronbahn $\perp\,\boldsymbol{q}$ auf der Fermifläche) definiert ist. Für eine kreisförmige Extremalbahn ist $l$ als Produkt des Bahnumfangs $\pi D$ und der mittleren Umlaufszahl $\frac{\omega_c \tau}{2\pi}$ direkt die mittlere freie Weglänge der Elektronen; im allgemeinen ist jedoch die korrekte freie Weglänge das Produkt von $l$ (2.6) und dem Verhältnis der Bogenlänge einer Extremalbahnperiode zu $\pi D$.

Ist $A_n$ kleiner als die relative Meßgenauigkeit $\varepsilon$, so ist die $n$-te Resonanz nicht mehr sichtbar. Somit gilt für die höchste beobachtbare Ordnung $N$

$$A_N = \frac{C}{\pi \sqrt{N + \frac{3}{8}} \cosh \frac{\pi^2}{q\,l}\left(N + \frac{3}{8}\right)} = \frac{\varepsilon}{100} \qquad (\varepsilon \text{ in } \%)\,. \tag{2.7}$$

Hieraus kann $q\,l$ in Abhängigkeit von $N$ für vorgegebene $\varepsilon$ berechnet werden. Das Ergebnis für $C = 1$ zeigt Fig. 2.

Aus (2.4) ergibt sich für den Koeffizienten $d_n$ des logarithmischen Dekrements

$$d_n = \frac{q\,l}{\pi^2} \log \frac{A_n}{A_{n+1}} =$$

$$= 1 + \frac{q\,l}{\pi^2}\left[\frac{1}{2}\log\left(1 + \frac{1}{n + \frac{3}{8}}\right) - \log\left(1 + \frac{1 - \mathrm{e}^{-\frac{2\pi^2}{q l}}}{\mathrm{e}^{\frac{2\pi^2}{q l}\left(n + \frac{3}{8}\right)} + \mathrm{e}^{-\frac{2\pi^2}{q l}}}\right)\right]. \tag{2.8}$$

In Fig. 3 ist $d_n$ für verschiedene $q\,l$-Werte graphisch dargestellt.

Wir kommen nun zum Vergleich der Ultraschallabsorption $\alpha_0$ ohne Magnetfeld und dem Sättigungswert $\alpha_\infty$ in hohen Magnetfeldern. Für $q\,l \gg 1$ erhält man $\alpha_0$ leicht aus (2.3), wenn man den geometrischen Resonanzterm streicht und im ersten Term den coth-Faktor gleich 1 setzt. Als Integral über die effektive Zone $\boldsymbol{q} \cdot \boldsymbol{v} \approx 0$ der Punkte stationärer Phase geschrieben ist [5]

$$\alpha_0 \approx \frac{\omega^2}{4\,\pi^2\,\hbar\,\varrho\,v_s^3} \oint\limits_{\boldsymbol{q}\cdot\boldsymbol{v}=0} \frac{\Lambda}{q\,v^2} R\,\mathrm{d}\varphi \qquad (q\,l \gg 1)\,, \tag{2.9}$$

wobei $R$ die reziproke Gaussche Krümmung der Fermifläche und $\varphi$ der Normalen-Polarwinkel auf der effektiven Zone ist. Ferner folgt für $\omega_c \to \infty$ aus (2.1)

$$\alpha_\infty = \frac{\omega^2}{2\,\pi^2\,\hbar^2\,\varrho\,v_s^3} \int m_c\,\widetilde{\Lambda}^2 \left(\widetilde{\frac{1}{\tau}}\right)^{-1} \mathrm{d}k_z\,, \tag{2.10}$$

worin die Schlange Mittelung über die Winkelvariable $\Theta$ bedeutet und die Relaxationszeit $\tau$ korrekterweise durch $\left(\widetilde{\frac{1}{\tau}}\right)^{-1}$ ersetzt ist.

Beide Absorptionswerte sind gleich ($\alpha_\infty = \alpha_0$), wenn die Bedingung

$$\int m_c\,\widetilde{\Lambda}^2 \left(\widetilde{\frac{1}{\tau}}\right)^{-1} \mathrm{d}k_z = \frac{\hbar}{2} \oint\limits_{\boldsymbol{q}\cdot\boldsymbol{v}=0} \frac{\Lambda^2}{q\,v^2} R\,\mathrm{d}\varphi \tag{2.11}$$

erfüllt ist. Im Modell freier Elektronen folgt daraus $q\,l = \frac{5\pi}{2}$ (oder wenn man $\alpha_0$ genauer als in (2.9) unter Berücksichtigung von Termen der relativen Größenordnung $O\left(\frac{1}{q\,l}\right)$ berechnet: $q\,l = 6{,}8$ [7]). Im allgemeinen läßt sich jedoch aus (2.11) nur dann ein zuverlässiger Wert für eine mittlere freie Weglänge gewinnen, wenn die Anisotropie des Deformationspotentials $\Lambda$, der reziproken Relaxationszeit und der Fermigeschwindigkeit sowie die Gaussche Krümmung der effektiven Zone bereits irgendwie bekannt sind.

Abschließend weisen wir darauf hin, daß für transversalen Schall mit $\boldsymbol{q} \perp \boldsymbol{B}$ eine zu (2.10) analoge Formel mit dem transversalen Deformationspotential $\Lambda_t$ gilt. Im Modell freier Elektronen ist $\widetilde{\Lambda}_t = 0$, also $\alpha_{t\,\infty} = 0$. Diese Tatsache wird (für nicht supraleitende Metalle) gelegentlich dazu benutzt, die nichtelektronische Restabsorption in hohen Magnetfeldern zu bestimmen [7]. Für beliebige Fermiflächen ist jedoch im allgemeinen $\widetilde{\Lambda}_t$ und damit $\alpha_{t\,\infty} \neq 0$. Daher ist diese Methode zur Bestimmung der Restabsorption nur für transversalen Schall in speziellen Symmetrierichtungen anwendbar, für die $\widetilde{\Lambda}_t = 0$ gilt.

### 3. Diskussion der Gültigkeit der Näherungen

In diesem Abschnitt sollen kurz die Bedingungen diskutiert werden, unter denen (wie im 2. Abschnitt) die Ultraschallwelle als stationär betrachtet und das selbstkonsistente Feld der Ultraschallwelle vernachlässigt werden kann.

1. Die Absorptionsformel für endliche $\omega\,\tau$ erhält man aus (2.3) bzw. (2.10), wenn man dort $1/\tau$ durch $1/\tau - i\,\omega$ ersetzt und dann den Realteil des Gesamtausdrucks bildet.

Die geometrische Resonanz für $\omega\,\tau \lesssim 1$ wurde von Gavenda und Chang [9] untersucht. Bemerkenswert dabei ist, daß im geometrischen Resonanzterm von (2.3) ein zusätzlicher Faktor

$$\cos\pi\frac{\omega}{\omega_c} = \cos\pi^2\frac{v_s}{v_F}\left(n + \frac{3}{8}\right) \tag{3.1}$$

auftritt. Da die Fermigeschwindigkeit $v_F$ für typische Metalle $\approx 300$ mal größer als die Schallgeschwindigkeit $v_s$ ist, spielt dieser Faktor für die Oszillationen niedriger Ordnung $n$ keine Rolle. Mit wachsender Ordnung bewirkt er eine zusätzliche Dämpfung, bis für $n \approx \frac{1}{2\pi}\frac{v_F}{v_s} \approx 50$ die Oszillationen völlig verschwinden. Für noch größere $n$ wird der cos-Faktor negativ, so daß die Oszillationen um $\pi$ phasenverschoben sind.

Es ist klar, daß die Oszillationen in unmittelbarer Nähe des Phasenumkehrpunktes zur Bestimmung der freien Weglänge ungeeignet sind. Ansonsten bringt die zusätzliche Dämpfung nur eine geringe Korrektur zum logarithmischen Dekrement (2.8), die z. B. für $n = 25$ nur 3% beträgt.

2. Der Beitrag des vom Ultraschall erzeugten elektrischen Feldes zur Ultraschallabsorption ist z. B. in [5] angegeben. Dieser Beitrag ist für transversalen Schall zumeist ausschlaggebend, dagegen für longitudinalen Schall oft nur eine kleinere Korrektur. Eine Abschätzung zeigt, daß der Koeffizient $d_n$ des logarithmischen Dekrements (2.8) um $\approx \frac{\pi}{2\,q\,l}\left(\cosh\frac{\pi^2\,n}{q\,l}\right)^{-2}$ sowie um weitere ähnliche Terme zu korrigieren ist. Diese Korrektur ist für größere $q\,l$-Werte und höhere Ordnungen unwesentlich und beträgt z. B. für $q\,l = 30$, $n = 5$ nur $\approx 1\%$.

Die Feldabsorption kann nur in sehr hohen Magnetfeldern wesentlich werden. Dies ist dadurch bedingt, daß bei der Bewegung des Gitters mit der Geschwindigkeit $\boldsymbol{u}$ ein elektrisches Feld $\boldsymbol{E} = \frac{u}{c} \times \boldsymbol{B}$ induziert wird. Da die Absorption $\sim \boldsymbol{E}^2$ ist, führt dies jenseits des Sättigungsgebietes zu einem quadratischen Anstieg der Absorption mit dem Magnetfeld.

Im Modell freier Elektronen haben Cohen, Harrison und Harrison [10] gezeigt, daß dieser Anstieg beginnt, wenn $q\,D \lesssim (2\,q\,\delta_n)^2$ wird, wobei $\delta_n = c\,(2\,\pi\,\omega\,\sigma_0)^{-1/2}$ die normale Skintiefe und $\sigma_0$ die Gleichstromleitfähigkeit ist. Zur Bestimmung des Sättigungswertes (2.10) der Ultraschallabsorption darf also das Magnetfeld $B$ nicht zu hoch sein, so daß der maximale Bahndurchmesser $D$ der Bedingung

$$(2\,q\,\delta_n)^2 \ll q\,D \ll 1 \tag{3.2}$$

genügt.

## 4. Zusammenfassung

Wir haben gezeigt, daß die in der Literatur gebräuchliche Auswertung von Messungen der geometrischen Resonanz in der longitudinalen Ultraschallabsorption inkorrekt ist und zu große freie Weglängen ergibt. Aus der Theorie der Ultraschallabsorption folgt, daß der $q\,l$-Wert pro beobachtbarer Oszillation bei 1 bis 2% relativer Meßgenauigkeit zwischen 3 und 5 liegt (Fig. 2). Ferner klingen die relativen Oszillationsamplituden annähernd exponentiell mit einem logarithmischen Dekrement $d_n \frac{\pi^2}{q\,l}$ ab, wobei der Koeffizient $d_n$ Werte zwischen 1,1 und 1,3 annimmt (Fig. 3). Schließlich läßt sich aus dem Vergleich der Ultraschallabsorption im verschwindenden bzw. hohen Magnetfeld nur dann zuverlässig die freie Weglänge bestimmen, wenn die Anisotropie des Deformationspotentials und der Fermigeschwindigkeit sowie die Gaussche Krümmung der effektiven Zone auf der Fermifläche bekannt sind.

Abschließend danke ich Herrn Dr. Stolz für wertvolle Diskussionen sowie eine kritische Durchsicht des Manuskripts.

## Literatur

[1] A. A. Galkin und A. P. Korolyuk, Zh. eksper. teor. Fiz. **38**, 1688 (1960).
[2] J. A. Rayne und B. S. Chandrasekhar, Phys. Rev. **125**, 1952 (1962).
[3] V. J. Easterling und H. V. Bohm, Phys. Rev. **125**, 812 (1962).
[4] T. Kjeldaas und T. Holstein, Phys. Rev. Letters **2**, 340 (1959).
[5] J. Mertsching, phys. stat. sol. **14**, 3 (1966).
[6] H. V. Bohm und V. J. Easterling, Phys. Rev. **128**, 1021 (1962).
[7] B. C. Deaton und J. D. Gavenda, Phys. Rev. **129**, 1990 (1963).
[8] B. C. Deaton, Phys. Rev. **140**, A 2051 (1965).
[9] J. D. Gavenda und F. H. S. Chang, Phys. Rev. Letters **16**, 228 (1966).
[10] M. H. Cohen, M. J. Harrison und W. A. Harrison, Phys. Rev. **117**, 937 (1960).

*(Received May 5, 1966)*

phys. stat. sol. 16, 277 (1966)

*3. Physikalisches Institut der Technischen Hochschule Stuttgart*

# Direct Measurement of the Diffusion Length of Singulet Excitons in Solid Phenanthrene

By

G. GALLUS and H. C. WOLF

The average radial diffusion length of excitons in evaporated layers of solid phenanthrene is measured directly. Assuming surface-quenching of the excitons, it is found, that $r = 263 \pm 85$ Å.

Die mittlere radiale Diffusionslänge von Exzitonen wird für aufgedampfte Schichten von festem Phenanthren direkt bestimmt. Unter der Annahme von Exzitonentilgung an der Oberfläche wird $r = 263 \pm 85$ Å gefunden.

Excitons in organic molecular crystals are able to transfer energy over appreciable distances. The diffusion length can be estimated from the concentration dependence of sensitized fluorescence [1]. Simpson [2] measured the diffusion length of singulet excitons in evaporated layers of anthracene directly. Competing energy transfer by reabsorption of fluorescence light could not be excluded completely in this system.

We used Simpsons method to measure the diffusion length of singulet excitons in polycrystalline layers of phenanthrene. Using this substance, we were able to exclude energy transfer by reabsorption of fluorescence light.

The principle of measurement is given in Fig. 1. P is a layer of highly purified phenanthrene, P + A is a layer of phenanthrene containing anthracene in a concentration lower than $10^{-2}$. The exciting light is absorbed mainly in the layer P which has a quantum efficiency appreciably lower than 1 [3]. Diffusion of excitons from P to P + A causes an increase of the total fluorescence quantum yield, since the quantum yield of the layer P + A is approximately 1. Rediffu-

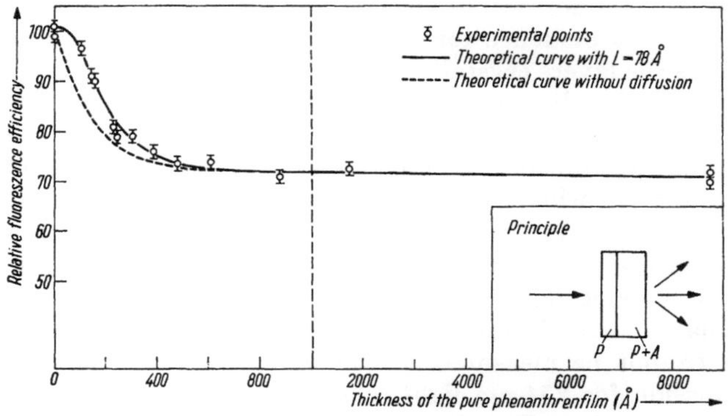


Fig. 1. Fluorescence efficiency as a function of the thickness of layer P

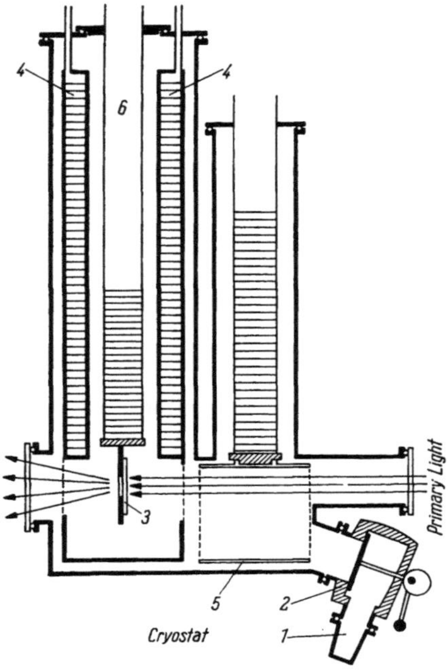


Fig. 2. Cryostat. 1 — container, 2 — valve, 3 — quartz plate, 4 — $N_2$-bath, 5 — cooled surface, 6 — $N_2$-bath

sion of excitons from P + A to P is excluded by the presence of trapping anthracene molecules in P + A. Diffusion of excitons from P to P + A increases the total quantum yield of the two-layer system. By measuring the total quantum yield as a function of the thickness of the layer P one gets evidence for exciton diffusion and a value for the diffusion length.

We used the 254 nm line of a low pressure mercury arc isolated by a grating monochromator for excitation and an EMI 6256 B multiplier for detection of the fluorescence. The exciting light is completely absorbed in the two-layer system, which has a constant total thickness of 8750 Å.

The cryostat is shown in Fig. 2. The two substances (P and P + A) are in two containers. Only one of them is shown in Fig. 2. They can be closed by valves. If the valves are open, the substance is evaporated on a cool surface of fused quartz.

The phenanthrene was purified chemically as described in [3] and [4] and by zone refining using 235 passes. Anthracene was not detectable in the purest phenanthrene P. The layers of P + A were evaporated from P doped with $10^{-2}$ parts of anthracene.

The quartz disk is held at 77 °K during the evaporation. The layer P was evaporated on the quartz in the desired thickness. 30 s afterwards the layer P + A was evaporated on the layer P. The total thickness of the two layers was always the same (8750 Å) in order to avoid errors caused by scattering of fluorescence light in the layers. After 30 min the evaporation process was complete. After this process the fluorescence spectrum of the layers is structureless. Electron diffraction gives evidence for an amorphous state. Annealing at 230 °K for 15 min makes the layers crystalline with a fluorescence spectrum very similar to the fluorescence of melt grown crystals (Fig. 3). These layers at 230 °K were used for the diffusion measurement. Careful annealing was very important, since by changing the annealing conditions one gets at least two different kinds of fluorescence spectra which correspond to different crystalline phases [4].

The thickness of the layers was measured during the evaporation by counting the number of interference maxima for the transmission of 365 nm light. In order to calibrate the method we dissolved layers with known number of interference maxima in alcohol and measured the optical absorption. Thicknesses below 1000 Å cannot be measured in this way. For such layers we used the optical transmission for 254 nm light, checked against the interference fringes and applying the Lambert-Beer law.

Fig. 3. Fluorescence spectra of evaporated phenanthrene layers P

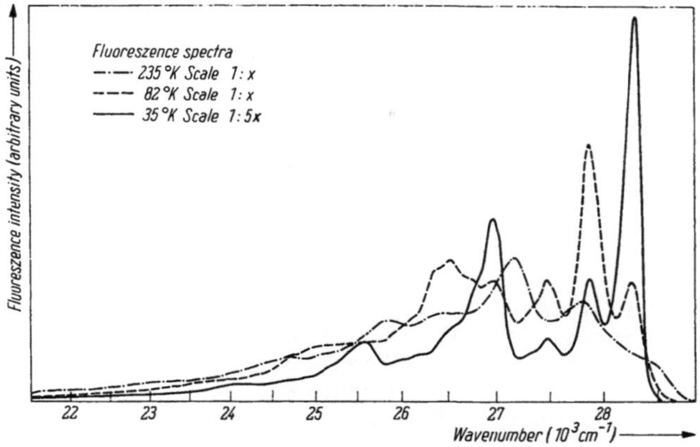


The measured dependence of the total quantum yield from the thickness $d$ of the layer P is shown in Fig. 1. The asymptotic behavior of the curve for large $d$ gives the absorption constant ($\varepsilon = 4.2 \times 10^4$ l/mol cm or $k = 6.6 \times 10^5$ cm$^{-1}$) needed for the evaluation. In this way one has not to take into account reflection of the primary light from the surface of the layer.

The mathematical treatment of the problem is simple [2]. For stationary excitation the exciton concentration $n$ is given by the differential equation

$$\frac{\mathrm{d}n}{\mathrm{d}t} = J_0\, k\, \mathrm{e}^{-kx} - \frac{n}{\tau} + D\,\frac{\partial^2 n}{\partial x^2} = 0\ ;$$

$J_0$ intensity of primary light,
$k$ absorption coefficient of exciting light,
$x$ penetration depth,
$\tau$ lifetime of excitons,
$D$ diffusion coefficient.

As boundary condition for the interface between both layers we used $n = 0$. The boundary condition at the front surface is open to discussion. Reflection of excitons at the surface [2, 5] gives $\mathrm{d}n/\mathrm{d}x = 0$, destruction at the surface [6 and 7] gives $n = 0$. We have no evidence for or against one of these two assumptions.

We get the following values for the diffusion length $L = \sqrt{D\,\tau}$ and the average radial diffusion length $\bar{r}$ in solid phenanthrene at 230 °K ($L$ is the projection of $\bar{r}$ on a certain direction):

For $\frac{\mathrm{d}n}{\mathrm{d}x} = 0$ at the front surface

$L = 78 \pm 25$ Å, $\bar{r} = 135 \pm 45$ Å.

For $n = 0$ at the front surface

$L = 152 \pm 48$ Å, $\bar{r} = 263 \pm 85$ Å.

The measured diffusion length in phenanthrene is appreciably shorter than in anthracene [2]. This agrees with the lower quantum yield and the lower energy transfer efficiency of phenanthrene as compared with anthracene [3].

It has to be mentioned that we measured the average diameter of the crystallites in the phenanthrene layers by electron diffraction. The measured value of 60 Å is close to the exciton diffusion length. If grain boundaries are hindering exciton diffusion the values for $L$ given above are lower limits for the diffusion length in perfect crystals.

We acknowledge financial support of the Deutsche Forschungsgemeinschaft.

## References

[1] H. C. WOLF, Festkörperprobleme **4**, 57 (1965) (review article).
[2] O. SIMPSON, Proc. Roy. Soc. **A 238**, 402 (1956).
[3] K. W. BENZ and H. C. WOLF, Z. Naturf. **19a**, 181 (1964).
[4] G. GALLUS, Diplomarbeit, Stuttgart 1965.
[5] R. E. MERRIFIELD, J. chem. Phys. **28**, 647 (1958).
[6] R. G. KEPLER and A. C. SWITENDICK, Phys. Rev. Letters **15**, 59 (1965).
[7] V. V. EREMENKO and V. S. MEDVEDEV, Fiz. tverd. Tela **2**, 1572 (1960).

*(Received May 6, 1966)*

phys. stat. sol. **16**, 281 (1966)

*Centre d'Études Nucléaires de Grenoble*

# Étude et interprétation du stade I dans le nickel irradié aux électrons à 20 °K

Par

P. Peretto, J. L. Oddou, C. Minier-Cassayre, D. Dautreppe et P. Moser

Des expériences couplées de traînage magnétique et de restauration de la résistivité électrique sur du Nickel pur et dopé de fer, irradié aux électrons à 20 °K, permettent de proposer un modèle détaillé de la structure des défauts ponctuels évoluant au cours du stade I. Chaque sous-stade est attribué à l'annihilation d'interstitiels placés sur des groupes de sites équivalents par rapport à la lacune sur lesquels ils peuvent éventuellement migrer avant da disparaître.

Magnetic after effect and electrical resistivity recovery experiments on electron-irradiated, pure or iron doped nickel at 20 °K give information for a refined model of the point defects which anneal out during stage I. The sub-stages are ascribed to interstitials on groups of equivalent sites. The interstitials can migrate via the different sites of a particular group before being trapped by a vacancy.

## 1. Introduction

Le cuivre est sans aucun doute le métal qui a suscité le plus grand nombre d'expériences d'irradiation, et il est pris à juste titre comme élément de comparaison pour les études sur les métaux cubiques à faces centrées. C'est pourquoi nous rappelons brièvement les résultats et les modèles proposés, dans le but de faire un parallèle entre le nickel et le cuivre.

Lucasson et Walker [1] ont montré l'existence à basse température d'un stade de restauration de la résistivité dans le cuivre irradié à 4 °K aux électrons. Une étude expérimentale fine révèle que ce stade est en fait composé de cinq sous-stades bien différenciés: $I_A$, $I_B$, $I_C$, $I_D$ et $I_E$. D'après Corbett, Smith et Walker [2], l'annihilation de paires proches-lacunes interstitiels serait responsable des sous-stades $I_A$, $I_B$, et $I_C$; quant aux sous-stades $I_D$ et $I_E$ ils correspondraient respectivement à l'annihilation avec et sans corrélation d'interstitiels dans les lacunes. L'annihilation avec corrélation au cours de $I_D$ est suggérée par une comparaison des résultats expérimentaux avec la théorie de Waite [3].

D'après Granato et Nilan [4] tous les sous-stades peuvent s'interpréter en termes d'interaction de l'interstitiel avec la lacune. A chaque position de l'interstitiel par rapport à la lacune est associée une énergie d'interaction, décroissante avec la distance séparant les deux composantes du défaut de Frenkel. Les sous-stades $I_A$, $I_B$ et $I_C$ correspondent aux positions de proches voisins: le sous-stade $I_D$ englobe l'ensemble des interstitiels des sites plus éloignés. Lorsque l'interaction devient négligeable l'interstitiel est libre et s'annihile au sous-stade $I_E$ dans une lacune quelconque.

D'après Seeger [5] les sous-stades $I_A$, $I_B$, $I_C$ et $I_D$ sont dûs à la guérison de paires proches. Au stade $I_E$ s'annihile une forme spéciale de l'interstitiel, le crowdion, et l'interstitiel libre (en configuration d'interstitiel dissocié) s'annihile au stade III.

Les expériences de traînage magnétique sur le nickel, comparées aux expériences de résistivité, permettent d'apporter des éléments nouveaux pour l'interprétation du stade I.

## 2. Étude expérimentale

Après irradiation du nickel de fusion de zone à 20 °K avec des électrons de 2,5 MeV, les études de résistivité mettent quatre sous-stades de restauration en évidence au-dessus de la température de l'hydrogène liquide: $I_B$, $I_C$, $I_D$ et $I_E$, notés par analogie avec le cuivre (le stade $I_A$, ou n'existe pas, ou se trouve à une température inférieure à 20 °K). La température des sous-stades $I_B$, $I_C$ et $I_D$ ne dépend pas de la concentration en défauts mais la température du sous-stade $I_E$ semble en dépendre. Ces résultats ont déjà été trouvés dans le cuivre.

L'addition de fer, en diminuant l'amplitude du stade $I_D$ fait apparaître un nouveau stade noté $II_A$ (Fig. 1 et 2).

Nous avons irradié ensemble à 20 °K deux anneaux de nickel de fusion de zone tirés du même lingot. Ces anneaux étaient très près l'un de l'autre pendant le recuit isochrone, et leurs températures étaient identiques; ils ont ainsi été soumis exactement au même traitement thermique. L'un servait aux mesures de traînage magnétique, l'autre aux mesures de résistivité. Nous avons pu ainsi associer de façon non équivoque les sous-stades de restauration de la résistivité avec la disparition des bandes de traînage magnétique.

Parmi les zones de traînage déjà observées antérieurement [6], trois disparaissent exactement pendant les sous-stades $I_B$, $I_C$ et $I_E$ détectés par la résistivité électrique. Aucune bande ne peut être associée au sous-stade $I_D$ (Fig. 3 et 4). Une zone supplémentaire apparaît dans le nickel pollué de fer: sa disparition correspond au stade $II_A$ (Fig. 5).

Le défaut responsable de la bande $I_E$ s'annihile suivant une cinétique d'ordre 2; la restauration dépend de la concentration, et le nombre de sauts de guérison est de l'ordre de 300 pour la dose utilisée ($\approx 10^{18}$ électrons/cm²). L'énergie de réorientation du défaut est de l'ordre de 0,15 eV, égale à l'énergie de guérison. La symétrie du défaut n'est pas [111] (en supposant l'existence de parois à 71°). Les énergies de réorientation des défauts responsables des zones de traînage

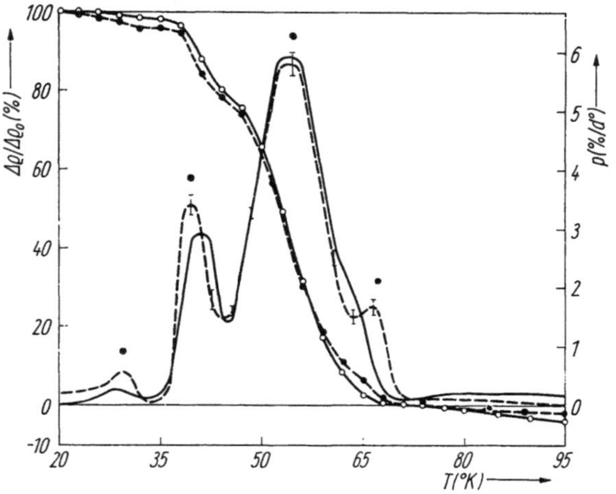


Fig. 1. Restauration, au cours d'un recuit isochrone de la résistivité électrique, d'un nickel préparé par la méthode de la zone fondue, irradié aux électrons à 20 °K. En traits pleins: Irradiation avec les électrons de 3 MeV, dose de $1,4 \times 10^{18}$ électrons/cm² — isochrone 3° — 10 min. En traits pointillés: Irradiation avec les électrons de 2,5 MeV, dose de $10^{18}$ électrons/cm² — isochrone 3° — 20 min. (Les quatre points signifient de gauche à droite $I_B$, $I_C$, $I_D$ et $I_E$)

Fig. 2. Restauration de la résistivité électrique d'un nickel dopé de 0,5% de fer dans des conditions d'irradiation et de recuit identiques à celles choisies pour l'expérience d'irradiation du nickel pur, à 3 MeV. (Les cinq points signifient de gauche à droite $I_B$, $I_C$, $I_D$, $I_E$ et $II_A$)

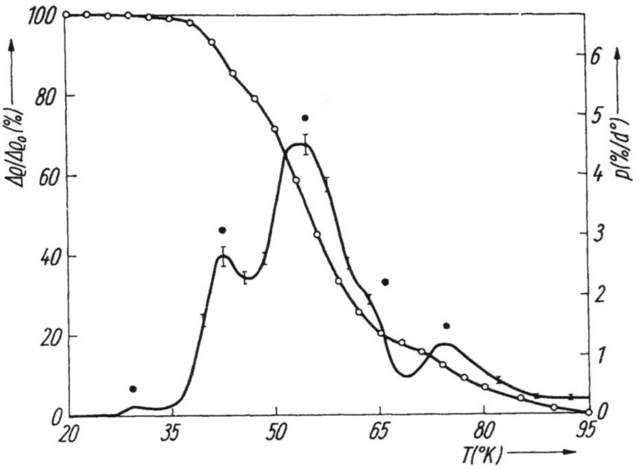


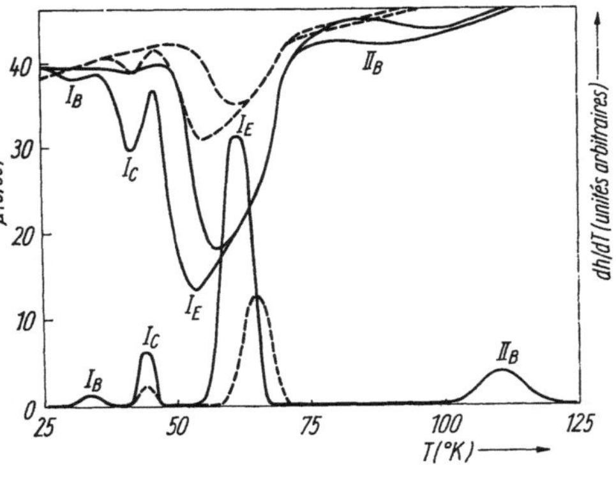


Fig. 3. Perméabilité initiale du nickel irradié aux électrons à 20 °K (dose: $10^{18}$ électrons/cm²) au cours d'une montée linéaire en température (0,4°/min).
En traits pleins: Perméabilité du nickel irradié à 2,5 MeV pour des durées de stabilisation de 3 s (trait supérieur) et de 166 s (trait inférieur) après désaimantation.
En traits pointillés: Perméabilité du nickel irradié à 1,2 MeV pour les mêmes durées de stabilisation. Nous avons tracé les courbes de vitesse de disparition des champs de traînage magnétique pour les situer par rapport aux pics de restauration de la résistivité. Leurs amplitudes n'ont qu'une valeur indicative

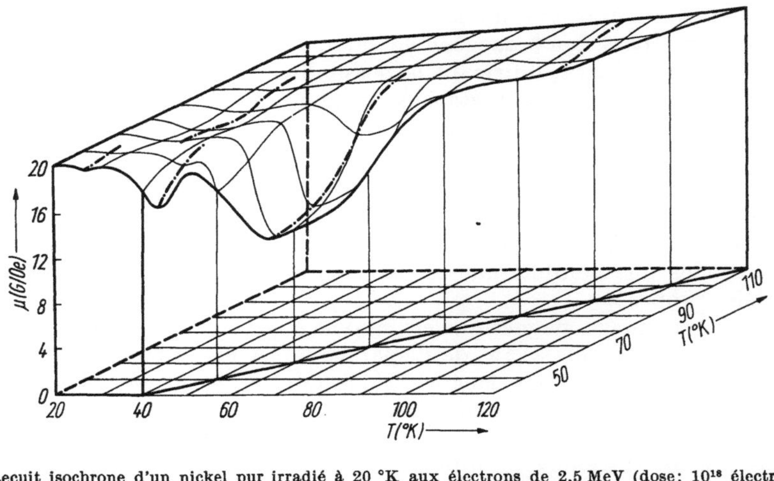


Fig. 4. Recuit isochrone d'un nickel pur irradié à 20 °K aux électrons de 2,5 MeV (dose: $10^{18}$ électrons/cm²). En coordonnées obliques: Les températures de fin de recuit. Chaque dépression représente un défaut

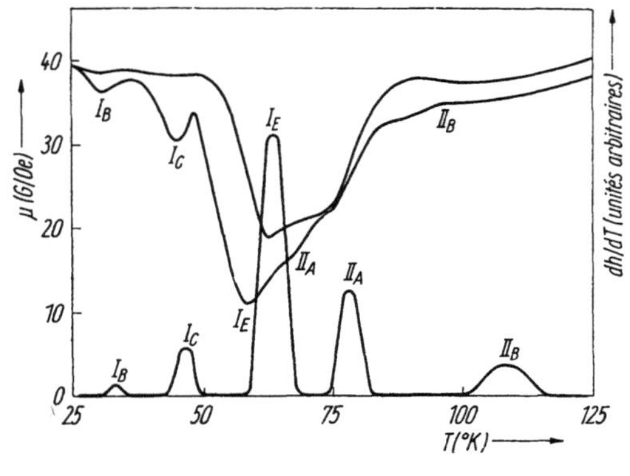


Fig. 5. Perméabilité initiale du nickel dopé à 0,3% de fer irradié aux électrons de 2,5 MeV (dose: $10^{18}$ électrons/cm²)

$I_B$ et $I_C$ obtenues en appliquant la règle de Wert et Marx sont estimées à 0,08 et 0,13 eV respectivement, et leurs énergies de guérison sont très légèrement supérieures à celles-ci; le nombre de sauts est faible, de l'ordre de 4 à 5 pour la bande $I_C$; il est sans doute encore plus petit pour la bande $I_B$.

### 3. Interprétation

Rappelons tout d'abord que l'apparition de traînage magnétique est liée à l'existence de défauts de symétrie inférieure à celle du cristal et migrant sur des sites équivalents (ou du moins dont les énergies diffèrent au plus de $kT = 5 \times 10^{-3}$ eV). Toute annihilation de défaut sera supposée s'accompagner d'une diminution de résistivité.

#### *3.1 Stade* $I_E$

Nous attribuons le stade $I_E$ à la disparition par migration d'interstitiels libres: toutes les données expérimentales confirment cette hypothèse (de plus la zone de traînage $I_E$ est la plus importante des zones).

Puisque sa symétrie est inférieure à la symétrie cubique, c'est que l'interstitiel est en configuration d'interstitiel dissocié; le nombre de sauts est compatible avec une répartition homogène de défauts se recombinant dans les lacunes par migration au hasard, infirmant la possibilité d'une réorientation sur place de l'interstitiel. Notre interprétation est donc en contradiction sur ce point avec le modèle proposé par Seeger [7].

#### *3.2 Stades* $I_B$ *et* $I_C$

Le faible nombre de sauts, les valeurs des énergies de réorientation ainsi que l'indépendance de la restauration vis-à-vis de la concentration, suggèrent pour ces sous-stades une annihilation de paires proches, donc en accord avec tous les modèles proposés.

Puisque un traînage magnétique est associé à ces défauts, c'est qu'ils migrent sur des sites équivalents avant de disparaître. La Fig. 6 montre les sites équivalents possibles près d'une lacune. Nous appelerons groupe, l'ensemble des sites qui sont à une même distance donnée de la lacune. Dans un groupe donné nous distinguerons des sous-groupes rassemblant les sites équivalents qui sont premiers voisins entre eux. L'interstitiel étant supposé sous la forme dissociée, on définit aisément l'axe de cet interstitiel. Le traînage sera dû à la migration de l'interstitiel dissocié sur les sites d'un sous-groupe, si les positions de l'axe de l'interstitiel sur les sites du sous-groupe sont équivalentes par rapport à la lacune. Nous supposerons d'autre part que l'interaction lacune-interstitiel est fortement anisotrope en accord avec l'étude de Vineyard et al. [8] sur le cuivre. Nous admettrons également que cette interaction décroît avec la distance lacune-interstitiel.

Tous les sites du groupe I forment un sous-groupe mais les calculs montrent leur instabilité [8]. La migration d'un défaut partant d'un site 2 le fait passer par un site 1. Aux sites 2 ne pourra donc pas être rattachée une zone de traînage magnétique. Mais la disparition d'interstitiels des sites 2 engendre, au cas où ces sites sont métastables, un stade de restauration de la résistivité. Or le sous-stade de restauration de la résistivité 1 B s'accompagne de traînage magnétique et ne peut donc être attribué aux interstitiels placés sur les sites 2. Donc les défauts associés à ce groupe sont instables au-dessus de 20 °K. Aux sites du groupe 3 qui par contre présente des sous-groupes nous rattacherons le stade $I_B$. Les sites du groupe 4 de même que les sites du groupe 2 ne peuvent donner de traînage magnétique mais devraient conduire à un stade de résistivité. Or, on ne voit aucun stade de restauration de résistivité entre $I_B$ et $I_C$; donc les sites du groupe 4 seraient instables (ce qui est conforme aux calculs de Vineyard pour le cuivre), ou peu peuplés. De même les sites du groupe 5 seraient rattachés au stade 1 C. (Pour le cuivre Vineyard les trouve statiquement stables mais instables dynamiquement.)

La migration fugitive des interstitiels sur les groupes de sites 3 et 5 nécessite l'existence d'une barrière de potentiel entre sites équivalents légèrement inférieure

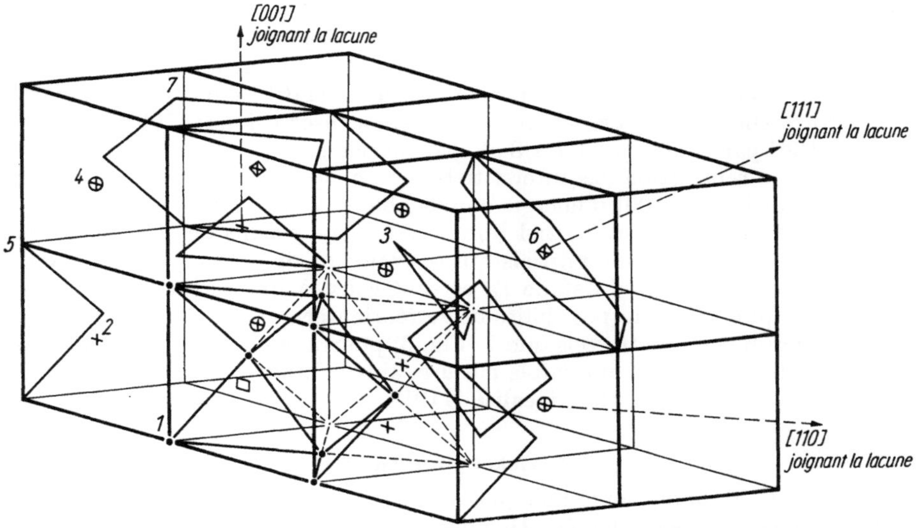


Fig. 6. Groupes de sites équivalents autour d'une lacune. Au sommet d'une même figure géométrique se trouvent les sites d'un même sous-groupe.

Les points noirs sont les premiers voisins: $\frac{a}{2}(1, 1, 0)$, les croix — les deuxièmes: $\frac{a}{2}(2, 0, 0)$, les triangles — les troisièmes: $\frac{a}{2}(2, 1, 1)$, les croix entourées d'un cercle — les quatrièmes: $\frac{a}{2}(2, 2, 0)$, les carrés — les cinquièmes: $\frac{a}{2}(3, 1, 0)$, les croix entourées d'un carré — les sixièmes: $\frac{a}{2}(2, 2, 2)$, les hexagones — les septièmes: $\frac{a}{2}(3, 2, 1)$

à la barrière entre les sites des groupes 3 et 2 d'une part et ceux des groupes 5 et 4 d'autre part, dont le franchissement signifie l'annihilation de l'interstitiel.

Le fer irradié aux électrons ne présente pas de zones de traînage identiques aux zones 1b et 1c (ou alors elles sont très faibles) [6]. Or, dans le fer, il n'y a pas parmi les sites à égale distance de la lacune, des sites qui soient premiers voisins entre eux, autrement dit, les sous-groupes ne comportent qu'un seul site. La migration sur les sites d'un groupe nécessite donc ou un saut direct de 2ème voisinage ou bien un saut intermédiaire sur un site d'un autre groupe ce qui peut conduire à une annihilation selon la structure des barrières de potentiel au voisinage de la lacune.

Dans un article à paraître dans Physical Review, R. A. Johnson utilisant un potentiel central pour décrire l'interaction atomique fait une étude numérique de la stabilité de l'interstitiel de nickel sur les différents sites voisins de la lacune. L'accord entre nos expériences et ses calculs est bon en ce qui concerne l'instabilité des sites, seuls résultats parmi ceux qu'il propose, pouvant être testés par traînage magnétique ou résistance électrique.

### 3.3 *Stade* $I_D$

L'absence de traînage magnétique au cours du stade $I_D$ est en contradiction avec le modèle proposé par Corbett et al. [2] pour le cuivre. Ce défaut ne se réoriente pas. On pourrait penser que le défaut correspondant est le ,,crowdion", mais la guérison du crowdion, si proche de la migration de l'interstitiel, serait un bien grand hasard (d'autant que tel devrait être le cas pour des métaux différents comme le cuivre et le nickel). On ne peut interpréter les expériences sur les défauts créés par irradiation sans tenir compte de la dynamique de création des défauts, laquelle a été étudiée par Vineyard et al. [8] pour le cuivre. Essentiellement ces auteurs ont trouvé que les chocs focalisés [100] et [110] ont des énergies seuils comparables: donc pour la création des défauts ces deux processus sont d'égale importance. Le choc focalisé [110] perd $^2/_3$ eV par marche (à trop forte énergie il se défocalise pour se refocaliser suivant [100]), et le choc focalisé [100] par contre perd 7 à 8 eV par marche.

Par suite la répartition des interstitiels suivant les sites est très inhomogène. Les chocs voisins de [100] d'énergie convenable envoient l'interstitiel très loin, c'est-à-dire forment un interstitiel libre. Les chocs voisins de [100] s'amortissent vite et laissent des interstitiels qui se trouvent dans le rayon d'attraction de la lacune. Au cours du stade $I_D$ ce sont ceux-là qui disparaissent dans la lacune sans qu'ils aient la possibilité de se réorienter.

D'autres groupes de sites tels que 7 (Fig. 6) pourraient participer au stade $I_D$, mais de toute façon dans une faible mesure, ainsi d'ailleurs que les plus proches des interstitiels créés par choc focalisé [110].

### 3.4 *Nickel pollué de fer*

Le stade $I_D$ diminue lorsqu'on pollue le nickel avec du fer, par contre les stades $I_B$ et $I_C$ sont peu affectés. De plus, apparaît une zone de traînage supplémentaire. Aux concentrations utilisées (0,3%), un atome de fer a peu de chance de se trouver à proximité des sites $I_B$ ou $I_C$. Ceci est vérifié expérimentalement puisque les zones de traînage magnétique $I_B$ et $I_C$ dans le nickel pollué sont peu modifiées par rapport à celles du nickel pur. Par contre, l'interstitiel responsable du stade $I_D$ explore au cours de son annihilation une plus grande fraction du cristal et peut de cette façon être piégé par un atome de fer. Ce nouveau défaut complexe impureté-interstitiel au voisinage d'une lacune ne serait pas capable de se réorienter à cause de l'absence pour lui de sites équivalents. De même l'interstitiel libre peut être piégé pendant sa migration; le défaut ainsi créé est susceptible de se réorienter (zone de traînage $II_A$ d'énergie de réorientation: 0,17 eV). La zone de traînage disparaît au stade $II_A$ de restauration de la résistance quand l'interstitiel échappe à l'action de l'atome d'impureté (énergie de guérison: 0,21 eV).

## 4. Conclusion

Nous avons interprété le stade I par la disparition de l'interstitiel dans les lacunes. Les stades $I_B$ et $I_C$ sont attribués à l'annihilation d'interstitiels occupant des sites bien particuliers vis-à-vis de la lacune. Pour sa plus grande partie le stade $I_D$ est attribué à l'annihilation des interstitiels placés sur les sites distribués suivant les directions [100] issues de la lacune. Le stade $I_E$ est attribué à l'annihilation des interstitiels dissociés libres de toute interaction avec les lacunes.

Les expériences couplées de mesures de résistivité électrique et de traînage magnétique, associées aux conclusions des calculs de Vineyard, nous ont permis de préciser un modèle qui dans son principe s'accorde avec le modèle discuté par Granato et al. [4] pour le cuivre.

## Bibliographie


[1] P. G. Lucasson et R. M. Walker, Disc. Faraday Soc. **31**, 57 (1961).

[2] J. W. Corbett, R. B. Smith et R. M. Walker, Phys. Rev. **114**, 1452 (1959).

[3] T. R. Waite, Phys. Rev. **107**, 463 (1957).

[4] A. V. Granato et T. G. Nilan, Phys. Rev. Letters **7**, 171 (1961).

[5] A. Seeger, Symposium on Radiation in Solids, Venice (Italy) 1962, I.A.E.A., Vienna 1962 (p. 101).

[6] J. Verdone, P. Peretto, P. Moser, D. Dautreppe et J. Verdier, C. R. Acad. Sci. (France) **260**, 5209 (1965).

[7] A. Seeger, H. Kronmüller et H. Rieger, Z. angew. Phys. **18**, 377 (1965).

[8] J. B. Gibson, A. N. Goland, M. Milgram et G. H. Vineyard, Phys. Rev. **120**, 1229 (1960).


*(Received May 9, 1966)*

phys. stat. sol. **16**, 289 (1966)

*Department of Applied Physics, Welsh College of Advanced Technology, Cardiff*

# Symmetry Orbitals and UV Selection Rules for the Polymethylene Chain

By

W. L. McCubbin

Basis functions for irreducible representations of the line group of polymethylene chain are presented. This work confirms the general features of a semi-quantitative band structure given previously. The UV selection rules are derived and are found to support the earlier conjecture that the transition probability for optical transitions to the first unoccupied band is negligible. It is pointed out that the selection rules may also be used to predict the dichroism of the polymethylene crystal.

Es werden die Basisfunktionen für die irreduzible Darstellung der Liniengruppe der Polymethylenkette gegeben. Die Arbeit erhärtet die allgemeinen Züge der schon früher gegebenen halb-quantitativen Bandstruktur. Die UV Auswahlregeln werden hergeleitet. Sie unterstützen die frühere Vermutung, daß die Übergangswahrscheinlichkeit für optische Übergänge zum ersten unbesetzten Band vernachlässigbar ist. Es wird weiterhin gezeigt, daß die Auswahlregeln auch den Dichroismus der Polymethylenkristalle voraussagen können.

The present work was undertaken to confirm certain aspects of the band structure of the polymethylene chain [1] and to assist more rigorous calculations now being set up. Owing to the difficulty encountered with boundary conditions normal to the chain axis, quantitative results are not likely to appear in the near future, and for this reason a semi-quantitative theory is a valuable short-cut.

For several problems of current interest a qualitative, but formally exact, description of the eigenfunctions in each energy band is at least as important as numerical results for the eigenvalues.

## 1. Symmetry Orbitals for the Polymethylene Chain

For the purpose of obtaining selection rules it is necessary only to use orbitals which exhibit the correct symmetry; solutions (exact or otherwise) of the Hartree-Fock equations are not required. Accordingly, we shall use below Bloch-type symmetry orbitals, our main task being to allocate them to the appropriate irreducible representations of the space group of the chain.

Since translational symmetry exists in one dimension only, the space group is termed Line group.[2]) Apart from introducing essential notation, we shall demonstrate the advantages of using a reduced set of symmetry operations.

The well-known paraffin structure possesses the symmetry elements shown in Table 1. Here, $R_n = n\,t$ and $\tau = t/2$, where $t$ is the basic primitive translation.[1]) Under cyclic boundary conditions, the ensemble of operations

---

[1]) We shall dispense with vector notation except where essential since the system is one-dimensional

[2]) For a systematic study of Line groups see [2].

Table 1

Symmetry operations of the polymethylene chain. Axes are rectangular Cartesian, $0x$ being coincident with the chain axis and $0z$ normal to the plane of the carbon atoms. The symbols in paranthesis refer to the axis of rotation or to the normal to the reflection plane, whichever is relevant. There are $2N$ possible planes $\sigma(x)$ and axes $C_2(y)$, $C_2(z)$ in an array of $N$ unit cells cyclicly disposed. Here, $\sigma(x)$ is chosen to contain a (particular) carbon atom, thus fixing the origin to which the symmetry operations are referred

| Symmetry elements | Typical symmetry operations | Basic operations |
|---|---|---|
| Screw axis $C_2(x)$ | $\tau\, C_2(x)$ | |
| Axes $C_2(y)$ | $C_2(y)$ | |
| Axes $C_2(z)$ | $\tau\, C_2(z)$ | |
| Inversion centres $i$ | $\tau\, i$ | |
| Planes $\sigma(x)$ | $\sigma(x)$ | $\{\sigma(x) \mid 0\}$ |
| Glide plane $\sigma(y)$ | $\tau\, \sigma(y)$ | $\{\sigma(y) \mid \tau\}$ |
| Plane $\sigma(z)$ | $\sigma(z)$ | $\{\sigma(z) \mid 0\}$ |
| Primitive translations $T$ | $T^n$ | $\{\varepsilon \mid R_n\}$ |

$T, T^2, \ldots T^N = E$ constitute an invariant subgroup of the line group $G$. It can be shown that there are $N$ one-dimensional representations of $G$ with matrices of the form $(e^{ik R_n})$ where $k = 2\pi m/N t$ and $m$ is an integer satisfying $0 \leqq m < < N$. The remaining operations plus the identity comprise the factor group $I$, in this case isomorphous with the point group $D_{2h}$ of order 8. The four operations given a Koster symbol [3] in the final column of Table 1 constitute the reduced set referred to above. By constructing the group multiplication table for $I$ we can express the other operations as products of those belonging to the reduced set:

$$\begin{aligned} \tau\, C_2(x) &= \{\sigma(y) \mid \tau\}\ \{\sigma(z) \mid 0\}\ , \\ C_2(y) &= \{\sigma(x) \mid 0\}\ \{\sigma(z) \mid 0\}\ , \\ \tau\, C_2(z) &= \{\sigma(x) \mid 0\}\ \{\sigma(y) \mid \tau\}\ , \\ \tau\, i &= \{\sigma(x) \mid 0\}\ \{\sigma(y) \mid \tau\}\ \{\sigma(z) \mid 0\}\ . \end{aligned}$$

Consequently, all the group-theoretical information is contained in the representation matrices of the basic operations.[3])

A method for obtaining these matrices has been described by Heine [4]. The central theorem states that the ensemble of operations which leave $k$ invariant itself constitutes a group, $K$, the importance of which lies in the fact that irreducible representations of $G$ can be derived from those of $K$. For $0 < k < \pi/t$, only one of the basic operations, $\{\sigma(x) \mid 0\}$, transforms $k \rightarrow -k$. Thus the elements of $K$ are all powers and products of $\{\sigma(y) \mid \tau\}$ and $\{\sigma(z) \mid 0\}$, viz., $E$, $\{\sigma(y) \mid \tau\}$, $\{\sigma(z) \mid 0\}$, and $\{\sigma(y) \mid \tau\} \mid \{\sigma(z) \mid 0\}$. $K$ is, therefore, isomorphous with the point group $C_{2v}$. Out of some function $\Psi(k, r)$ which transforms as any one of the representations of $K(C_{2v})$ we can generate a partner $\Psi(-k, r)$ by

[3]) There is, of course, nothing unique about the reflection operations; other operations could have been chosen as basic.

operating with $\{\sigma(x) \mid 0\}$. The two-dimensional representations of $G$ can thus be constructed from those of $K$ by supplying a second term similar to the entries of $K(C_{2v})$ but with $k$ replaced by $-k$. The matrices for the operation $\{\sigma(x) \mid 0\}$ itself are obtained by noting that it transforms $\Psi(k, r) \to \pm \Psi(-k, r)$. At $k = 0$ the irreducible representations are those of the point group $D_{2h}$.

For general $k$, unnormalised symmetry orbitals are

$$\Psi_k^{\pm}(\Phi) = \sum_{R_n} \mathrm{e}^{ik R_n} \{\Phi(\boldsymbol{r} - \boldsymbol{R}_n) \pm \mathrm{e}^{-ik\tau} \Phi(\boldsymbol{r} - \boldsymbol{R}_n - \boldsymbol{a})\} ,$$

$$\Psi_{-k}^{\pm}(\Phi) = \sum_{R_n} \mathrm{e}^{-ik R_n} \{\Phi(\boldsymbol{r} - \boldsymbol{R}_n) \pm \mathrm{e}^{ik\tau} \Phi(\boldsymbol{r} - \boldsymbol{R}_n - \boldsymbol{a})\} ,$$

where $\Phi$ is one of the carbon atomic functions. $\boldsymbol{R}_n$ locates the first atom in the $n$-th unit cell and $\boldsymbol{R}_n + \boldsymbol{a}$ the second. For $0 < k < \pi/t$ only two operations — $\{\sigma(y) \mid \tau\}$ and $\{\sigma(z) \mid 0\}$ — are required to allocate the $\Psi_k(\Phi)$ to their respective representations. (This is so because the characters of the remaining operations are identical in each representation). By the method outlined above it can be shown that the matrix expressions for these operations are of the form $\pm\begin{pmatrix}\mathrm{e}^{ik\tau} & 0\\ 0 & \mathrm{e}^{-ik\tau}\end{pmatrix}$ and $\pm\begin{pmatrix}1 & 0\\ 0 & 1\end{pmatrix}$ respectively. Applying them to the $\Psi_k(\Phi)$, one obtains Table 2 for $0 < k < \pi/t$. The entries + and — refer to the combination of atomic functions belonging to the representation in question.

For $k = 0$, the symmetry orbitals simplify in the obvious way and belong to the factor group representations shown in Table 3. Orbitals for $k = \pi/t$ are of no particular interest and will not be developed here.

Table 2

Symmetry of functions $\Psi_k^{\pm}(\Phi)$, where $0 < |k| < \pi/t$. The entry 0 implies that neither + or — combinations of the atomic function at the top of the column belong to that representation

| $\Phi$ | $s$ | $p_x$ | $p_y$ | $p_z$ |
|---|---|---|---|---|
| ${}^2\Gamma_1$ | + | + | — | 0 |
| ${}^2\Gamma_2$ | 0 | 0 | 0 | — |
| ${}^2\Gamma_3$ | — | — | + | 0 |
| ${}^2\Gamma_4$ | 0 | 0 | 0 | + |

Table 3

Symmetry of functions $\Psi_0^{\pm}(\Phi)$

| | $\{\varepsilon \mid R_n\}$ | $\{\sigma(x) \mid 0\}$ | $\{\sigma(y) \mid \tau\}$ | $\{\sigma(z) \mid 0\}$ | $\Psi_0^{\pm}(\Phi)$ |
|---|---|---|---|---|---|
| $\Gamma_1(A_g)$ | 1 | 1 | 1 | 1 | $\Psi_0^+(s)$, $\Psi_0^-(p_y)$ |
| $\Gamma_2(B_{3u})$ | 1 | 1 | −1 | 1 | $\Psi_0^-(s)$, $\Psi_0^+(p_y)$ |
| $\Gamma_3(B_{1u})$ | 1 | −1 | 1 | 1 | $\Psi_0^+(p_x)$ |
| $\Gamma_4(B_{2g})$ | 1 | −1 | −1 | 1 | $\Psi_0^-(p_x)$ |
| $\Gamma_5(B_{2u})$ | 1 | 1 | 1 | −1 | $\Psi_0^+(p_z)$ |
| $\Gamma_6(B_{1g})$ | 1 | 1 | −1 | −1 | $\Psi_0^-(p_z)$ |

## 2. Selection Rules for Optical Absorption

It is now necessary to be more precise about the nature of the eigenstates involved in the electronic transitions we are considering. A semi-quantitative band structure for the paraffin chain has already been presented [1]. The uppermost ($k = 0$) valence state was shown to be the bonding combination of 2 $p_x$ orbitals. This is the initial state from which we envisage a photon-induced transition to occur. Table 3 shows that it belongs to the representation $\Gamma_4(B_{2g})$. For the final state there are two possibilities: the lowest levels of the first two empty bands respectively. Both of these occur at $k = 0$, but are composed of different combinations of carbon wave functions. From Table 3 and Fig. 1 of reference [1], duly rearranged to take account of the fact that the Brillouin boundary occurs at $\theta = \pi/2$, it is clear that the lower edge of the first empty band belongs to the representation $\Gamma_1(A_g)$ and that of the second to $\Gamma_2(B_{3u})$.

Dipole transitions are allowed for those components of the vector $\boldsymbol{r}$ for which the integrals $\langle ij | \boldsymbol{r} | f\, k \rangle \neq 0$, where $i$ and $f$ signify "initial" and "final" and where $j$ and $k$ label the representations. As is well known this is so when the representation of the product $\Psi_{ij}\, \Psi_{fk}$ is the same as for at least one component of of $\boldsymbol{r}$. Now, the $x$-, $y$-, and $z$-coordinates transform according to $\Gamma_3$, $\Gamma_2$, and $\Gamma_5$, respectively. By utilising Table 3 to find the characters of the product representations $\Gamma_{jk}$ one may show that $\Gamma_{24} = \Gamma_3$ and $\Gamma_{14} \neq \Gamma_3$, $\Gamma_2$, or $\Gamma_5$.

Therefore, the $k = 0$ dipole transition to the lowest unoccupied state is forbidden, while that to the lowest state in the second unoccupied band is allowed for $x$-polarised radiation. The highest state of the first unoccupied band also occurs at $k = 0$ and belongs to $\Gamma_2$. As previously stated [1], however, the orbitals in this case have lobes directed away from the chain and the corresponding transition moment will be much smaller than that for which the above allowed transition is predicted.

## 3. Discussion

The first use that we make of the above results is to dispose of one possible objection to the band structure of reference [1], which was: If an unoccupied band is situated $\approx 3$ eV above the valence band, why is no intrinsic absorption detected in the near ultraviolet. As shown above for $k = 0$[4]) the transition moments are either identically zero or negligible, while those for transition to the second unoccupied band are not. The identity of the thermal and optical transitions proposed in [1] are thus confirmed.

Looking ahead to the time when quantitative estimates of the band edge eigenvalues will be available, we consider that more should be done to collect reliable experimental information on the uv absorption bands of polymethylene and other polymers. The present lack of such information for polymethylene is related to the complicating effects of scattering and impurity absorption. However, on the basis of the above selection rules, one can immediately predict the uv dichroism of the polymethylene crystal and thereby offer a way round the practical difficulties [5].

---

[4]) Exactly the same situation can be shown to hold for $k \neq 0$.

### *Acknowledgement*

It is a pleasure to acknowledge the assistance of Dr. M. Boon in the initial stages of this work.

## References

[1] W. L. McCubbin and I. D. C. Gurney, J. chem. Phys. **43**, 983 (1965).
[2] K. E. Lawson and G. A. Crosby, Sandia Corp. Res. Rep. SC-4655 (RR), University of New Mexico 1961.
[3] G. F. Koster, Solid State Phys. **5**, 174 (1957).
[4] V. Heine, Group Theory in Quantum Mechanics, Pergamon Press 1960.
[5] W. L. McCubbin and J. C. Weeks, to be published in J. appl. Phys.

*(Received May 9, 1966)*

phys. stat. sol. **16**, 295 (1966)

*Laboratoire d'Electrostatique et de Physique du Métal, Grenoble*

# Structural and Magnetic Properties of Epitaxial Permalloy Films[1])

By

P. ESCUDIER, F. BIRAGNET, J. DEVENYI, and A. YELON [2])

A study is made of the structural and magnetic properties of (111) films of permalloy (80% Ni, 20% Fe), produced epitaxially on epitaxial (111) Ag on mica. The orientation of the NiFe is parallel to that of the Ag. The anisotropy field is about 2 Oe, and the coercive force is slightly less than this value. The dispersion, $\alpha_{90}$, is of the order of a degree. The films are largely uniaxial, with a superimposed triaxial component. The presence of a triaxial component is demonstrated by the slow switching properties of the permalloy films, and by their domain structures. It seems that the reversal process, away from the easy axis, is a partial rotation.

Nous avons étudié les propriétés structurales et magnétiques des lames (111) de permalloy (80% Ni, 20% Fe) en épitaxie sur de l'argent (111) lui-même en épitaxie sur du mica. L'épitaxie obtenue est une épitaxie parallèle. Le champ d'anisotropie est d'environ 2 Oe, le champ coercitif lui est légèrement inférieur. Quant à la dispersion $\alpha_{90}$, elle est de l'ordre du degré. Les lames de permalloy sont principalement uniaxiales mais présentent néanmoins une composante triaxiale superposée. Ce caractère triaxial est mis en évidence par les études du basculement lent de l'aimantation et de la structure en domaines. Nous pensons que le processus de renversement de l'aimantation par application d'un champ en dehors de l'axe facile est un phénomène de rotation partielle.

We have studied the structure and the magnetic behaviour of 80 Ni-20 Fe permalloy films produced by epitaxy on Ag (111). The silver support is itself an epitaxial film on mica, whose surface is known to be very smooth [1]. The films are made in the same way as those of iron [2] or nickel [3] on silver on mica. The temperature of the support during evaporation is 315 °C, slightly lower than in the preceding cases. Some of these permalloy films are evaporated and cooled under the influence of an applied field in the plane of the film. The epitaxial relationship is the same as that for nickel [3], that is, the film has a (111) orientation, parallel to that of the substrate.

Fig. 1 shows this relationship, and in addition to the normal spots for the (111) orientation, we observe extra irrational spots of low intensity. The spots are situated at one third and at two thirds the distance of the spots belonging to the family {422}. This phenomenon has been observed for gold films, by Pashley [4], who proposes that these spots result from a double diffraction by the matrix, with normal orientation, and a system of twins. The twinning takes place about one of the ⟨111⟩ axes. We suppose that this explanation is correct for the permalloy films examined here. From electron micrographs, we find that the subcrystallites are of the order of 1000 Å.

---

[1]) Contribution to the International Colloquium on Magnetic Thin Films, 25th to 28th April 1966 in Jena.

[2]) The authors wish to thank the Direction des Recherches et Moyens d'Essais du Ministère des Armées for its financial support, Contract 64—34—063—00—480—75.

Fig. 1. Electron diffraction pattern in transmission of a (111) permalloy film

The investigations of the magnetic properties of the permalloy are performed with a hysteresiscope, with Kerr effect observations, and by Lorentz microscopy.

Theoretically, the (111) oriented films should have a triaxial anisotropy, with an anisotropy energy per unit volume given by

$$E_K = \frac{K_2}{108} \sin 6\,\theta \,.$$

The value of $K_2/108$ for the composition studied is of the order of 150 erg/cm³ [5]. We find, in fact, that the films which are evaporated without applied field, like those evaporated in a magnetic field are uniaxial. For those without applied field the coercive force, $H_c$, is of the order of 2.2 Oe, and the anisotropy field is of the order of 1.8 Oe. The dispersion is large, and may attain 16°. We assume that the uniaxial anisotropy and the dispersion are due to stray fields, or to strains acting on the support.

For the films evaporated in an applied field, we give some typical values of $H_c$, $H_K$, and $\alpha_{90}$, the dispersion.

The values of $H_c$, $H_K$, and $\alpha_{90}$ tend to be somewhat better for possible memory applications than are normally obtained with polycrystalline films.

We have studied the switching behaviour of the films with the lowest dispersion. The sample considered here is film 1, of Table 1, with $H_c = 1.76$ Oe, $H_K = 2.16$ Oe and $\alpha_{90} = 42'$. We have observed its quasistatic switching behaviour

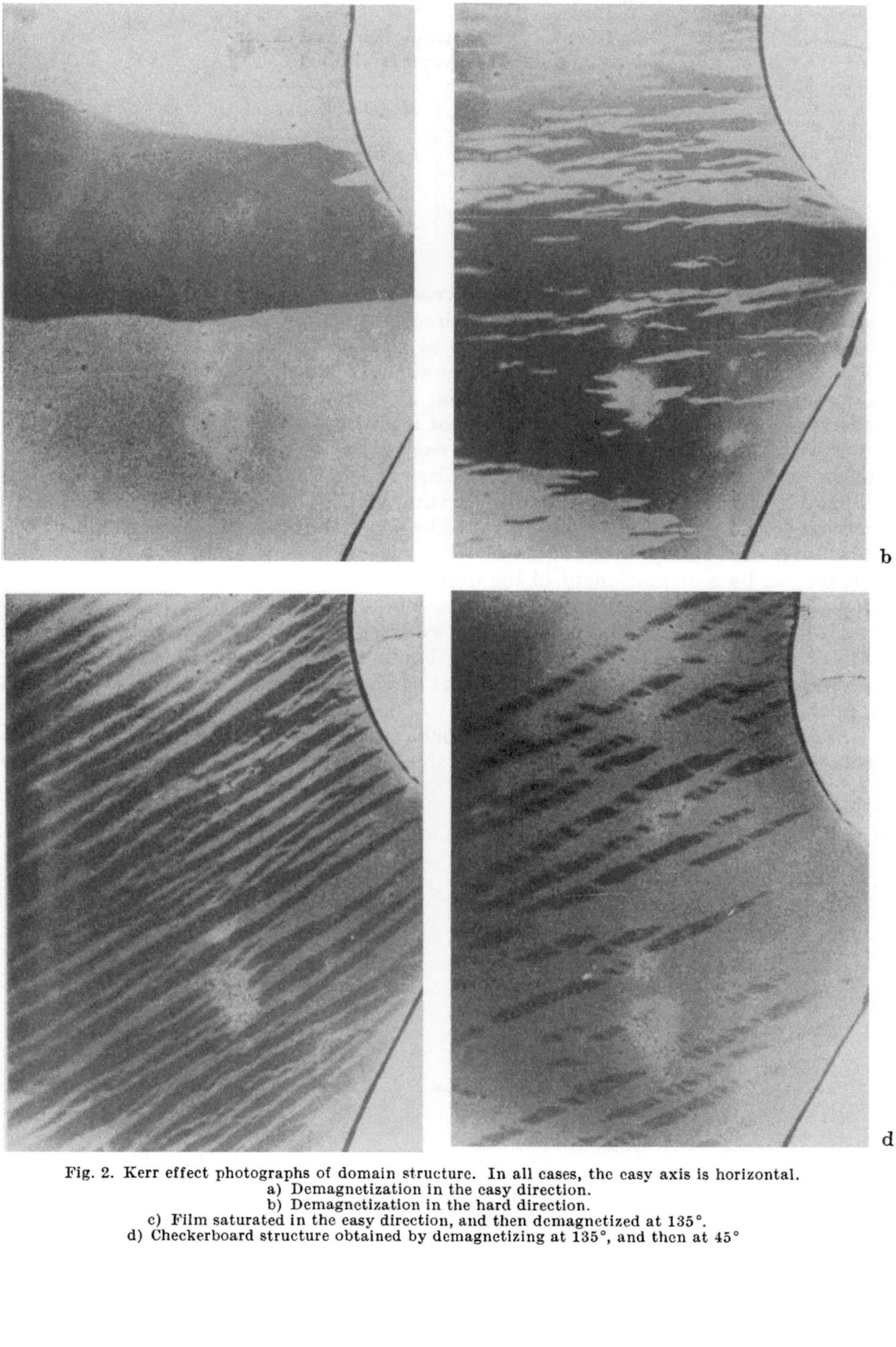


Fig. 2. Kerr effect photographs of domain structure. In all cases, the easy axis is horizontal.
a) Demagnetization in the easy direction.
b) Demagnetization in the hard direction.
c) Film saturated in the easy direction, and then demagnetized at 135°.
d) Checkerboard structure obtained by demagnetizing at 135°, and then at 45°

Table 1

Coercive force, anisotropy field, and dispersion of some epitaxial films

| Film | $H_c$ (Oe) | $H_K$(Oe) | $\alpha_{90}$ |
|---|---|---|---|
| 1 | 1.76 | 2.16 | 42′ |
| 2 | 1.92 | 2.1 | 3° |
| 3 | 1.75 | 1.75 | 1°45′ |
| 4 | 1.5 | 1.8 | 1°30′ |

with Kerr effect and with Lorentz microscopy. With Kerr effect, applying an alternating field applied in the easy direction, we obtain, by domain wall motion, domains whose walls are parallel to the easy direction. This is shown in Fig. 2a. If we demagnetize the film in the hard direction, we observe narrow domains, parallel to the easy direction, as illustrated in Fig. 2b. In both of these cases, the behaviour is like that of a uniaxial polycrystalline film.

If we saturate the film parallel to the easy direction, at 0°, and we apply a field at an angle in the vicinity of 135°, we observe what appears to be a partial rotation. The domains exhibit a band structure, which is stable when the field is removed (Fig. 2c). This does not appear to be a labyrinth propagation, as the bands are long, and rarely broken. If the field is again increased, the film attains saturation by a displacement of the domain walls limiting the bands, and not by a wall parallel to the easy axis.

If, starting from the band structure we have just discussed, a field is applied at 45° to the easy axis, a second set of bands may be produced, approximately perpendicular to the first, yielding a sort of checkered pattern. This is shown in Fig. 2d.

We have attempted to explain the behaviour observed in terms of a partial rotation model [6, 7]. In order to do this, we have traced two critical curves

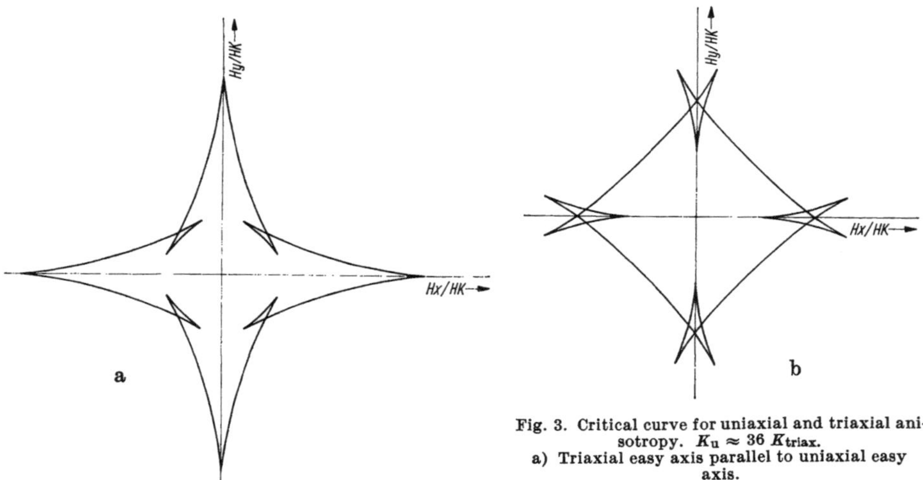


Fig. 3. Critical curve for uniaxial and triaxial anisotropy. $K_u \approx 36\ K_{triax.}$
a) Triaxial easy axis parallel to uniaxial easy axis.
b) Triaxial easy axis parallel to uniaxial hard axis

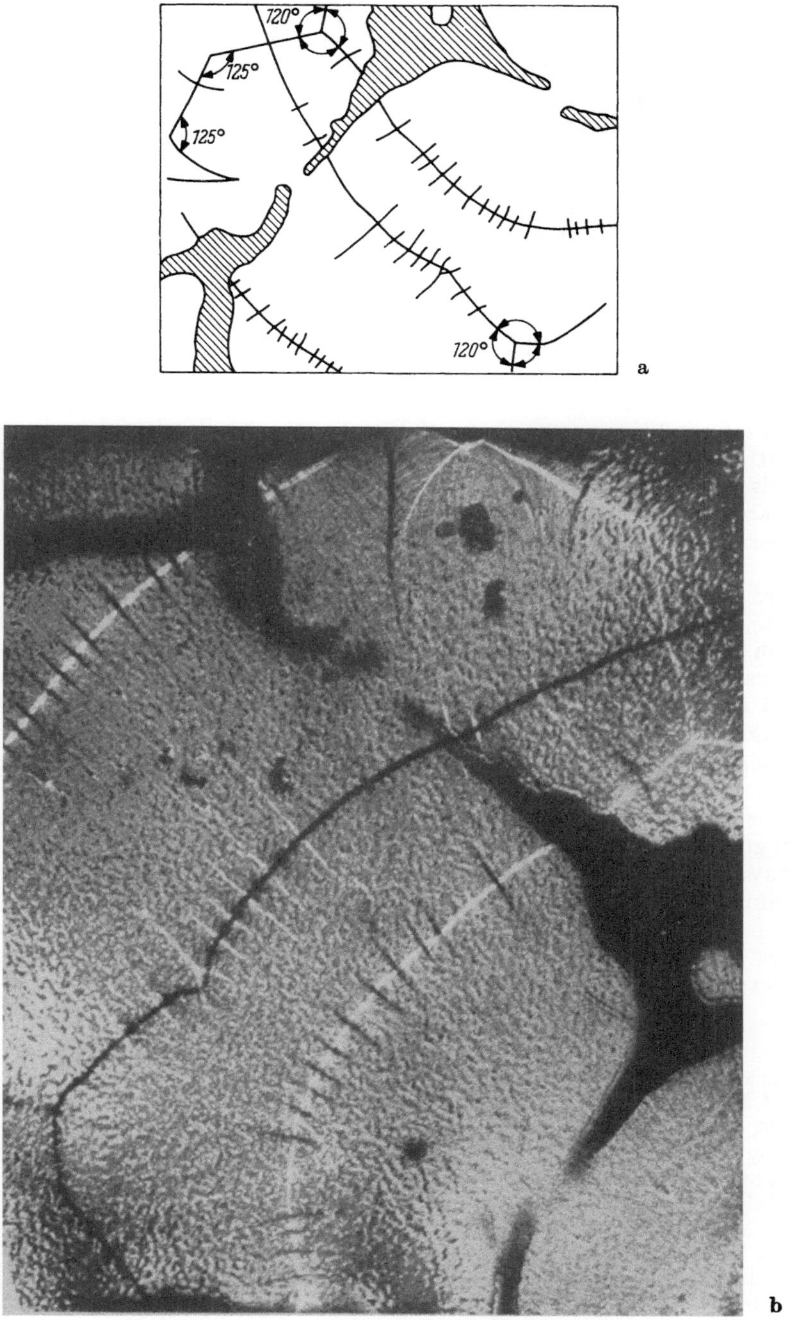


Fig. 4. Domain structure shown by Lorentz microscopy.
a) Diagram of the domain structure.
b) Photograph of the same region

for combined uniaxial and triaxial anisotropy, with $K_{\text{triax}} \approx K_u/36$. We have taken two extreme cases: a triaxial easy direction parallel to the uniaxial easy direction, as shown in Fig. 3a, and a triaxial easy direction parallel to the uniaxial hard direction, as shown in Fig. 3b. At zero applied field, there is only one easy axis in both cases, but the critical curves are very different from that of a simple uniaxial film. Even though the triaxial anisotropy is small compared to the uniaxial one when we take two derivatives of the energy to calculate the curve, the $6\,\theta$ effect becomes significant.

It has been shown [7, 8] that the direction of the partial rotation bands should be perpendicular to the tangent at the point on the critical curve where the switching takes place. As we have found from the experiment illustrated by Fig. 2c, when a field is applied in a reasonable range of angles around 135°, the bands are roughly at 135°. This is what we should expect if the experimental situation resembles that shown in Fig. 3b, that is, if an easy direction of the triaxial anisotropy is not far from a hard direction of the uniaxial anisotropy. If the experimental conditions were those of Fig. 3a, the bands should make a small angle with the easy axis for fields applied between 90° and 135°, and a large angle for fields applied between 135° and 180°. In the second case, the bands would be unstable, and should tend to break up into labyrinths. Such a behaviour should be observable, but we have not found it in our samples up to the present time.

The study of these films by Lorentz microscopy also shows their triaxial character. We note, in the diagram of Fig. 4a, which is a schematic representation of the micrograph of Fig. 4b, taken at $H = 0$, that the domains are limited by parallel domain walls which appear to be in good agreement with our observation that the films are uniaxial. However we note that the crossings of the walls frequently make an angle of the order of 120°. This triaxial effect is probably due to a blocking of the magnetization as it returns toward the easy direction, from one of the minor easy directions. We also note that although there is a dispersion of three quarters of a degree, we observe no ripple, except in the regions where the magnetization changes abruptly.

Thus, we have seen that although the triaxial anisotropy energy is small compared to the uniaxial component, its effect on the magnetic properties is not negligible.

The authors wish to thank J. C. Bruyère for his help in obtaining the Kerr effect photographs.

## References

[1] D. W. Pashley, Phil. Mag. **5**, 316 (1959).
[2] R. Cinti, J. Devenyi, P. Escudier, R. Montmory, and A. Yelon, C.R. Acad. Sci. (France) **260**, 6849 (1965).
[3] F. Biragnet, J. Devenyi, P. Escudier, R. Montmory, D. Paccard, and A. Yelon, Proceedings of the International Symposium on Basic Problems in Thin Film Physics, Clausthal-Zellerfeld, 1965 (to be published).
[4] D. W. Pashley, Phil. Mag. **8**, 1606 (1963).
[5] E. Kneller, Ferromagnetismus, Springer Verlag 1962 (p. 194).
[6] S. Methfessel, S. Middelheok, and H. Thomas, J. appl. Phys. **32**, 1959 (1961).
[7] S. Middelhoek, I.B.M. J. Res. Development **6**, 394 (1962).
[8] W. Metzdorf, Z. angew. Phys. **14**, 79 (1962).

*(Received May 9, 1966)*

phys. stat. sol. **16**, 301 (1966)

*Laboratoire d'Electrostatique et de Physique du Métal, Grenoble*

# A New Property of Ferromagnetic-Antiferromagnetic Coupling[1])

By

D. PACCARD, C. SCHLENKER, O. MASSENET, R. MONTMORY, and A. YELON

Three systems exhibiting ferromagnetic-antiferromagnetic coupling, Co–CoO, NiFe–NiFeMn, NiFe–$Cr_2O_3$, are studied using uniaxial thin films. A new property of such systems is found, i.e., a concentration of the hysteresis loops as a function of the number of cycles traced ($n$). This contraction follows a $1/\sqrt{n}$ law for the first 50 cycles and is related to the rotational hysteresis. This phenomenon is attributed to fluctuations in the ferromagnetic-antiferromagnetic coupling, which modify the magnetic interactions between differently coupled regions with each reversal.

On a étudié sur des couches minces à anisotropie uniaxiale trois systèmes présentant les propriétés caractéristiques des couplages ferromagnétiques-antiferromagnétiques, Co-CoO, NiFe-NiFeMn, NiFe-$Cr_2O_3$. On a mis en évidence une propriété nouvelle de contraction des cycles d'hystérésis avec l'ordre $n$ du cycle décrit. Cette contraction suit une loi en $1/\sqrt{n}$ pour les 50 premiers cycles et est liée à l'hystérésis de rotation. On attribue ce phénomène à des fluctuations dans l'intensité du couplage ferromagnétique-antiferromagnétique entraînant lors des différents basculements, des modifications des interactions magnétiques entre des régions à couplage différent.

## 1. Introduction

The phenomenon of ferromagnetic-antiferromagnetic coupling [1] has been the object of a substantial number of studies on various materials, including the systems of metal-metal oxides and of disordered alloys containing manganese [2]. The properties characteristic of this phenomenon include

a hysteresis loop displaced along the field axis,
a unidirectional component in the torque curve,
rotational hysteresis persisting to high fields.

Normally, the existence of rotational hysteresis has been attributed to a weak anisotropy of the antiferromagnetic component, permitting a discontinuous spin reorientation in this material, creating energy losses. Thus, one should not observe rotational hysteresis and unidirectional anisotropy simultaneously in the same sample. In the case of inhomogeneous Ni-Mn alloys, one finds, effectively, that rotational hysteresis begins to appear as the sample is heated, while the displacement of the hysteresis loop disappears. By contrast, fine particles of partially oxidized cobalt exhibit all three properties simultaneously. This has never been fully understood.

Bean [3] proposed that there must be some grains with an extremely thin oxide layer, and that this thin layer might have an anisotropy well below that of bulk. But the study of oxidized films showed that the unidirectional aniso-

[1]) Contribution to the International Colloquium on Magnetic Thin Films, 25th to 28th April 1966 in Jena.

tropy and the hysteresis appear at about the same time, for oxide layers less than 30 Å. It was therefore proposed [4] that the successive creation and destruction of domain walls at the interface could produce such a loss, but this has not been verified.

We have continued the studies of Co–CoO in thin films, and we have studied two other systems, NiFe–NiFeMn and NiFe–$Cr_2O_3$. We have demonstrated the existence of the properties considered characteristic of ferromagnetic-antiferromagnetic coupling in these three systems, as well as a new property.

## 2. Sample Preparation and Measurement Techniques

The ferromagnetic films are evaporated in classical vacuum ($10^{-6}$ Torr), in the presence of a magnetic field. They all possess uniaxial anisotropy.

The cobalt films, between 500 and 1000 Å in thickness, are oxidized superficially in air between 150 and 250 °C, for times ranging from 30 minutes to several hours. The antiferromagnetic material is the oxide CoO which has a Néel point of 18 °C.

The system NiFe–NiFeMn [5] is produced by evaporating a film of FeNi (80% Ni, 20% Fe) onto a film of Mn at 350 °C. The Mn diffuses into the NiFe, forming a transition layer, consisting of a series of alloys of $Ni_xFe_yMn_z$, which are antiferromagnetic and have Néel temperatures ranging from 150 to 250 °C. The films are cooled in the evaporation system or in the measuring apparatus in the presence of an applied magnetic field.

The films of NiFe–$Cr_2O_3$ are prepared in the following manner: A very thin film of chromium (200 Å) is evaporated and then oxidized for about one hour at temperatures near 400 °C under oxygen. A film of NiFe of about 400 Å is then evaporated onto this layer, at 300 °C. The oxide of chromium which is formed is $Cr_2O_3$, which is antiferromagnetic, with a Néel temperature of 35 °C.

We have performed three series of measurements on all our samples. The hysteresis loops have been measured with either a 1 kHz hysteresiscope or a Kerr apparatus operating at 50 Hz. We have measured the magnetic torque curves on a torque balance. With the same instrument and two power supplies commanded by a low frequency generator, hysteresis loops [6] have been obtained at $5\times10^{-3}$ Hz.

In all of these measurements, the samples are studied at 20 °C and −196 °C. They are cooled in a magnetic field applied in the easy direction of the ferromagnetic film, from a temperature above or very slightly below the Néel temperature.

Thus, one expects that the two anisotropies, uniaxial and unidirectional, have the same axes. The hysteresis loops are traced along this common direction. For NiFe–NiFeMn, the exchange anisotropy properties are present at 20 °C and −196 °C, and for the two other systems, only at −196 °C.

## 3. Results

### *3.1 50 Hz or 1000 Hz hysteresis loops*

We have observed a hysteresis loop displacement measured at 50 or 1000 Hz for all three systems. We define the real coercive force, $H_c$, as the half width of the hysteresis loop, and the coupling field, $H_e$, the displacement of the hysteresis loop along the field axis. The coupling field is theoretically related to the

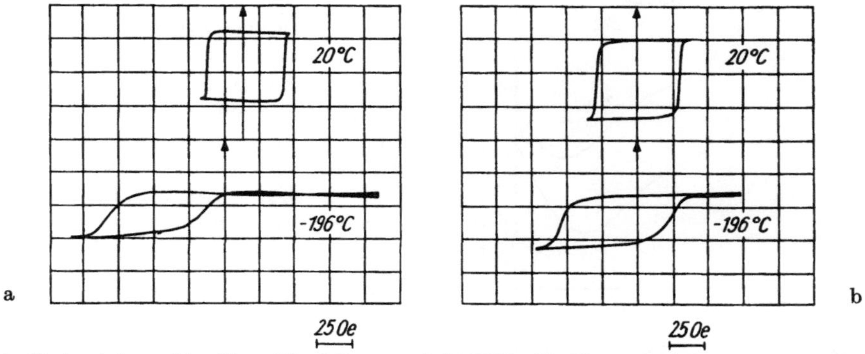


Fig. 1. Hysteresis loops of two films of Co-CoO measured at 1000 Hz. The films are cooled from room temperature to liquid nitrogen temperature, in the presence of a magnetic field sufficient to saturate the Co. a) 20 °C, b) — 196 °C

unidirectional anisotropy constant, $K_e$, by the relationship $H_e = \frac{K_e}{M_s}$. The measured coercive fields are

$$|H_\alpha| = H_c + |H_e| \quad \text{and} \quad |H_\beta| = -H_c + |H_e| .$$

In the case of Co-CoO, $H_c$ and $H_e$ may have values of several tens of Oersteds. We note that there is always a diminution of the flux detected at liquid nitrogen temperature compared with that measured at room temperature, which indicates that a part of the magnetization is blocked at low temperatures, at the measuring frequency, and that the measurements are characteristic of only a fraction of the film. This is illustrated in Fig. 1.

For NiFe-NiFeMn, the measured fields are smaller, and vary relatively little between 20 °C and −196 °C. In both of these cases, the coupling may be suffi-

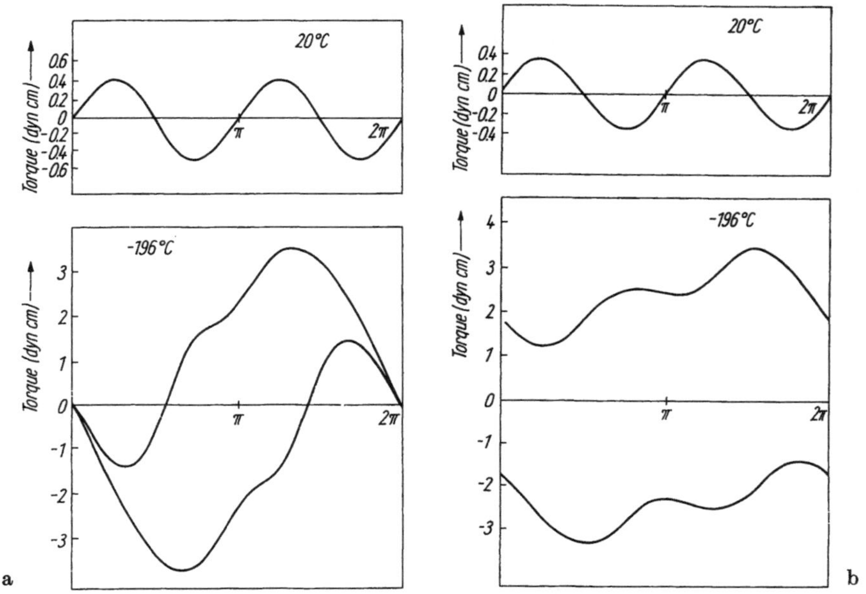


Fig. 2. Torque curves of Co-CoO films as function of angle.
a) Co-CoO I (at 360 Oe) exhibits a great deal more rotational hysteresis than b) CoO II (at 360 Oe)

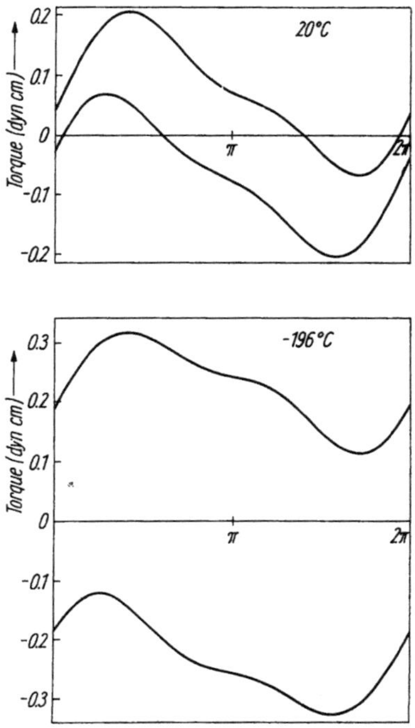


Fig. 3. Torque curves of NiFe-NiFeMn (at 180 Oc) as function of angle

ciently large so that $H_\alpha$ and $H_\beta$ may have the same sign. There is then only one equilibrium position for the magnetization at $H = 0$.

For the system NiFe-$Cr_2O_3$ the loop displacements, if they exist at all, are small (a few Oersteds), the loop may even be symmetrical, giving no indication of exchange anisotropy.

### 3.2 *Magnetic torque curves*

In the three cases, the torque curves consist of a uniaxial component $K_u \sin 2\theta$ and a unidirectional component $K_e \sin \theta$. Examples of this are shown in Fig. 2 and 3. Table 1 gives some typical numerical values of these quantities.

We note that the anisotropy constants for the NiFe-NiFeMn system vary very little between 20 °C and −196 °C.

The coupling field $H_e$, measured at 50 or 1000 Hz is always smaller than $K_c/M_s$, its theoretical value, as shown in Table 2.

Table 1

$K_u$, $K_e$, and $W_r$ per unit volume of the ferromagnetic material.
$K_u$ is the uniaxial anisotropy constant, $K_e$ the unidirectional anisotropy constant. $W_r$ is the rotational hysteresis at fields much greater than the ferromagnetic anisotropy field.

| | | $10^{-3}$ $K_u$ (erg/cm³) | $10^{-3}$ $K_e$ (erg/cm³) | $10^{-5}$ $W_r$ (erg/cm³) |
|---|---|---|---|---|
| Co–CoO I (228 f) | −196 °C | 34.8 | 56.4 | 21.6 |
| Co–CoO II (227 e) | −196 °C | 14.2 | 131 | 8.7 |
| NiFe–NiFeMn (L 13 II) | 20 °C | 0.41 | 1.20 | 0.78 |
| | −196 °C | 0.34 | 0.92 | 2.80 |
| NiFe–$Cr_2O_3$ (L 12 IV) | −196 °C | 2.84 | 2.18 | 0.11 |

Table 2

Comparison of coupling field, $H_e$, measured at 50 or 1000 Hz, and of $K_e/M_s$

| | | $H_e$ (Oe) | $\frac{K_e}{M_s}$ (Oe) |
|---|---|---|---|
| Co–CoO | (227 e) | 60 | 92 |
| | (197) | 15 | 32 |
| NiFe–$Cr_2O_3$ | (L 12 IV) | 0.85 | 2.05 |

### *3.3 Rotational hysteresis*

Films of all three systems of composition exhibit rotational hysteresis (Fig. 2 and 3). This hysteresis persists to fields well above 1000 Oe, and varies very little with fields above a certain value, as is shown in Fig. 4. For NiFe-NiFeMn films, we find that the rotational hysteresis is much stronger at low temperature than at room temperature. Some NiFe-$Cr_2O_3$ samples did not show any displacement of the hysteresis loop at 50 Hz, but all the torque curves for this system exhibit unidirectional anisotropy and rotational hysteresis, showing that there is a ferromagnetic-antiferromagnetic coupling. Several values of rotational hysteresis are given in Table 1.

### *3.4 Quasi-static hysteresis loops*

The disagreement between the values of $H_e$ measured at 50 or 1000 Hz and the calculated values $K_e/M_s$ have led us to examine the hysteresis loops by a quasi-static method, at $5 \times 10^{-3}$ Hz.

This study has brought to light a new phenomenon: The successive cycles are not superimposed. One observes a contraction of the hysteresis loop, corresponding to a diminution of $H_\alpha$ and of $H_\beta$, the measured coercive fields in each cycling. This is illustrated in Fig. 5 and 6. The variations become slower as the number $n$ of the cycle increases. The difference between the first and second cycles is much larger than the succeeding differences. One observes on Co-CoO films, where the changes are the greatest, that the variations persist for a great number of cycles, for example, $n = 100$. At this point they are very small.

The field $H_\alpha$ varies more rapidly than $H_\beta$, thus, not only the real coercive force $H_c$, but also the coupling field, $H_e$, decreases. In certain films of NiFe-$Cr_2O_3$, $H_e$, which is nonzero for the first cycle, disappears after several cycles. This corresponds to the fact that we measure no displacement at 50 Hz.

We note also that the switching is slower than it would be in the absence of ferromagnetic-antiferromagnetic coupling. Thus, for Co-CoO the switching is much slower at $-196$ °C than at 20 °C. From this we conclude that the diminution of flux in the hysteresis loops obtained at 1000 Hz is due to the fact that at this frequency a part of the magnetization does not have time to switch, and thus remains effectively blocked. These phenomena appear to be due to the magnetic viscosity, shown by Jacobs and Kouvel [7] to exist in systems with exchange anisotropy, in particular the manganites with Hausmannite structure. We may also remark that the switching in the first cycle may be substantially more rapid than in the subsequent ones, as is noticeable in Fig. 5.

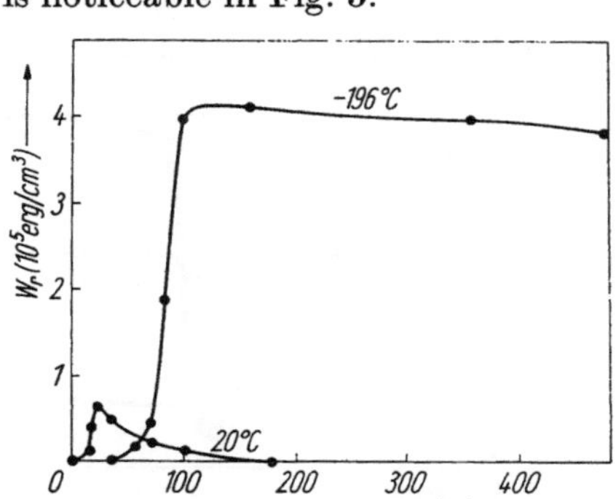


Fig. 4. Variation of rotational hysteresis, $W_r$, with field, for a Co-CoO film

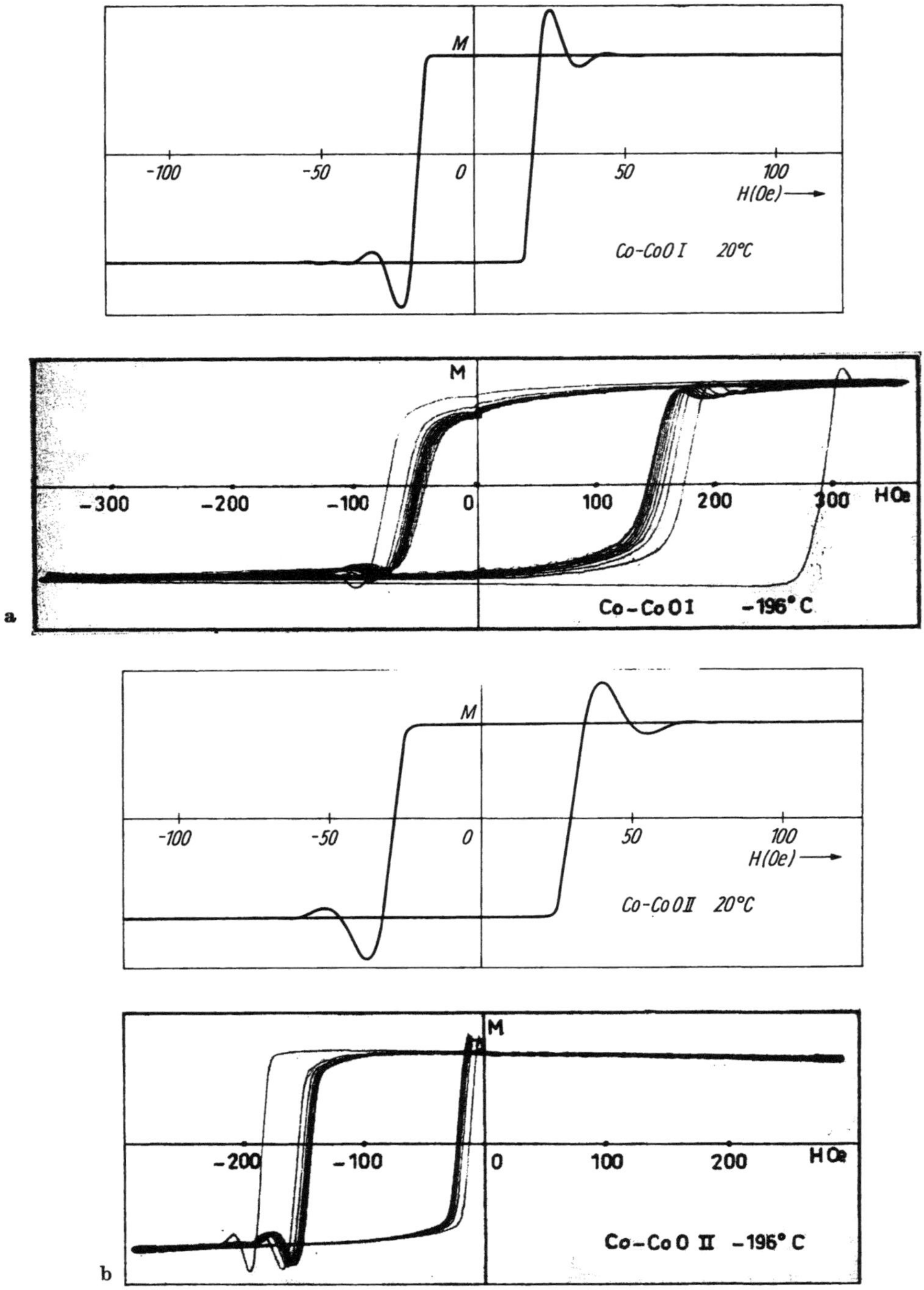


Fig. 5. Hysteresis loops measured with a torque balance ($5 \times 10^{-3}$ Hz). a) Co-CoO I, b) Co-CoO II. The overshoot is an instrumental effect

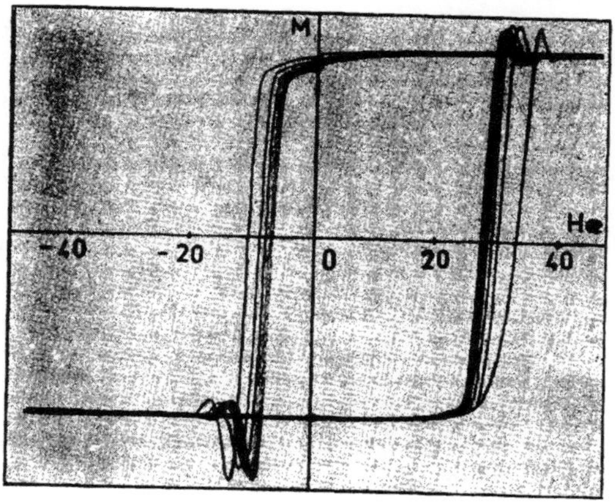


a

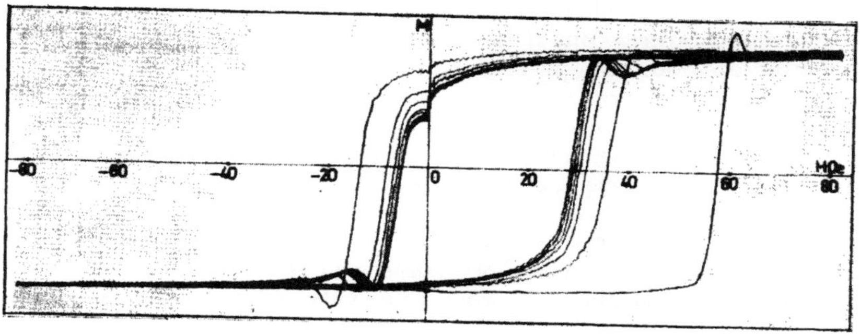


b

Fig. 6. Quasistatic hysteresis loops for a film of NiFe-NiFeMn. a) 20 °C, b) —196 °C

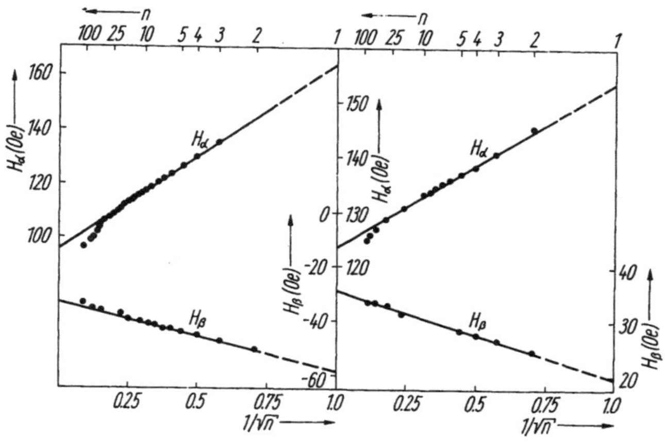


a b

Fig. 7. Coercive forces $H_{\alpha n}$ and $H_{\beta n}$ vs. $1/\sqrt{n}$, where $n$ is the number of the cycles. The values for $n = 1$ are not shown. a) Co–CoO I, b) Co-CoO II

To obtain a quantitative law governing the variation of $H_e$, we must ignore the first cycle. We then find that for $n > 1$,

$$H_{\alpha n} - H_{\alpha\infty} = \frac{A}{\sqrt{n}} \quad \text{and} \quad H_{\beta n} - H_{\beta\infty} = \frac{B}{\sqrt{n}} .$$

$H_{\alpha n}$ and $H_{\beta n}$ are the coercive forces measured in the $n$-th cycle. $A$ and $B$ are experimental constants. By drawing the curves $H_{\alpha n}$ vs. $\frac{1}{\sqrt{n}}$ and $H_{\beta n}$ vs. $\frac{1}{\sqrt{n}}$ one obtains straight lines (Fig. 7, 8, and 9), so that the limiting values $H_{\alpha\infty}$ and $H_{\beta\infty}$ may be deduced.

This law is sometimes not obeyed too well for $n = 2$ or 3. For large values of $n$ ($> 40$ or 50), the variation is faster. But the measurements are well described by such a law for the first 50 cycles.

We have found two methods for returning to the initial conditions (the state which precedes the first cycle). One can warm the film to a temperature above $T_N$, the Néel temperature, and again cool in the saturated state, or one can saturate the film perpendicular to the easy direction in the film plane, at the measuring temperature. But if the magnetization is parallel or antiparallel to the preferred direction, the phenomenon is largely irreversible.

However, if after a large number of cycles one leaves the magnetization parallel to the preferred direction for about one hour, the first cycle after this is somewhat larger than the one which preceded it; the next is very close to the undisturbed size. Conversely, if the magnetization is antiparallel to the preferred direction for about one hour, the process seems to be slightly accelerated but after a few cycles we return to the undisturbed variation. This indicates that a small fraction of the process is reversible, and seems to be explicable by the phenomenon of extraordinary magnetic viscosity, as described by Jacobs and Kouvel [7].

However, we do not believe that the contractions of the hysteresis loop can be explained by magnetic viscosity. There exists presumably a superposition of these effects and of a process depending upon the reversal of the magnetization. Our results seem related to the changes of the initial magnetization curves obtained at low temperatures by Vlasov et al. [8] on disordered NiMn alloys.

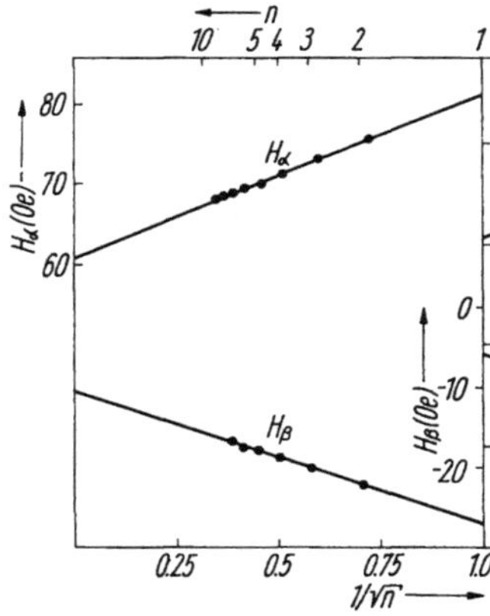


Fig. 8. $H_{\alpha n}$ and $H_{\beta n}$ vs. $1/\sqrt{n}$ for NiFe-NiFeMn at 20 °C

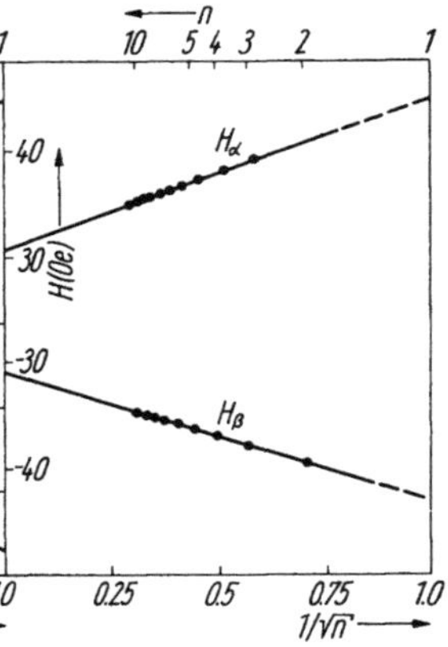


Fig. 9. $H_{\alpha n}$ and $H_{\beta n}$ vs. $1/\sqrt{n}$ for NiFe-$Cr_2O_3$ at 77 °K

## 4. Discussion

We propose that the origin of the variation of the hysteresis loops is connected with fluctuations in the intensity of exchange anisotropy. Each reversal of the magnetization of the film as a whole produces modifications in the magnetic interactions between regions of different coupling intensity, giving rise to variations in $H_\alpha$ and $H_\beta$. Such variations of the coupling over the surface may arise from the variations in the crystallographic orientations at the interface between the ferromagnetic and the antiferromagnetic.

One might imagine that the contraction of the loops would be reflected by an analogous change in the torque curves. After the second cycle, no noticeable change in these curves takes place. We suppose that this is due to the fact that in the course of plotting the torque curves, a high field is applied perpendicular to the easy axis, and that this returns the film to its initial state.

It is interesting to compare the coupling fields obtained for different numbers of cycles with $K_e/M_s$. This is done in Table 3.

Table 3

Comparison of different coupling fields

| | | $\frac{K_e}{M_s}$ (Oe) | $H_{e1}$(Oe) | $H_{e2}$(Oe) | $H_{e\infty}$(Oe) |
|---|---|---|---|---|---|
| Co–CoO | | | | | |
| (228 f) | | 40 | 48.5 | 43.5 | 30 |
| (227 e) | | 92 | 97 | 85 | 79 |
| (197) | | 32 | 44 | 25 | 18 |
| NiFe–NiFeMn (L 13 II) | 20 °C | 13.2 | 11.8 | 11.5 | 10.6 |
| | −196 °C | 10.1 | 21.9 | 13.6 | 10.8 |
| $NiFe–Cr_2O_3$ (L 12 IV) | | 2.05 | 2.7 | 2.4 | 2.05 |

$K_e/M_s$ generally falls between $H_{e1}$ and $H_{e\infty}$. This is not surprising, as the mechanism of modification of local interactions is probably different in the different measurements. Furthermore, the effect of inhomogeneities is important, and so we cannot speak of an average surface energy. In this case $H_e = \frac{K_e}{M_s}$ is not necessarily valid.

The phenomenon which produces variation of the hysteresis loops appears to be related to the hysteresis of rotation rather than to any gradual change of the torque curves. Whatever the processes taking place during the measurement of the torque curves, they should be accompanied by irreversible changes in the local interactions, during one cycle. Thus there should be energy losses.

We find, in fact, for the NiFe-NiFeMn system that the rotational hysteresis, $W_r$, increases at liquid nitrogen temperature, along with the changes in $H_c$ and $H_e$, while $K_e$ and $K_u$ vary very little. This is shown in Table 4.

Table 4

Comparison of the changes in $H_c$, $H_e$, and $W_r$ as a function of temperature for NiFe–NiFeMn

| | $H_{c1} - H_{c\infty}$ (Oe) | $H_{e1} - H_{e\infty}$ (Oe) | $10^{-5}$ $W_r$ (erg/cm³) |
|---|---|---|---|
| 20 °C | 8.1 | 1.1 | 0.78 |
| −196 °C | 21.3 | 11.1 | 2.80 |

We have been able to demonstrate a similar behavior for Co–CoO. For a series of films made under similar conditions, the variation ($H_{c1} - H_{c\infty}$) is approximately proportional to the rotational hysteresis, as is shown in Fig. 10. We therefore conclude that the rotational hysteresis observed in systems which simultaneously exhibit exchange anisotropy and rotational hysteresis, has the same origin as the change of the hysteresis loops.

To explain this contraction, one can imagine that as the hysteresis loops are traced, a spiral domain wall might be built up around a point where the coupling is so strong that the spins are blocked. But one would expect that such a process would produce an increase, rather than a decrease in the coercive force. One might also propose the creation of a helicoïdal domain wall in the thickness of the film, at the ferromagnetic-antiferromagnetic interface. But the state obtained by such a process would be very unstable.

In fact, it is quite likely that reorientations of the spins of the antiferromagnet take place at each reversal. But we do not know what the mechanism of such reorientations might be.

It might be expected that some of these processes would also result in a change in the total flux switched, from one cycle to another, In fact, for the Co-CoO films, we have observed a decrease of the order of 2 to 4% in the magnetization, in the course of a large number of cycles.

Our results may be compared with those of Nguyen van Dang [9] on the development of minor hysteresis loops, which have been discussed theoretically by Néel [10]. We note, however, that in our case there is a large change in $H_c$ and that the fields applied should be large enough to saturate the ferromagnetic material.

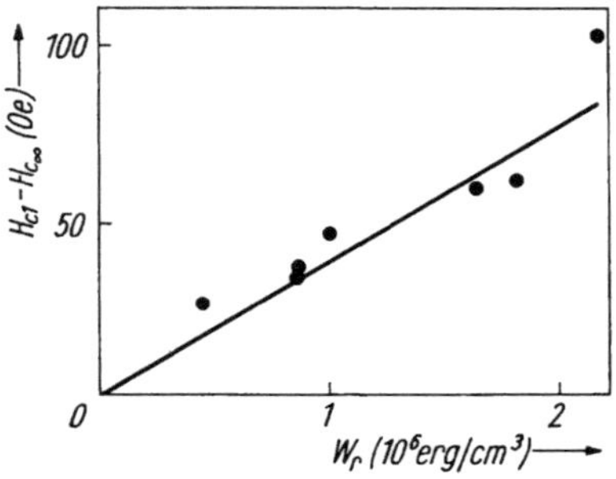


Fig. 10. Rotational hysteresis, $W_r$, vs. the contraction of the hysteresis loop ($H_{c1} - H_{c\infty}$) for a series of Co–CoO films

## 5. Conclusion

We have demonstrated the existence of a new effect in systems exhibiting ferromagnetic-antiferromagnetic coupling, the contraction of the hysteresis loops. We do not yet understand the origin of this phenomenon, but we have shown that it is related to the rotational hysteresis. As this effect seems to be present in a large number of exchange anisotropy systems, we believe that it is important for an understanding of such systems, generally.

## References

[1] W. H. Meiklejohn, and C. P. Bean, Phys. Rev **102**, 1413 (1956); **105**, 904 (1957).
[2] J. S. Kouvel, C. D. Graham Jr., and J. J. Becker, J. appl. Phys. **29**, 518 (1958).
J. S. Kouvel and C. D. Graham Jr., J. appl. Phys. **30**, 3125 (1959).
J. S. Kouvel, J. Phys. Chem. Solids **24**, 795 (1963).
[3] C. P. Bean, Structure and Properties of Thin Films, Wiley 1959 (p. 331).
[4] W. H. Meiklejohn, J. appl. Phys. **33**, 1328 S (1962).
[5] O. Massenet and R. Montmory, C. R. Acad. Sci. (France) **258**, 1752 (1964).
O. Massenet, R. Montmory, and L. Néel, IEEE Trans. Magnetics MAG **1**, **63** (1965).
[6] E. J. Torok, D. G. Agouridis, A. L. Olson, and H. N. Oredson, Rev. sci. Instrum. **35**, 1039 (1964).
[7] I. S. Jacobs and J. S. Kouvel, Phys. Rev. **122**, 412 (1961).
[8] K. B. Vlasov, N. V. Volkenshtein, S. V. Vonsovskii, A. I. Mitsek, and M. I. Turchinskaya, Izv. Akad. Nauk SSSR, Ser. fiz. **28**, 423 (1964).
[9] Nguyen van Dang, J. Phys. Radium **20**, 222 (1959).
[10] L. Néel, C. R. Acad. Sci. (France) **246**, 2313 (1958); **246**, 2963 (1958).

*(Received May 9, 1966)*

phys. stat. sol. **16**, 313 (1966)

*Instytut Podstawowych Problemów Techniki PAN, Warszawa*

# Ramp Field Switching in Thin Magnetic Films[1])

By

H. LACHOWICZ

A model of magnetization reversal by the domain wall movement is proposed. This model is formulated in general for polycrystalline square loop ferromagnets which are magnetized by a linearly increasing magnetic field. The switching time and dynamic coercivity of the material are expressed as a function of the rate of increase of the field. These calculation are based on the expression for the velocity of a domain wall obtained from the well-known Landau-Lifshits equation. Experimental results, for thin permalloy films are in good agreement with theoretical predictions. A practical aspect of the present work is a new method of determining the static coercivity of the material.

Ein Modell für die Magnetisierungsumkehr durch Bewegung von Domänenwänden wird vorgeschlagen. Dieses Modell wird allgemein für polykristalline Ferromagnete mit rechteckiger Schleife, die durch ein linear ansteigendes Feld magnetisiert werden, formuliert. Die Schaltzeit und die dynamische Koerzitivkraft des Materials werden als Funktionen der Anstiegsrate des Feldes ausgedrückt. Die Berechnungen beruhen auf der Lösung der bekannten Landau-Lifshits Gleichung für die Geschwindigkeit einer Domänenwand. Die experimentellen Ergebnisse für dünne Permalloyschichten befinden sich in guter Übereinstimmung mit theoretischen Voraussagen. Ein praktischer Aspekt dieser Arbeit ist eine neue Methode zur Bestimmung der statischen Koerzitivkraft des Materials.

## 1. Introduction

The *B-H* characteristics of thin ferromagnetic films for storage applications are very often measured by means of a well-known oscilloscope method. The measurements are carried out using a low-frequency sinusoidal driving field. An increase of amplitude of this field is generally accompanied by an increase of dynamic coercivity i.e. an increase of *B-H* loop width (Fig. 1). The most important parameters of the *B-H* loop of thin anisotropic films are the easy-axis coercivity $H_c$ and anisotropy field $H_k$. The latter is defined as the field required to obtain the saturated state in the hard direction of magnetization. In the previous works [1, 2] it was shown that the changes of dynamic coercivity are due solely to changes in the rate of rise of the driving field. It was also concluded that the above effects are observed only in those cases when magnetization reversal takes place by the motion of Bloch walls. As confirmed by theory and numerous experiments (see e.g. [3]), the easy axis magnetization reversal in thin anisotropic permalloy films of a thickness greater than approximately 500 Å takes place by the motion of Bloch walls and by the rotation in the hard direction. Since the easy axis coercivity depends upon the conditions of measurement, the coercivity, as well as the often quoted ratio $H_c/H_k$, have values which are difficult to interpret.

---

[1]) Contribution to the International Colloquium on Magnetic Thin Films, 25th to 28th April 1966 in Jena.

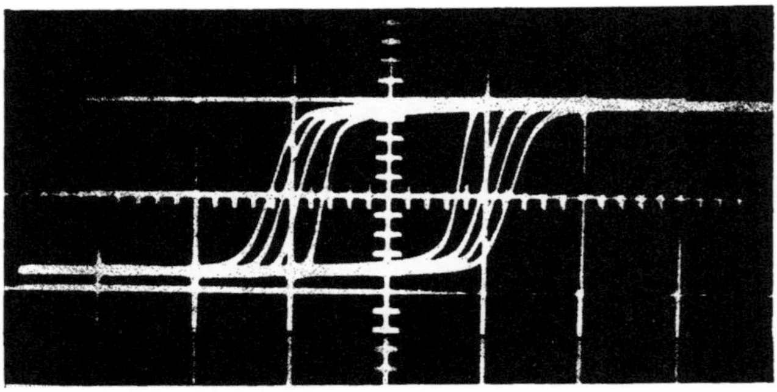

**Fig. 1.** *B-H* characteristics of a thin magnetic film with increasing amplitude sine-wave drive at 1000 Hz. Horizontal calibration 2.5 Oe/division

## 2. Domain Walls Motion Model of the Magnetization Reversal

We consider an anisotropic thin film sample which is magnetized in the easy direction by a field increasing linearly with time, i.e. $h = k \cdot t$. The initial state of the sample is the remanence state. If the film is sufficiently thick, the magnetization reversal takes place by the motion of Bloch walls which are, generally, parallel to the easy direction. The velocity of an isolated wall, as confirmed both by theory [4] and experiment [5], is proportional to the net magnetic field

$$\vartheta_i = R\,(h - H_{0_i})\,, \tag{1}$$

where $R$ is the domain wall mobility, $H_{0_i}$ the starting field for the irreversible wall motion, and $h$ the applied magnetic field. In the case considered, i.e. $h = k \cdot t$, where $k$ is the rate of rise of the applied field, and $t$ the time, the isolated wall is uniformly accelerated as can be easily seen from (1).

In any real ferromagnetic films there are many nuclei for the formation of domain walls such as crystal boundaries, lattice defects, and imperfections in the film, substrate and surface irregularities. We assume that the thin film sample can be devided into areas within which the wall which is freed from an obstacle will continue to move freely. A particular distribution of starting field strengths $H_{0_i}$ of the nucleated walls can be attributed to each of such areas with respect to the value of the applied field. Let us further assume that the minimum value of $H_{0_i}$ in such a distribution is $(H_{0_i})_{\min}$ and that the succeeding value is greater by $\Delta H_i$. Then, during the time interval in which the applied field increases from $(H_{0_i})_{\min}$ to $[(H_{0_i})_{\min} + \Delta H_i]$, the wall which started at $(H_{0_i})_{\min}$ passes a distance

$$\Delta d = \int\limits_{\frac{(H_{0_i})_{\min}}{k}}^{\frac{(H_{0_i})_{\min} + \Delta H_i}{k}} \vartheta_i \, \mathrm{d}t = R \cdot k \int\limits_{0}^{\frac{\Delta H_i}{k}} t_1 \, \mathrm{d}t_1 = \frac{(\Delta H_i)^2 \cdot R}{2\,k}\,. \tag{2}$$

From (2) it can be easily seen that for a certain range of $k$, beginning from $k = 0$, the wall of minimum starting field can pass a full free path length within the corresponding area before the applied field reaches the value $[(H_{0i})_{\min} + \Delta H_i]$. We can then expect that for a suitable range of $k$, the mode of magnetization reversal will be the same in each of such areas. Therefore, the reversal process of the film sample will be uniform i.e. changes of magnetization will occur at a constant number of moving walls. In the range of $k$ where the process is uniform, an average Bloch wall can be introduced. The starting field of this wall is assumed to be equal to the static coercive force $H_{c_0}$ since the value of the static coercivity can be evaluated as the mean value of the minimum starting fields. The velocity of the average Bloch wall is then

$$\langle \vartheta \rangle = R\,(k\,t - H_{c_0})\,. \tag{3}$$

Introducing now a mean free path length $\langle d \rangle$, which passes an average wall during the reversal time $\tau$, the latter can be evaluated by solving the integral

$$\langle d \rangle = \int_{H_{c_0}/k}^{H_{c_0}/k+\tau} \langle \vartheta \rangle \, \mathrm{d}t\,, \tag{4}$$

hence

$$\tau = \sqrt{2\,\langle d \rangle / R\,k}\,. \tag{5}$$

The switching coefficient, as it can be easily proved, is $S_w = \langle d \rangle / R$. Introducing $S_w$ to (5) the reversal time is finally expressed by

$$\tau = \sqrt{2\,S_{w_1}/k}\,. \tag{6}$$

It must be noted that the numerical value of $S_{w_1}$ in (6) can differ considerably from the value of $S_w$ which is evaluated as the reciprocal of the slope of the first linear part of a well-known step-field switching characteristic $1/\tau = f(H_m)$ where $H_m$ is the amplitude of the applied field. This is obvious because the value of $S_{w_1}$ in (6) defines the lowest speed of the magnetization reversal, whereas the second value characterizes the highest speed which is attainable, in the considered sample by the motion of Bloch walls. The ratio of these values gives a measure of the increase in number of moving walls while the rate of rise of the driving field is changed from $k = 0$ to $k = \infty$.

The easy axis dynamic coercive force $H_{cd}$, defined as a half of the displayed $B$-$H$ loop width, can be expressed as follows

$$H_{cd} = H_{c_0} + c\,k\,\tau\,, \tag{7}$$

where $c$, defined in Fig. 2, is a constant within the range of $k$ where the reversal

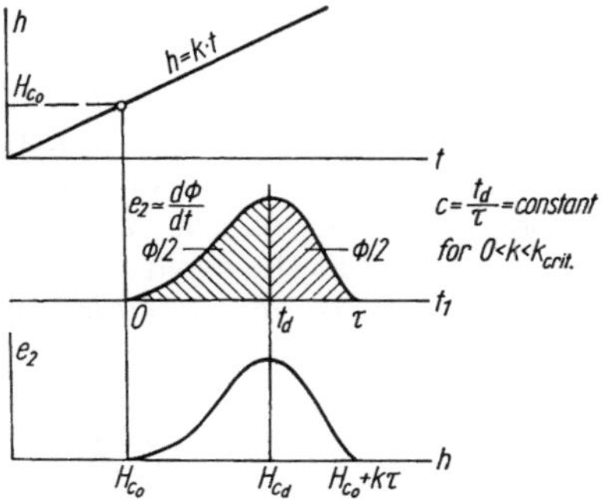


Fig. 2. Rate of change of flux vs. time

process is uniform. Substitution of the previously expressed reversal time (6) into (7) gives

$$H_{cd} = H_{c_0} + c\sqrt{2\,S_{w_1}\,k}\ . \tag{8}$$

It is interesting to note that an expression identical with (6) can be found in a quite formal way, the assumption being made that the reversal process by the irreversible wall motion can be described by the following differential equation

$$\mathrm{d}M_z/\mathrm{d}t = f(M_z)\,[h(t) - H_{c_0}]\,, \tag{9}$$

where $M_z$ is the easy-direction component of magnetization and $h(t)$, as previously, the applied field.

The reversal time $\tau$ is obtained by integrating (9)

$$\int\limits_{-M_s}^{+M_s} \frac{\mathrm{d}M_z}{f(M_z)} = \int\limits_{t_0}^{t_0+\tau} [h(t) - H_{c_0}]\,\mathrm{d}t\,, \tag{10}$$

where $h(t_0) = H_{c_0}$.

In step-field measurements the relationship between the reversal time and switching field is usually given by

$$S_w = \tau\,(H_m - H_0)\,, \tag{11}$$

where $S_w$ is the switching coefficient.

Equation (11) follows directly from (10) when the left-hand side of (10) is identified with $S_w$ and when $h(t) = H_m$ and $H_{c_0} = H_0$. In the case considered, when $h(t) = k\,t$, (10) becomes

$$S_w = \int\limits_{t_0}^{t_0+\tau} [k\,t - H_{c_0}]\,\mathrm{d}t\,. \tag{12}$$

For the sake of simplicity, this integral can be solved in coordinates shifted along the time axis by the value $H_{c_0}/k$. Therefore

$$S_w = \int\limits_0^{\tau} k\,t_1\,\mathrm{d}t_1 \tag{13}$$

and finally the reversal time is given as

$$\tau = \sqrt{2\,S_w/k}\ . \tag{14}$$

It is easily seen that (14) is identical with the previously found (6).

The function $f(M_z)$ in (9) defines the instantaneous state of magnetization. It is generally assumed that this function does not depend on the way in which this state is reached. However, the considerable differences in numerical values of $S_w$ obtained by the author [6, 7] for a ferrite and tape core in conditions of a linearly rising and step-field indicate that the function $f(M)$ does depend on the way in which the material is magnetized.

A similar formal analysis has been made by Hatfield [8] in the case of thin anisotropic films magnetized by a sine-wave field. Use of a sinusoidal driving field causes that the separation of the effects of amplitude and the rate of change of amplitude is not possible to achieve. The conclusions made by Hatfield that the switching depends on the time rate of change of the switching field as well as on its amplitude are, therefore, not completely true. The experiments carried out by the author on thin permalloy films as well as ferrites and tape metallic cores [1] under the conditions of a triangular driving field (Fig. 3)

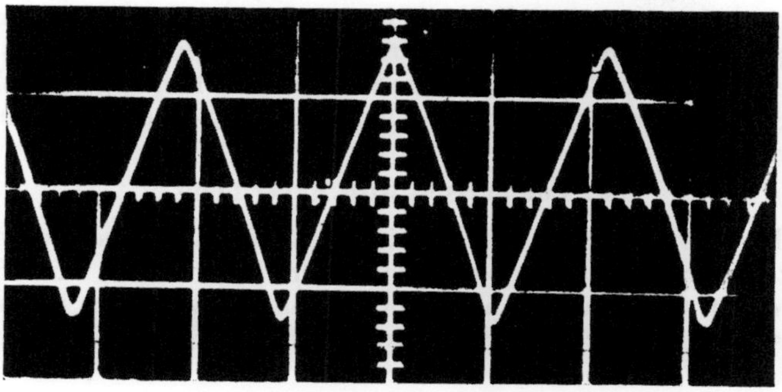

Fig. 3. Triangular wave-form of driving field

show that the switching depends only on the rate of rise of the driving field. The amplitude effect is perceptible only in the range of small field amplitudes within which the film sample is partially switched.

When the rate of rise of the applied field $k$ is sufficiently large, a certain number of walls, which have starting fields greater than the smallest in the individual areas, take part in the magnetization reversal process. Assuming that these starting fields are on the average equal to $H_{c_0} + \langle \Delta H \rangle$, the magnetization reversal can be now considered as a movement of two average domain walls having the respective velocities

$$\vartheta_1 = R\,k\,t_1\,, \qquad \vartheta_2 = R\,k\left[t_1 - \frac{\langle \Delta H \rangle}{k}\right], \tag{15}$$

where $t_1$ is the time required to shift the coordinates along the $t$ axis by a value of $H_{c_0}/k$.

During the reversal time, the two average walls, before colliding with each other, will pass the total path length denoted by $\langle s_0 \rangle$. The latter can be interpreted as the average value of the distances in each area between the walls of minimum and greater than minimum starting fields. The total path length can be expressed as

$$\langle s_0 \rangle = \int_0^\tau \vartheta_1\,\mathrm{d}t_1 + \int_{\langle \Delta H \rangle/k}^{\tau} \vartheta_2\,\mathrm{d}t_1\,. \tag{16}$$

The reversal time can be now evaluated by solving the integral (16) with the velocities previously expressed (15). It gives a square algebraic equation with respect to $\tau$. After some simplifications, the reversal time is finally determined as

$$\tau = \frac{\langle \Delta H \rangle}{2}\,x^2 + \sqrt{2\,S_{w_2}}\,x\,, \tag{17}$$

where $x = 1/\sqrt{k}$ is an introduced new variable, and $S_{w_2}$ the switching coefficient.

The numerical value of $S_{w_2}$ defines the highest reversal speed which can be obtained in the considered sample by the wall motion. It can be expected that this value should be equal to the value of $S_w$ calculated as the reciprocal of the slope of the first linear part of a well-known switching characteristic $1/\tau = f(H_m)$.

We must note that (17) is valid in the range of $k$ where the reversal process is non-uniform i.e. the number of moving walls increases with the increase of the rate of rise of the applied field. This range extends from a certain critical value of $k$ up to $k = \infty$ while the uniformity of the reversal occurs from $k = 0$ up to this critical value. At the boundary case, when $k = \infty$, the proposed model predicts that the reversal process will proceeds with the maximum number of moving walls.

## 3. Comparison with Experiment

A triangular driving field (as shown in Fig. 3) is used to switch the thin film samples in the easy direction of magnetization. The rate of rise of this field is changed in such a way that the amplitude remains constant. In each case this amplitude is large enough to guarantee that the reversal process is always completed.

The reversal time characteristic for an anisotropic permalloy thin film sample plotted along the variable $x = 1/\sqrt{k}$ is shown in Fig. 4. The reversal time is defined as the time during which the voltage response is greater than 10% of the peak value. The film is deposited electrolytically on a circular copper substrate which has a diameter of 1 cm. Before deposition, the substrate is polished mechanically and electrically. The film composition is 80% Ni and 20% Fe, as measured by means of spectral analysis. During the deposition a dc magnetic field of 100 Oe is applied in the plane of the film substrate. The thickness of the sample is approximately 7000 Å.

As it can be expected from the model, there are two distinct parts of the curve. The first, with the constant slope corresponds to the uniform process of magnetization reversal and therefore can be described by (6), whereas the second, non-linear part can be connected with the non-uniform mode of magnetization reversal (see (17)). A numerical approximation of the non-linear part of the time reversal curve to (17) gives $\tau = (4.43\,x^2 + 0.069 \cdot x)$ ms. The plot of the numerical expression is shown on the same graph. It can be seen that the experimental curve fits the parabolic formula quite well. Equating the appropriate coefficients of both the theoretical and numerical expression, the values of $S_{w_2}$ and $\langle \Delta H \rangle$ are calculated giving $S_{w_2} = 2.38$ Oe µs and $\langle \Delta H \rangle = 8.86$ Oe. The slope of the linear part gives the value of $S_{w_1} = 2.53$ Oe ms. Therefore, the increase in number of moving walls can be determined from the ratio $S_{w_1}/S_{w_2}$ and gives 1063.

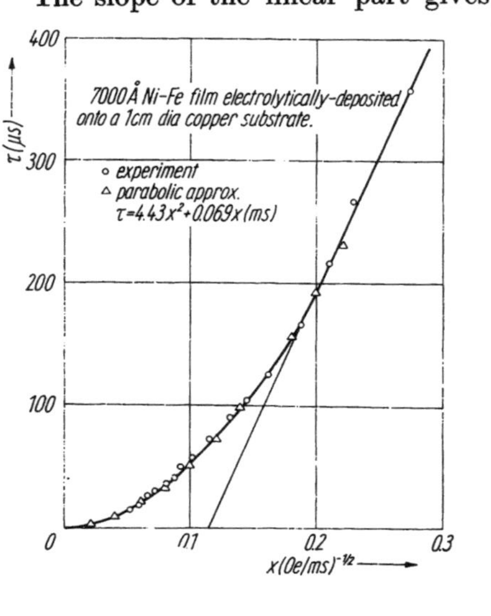


Fig. 4. Reversal time characteristic of an electrolytically-deposited thin magnetic film

An experimental plot (after [1]) of the variation of the dynamic easy-axis coercivity with the rate of rise of the applied field for various values of dc hard direction field is shown in Fig. 5. The coercivity is measured directly from the oscilloscope trace of the displayed $B$-$H$ loop of the sample using an apparatus which is described in detail in [2].

As it can be predicted from the assumed model (equation (8)), $H_{cd}$ is a linear function of $k$. The slopes of the curves plotted at different values of dc hard direction field are almost identical. This indicates that the simple domain

Fig. 5. Easy-axis dynamic coercivity dependence of a vacuum-deposited thin magnetic film

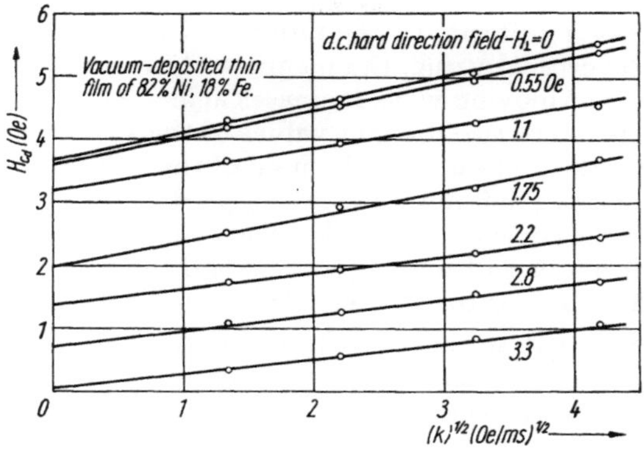


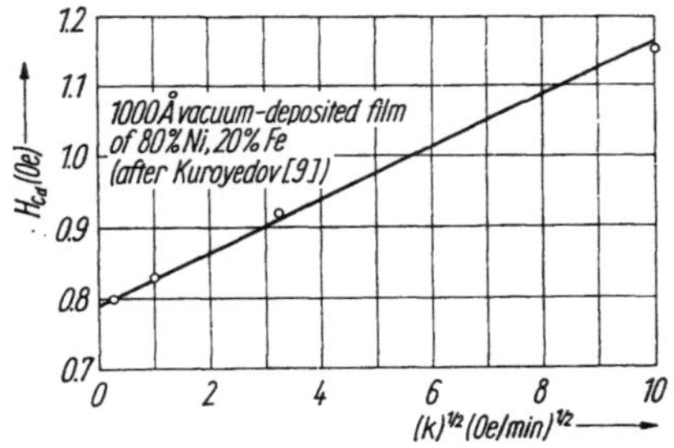


Fig. 6. Easy-axis dynamic coercivity characteristic of a vacuum-deposited thin film measured in the range of low rate of rise of the applied field (after Kuroyedov [9])

rotation does not occur since this would imply zero slope. Simultaneous application of a triangular driving field in the easy direction and of a dc field in the hard direction should cause the magnetization vector to rotate from one state of equilibrium to the other. The experiment shows that this is not the case.

Extrapolation of these curves, made to obtain values of $H_{cd}$ under conditions of zero rate of rise of applied field gives the static values of coercivity.

The above effects have been observed with vacuum deposited magnetic films of 5 mm diameter, composition 82% Ni, 12% Fe, in the thickness range 500 to 2500 Å.

The influence of the rate of rise of the applied field on the coercivity was investigated also by Kuroyedov [9]. It is interesting for the author that the experiment made in [9] was carried on in the range of a relatively low rate of rise (from 0.05 to 100 Oe/min) as compared to the values of $k$ at which the presented measurements have been made. The experimental curve, being originally plotted as $H_{cd} = f(k)$ is shown in Fig. 6 as a function $H_{cd} = f(\sqrt{k})$. As it is seen, this function is still linear even for the range of a very low rate of rise of the applied field.

## 4. Summary

Thin anisotropic films, (sufficiently thick, i.e.: $D > 500$ Å), which are magnetized in the easy direction by a field increasing linearly with time change their state of magnetization mainly by the movement of domain walls. The motion of individual walls is uniformly accelerated since the acceleration is proportional to the rate of rise of the applied field $k$ and depends on the properties of the sample. In the range of $k$, from $k = 0$ to $k = k_c$, the latter defined as the critical value of the rate of rise depending also on the film properties, the reversal process is uniform, i.e the number of moving walls is constant. If $k$ increases above the critical value, the magnetization reversal become non-uniform, i.e. the number of moving walls increases achieving the maximum possible value at $k = \infty$. There are two distinct values of the switching coefficient: $S_{w_2}$, which characterizes the highest possible speed of the switching process by the domain-wall motion, and $S_{w_1}$, defining the lowest speed of the reversal process which can be achieved in the considered sample. The ratio of both values gives a measure of the increase in number of moving walls when the rate of rise of the applied field is changed from $k = 0$ to $k = \infty$. The numerical values of $S_{w_1}$ and $S_{w_2}$ can be evaluated from a time reversal characteristic plotted along the variable $x = 1/\sqrt{k}$.

The easy-axis dynamic coercivity $H_{cd}$, defined as a half of the displayed $B$-$H$ loop width, increases with the increase of the rate of rise of the applied field. In the range of $k$, within which the reversal process is uniform ($0 < k < k_c$), $H_{cd}$ is proportional to $(k)^{1/2}$. Extrapolation of the curve $H_{cd} = f(k)^{1/2}$, made to obtain the value of coercivity under the conditions of zero rate of rise of the applied field, gives the static value of the coercive force.

Additionally the experiments show that, even under conditions in which a uniform rotation is expected to occur, domain wall motion is the resulting way by which a magnetization change takes place.

## References


[1] G. R. HOFFMAN, J. A. TURNER, and H. K. LACHOWICZ, J. appl. Phys. **34**, 2708 (1963).
[2] H. LACHOWICZ, J. A. TURNER, and G. R. HOFFMAN, Arch. Elektrotechn. **11**, 798 (1962).
[3] M. PRUTTON, Thin Ferromagnetic Films, Butterworths, London 1964.
[4] L. LANDAU and E. LIFSHITS, Phys. Z. Sowjetunion **8**, 153 (1935).
[5] K. J. SIXTUS, Phys. Rev. **48**, 425 (1935).
[6] H. LACHOWICZ, Arch. Elektrotechn. **15**, 191 (1966).
[7] H. LACHOWICZ, Arch. Elektrotechn., in print (1966).
[8] W. B. HATFIELD, J. appl. Phys. **36**, 2662 (1965).
[9] K. A. KUROYEDOV, Izv. Akad. Nauk SSSR, Fiz. Ser. **29**, 599 (1965).


*(Received May 9, 1966)*

phys. stat. sol. **16**, 321 (1966)

*Department of Physical Chemistry, University of Nottingham*

# Thermally Activated Domain Wall Movement[1])

By

D. J. Craik and M. J. Wood

Magnetization reversal by slow domain wall motion under constant applied field has been observed by the Faraday effect in permalloy films. A quantitative study (Craik and Wood 1965) of the dependence of wall velocity on temperature and field has been made in the temperature range −47 to +22 °C. The linear relationship between the field $H$ and the logarithm of velocity is consistent with previous work, but in addition the variation of the slope of this line with temperature is noted. The analysis of the results is based on Stacey's theory for uniform potential barriers, and an activation volume deduced which is approximately equal to the cube of the film thickness. An estimate is also made of the frequency factor incorporated in the theory, which is consistent with theoretical values. The validity of these calculations is discussed.

Magnetisierungsumkehr durch langsame Bewegung von Domänenwänden bei konstantem äußeren Feld wird mit Hilfe des Faraday-Effekts an Permalloyschichten beobachtet. Eine quantitative Untersuchung (Craik und Wood 1965) der Abhängigkeit der Wandgeschwindigkeit von der Temperatur und dem Feld wurde im Temperaturbereich von −47 bis +22 °C durchgeführt. Die lineare Beziehung zwischen dem Feld $H$ und dem Logarithmus der Geschwindigkeit stimmt mit der früheren Arbeit überein, zusätzlich ändert sich aber der Anstieg der Kurve mit der Temperatur. Die Auswertung der Ergebnisse beruht auf einer Theorie von Stacey für gleichmäßige Potentialbarrieren. Es wird ein Aktivierungsvolumen von ungefähr der dritten Potenz der Schichtdicke hergeleitet. Eine Abschätzung des in der Theorie enthaltenen Frequenzfaktors befindet sich in Übereinstimmung mit dem theoretischen Wert. Die Gültigkeit dieser Rechnungen wird diskutiert.

## 1. Introduction

If the magnetic field being applied to a ferromagnetic material is suddenly changed, the corresponding change in magnetization is not instantaneous. This magnetic "after effect" is of two kinds at temperatures where no magnetic annealing occurs; a short term process in which domain wall velocities are limited by eddy currents, and a much longer term effect which can be accounted for by thermal activation. The process of magnetization reversal involves the surmounting of a series of energy barriers, giving rise to discrete changes of magnetization, or Barkhausen jumps. The necessary energy derives partly from the applied field and partly from thermal fluctuations, giving rise to time dependent effects (magnetic viscosity).

This paper presents the results of some studies of wall velocities in NiFe films, and attempts to correlate them with existing theories of viscosity.

## 2. Theories of Magnetic Viscosity

Almost any theory of magnetic viscosity begins with a formula of the type

$$P = 1/\tau = C \cdot \exp\,[-E/kT]\,, \tag{1}$$

[1]) Contribution to the International Colloquium on Magnetic Thin Films, 25th to 28th April 1966 in Jena.

where $P$ is the probability per unit time that an activation process will take place, $E$ is the activation energy, $\tau$ the relaxation time for the process, and $C$ is a frequency factor. Different theories give different values for $C$, but it is of order $10^9$ to $10^{12}$ s$^{-1}$ [1, 2, 3].

In films two basic types of model have been considered, corresponding to two different mechanisms of magnetization reversal. In the thickness range greater than a few hundred Angstroms, wall motion is known to be responsible for switching. For the ideal case in which the film is "homogeneous" i.e., if its properties, in particular the coercive force, are the same in all parts of the film, then the energy barriers will be of equal height, the hysteresis loop will be extremely square, and under a constant field the domain wall will move with constant velocity. If, further, the wall does not change in length, the rate of change of magnetization remains constant.

This type of switching has been considered by Stacey [4], who derives an equation for the wall velocity $v$ of the form

$$\ln v = \ln C\,d + \frac{2\,I_s\,V}{k\,T}\,(H - H_0)\,, \tag{2}$$

where $d$ is the distance moved by the wall between barriers, $H$ the applied field, $H_0$ the coercive force at absolute zero, and $V$ the volume of material involved in one activation process (not to be confused with the volume of the subsequent jump).

If the film is very thin, Wohlfarth [5] has suggested that it may approximate to an assembly of single domain particles. Magnetization reversal occurs in this case by rotation against magnetocrystalline or shape anisotropy, and the application of Néel's treatment becomes possible [2]. In contrast to Stacey's model, the rate of change of magnetization is not constant, but decreases exponentially with time.

On account of its appearance in the exponent, the volume $V$ has a large influence on the relaxation time, both for wall motion, where $V$ represents the volume of material forming the energy barrier, and rotation in single domain particles, where $V$ is the effective particle volume. If $V$ is large, thermal activation will have little effect except when $H$ is very close to $H_0$, and observation of any magnetic after effect becomes difficult. On the other hand, if $V$ is small the barriers may be readily overcome even at $H = 0$. A change of $V$ by factor of 3 can change the relaxation time from a fraction of a second to many years. The range of particle size over which observation of magnetic viscosity is possible is thus small.

## 3. Survey of Previous Work

Slow domain wall movement with uniform velocity has been observed in NiFe films by Olmen and Mitchell [6]. They obtained a relationship between velocity and field of the form (2), but little information could be drawn about the temperature variation. However, Stacey [4] applied his theory to their results and obtained an estimate of $C\,d$ of $10^{10}$ cm/s.

Similar results have been obtained on cobalt films by Spivey [7]. In this case, however, the wall velocity was not constant, but continually decreased, and instead of velocity, the time $t^*$ to reach $I = 0$ was noted. These results further yielded a linear relationship between $H$ and $T$ for a given switching time, as in (2), but the slope of the $\ln t^*$ versus $H$ curve was found to be independent of temperature.

Hellenthal has studied slow switching in very thin nickel films [8, 9]. Again the rate of change of magnetization continually decreased. Contrary to expectation, he finds that it is ln $H$ which varies linearly with ln $t^*$. However, Wohlfarth [5] has analysed some of his data [8] on the basis of Néel's theory, and calculates an effective particle size (activation volume). This turns out to be much larger than the actual size, a discrepancy he attributes to the interactions between crystallites.

The influence of particle size is clearly demonstrated by Hellenthal [9] where, in very thin films, one might expect the crystallite size to be a function of film thickness. Thermal activation becomes increasingly effective as the film thickness is decreased, as shown by a decrease in the switching field for a given time $t^*$ and an increase in the variation of $H$ with $t^*$.

## 4. Techniques

The apparatus for observing domain structures and recording the magnetization has previously been described in detail [10]. The magnetization, as revealed by the Faraday effect, was detected by a photomultiplier, and a continuous record of magnetization with time displayed on a pen recorder.

Heating and cooling was effected by circulating respectively warm water from a thermostated bath, or nitrogen vapour from the boiling liquid, through a flat brass tube on which the specimen was placed. In order to avoid the formation of frost on the specimen, a well lagged chamber was constructed, extending from the condenser to the objective lenses of the microscope, so that the possible misting up of windows was eliminated. This was filled with dry nitrogen prior to cooling.

The construction of the specimen chamber was satisfactory in that no frost was observed to form even at the lowest temperatures, but unfortunately fluctuations in the transmitted intensity began to appear below about −50 °C. These seemed to be due to convection currents occurring within the specimen chamber, and so far have prevented observations of slow switching below this temperature. The temperature could be controlled to within about 3 °C.

Magnetic fields were produced by a pair of coils energized from a stabilized power supply. A potentiometer was employed for measurement of the current, enabling fields to be measured to 0.001 Oe. The fields were calibrated by a Hall effect gaussmeter and were uniform to within 0.15% over the region occupied by the specimen.

## 5. Results

### *5.1 The I versus t curve*

All films studied so far have been in the thickness range 500 to 1500 Å, and in all cases the magnetization reversed by wall motion. Thus, in the ideal case, where the film is perfectly uniform, Stacey's treatment should be applicable.

In fact, very few films which have been grown approach this ideal case of uniform energy barriers. Because the wall velocity depends exponentially on $(H - H_0)$, only a small change of $H_0$ is needed to give a large change in wall velocity. Thus under a constant field, walls will move into the film from a region where $H_0$ is lowest, will initially have a high velocity, but will slow down and possibly effectively stop altogether as they move to another part of the film.

The importance of obtaining "homogeneous" specimens was briefly mentioned by Olmen and Mitchell [6], but does not seem to have received attention elsewhere. It would seem that there is always a danger that a non-uniform switching speed could be ascribed to lack of homogeneity of the film, rather than having any intrinsic connection with the mode of reversal.

In all present work, specimens were 4 mm in diameter, evaporated onto glass cover slips from an 80:20 NiFe melt, in a vacuum of about $10^{-5}$ Torr. Several possible causes of the inhomogeneities are:

i) Long range thickness variations. This possibility can be readily dismissed by a simple calculation based on the geometry of the evaporating system, for any reasonable estimate of the thickness dependence of coercivity.

ii) The existence of random stresses, set up as the film cools down from its deposition temperature. In this case, greater homogeneity should be observed for films either grown at room temperature, or having the zero magnetostriction composition, neither of which was observed.

iii) Inhomogeneities in the substrate. The cover slips were teepol washed, then rinsed in ethanol, and baked at 400 to 500 °C prior to evaporation, but the absence of baking did not increase the inhomogeneity of the film.

The exact cause of the inhomogeneity is still unknown, although insufficient substrate preparation seems most likely. However, some films have been obtained in which the domain wall velocity was constant over a sufficiently large region for quantitative measurements to be made, and the results for two such films are given below.

### *5.2 Slow domain wall movement under constant applied field*

Magnetization curves for two films, A and B, with constant applied field, were recorded for a range of switching times from three seconds, a limitation imposed by the time constant of the apparatus, to about half an hour, although

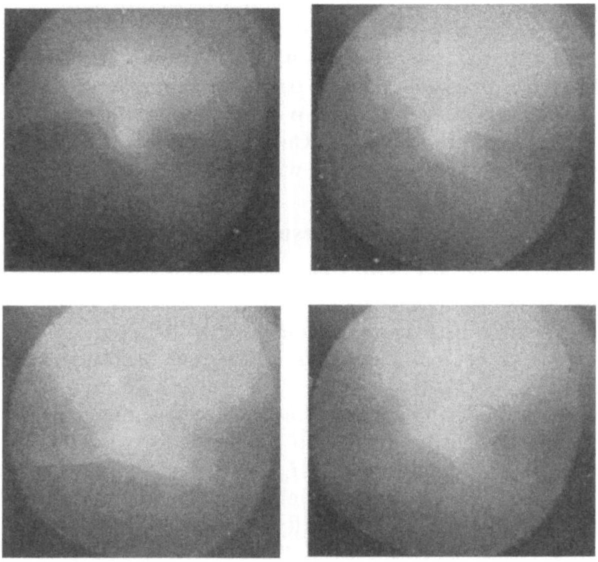

Fig. 1. Domain wall movement under constant applied field in a nickel-iron film

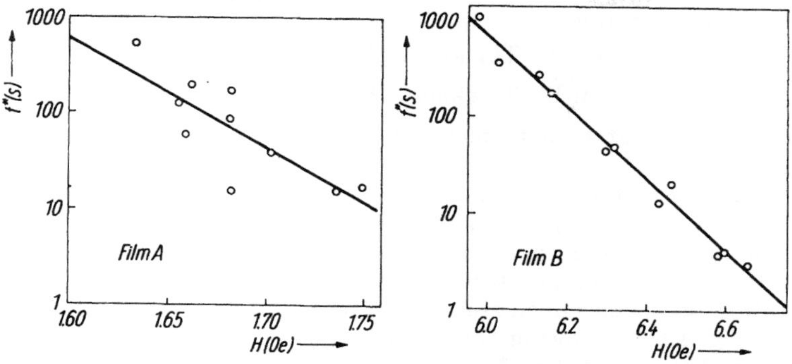


Fig. 2. Dependence of switching time on applied field in NiFe films at room temperature

in theory no upper limit exists. Fig. 1 illustrates the mode of reversal of film A. A plot of $H$ against log $t^*$, where $t^*$ now represents the time to reverse a given area of film, is shown in Fig. 2a and b for the room temperature measurements, where a linear relationship, for the film B anyway, is clearly indicated. The much larger scatter for film A is not due to experimental error, and can be ascribed to differences in the detailed modes of reversal. Although in both films the *average* wall velocity was constant, the magnetization in film B was reversed in very small jumps, unresolvable at the low magnification used, whereas in film A the wall was seen to move in a series of large jumps. This is illustrated in Fig. 3. Since we are dealing with a statistical effect here, only an average over a large number of energy barriers of equal height can lead to a $t^*$ which is a smoothly varying function of $H$.

For the purpose of the present analysis, equation (2) will be written

$$\log t^* = -\log C\,m\,p + \frac{2\,I_s\,V}{2.303\,kT}(H_0 - H)\,, \tag{3}$$

where $m$ is the change in magnetization per jump, expressed as a fraction of the total change over the area under consideration, and $p$ is the number of effectively independent sections of wall.

Room temperature measurements alone cannot determine $H_0$, but from equation (3) one can evaluate $V$ from the slope of the graph in Fig. 2. For the two

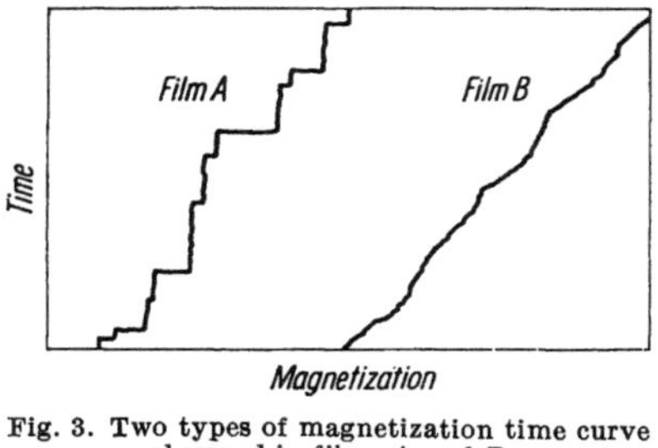


Fig. 3. Two types of magnetization time curve observed in films A and B

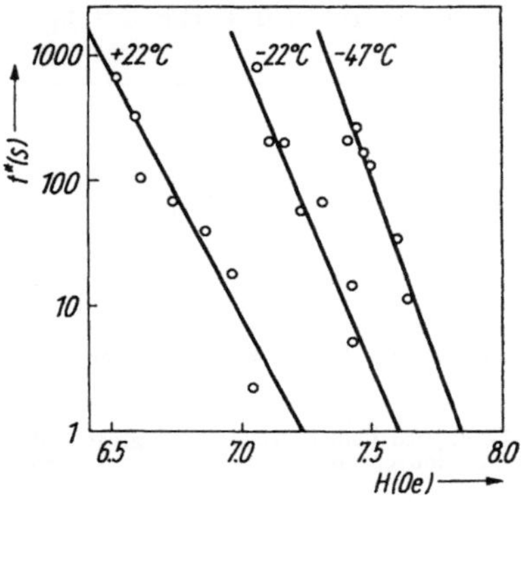


Fig. 4. Dependence of switching time on applied field in a NiFe film (film B) at various temperatures

films A and B we obtain

$$\text{Film A: Thickness } 1100\ \text{Å},\ V = (910\ \text{Å})^3\,,$$

$$\text{Film B: Thickness } 670\ \text{Å},\ V = (620\ \text{Å})^3\,.$$

One notes a qualitative agreement at least between film thickness and activation volume.

Curves similar to those in Fig. 2 have also been plotted at different temperatures for film B (Fig. 4). Only for three representative temperatures were results obtained, so the conclusions can only be regarded as tentative. The increase in the room temperature coercive force over that in Fig. 2 was due to deterioration of the specimen since those measurements were taken. From the results in Fig. 4 the following observations can be made:

i) The gradient $\mathrm{d}H/\mathrm{d}\,(\log t^*)$ decreases in magnitude with decreasing temperature. Fig. 5 shows this variation, and although a linear relation is possible within the limits of experimental error, little conclusive support of equation (3) can be drawn from this graph.

ii) $H_0$ can be estimated by plotting $H$ against $T$ for a fixed value of $t^*$, and extrapolating to $T = 0$. This has been done for two values of $t^*$ (Fig. 6), giving

$$H_0 = 10.1\ \text{Oe (using } \log t^* = 3)\,,$$

$$H_0 = 9.9\ \text{Oe (using } \log t^* = 0)\,.$$

iii) The slope of the graph in Fig. 6 can be used to give a value of $Cmp$. Taking an average of the two curves we obtain

$$\log Cmp = 10.5\,.$$

The quantities $m$ and $p$ were not measured, but $m$ must be less than 0.005, this being the limit of resolution of the apparatus. $p$ can be estimated from the size of the activation volume, assuming that this is spherical, which, for a wall length of 3 mm, gives $p = 5 \times 10^4$. Hence, one obtains a lower limit for $C$ of $10^8\ \mathrm{s}^{-1}$ which is consistent with theoretical considerations.

It should be noted that the deductions made from equation (3) assume that $V$ is independent of $H$, an assumption made by Stacey. In fact, the equilibrium position of the wall relative to an energy barrier will alter as the applied field is altered. The nature of this variation depends on the exact form of the energy barriers, which is unknown. Over the small range of experimental points it is probably safe to assume it is constant, and the linearity of the graphs in Fig. 2 and 4 bears this out, but it must be born in mind that the estimates of $V$, $H_0$, and $C$ may be in error if $\mathrm{d}V/\mathrm{d}H$ is large. It is interesting to note a slight de-

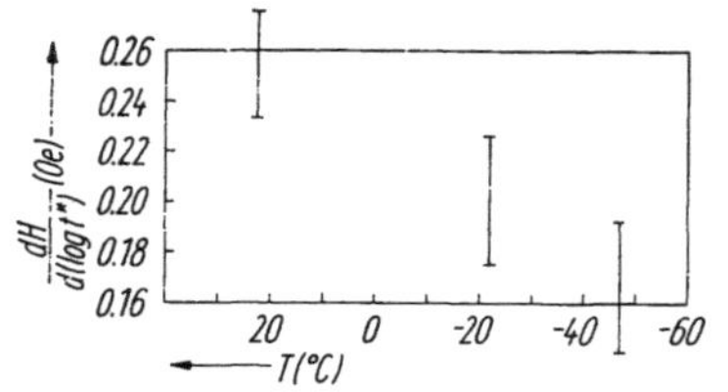


Fig. 5. Slope of curves in Fig. 4 as a function of temperature

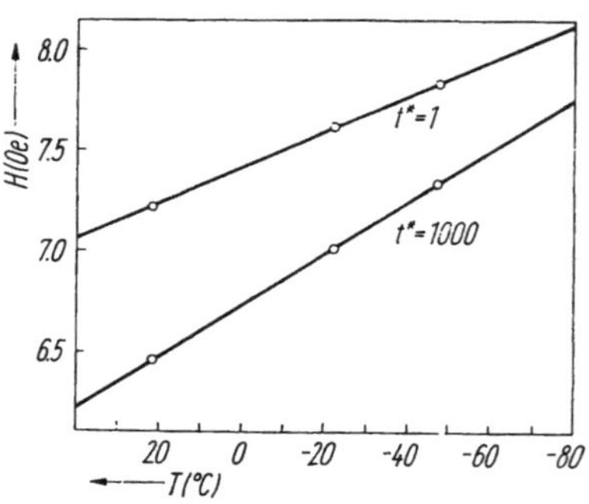


Fig. 6. The applied field necessary for a given switching time as a function of temperature in a NiFe film (film B)

parture from linearity over a large range of $t^*$, quoted by Bates et al. [11], which might possibly be accounted for in this way.

## 6. Validity of the *H* versus *T* Curve

A knowledge of the variation of the relevant parameters with temperature is essential to any phenomenon involving thermal activation processes. In the present case, the height of the energy barriers ($H_0$) and hence the frequency factor $C$, can only be determined from a plot of switching field against temperature. It is thus important to consider whether the observed variation can be attributed to thermal activation alone, or whether there exist any other parameters, whose temperature variation would likewise lead to a change in coercive force, i.e., to a change in the barrier height $H_0$ itself.

The question thus arises; on what parameters does the coercive force depend? It is difficult to supply an adequate answer to this problem, since little is known about the origin of the wall coercive force in films. One obvious temperature dependent parameter is the saturation magnetization $I_s$. The Curie point of permalloy is about 900 °K, and the range of study in the present case is around 200 to 300 °K. An inspection of the $I_s$ versus $T$ curve for ferromagnetic materials shows that the temperature variation of magnetization can be neglected here.

Another possible temperature-sensitive parameter is the anisotropy constant $K$. Indeed, in the case of coherent rotation, the observed coercive force would be $2\,K/I_s$. In Fig. 7 the anisotropy field $H_k$ as measured by the hard direction susceptibility, is plotted against temperature for two films. Although the errors in measurement are large the results indicate a variation of $H_k$ which is significant when compared with Fig. 6. Thus, if the anisotropy plays a predominant part in wall coercivity, no information about thermal activation mechanisms can be obtained from the temperature dependence of the switching field.

The presence of a wall coercive force derives from the variation of wall energy as it moves across the film. In films in the thickness range greater than about 50 Å, the anisotropy contribution to the total wall energy is negligible as compared with the magnetostatic energy, which latter gives rise to a term proportional to $I_s$ [12]. This condition applies to either Bloch or Néel walls. One might then expect that variations of $H_0$ with temperature would follow those of $I_s$ and thus be negligible. However, Middelhoek [12], points out that due to the anisotropy dispersion, the effective value of $I_s$ which must be included in the wall energy will vary from crystallite to crystallite, so that the anisotropy constant, in as much as it affects the dispersion, will enter into the formulae for the coercive force. The actual form of the dependence is difficult to assess;

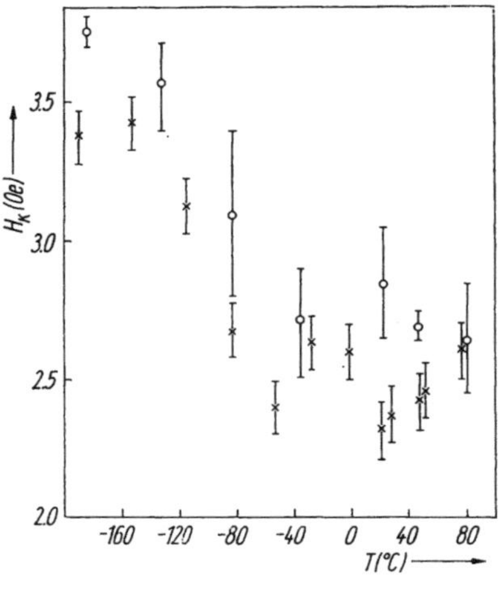


Fig. 7. Dependence of anisotropy field on temperature in two NiFe films. Circles and crosses differentiate between the two films

Middelhoek calculates that the coercivity due to this mechanism varies as $1/K^2$.

However, it is unlikely that this is the only origin of coercivity. The wall energy may also be affected by the presence of inclusions or film thickness variations, and the influence of $H_k$ on coercivity can really only be assessed experimentally. Middelhoek varied $H_k$ by bending a film, and found a dependence of $H_c$ which, for the films shown in Fig. 7, would give a negligible contribution to the total coercivity change. Further, a large number of films have been grown in this laboratory, and although any individual film may show large departures from the average behaviour, consideration of all films failed to indicate any correlation between $H_c$ and $H_k$.

The above observations seems to indicate that the influence of $H_k$ on the temperature dependence of coercivity is probably small, although the evidence is by no means conclusive.

## References

[1] F. D. Stacey, Proc. Phys. Soc. **73**, 136 (1959).
[2] L. Néel, Ann. Geophys. **5**, 99 (1949).
[3] W. F. Brown, J. appl. Phys. **30**, 130 S (1959).
[4] F. D. Stacey, Austral. J. Phys. **13**, 599 (1960).
[5] E. P. Wohlfarth, J. Electronics and Control **10**, 33 (1961).
[6] R. W. Olmen and E. N. Mitchell, J. appl. Phys. **30**, 258 S (1959).
[7] Sylvia M. Spivey, Thesis, University of Nottingham 1963.
[8] W. Hellenthal, Naturwissenschaften **47**, 371 (1960).
[9] W. Hellenthal, Z. Phys. **184**, 39 (1965).
[10] D. J. Craik and M. J. Wood, J. sci. Instrum. **42**, 410 (1965).
[11] L. F. Bates, D. J. Craik, and S. M. Rushton, Proc. Phys. Soc. **80**, 768 (1962).
[12] S. Middelhoek, Thesis, University of Amsterdam 1961.

*(Received May 9, 1966)*

phys. stat. sol. 16, 329 (1966)

*Institute of Atomic Physics, Bucharest*

# Influence of the Perpendicular and Parallel Anisotropy on the Ferromagnetic Properties of Thin Films[1])

By

A. CORCIOVEI and G. COSTACHE

A calculation is given of the dependence of the spontaneous magnetization in a thin film on the temperature and the number of monoatomic layers in the film. The dependence of Curie temperature on the number of monoatomic layers is also calculated, taking into account the effects of perpendicular and parallel magnetic anisotropy. This work is an extension of a previous theoretical study by Corciovei.

Die Abhängigkeit der spontanen Magnetisierung von der Temperatur und der Zahl der monoatomaren Ebenen in dünnen Schichten und die Abhängigkeit der Curietemperatur von der Zahl der monoatomaren Ebenen werden unter Berücksichtigung senkrechter und paralleler magnetischer Anisotropie berechnet. Die Arbeit ist eine Erweiterung einer früheren theoretischen Untersuchung von Corciovei.

## 1. Introduction

Theoretical studies of single domain thin films have been carried out using the Heisenberg trace method at high temperatures and the Bloch spin wave method at low temperatures.

Valenta [1] and Corciovei [2, 3] applied the trace method and Klein et al. [4, 5] and Döring [6] applied the spin wave method in the Bloch formulation. Corciovei [7] applied the spin wave theory in the Holstein-Primakoff formulation. Recently, the Green function method expressed its advantages in the study of ferromagnetic thin films [8, 9].

In the present paper we shall attempt to develop further a previous theoretical study by Corciovei [3]. We shall calculate the dependence of the spontaneous magnetization on temperature and number of monoatomic layers of thin films. The dependence of Curie temperature vs. number of monoatomic layers will be also calculated, taking into account the influence of magnetic anisotropy on the magnetic properties of thin films.

For this purpose, a magnetic anisotropy term, which includes the possibility both of a parallel and perpendicular anisotropy, was introduced in the Hamiltonian of the system. We shall use the trace theory of Heisenberg, as it was applied in [2] and [3].

## 2. Matrix Elements of the Hamiltonian

The values of all projections of the spins on the magnetization axis completely determine the quantum state of the atomic spin system. All atoms are supposed to have spins 1/2. Hence, we may label a quantum state $|n\rangle$ by the set $n$ of

---

[1]) Contribution to the International Colloquium on Magnetic Thin Films, 25th to 28th April 1966 in Jena.

the order numbers of all the atoms having their spins directed upwards.

We assume that the ferromagnetic thin film consists of $q$ monoatomic layers, parallel to the surface of the film, $N$ atoms being in each layer. Let $r_i$ be the number of atoms in a layer $i$, having their spin directed upwards. Then, the total spin of the layer $i$ will be

$$m_i = \frac{1}{2} r_i - \frac{1}{2} (N - r_i) = r_i - \frac{1}{2} N . \tag{1}$$

By specifying all quantities $m_i$ ($i = 1, 2, \ldots, q$), the thermodynamic state of the film is determined.

We shall calculate the partition function for a given set of numbers $m_i$ which we shall simply denote by $m$. This can be done by computing the traces of the Hamiltonian of the system for a given $m$.

Firstly, we note that the number of microscopic states corresponding to a given thermodynamic state $m$ is given by

$$f(m) = f(m_1, \ldots, m_q) = \binom{N}{r_1} \cdots \binom{N}{r_q} = \prod_{i=1}^{q} \binom{N}{m_i + \frac{1}{2} N} . \tag{2}$$

Now, we must know the matrix elements of the Hamiltonian between two states $|n\rangle$ and $|n'\rangle$. The Hamiltonian contains two terms: the exchange term $H^{\text{ex}}$ and the anisotropy term $H^{\text{anis}}$:

$$H = H^{\text{ex}} + H^{\text{anis}} , \tag{3}$$

where $H^{\text{ex}}$ is the well-known Heisenberg spin Hamiltonian:

$$H^{\text{ex}} = -\frac{1}{2} \sum_{k, k'} A_{k k'} (1 + \sigma_k \sigma_{k'}) . \tag{4}$$

Here $A_{kk'}$ is the exchange integral of the atoms $k$ and $k'$. The diagonal matrix elements of the exchange term are the following:

$$H^{\text{ex}}_{nn} = - \Sigma^* A_{k k'} , \tag{5}$$

where we indicate by $\Sigma^*$ that the summation must be performed over atoms $k$ and $k'$ so that either $k, k' \in n$ or $k, k' \in C\, n$. (We shall adopt the notation used in [2].) The nondiagonal matrix elements $H^{\text{ex}}_{nn'}$ are

$$H^{\text{ex}}_{nn'} = - A_{k k'} \tag{6}$$

when $n$ differs from $n'$ by the interchange of spins $k$ and $k'$ and, otherwise, equals zero.

Let us now study the anisotropy term. We shall work with an anisotropy Hamiltonian of the form

$$H^{\text{anis}} = - K_{\perp} \sum_{k, k'} S^x_k S^x_{k'} - K_{||} \sum_{k, k'} S^z_k S^z_{k'} , \tag{7}$$

where the axes of anisotropy are $0x$ and $0z$ (we choose such a coordinate system that the $0x$ axis is perpendicular to the surface of the thin film) and $K_{\perp}$, $K_{||}$ are the anisotropy constants. The summation must be performed only over nearest neighbours.[2])

If we denote by $A$ the exchange integral between two neighbours of first order, we can introduce the quantities

$$\eta_{\parallel} = \frac{K_{\parallel}}{A}, \qquad \eta_{\perp} = \frac{K_{\perp}}{A}. \tag{8}$$

Now we are able to calculate the matrix elements of the anisotropy term. The diagonal matrix elements are

$$H_{nn}^{\text{anis}} = -\frac{K_{\parallel}}{4} \sum_{k,k'} \langle n|\, \sigma_k^z \sigma_{k'}^z \,|n\rangle . \tag{9}$$

We know that

$$H_{nn}^{\text{ex}} = -\sum_{k,k'} A_{kk'} - \frac{1}{2} \sum_{k,k'} A_{kk'} \langle n|\, \sigma_k^z \sigma_{k'}^z \,|n\rangle$$

or

$$H_{nn}^{\text{ex}} = -\frac{A\,N}{4}\Big(\sum_i z_{ii} + \sum_{i \neq j} z_{ij}\Big) - \frac{A}{2} \sum_{k,k'} \langle n|\, \sigma_k^z \sigma_{k'}^z \,|n\rangle ,$$

where $z_{ij}$ denotes the number of neighbours of first order in layer $j$ of any atom in layer $i$.

So we find that

$$H_{nn}^{\text{anis}} = \frac{\eta_{\parallel}}{4} \frac{A\,N}{2}\Big(\sum_i z_{ii} + \sum_{i \neq j} z_{ij}\Big) + \frac{1}{2} \eta_{\parallel} H_{nn}^{\text{ex}} . \tag{10}$$

The matrix elements between two states $|n\rangle$ and $|n'\rangle$ are non-zero only if we obtain the state $|n'\rangle$ from $|n\rangle$ by reversing two arbitrary nearest spins $k$ and $k'$, and then

$$H_{nn'}^{\text{anis}} = -\frac{K_{\perp}}{2} . \tag{11}$$

## 3. The Traces of the Hamiltonian

Let $H_m$ be the submatrix of the Hamiltonian matrix which contains all elements $H_{nn}$, where the quantum states $|n\rangle$ and $|n'\rangle$ correspond to the same thermodynamic state $m$. Similarly, we introduce $H_m^{\text{ex}}$ and $H_m^{\text{anis}}$. Then,

$$H_m = H_m^{\text{ex}} + H_m^{\text{anis}}$$

and

$$\text{tr}\, H_m = \text{tr}\, H_m^{\text{ex}} + \text{tr}\, H_m^{\text{anis}} .$$

---

[2]) Such an anisotropy term arises immediately from the generalized term of Orbach [10]. In fact, Orbach considers for the exchange term of a linear chain the expression.

$$\mathcal{H} = 2\,J \sum_{i=1}^{N} [S_i^z S_{i+1}^z + (1-\alpha)(S_i^x S_{i+1}^x + S_i^y S_{i+1}^y)]$$

in which two different constants have been chosen, because the linear chain has cylindrical symmetry. In our case, there is no such symmetry, because the magnetization axis is contained in the plane of the film. Thus, three different constants are necessary. We note that this term is similar to the corresponding term used by Corciovei in [3]. Also, we remark that, if the more usual form for the anisotropy term which introduces an internal magnetic anisotropy field would be used, the effect of the anisotropy would also arise, but by going to higher order terms in the Krikwood method.

Obviously, the trace of zero order of the $H_m$ matrix will be

$$\operatorname{tr} H_m^0 = f(m) . \tag{12}$$

The first order trace of the $H_m$ matrix is given by the expression

$$\operatorname{tr} H_m^1 = \left(1 + \frac{1}{2} \eta_{\|}\right) \operatorname{tr} H_m^{ex} + C\, f(m) , \tag{13}$$

where

$$C = \frac{\eta_{\|}}{2} \frac{NA}{4} \left( \sum_i z_{ii} + \sum_{i \neq j} z_{ij} \right) . \tag{14}$$

Thus, the mean energy is

$$\bar{E}_m = \left(1 + \frac{1}{2} \eta_{\|}\right) \overline{E_m^{ex}} + C . \tag{15}$$

The quantity $\overline{E_m^{ex}}$ was calculated in [2], formula (8).

The partition function for a determined thermodynamic state $m$, which we need, is in Kirkwood's second order approximation

$$Z_m = f(m) \exp\left(- \frac{\bar{E}_m}{k\,T}\right) \exp\left(\frac{\overline{\Delta E_m^2}}{2(k\,T)^2}\right) , \tag{16}$$

where

$$\overline{\Delta E_m^2} = \frac{\operatorname{tr} H_m^2}{\operatorname{tr} H_m^0} - \left(\frac{\operatorname{tr} H_m^1}{\operatorname{tr} H_m^0}\right)^2 .$$

By computing $H^2$ we obtain

$$H^2 = (H^{ex})^2 + (H^{anis})^2 + (H^{ex}\, H^{anis}) + (H^{anis})\, (H^{ex})$$

and

$$[(H^{anis})^2]_{nn} = C^2 + \eta_{\|}\, C\, H_{nn}^{ex} + \frac{1}{4} \eta_{\|}^2 (H_{nn}^{ex})^2 + C \frac{\eta_{\perp}^2}{\eta_{\|}} A , \tag{17}$$

$$(H^{ex}\, H^{anis})_{nn} = C\, H_{nn}^{ex} + \frac{1}{2} \eta_{\|} (H_{nn}^{ex})^2 + \frac{1}{2} \eta_{\perp}\, A \sum^{+} A_{\mu} . \tag{18}$$

A straightforward calculation leads to

$$\overline{\Delta E_m^2} = \left(1 + \frac{1}{2} \eta_{\|}\right)^2 \left\{\overline{\Delta E_m^{ex2}} + \alpha \frac{NA^2}{2} \left[ \sum_i z_{ii} \left(\frac{1}{2} - \frac{2\, m_i^2}{N^2}\right) + \sum_{i \neq j} z_{ij} \left(\frac{1}{2} - \frac{2\, m_i\, m_j}{N^2}\right)\right]\right\} , \tag{19}$$

where

$$\alpha = \frac{\eta_{\perp} - \eta_{\|} - \frac{1}{4} \eta_{\|}^2}{\left(1 + \frac{1}{2} \eta_{\|}\right)^2} . \tag{20}$$

In expression (19) we neglected the constant $C A\, \eta_{\perp}^2 / \eta_{\|}$, producing only a shift of the energy. $\overline{\Delta E_m^{ex2}}$ was calculated in [2], formula (19). Finally, we obtain

$$\overline{\Delta E_m^2} = \left(1 + \frac{1}{2} \eta_{\|}\right)^2 NA^2 \left\{ \sum_i z_{ii} \left[\frac{3}{8} - \frac{2\, m_i^2}{N^2} + \frac{2\, m_i^4}{N^4} + \alpha \left(\frac{1}{4} - \frac{m_i^2}{N^2}\right)\right] + \right.$$
$$\left. + \sum_{i \neq j} z_{ij} \left[\frac{3}{8} - \frac{(m_i + m_j)^2}{2\, N^2} + \frac{2\, m_i^2\, m_j^2}{N^4} + \alpha \left(\frac{1}{4} - \frac{m_i\, m_j}{N^2}\right)\right]\right\} . \tag{21}$$

If we take $\eta_{\|} = \eta_{\perp} = 0$, we obtain the expression (19) from [2] and for

$\eta_{\|} = 0$, $\eta_{\perp} = \alpha$ we obtain the expression (15) from [3]. According to (15) and [2], formula (8), we get

$$\overline{E}_m = -\left(1 + \frac{1}{2}\eta_{\|}\right) \times$$

$$\times \frac{NA}{2}\left\{\sum_i z_{ii}\left[\left(\frac{1}{2} + \frac{2\,m_i^2}{N^2}\right) + \left(-\frac{1}{2\,N} + \frac{2\,m_i^2}{N^3}\right)\right] + \sum_{i \neq j} z_{ij}\left(\frac{1}{2} + \frac{2\,m_i\,m_j}{N^2}\right)\right\} + C\,. \quad (22)$$

Thus, we see that the anisotropy term gives a significant contribution to the partition function due to parallel anisotropy, even in first order. Perpendicular anisotropy gives a significant contribution only in second order approximation of the partition function.[3])

## 4. The Equations of Magnetization

By equating to zero the expressions $\partial \log z_m/\partial m_i$ and using Stirling's formula

$$\log n! = n\,(\log n - 1)$$

after some simple transformations the following system of implicit equations is obtained:

$$y_i = \tanh\left\{\frac{1}{2}\beta\left(1 + \frac{1}{2}\eta_{\|}\right)\sum_j z_{ij}\,y_j - \frac{1}{4}\beta^2\left(1 + \frac{1}{2}\eta_{\|}\right)^2\sum_j z_{ij}\,(y_i + y_j + \alpha\,y_j - y_i\,y_j^2)\right\}, \quad (23)$$

where $y_i = m_i/(N/2)$ is the relative magnetization of the layer $i$ and $\beta = A/k\,T$. Equations (23) enable us to find out how the magnitudes $y_i$ depend on temperature and on $\eta_{\|}$, $\eta_{\perp}$.

The relative spontaneous magnetization of the film is given by

$$y = \frac{1}{q}\sum_{i=1}^{q} y_i \quad (24)$$

and we see that this is a function of temperature, number of layers $q$, and anisotropy constants.

If one wants to obtain the Curie temperature, one has to consider the equations

$$y_i = \frac{1}{2}\beta\left(1 + \frac{1}{2}\eta_{\|}\right)\sum_j z_{ij}\,y_i - \frac{1}{4}\beta^2\left(1 + \frac{1}{2}\eta_{\|}\right)^2\sum_j z_{ij}\,(y_i + y_j + \alpha\,y_j) \quad (25)$$

$$(i = 1, 2, \ldots, q)$$

and to require that the system of equations has nonvanishing solutions. The Curie temperature will depend on the number $q$ of layers and on $\eta_{\perp}$ and $\eta_{\|}$.

When $q$ is very great the thin film becomes a solid crystal. In this case $y_i = y$ for any $i$, and the system (24) reduces to a single equation:

$$y = \frac{1}{2}\beta\left(1 + \frac{1}{2}\eta_{\|}\right)z\,y - \frac{1}{2}\beta^2\left(1 + \frac{1}{2}\eta_{\|}\right)^2\left(1 + \frac{1}{2}\alpha\right)z\,y \quad (24')$$

wherefrom we obtain for $\eta_{\|} = \eta_{\perp} = 0$ the well-known Heisenberg relation

$$\beta = \frac{1}{2} - \frac{1}{2}\sqrt{1 - \frac{8}{z}}\,. \quad (26)$$

[3]) Our result is thus in agreement with the previous paper of Corciovei [3].

For crystallized cobalt in a hexagonal closed packed lattice, with the normal to the surface given by the crystal direction (0001), we have $z = 12$. Then $\beta$ has the value 0.2113.

Let us consider now the case that our thin film has only a single layer. In this case the system (25) becomes

$$y_i = \frac{1}{2}\beta\left(1 + \frac{1}{2}\eta_{\|}\right) z_{ii}\, y_i - \frac{1}{4}\beta^2\left(1 + \frac{1}{2}\eta_{\|}\right)^2 (2 + \alpha)\, z_{ii}\, y_i$$

wherefrom we obtain

$$\beta = \frac{1}{(2+\alpha)\left(1 + \frac{1}{2}\eta_{\|}\right)} - \frac{1}{(2+\alpha)\left(1 + \frac{1}{2}\eta_{\|}\right)}\sqrt{1 - \frac{8}{z_{ii}}\left(1 + \frac{1}{2}\alpha\right)}.$$

We know that $z_{ii} = 6$. In order to obtain a real $\beta$ it is necessary that

$$\alpha < -\frac{1}{2}.$$

In [4] $\eta_\perp$ was denoted by $\alpha$. In that case it was not possible to have $\alpha < -{}^1/_2$. Let us see whether in our case a solution of the inequality $\alpha < -{}^1/_2$ is possible, i.e. whether spontaneous magnetization occurs in a single layer. From $\alpha < -{}^1/_2$ we obtain that ferromagnetism exists even for a single layer if

$$\eta_{\|} \geqq 2\left(\sqrt{2\,(1 + \eta_\perp)} - 1\right). \tag{27}$$

If $\eta_\perp = 0$, then $\eta_{\|} \geqq 0.82$.[4]) This is the lowest value admitted for $\eta_{\|}$. Evidently, this value is too large for being acceptable and we can conclude that ferromagnetism does not exist for a single layer.

## 5. The Curie Temperature of Thin Films

The Curie temperature $T_c^{(\infty)}$ of the solid crystal without anisotropy field is given by (26). In the case $z = 12$, we obtain $\beta = 0.2113$. If the relative temperature $x = T/T_c^{(\infty)}$ is introduced, we get

$$\beta = \frac{0.2113}{x}. \tag{28}$$

The magnitude $T_c^{(\infty)}$ must be considered as equal to the Curie temperature of the solid metal without anisotropy field. If $\beta$, given by (28), is introduced in the determinant of the coefficients of the system (25), an equation of degree $2q$ in $x$ is obtained. Using the electronic computer CET 500, the solutions were calculated up to $q = 11$. We took only the highest values of $x$. We note that solutions $x > 1$ were obtained, because the Curie temperature for the solid metal was calculated without anisotropy field. If we shall define the Curie temper-

---

[4]) Generally $\eta_{\|} \geqq 4\sqrt{\frac{1}{z_{ii} - 4}} - 2 = a$, where we remark the restriction $z_{ii} > 4$. In order that the quantity $a$ be positive it is necessary that $z_{ii} < 8$.

Fig. 1

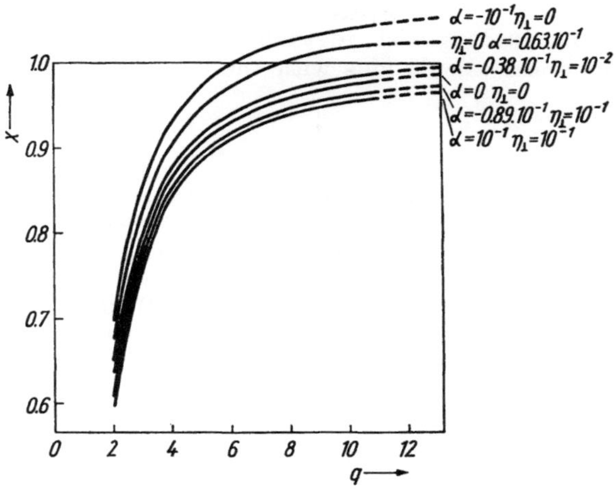


ature relative to a solid metal with the same anisotropy constants as in the case of thin films we should obtain $x < 1$. The results of the calculations are presented in Fig. 1. We remark that for increasing $\alpha$ the Curie temperature decreases. For negative values of $\alpha$, the Curie temperature increases and it becomes possible for every value of $q$, to find such a corresponding $\alpha$ denoted by $\alpha_q^{(c)}$ so that the Curie temperature of the thin film becomes equal to that of a solid with $\eta_{\parallel} = \eta_{\perp} = 0$. Such a situation is possible to arise sometimes if we obtain experimentally for thin films with very small $q$ the same Curie temperature as for the solid. In this manner, we can obtain the function $\alpha_q^{(c)} = \alpha^{(c)}(q)$ (or $\eta_{\parallel}^{(c)} = \eta_{\parallel}^{(c)}(q)$ for $\eta_{\perp} = 0$) for which $T_c^{(q)} = T_c^{(\infty)}$ (Fig. 2). We see that this situation arises only for $\alpha < 0$, and $|\alpha|$ increases for decreasing $q$.

As we noted in [3], when $\eta_{\parallel} = 0$ and $\eta_{\perp}$ is very large it is possible that the magnetization for thin films with two or more layers vanishes. But the corresponding values for $\eta_{\perp}$ are surely larger than the experimental ones.

The careful consideration of parallel and perpendicular anisotropy in thin films can give a full insight in a more realistic interpretation of the experimental data for thin films with $q < 10$.

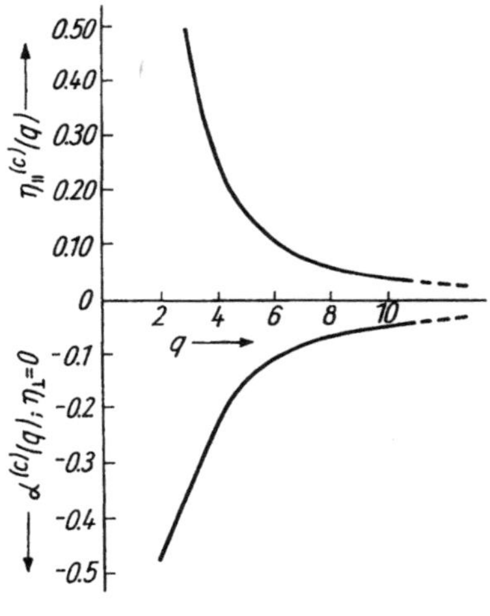


Fig. 2

The authors wish to thank Mr. J. Zamfirescu for his valuable interest in numerical calculations with the CET 500, and Mr. N. Angelescu for his assistance.

## References

[1] L. Valenta, Czech. J. Phys. **7**, 127 (1957); **7**, 136 (1957).
[2] A. Corciovei, Czech. J. Phys. **10**, 568 (1960).
[3] A. Corciovei, Czech. J. Phys. **10**, 917 (1960).
[4] M. J. Klein and R. S. Smith, Phys. Rev. **81**, 378 (1951).
[5] S. J. Glass and M. J. Klein, Phys. Rev. **109**, 288 (1958).
[6] W. Döring, Z. Naturf. **16a**, 1008 (1961); **16a**, 1146 (1961).
[7] A. Corciovei, Phys. Rev. **130**, 2223 (1963).
[8] A. Corciovei and G. Ciobanu, J. Phys. Chem. Solids **26**, 1939 (1965).
[9] W. Brodkorb and W. Haubenreisser, phys. stat. sol. **8**, K 21 (1965).
[10] R. Orbach, Phys. Rev. **112**, 309 (1958).

*(Received May 9, 1966)*

phys. stat. sol. **16**, 337 (1966)

*Institute of Applied Physics, Academy of Sciences of the Moldavian SSR, Kishinev*

# The Interaction of Two Excitons in a Crystal

By

A. I. BOBRYSHEVA

The energy of the exciton–exciton interaction is calculated as a function of the distance between their centres of mass for the atomic model of a semiconductor. Computed numerical results are presented for various values of the parameters of the atomic functions, relative-motion functions, and the electron–hole-effective mass ratio. The binding energy of two excitons is obtained by the Green's function method assuming that the exciton–exciton interaction is less than the interaction of the excitons with optical and acoustic phonons. The role of the direct exciton–exciton interaction and the indirect interaction, via optical or acoustic phonons, in the formation of a biexciton is studied. The results are given of calculations of the cross-section for elastic scattering between two excitons using the Born approximation.

Die Energie der Exziton–Exziton-Wechselwirkung wird als Funktion des Abstands zwischen ihren Massenzentren im Atommodell eines Halbleiters berechnet. Numerische Ergebnisse werden für verschiedene Werte der Parameter der Atomfunktionen, der Relativbewegungsfunktionen und des effektiven Elektronen–Löcher-Massenverhältnisses gegeben. Die Bindungsenergie zweier Exzitonen wird mit der Methode der Greenschen Funktionen erhalten, wobei angenommen wird, daß die Exziton–Exziton-Wechselwirkung kleiner ist als die Wechselwirkung der Exzitonen mit optischen und akustischen Phononen. Weiterhin wird die Rolle direkter Exziton–Exziton-Wechselwirkung und indirekter Wechselwirkung über optische und akustische Phononen während der Bildung des Bi-Exzitons untersucht. Die Ergebnisse werden durch die Berechnung des Wirkungsquerschnittes für elastische Streuung zwischen zwei Exzitonen gegeben, wobei die Bornsche Näherung benutzt wird.

## 1. Introduction

The character of the interaction between excitons is of interest in all cases when a study is made of an exciton gas of intermediate density. These problems include exciton-exciton scattering, Bose-Einstein condensation of excitons, and creation of biexcitons.

The possibility of the creation of biexcitons was first discussed by Lampert [1] and Moskalenko [2]. Further this idea was developed by Chessnut [4] and Trlifaj [5]

Two mechanisms are supposed to give rise to the binding between excitons:

1. direct interaction between excitons, which leads to an attraction,
2. interaction of the excitons with the lattice, which leads to an attractive exciton-exciton interaction.

Lampert [1] has presented estimates of the dissociation energy of a biexciton for limiting values of the electron to hole effective mass ratio: $\frac{m_h}{m_e} \gg 1$, $\frac{m_h}{m_e} = 1$, and $\frac{m_h}{m_e} \ll 1$. For these estimates Lampert made use of the fact that, in these limiting cases, the biexciton is analogous to a molecular or many-electron structure for which calculations are available.

Using a variational method Hylleraas and Ore [3] have calculated the dissociation energy of the positronium molecule. The variational function which was used depends only on four electron-positron distances. The two distances between the particles of the same kind were neglected. Apparently this leads to vastly reduced values of the dissociation energy. For example, from Hylleraas's function the dissociation energy of the hydrogen molecule is obtained to be 0.02 times the ionisation energy of the hydrogen atom instead of 0.33. Besides such a calculation does not give the possibility to calculate the interaction energy as a function of the distance between the excitons. Therefore this calculation cannot be used to estimate the dissociation energy of a biexciton.

Chessnut [4] has calculated the effective exciton-exciton interaction potential resulting from exciton-phonon coupling in a linear chain of dimers. In the two limiting cases of strongly and weakly bound dimers these calculations show that the virtual exchange of acoustic phonons leads to an attractive interaction while the virtual exchange of optical phonons leads to a repulsive potential. The total exciton-exciton interaction potential is probably attractive.

A paper of Trlifaj [5] points out that in a linear chain with two Frenkel excitons the excitation propagate as a biexciton.

It is of interest to study exciton-exciton interaction for a model which better describes real crystals and allows to evaluate the dissociation energy of a biexciton. It is obvious that when the distances between the excitons are smaller than two exciton radii $r_{ex}$ the biexciton must be regarded as an indivisible complex of four particles, which, for distances greater than 2 $r_{ex}$, breaks into two excitons. In order to study the exciton-exciton interaction in a unified way it is assumed that there are two excitons in the crystal, and the distance between their centres of mass is $R$. However, in solving the problem of biexciton creation in this scheme (in contradistinction to the indivisible complex) it is assumed that the relative motion of the two excitons is slower than the motion within a separate exciton (adiabatic approximation). This is true under the condition $I \gg D$ which is fulfilled when $m_{ex} > \mu$ ($I$ ionisation energy of an exciton; $D$ binding energy of the biexciton; $m_{ex}$ mass of an exciton, $\mu$ reduced mass of an exciton).

An experimental detection of biexcitons may be possible by observing phenomena which probably accompany the creation of biexcitons. Such phenomena can take place firstly at temperatures lower than the critical temperature which is determind by the relation $k\, T_{cr} = D$. Secondly, the formation of biexcitons can only be observed if the distances between the excitons are small, and this is most probable at sufficiently large concentrations of excitons. Owing to the existence of a temperature range in which biexcitons can be formed, it may be expected that at temperatures $T \geqq T_{cr}$ a phase transition from the biexciton state to the exciton state may occur. Owing to the radiative annihilation of the excitons which form the biexciton at constant concentration and $T < T_{cr}$ it may be expected that a luminescence band appears which differs from the exciton luminescence band by the biexciton dissociation energy. It must vanish at $T \geqq T_{cr}$. Besides, at constant temperature $T < T_{cr}$, the exciton luminescence band diminishes and the biexciton luminescence band increases as the concentration of excitons increases to a critical value. It may be expected in accordance to [6] that at this critical concentration the excitons decay, and therefore both bands vanish. Probably the biexciton band has a vibrational structure. The formation of biexcitons should affect the diffusion length because the mean free path of a biexciton for dominant biexciton-biexciton and biexciton-phonon scattering is smaller than that of an exciton. As an example we compare the mean free path for exciton-exciton scattering $l_{ex}$ and for biexciton biexciton scattering $l_{bex}$. $l_{ex}$ and $l_{bex}$ are assumed to have the same values as for an ideal gas

$$l_{ex} = \frac{1}{\sqrt{2}\,\pi\,(2\,r_{ex})^2\,n_{ex}}\,, \qquad l_{bex} = \frac{1}{\sqrt{2}\,\pi\,(2\,r_{bex})^2\,n_{bex}}\,.$$

Here $n_{ex}$ and $n_{bex}$ are the concentrations of excitons and biexcitons, respectively; $r_{ex}$ radius of an exciton; because of $r_{bex} = 2\,r_{ex}$ and $n_{bex} \leqq \frac{n_{ex}}{2}$, $l_{bex} \geqq \frac{1}{2}\,l_{ex}$. It may be possible to determine experimentally the dissociation energy of a biexciton if the crystal is irradiated with infrared radiation at corresponding concentrations of excitons. In addition, it should be noted that in [7] probably the interaction energy of excitons is determined experimentally.[1])

## 2. Direct Exciton-Exciton Interaction

To estimate the interaction potential between two excitons of arbitrary radius the model of an atomic semiconductor is used. For simplicity only the first two single-electron states 1 s and 2 s with two orientations of spin in each state are taken into account. The case is considered when two electrons from 1 s states at the sites $\boldsymbol{H}$ and $\boldsymbol{Q}$ make transitions to 2 s states at the sites $\boldsymbol{G}$ and $\boldsymbol{L}$, and the system consisting of two electrons and two holes forms two excitons. This simple band model is applicable, for example, for the yellow exciton series of $Cu_2O$. Coordinates of the centres of mass and relative motions are used

$$\left.\begin{aligned} \boldsymbol{R}_1 &= \frac{m_e\,\boldsymbol{G} + m_h\,\boldsymbol{H}}{m_e + m_h}\,, &\qquad \boldsymbol{\lambda}_1 &= \boldsymbol{G} - \boldsymbol{H}\,,\\ \boldsymbol{R}_2 &= \frac{m_e\,\boldsymbol{L} + m_h\,\boldsymbol{Q}}{m_e + m_h}\,, &\qquad \boldsymbol{\lambda}_2 &= \boldsymbol{L} - \boldsymbol{Q}\,. \end{aligned}\right\} \tag{1}$$

The wave function which corresponds to our model is

$$\begin{aligned} \Psi(\boldsymbol{R}_1, \boldsymbol{R}_2) = A \sum_{\lambda_1\,\lambda_2\,s\,t\,\pi\,T} \varphi(\boldsymbol{\lambda}_1)\,\varphi(\boldsymbol{\lambda}_2)\,a^+_{\boldsymbol{R}_1 + P_1\lambda_1, s}\,b^+_{\boldsymbol{R}_1 - P_2\lambda_1, t}\,a^+_{\boldsymbol{R}_2 + P_1\lambda_2, \pi} \times \\ \times\, b^+_{\boldsymbol{R}_2 - P_2\lambda_2, T}\,S(s, t, \pi, T)\,\psi_0\,, \end{aligned} \tag{2}$$

where

$$\begin{aligned} S = \frac{1}{2}\left[\delta\left(s, \frac{1}{2}\right)\delta\left(t, \frac{1}{2}\right) + \eta_0\,\delta\left(s, -\frac{1}{2}\right)\delta\left(t, -\frac{1}{2}\right)\right] \times \\ \times \left[\delta\left(\pi, \frac{1}{2}\right)\delta\left(T, \frac{1}{2}\right) + \eta_0\,\delta\left(\pi, -\frac{1}{2}\right)\delta\left(T, -\frac{1}{2}\right)\right]; \end{aligned} \tag{3}$$

$a_g^+$, $a_g$, $b_f^+$, $b_f$ creation and annihilation operators of electrons and holes, respectively; $\varphi(\boldsymbol{\lambda})$ function describing the relative motion of the electron and hole in the exciton;

$$P_1 = \frac{m_h}{m_h + m_e}\,; \qquad P_2 = \frac{m_e}{m_h + m_e}\,; \qquad P_1 + P_2 = 1\,;$$

$\eta_0 = 1$ for paraexcitons and $\eta_0 = -1$ for orthoexcitons; $A$ normalization coefficient;

$$A^2 = 1\left\{1 + \delta_{kr}(\boldsymbol{R}_1, \boldsymbol{R}_2) - \frac{1}{2}\sum_{\lambda}\varphi^2(\boldsymbol{\lambda})\left[\varphi^2\left(\frac{\boldsymbol{R}_2 - \boldsymbol{R}_1}{P_2} + \boldsymbol{\lambda}\right) + \varphi^2\left(\frac{\boldsymbol{R}_1 - \boldsymbol{R}_2}{P_1} + \boldsymbol{\lambda}\right)\right]\right\}. \tag{4}$$

[1]) The experiments are proposed by S. A. Moskalenko.

The specific character of the many-electron problem causes such a form of the normalisation coefficient, because the exchange between electrons and holes belonging to two separate excitons may occur through the vacuum state. The correction to the electron and hole effective mass in the expression for the energy of two excitons disappears because the centres of mass of the excitons are fixed.

The Hamiltonian which describes our model is averaged over the wave function (2). As a result we obtain the ground-state energy $E_0$, the energy of two separated excitons, and exciton–exciton interaction energy $W\,(|\boldsymbol{R}_1 - \boldsymbol{R}_2|)$ which depends on the distance between the centres of mass. The obtained expression is very lengthy and is not given here completely. We explain only the meaning of several terms of this expression:

1. Terms which describe the Coulomb and exchange interaction between the electrons of the two excitons:

a) When each of the electrons belongs to a separate exciton:

$$\sum_{\lambda_1 \lambda_2} \varphi^2(\boldsymbol{\lambda}_1)\,\varphi^2(\boldsymbol{\lambda}_2) \Big[F\,(\boldsymbol{R}_2 + P_1\,\boldsymbol{\lambda}_2, \boldsymbol{R}_1 + P_1\,\boldsymbol{\lambda}_1; \boldsymbol{R}_2 + P_1\,\boldsymbol{\lambda}_2, \boldsymbol{R}_1 + P_1\,\boldsymbol{\lambda}_1) - \\ - \frac{1}{2}\,F\,(\boldsymbol{R}_2 + P_1\,\boldsymbol{\lambda}_2, \boldsymbol{R}_1 + P_1\,\boldsymbol{\lambda}_1;\; \boldsymbol{R}_1 + P_1\,\boldsymbol{\lambda}_1, \boldsymbol{R}_2 + P_1\,\boldsymbol{\lambda}_2)\Big]. \quad (5)$$

b) When they belong to both excitons simultaneously:

$$\sum_{\lambda_1 \lambda_2} \varphi(\boldsymbol{\lambda}_1)\,\varphi(\boldsymbol{\lambda}_2)\,\varphi\left(\left|\frac{\boldsymbol{R}_1 - \boldsymbol{R}_2}{P_2} + \boldsymbol{\lambda}_2\right|\right)\varphi\left(\left|\frac{\boldsymbol{R}_2 - \boldsymbol{R}_1}{P_2} + \boldsymbol{\lambda}_1\right|\right)\Big[F\Big(\boldsymbol{R}_2 + \frac{P_1}{P_2}(\boldsymbol{R}_2 - \boldsymbol{R}_1) + \\ + P_1\,\boldsymbol{\lambda}_1, \boldsymbol{R}_1 + \frac{P_1}{P_2}(\boldsymbol{R}_1 - \boldsymbol{R}_2) + P_1\,\boldsymbol{\lambda}_2;\; \boldsymbol{R}_1 + P_1\,\boldsymbol{\lambda}_1;\; \boldsymbol{R}_2 + P_1\,\boldsymbol{\lambda}_2\Big) - \\ - \frac{1}{2}\,F\Big(\boldsymbol{R}_2 + \frac{P_1}{P_2}(\boldsymbol{R}_2 - \boldsymbol{R}_1) + P_1\,\boldsymbol{\lambda}_1, \boldsymbol{R}_1 + \\ + \frac{P_1}{P_2}(\boldsymbol{R}_1 - \boldsymbol{R}_2) + P_1\,\boldsymbol{\lambda}_2;\; \boldsymbol{R}_2 + P_1\,\boldsymbol{\lambda}_2, \boldsymbol{R}_1 + P_1\,\boldsymbol{\lambda}_1\Big)\Big]. \quad (6)$$

2. Analogous terms connected with the interaction between the two holes of the two excitons.

3. Terms which describe the Coulomb and exchange interaction between an electron and a hole of the two excitons:

a) When electron and hole belong to separate excitons:

$$\sum_{\lambda_1 \lambda_2} \varphi^2(\boldsymbol{\lambda}_1)\,\varphi^2(\boldsymbol{\lambda}_2)\,\Big[F(\boldsymbol{R}_2 + P_1\,\boldsymbol{\lambda}_2, \boldsymbol{R}_1 - P_2\,\boldsymbol{\lambda}_1; \boldsymbol{R}_2 + P_1\,\boldsymbol{\lambda}_2, \boldsymbol{R}_1 - P_2\,\boldsymbol{\lambda}_1) - \\ - \frac{1}{2}\,F(\boldsymbol{R}_2 + P_1\,\boldsymbol{\lambda}_2, \boldsymbol{R}_1 - P_2\,\boldsymbol{\lambda}_1;\; \boldsymbol{R}_1 - P_2\,\boldsymbol{\lambda}_1, \boldsymbol{R}_2 + P_1\,\boldsymbol{\lambda}_2)\Big]. \quad (7)$$

b) Electron and hole belong to both excitons simultaneously:

$$\sum_{\lambda_1 \lambda_2} \varphi(\boldsymbol{\lambda}_1)\,\varphi(\boldsymbol{\lambda}_2)\,\varphi\left(\left|\frac{\boldsymbol{R}_2 - \boldsymbol{R}_1}{P_1} + \boldsymbol{\lambda}_2\right|\right)\varphi\left(\left|\frac{\boldsymbol{R}_2 - \boldsymbol{R}_1}{P_2} + \boldsymbol{\lambda}_1\right|\right)\Big[F\Big(\boldsymbol{R}_2 + \frac{P_1}{P_2}(\boldsymbol{R}_2 - \boldsymbol{R}_1) + \\ + P_1\,\boldsymbol{\lambda}_1, \boldsymbol{R}_2 - P_2\,\boldsymbol{\lambda}_2; \boldsymbol{R}_1 + P_1\,\boldsymbol{\lambda}_1, \boldsymbol{R}_1 - \frac{P_2}{P_1}(\boldsymbol{R}_2 - \boldsymbol{R}_1) - P_2\,\boldsymbol{\lambda}_2\Big) - \\ - \frac{1}{2}\,F\Big(\boldsymbol{R}_2 + \frac{P_1}{P_2}(\boldsymbol{R}_2 - \boldsymbol{R}_1) + P_1\,\boldsymbol{\lambda}_1, \boldsymbol{R}_2 - P_2\,\boldsymbol{\lambda}_2; \\ \boldsymbol{R}_1 - \frac{P_2}{P_1}(\boldsymbol{R}_2 - \boldsymbol{R}_1) - P_2\,\boldsymbol{\lambda}_2, \boldsymbol{R}_1 + P_1\,\boldsymbol{\lambda}_1\Big)\Big]. \quad (8)$$

c) The electrons belong to separate excitons, the holes belong to both excitons simultaneously:

$$\sum_{\lambda_1 \lambda_2} \varphi^2(\boldsymbol{\lambda}_1)\, \varphi(\boldsymbol{\lambda}_2)\, \varphi\left(\left|\frac{\boldsymbol{R}_2 - \boldsymbol{R}_1}{P_2} + \boldsymbol{\lambda}_1\right|\right) \left[F\left(\boldsymbol{R}_2 + \frac{P_1}{P_2}(\boldsymbol{R}_2 - \boldsymbol{R}_1) + P_1 \boldsymbol{\lambda}_1, \boldsymbol{R}_2 - P_2 \boldsymbol{\lambda}_2;\right.\right.$$
$$\left. \boldsymbol{R}_2 + P_1 \boldsymbol{\lambda}_2, \boldsymbol{R}_1 - P_2 \boldsymbol{\lambda}_1\right) - (1 + \eta_0)\, F\left(\boldsymbol{R}_2 + \frac{P_1}{P_2}(\boldsymbol{R}_2 - \boldsymbol{R}_1) + \right.$$
$$\left.\left. + P_1 \boldsymbol{\lambda}_1, \boldsymbol{R}_2 - P_2 \boldsymbol{\lambda}_2;\ \boldsymbol{R}_1 - P_2 \boldsymbol{\lambda}_1, \boldsymbol{R}_2 + P_1 \boldsymbol{\lambda}_2\right)\right]. \tag{9}$$

d) Inverse case.

In these expressions $F(\boldsymbol{a}, \boldsymbol{b}; \boldsymbol{c}, \boldsymbol{d})$ are matrix elements of the type

$$F(\boldsymbol{a}, \boldsymbol{b};\ \boldsymbol{c}, \boldsymbol{d}) = e^2 \iint \frac{\psi_\alpha^* (\boldsymbol{r}_1 - \boldsymbol{a})\, \psi_\beta^* (\boldsymbol{r}_2 - \boldsymbol{b})\, \psi_\alpha (\boldsymbol{r}_1 - \boldsymbol{c})\, \psi_\beta (\boldsymbol{r}_2 - \boldsymbol{d})}{r_{12}}\, \mathrm{d}\boldsymbol{r}_1\, \mathrm{d}\boldsymbol{r}_2\,.$$

$F(\boldsymbol{a}, \boldsymbol{b};\ \boldsymbol{c}, \boldsymbol{d})$ are four-centre integrals, and a numerical calculation of $W(|\boldsymbol{R}_1 - \boldsymbol{R}_2|)$ is extremely difficult even with the help of a computer. Therefore the four-centre Coulomb and exchange integrals are approximated by expressions which contain two-centre integrals and depend exponentially on the distances between the centres

$$F(\boldsymbol{a}, \boldsymbol{b};\ \boldsymbol{c}, \boldsymbol{d}) \approx \mathrm{e}^{-\alpha |\boldsymbol{a} - \boldsymbol{c}|}\, \mathrm{e}^{-\beta |\boldsymbol{b} - \boldsymbol{d}|}\, F(\boldsymbol{a}, \boldsymbol{b};\ \boldsymbol{a}, \boldsymbol{b}).$$

In turn, the two-centre exchange integrals are calculated approximately by Gellmann's method using the function $\eta$ given in [9]. For simplicity, the wave functions of the ground state and excited state in cases when their orthogonality is not essential were approximated by the expressions

$$\psi_{2\mathrm{s}}(r) = \left(\frac{\beta^3}{\pi}\right)^{1/2} \mathrm{e}^{-\beta r}, \qquad \psi_{1\mathrm{s}}(r) = \left(\frac{\alpha^3}{\pi}\right)^{1/2} \mathrm{e}^{-\alpha r}$$

and the relative-motion function is $\varphi(r) = \left(\frac{\mu^3}{\pi}\right)^{1/2} \mathrm{e}^{-\mu r}$. Then the simplified expression of $W(|R_1 - R_2|)$ was tabulated by means of a computer for some parameters of atomic functions, relative-motion functions, and $P_1$, $P_2$. The values of these parameters are given in Table 1. The paraexciton-paraexciton interaction energy as a function of the distances between their centres of mass is given in Fig. 1, 2, and 3. A comparison of the obtained curves leads to the conclusion that the biexciton dissociation energy is large when the radius of the exciton is small and the electron to hole effective mass ratio is large.

Table 1

| No. of curve | $\frac{\mu}{10^8}$ (cm$^{-1}$) | $\frac{\alpha}{10^8}$ (cm$^{-1}$) | $\frac{\beta}{10^8}$ (cm$^{-1}$) | $P_1$ | $P_2$ |
|---|---|---|---|---|---|
| 1 | 0.1 | 2 | 1 | 3/4 | 1/4 |
| 2 | 0.2 | 2 | 1 | 3/4 | 1/4 |
| 3 | 0.1 | 1 | 0.5 | 3/4 | 1/4 |
| 4 | 0.2 | 1 | 0.5 | 3/4 | 1/4 |
| 5 | 0.1 | 2 | 1 | 2/3 | 1/3 |
| 6 | 0.2 | 2 | 1 | 2/3 | 1/3 |
| 7 | 0.1 | 1 | 0.5 | 2/3 | 1/3 |
| 8 | 0.2 | 1 | 0.5 | 2/3 | 1/3 |

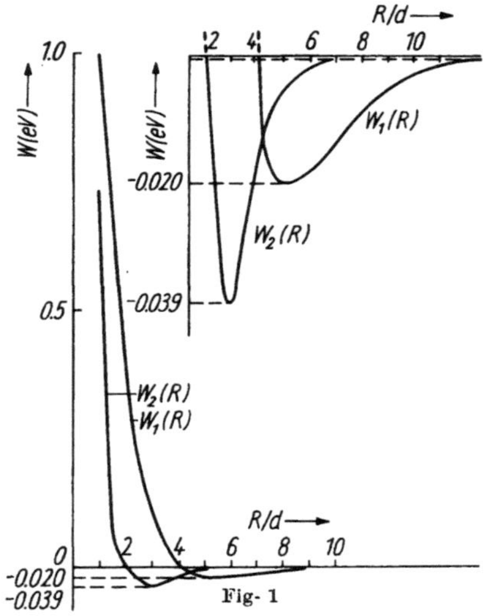


Fig. 1

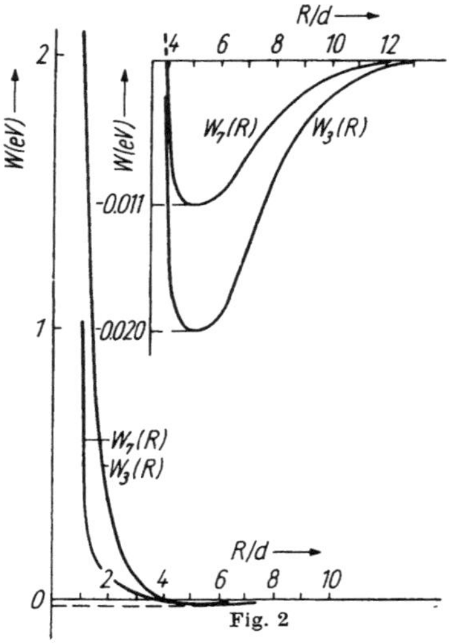


Fig. 2

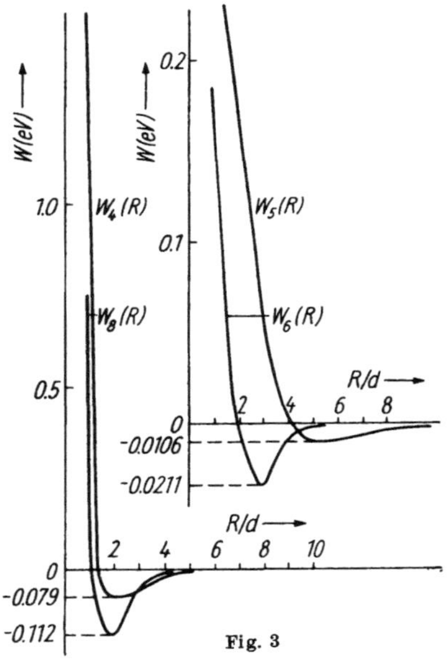


Fig. 3

Fig. 1, 2, and 3. Dependence of the paraexciton-paraexciton interaction potential on the distance between their centres of mass (the distance is given in units of $d = 3$ Å).

Fig. 1 for the parameter values of curves 1 and 2 in Table 1.

Fig. 2 for the parameter values of curves 3 and 7 in Table 1.

Fig. 3 for the parameter values of curves 4, 5, 6, and 8 in Table 1

The exciton-exciton interaction energy decreases exponentially at large distances, because in our calculation we ignore the polarization of excitons. The correction caused by the polarization of excitons may be calculated at least in one limiting case, namely in the adiabatic approximation, under the assumption that the exciton mass is larger than the reduced exciton mass. In this case it may be assumed that the polarization leads to a correction only in the relative-motion function.

In solving our problem it was assumed that there are four particles over the vacuum state which form two excitons. Therefore, after averaging the many-electron Hamiltonian over the operator part of the function (2), we obtain Hamiltonians of two excitons in centre-of-mass and relative-motion variables:

$$\hat{H} = \hat{H}_1 (\boldsymbol{R}_1 \boldsymbol{\lambda}_1) + \hat{H}_2 (\boldsymbol{R}_2, \boldsymbol{\lambda}_2) + \hat{U}(\boldsymbol{R}_1, \boldsymbol{R}_2, \boldsymbol{\lambda}_1, \boldsymbol{\lambda}_2)\,, \tag{10}$$

where $\hat{H}_1, \hat{H}_2$ are the Hamiltonians of two separate excitons, and $\hat{U}(\boldsymbol{R}_1, \boldsymbol{R}_2, \boldsymbol{\lambda}_1, \boldsymbol{\lambda}_2)$ is the interaction Hamiltonian.

The correction to the exciton-exciton interaction energy caused by polarization may be obtained by solving a variational problem with the relative-motion function

$$\varphi(\boldsymbol{\lambda}_1, \boldsymbol{\lambda}_2) = \varphi_0(\boldsymbol{\lambda}_1)\, \varphi_0(\boldsymbol{\lambda}_2) + \sum_{m,n}{}' C_{nm}\, \varphi_n(\boldsymbol{\lambda}_1)\, \varphi_m(\boldsymbol{\lambda}_2)\,. \tag{11}$$

Here the prime on the sum means that $n$ and $m$ are not simultaneously equal to zero; $\varphi_0(\boldsymbol{\lambda})$ relative-motion function of s-type; $\varphi_n(\boldsymbol{\lambda})$ relative-motion function of excited-state type. As a result we obtain the second-order correction to the exciton-exciton interaction energy

$$\varepsilon = -\sum_{n,m}{}' \frac{|U_{00,nm}|^2}{E_n + E_m - 2\,E_0}\,, \tag{12}$$

where $U_{00,nm}$ is the matrix element

$$U_{00,nm} = \iint \varphi_0^*(\boldsymbol{\lambda}_1)\, \varphi_0^*(\boldsymbol{\lambda}_2)\, \hat{U}(\boldsymbol{R}_1, \boldsymbol{R}_2, \boldsymbol{\lambda}_1, \boldsymbol{\lambda}_2)\, \varphi_n(\boldsymbol{\lambda}_1)\, \varphi_m(\boldsymbol{\lambda}_2)\, \mathrm{d}\boldsymbol{\lambda}_1\, \mathrm{d}\boldsymbol{\lambda}_2\,.$$

The correction (12) leads to a correct behaviour of $W(|\boldsymbol{R}_1 - \boldsymbol{R}_2|)$ at large distances. Indeed $U_{00,nm}$ contains the matrix elements $F$. A multipole expansion of $1/r_{12}$ from $F$ leads to terms $1/R^6$ in (12) which are typical of dispersion forces. The polarization of the medium connected with the creation of two excitons can be taken into account by introducing into the dielectric constant $\varepsilon$.

## 3. Calculation of the Dissociation Energy of a Biexciton

To take into account two mechanisms of biexciton formation the Hamiltonian of [10] is used:

$$\begin{aligned}\hat{H} = {} & \sum_{\boldsymbol{p}} (\Delta + T_{\boldsymbol{p}})\, a_{\boldsymbol{p}}^+ a_{\boldsymbol{p}} + \sum_{\boldsymbol{p}} B_{\boldsymbol{p}}\, b_{\boldsymbol{p}}^+ b_{\boldsymbol{p}} + \frac{1}{2V} \sum_{\boldsymbol{p}\boldsymbol{p}'\boldsymbol{q}\boldsymbol{q}'} \nu(\boldsymbol{p} - \boldsymbol{p}') \times \\ & \times \delta_{kr}(\boldsymbol{p} + \boldsymbol{q} - \boldsymbol{p}' - \boldsymbol{q}')\, a_{\boldsymbol{p}}^+ a_{\boldsymbol{q}}^+ a_{\boldsymbol{p}'} a_{\boldsymbol{q}'} + \\ & + \frac{i}{\sqrt{N_a}} \sum_{\boldsymbol{p}\boldsymbol{p}'} \theta(\boldsymbol{p} - \boldsymbol{p}')\, a_{\boldsymbol{p}}^+ a_{\boldsymbol{p}'} (b_{\boldsymbol{p}-\boldsymbol{p}'} - b_{\boldsymbol{p}'-\boldsymbol{p}}^+)\,,\end{aligned} \tag{13}$$

where $\Delta$ exciton creation energy; $T_{\boldsymbol{p}}$ kinetic energy of the exciton; $B_{\boldsymbol{p}}$ energy of a phonon (acoustic or optical); $\nu(\boldsymbol{p})$ Fourier-component of the exciton-exciton interaction energy which was studied in Section 2; $\theta(\boldsymbol{p})$ Fourier-component of the exciton-phonon interaction energy; $V$ volume of the crystal; $N_a$ number of atoms in the lattice; $a_{\boldsymbol{p}}^+$, $a_{\boldsymbol{p}}$, $b_{\boldsymbol{p}}^+$, $b_{\boldsymbol{p}}$ exciton and phonon creation and annihilation operators, respectively.

Three limiting cases of interaction in the exciton-phonon system may be investigated:

1 and 2. The exciton-phonon interaction is stronger than the exciton-exciton interaction. This approach corresponds to the assumption that the excitons first interact with phonons and this leads to changes of the exciton masses and the positions of the exciton levels. The two cases of weak and strong coupling

may be considered. In this paper only the case of weak exciton-phonon coupling is taken into account.

3. The exciton-phonon interaction is weaker than the exciton-exciton interaction. This assumption means that first the interaction occurs between excitons so that a biexciton may be created, and then changes of level position and biexciton effective mass occur.

To obtain the contribution to the exciton energy due to exciton-phonon interaction, the perturbation is written in canonical form: $\hat{H} = \mathrm{e}^{-S}\,\hat{H}\,\mathrm{e}^{S}$. The function $S$ is given by the expression

$$S = \sum_{pp'}{}' \{S^{(1)}(p - p')\, a_p^+ a_{p'} b_{p-p'} - S^{(2)}(p - p')\, a_p^+ a_{p'} b_{p'-p}^+\} \,. \tag{14}$$

As a result, to first order of perturbation theory we obtain a Hamiltonian which includes excitons with changed masses, phonons with changed frequencies, direct and effective exciton-exciton interaction.

To calculate the energy of the bound state of two excitons we use the Green's function method. The time-dependent two-particle advanced and retarded exciton Green's functions are defined by the relations

$$\left.\begin{aligned} G^{\mathrm{r}}_{f_1 f_2 g_1 g_2}(t - t') &= -i\,\theta(t - t')\,\langle[a_{f_1}^+(t)\, a_{f_2}^+(t), a_{g_1}(t')\, a_{g_2}(t')]\rangle\,,\\ G^{\mathrm{a}}_{f_1 f_2 g g_2}(t - t') &= i\theta(t' - t)\,\langle[a_{f_1}^+(t)\, a_{f_2}^+(t), a_{g_1}(t')\, a_{g_2}(t')]\rangle\,. \end{aligned}\right\} \tag{15}$$

Let us derive the equations of motion for the Green's functions. Unlinking the three-particle Green's functions in the first equation the infinite set of equations is found to reduce to the first equation. In this case for the Fourier-component of the Green's function we obtain the equation

$$\begin{aligned} &E\,\langle\langle a_{f_1}^+(t)\, a_{f_2}^+(t);\, a_{g_1}(t')\, a_{g_2}(t')\rangle\rangle_E = -\frac{1}{2\pi}\{\delta_{f_1 g_2}\langle a_{g_1} a_{f_2}^+\rangle + \delta_{f_2 g_2}\langle a_{g_1} a_{f_1}^+\rangle +\\ &+ \delta_{f_2 g_1}\langle a_{f_1}^+ a_{g_2}\rangle + \delta_{f_1 g_1}\langle a_{f_2}^+ a_{g_2}\rangle\} - \Bigg(2\Delta + T_{f_1} + T_{f_2} - \frac{1}{N_{\mathrm{a}}}\times\\ &\times \sum_p\left[\frac{\theta^2(p - f_1)}{T_p - T_{f_1} + B_{p-f_1}} + \frac{\theta^2(p - f_2)}{T_p - T_{f_2} + B_{p-f_2}}\right] + 2\,\nu(0)\,\frac{N}{V} + \frac{1}{N_{\mathrm{a}}}\times\\ &\times \sum_p\Bigg\{\left[\frac{\nu(f_1 - p)}{V_0} + \frac{2\,\theta^2(f_1 - p)\,B_{f_1-p}}{(T_{f_1} - T_p)^2 - B^2_{f_1-p}} + \frac{\nu(f_2 - p)}{V_0} + \frac{2\,\theta^2(f_2 - p)\,B_{f_2-p}}{(T_{f_2} - T_p)^2 - B^2_{f_2-p}}\right] n_p -\\ &-\left[\frac{2\,\bar{\xi}_{p-f_1}(T_p - T_{f_1})}{(T_p - T_{f_1})^2 - B^2_{p-f_1}} + \frac{2\,\bar{\xi}_{p-f_2}(T_p - T_{f_2})}{(T_p - T_{f_2})^2 - B^2_{p-f_2}}\right]\Bigg\}\,\langle\langle a_{f_1}^+(t)\, a_{f_2}^+(t);\, a_{g_1}(t')\, a_{g_2}(t')\rangle\rangle_E -\\ &-\frac{1}{N_{\mathrm{a}}}\sum_p\left[\frac{\nu(p - f_2)}{2\,V_0} + \frac{\theta^2(p - f_2)\,B_{p-f_2}}{(T_p - T_{f_2})^2 - B^2_{p-f_2}} + \frac{\nu(p - f_1)}{2\,V_0} + \frac{\theta^2(p - f_1)\,B_{p-f_1}}{(T_p - T_{f_1})^2 - B^2_{p-f_1}}\right]\times\\ &\times(1 + \bar{n}_{f_1} + \bar{n}_{f_2})\,\langle\langle a_p^+(t)\, a_{f_1+f_2-p}^+(t);\, a_{g_1}(t')\, a_{g_2}(t')\rangle\rangle_E\,, \end{aligned} \tag{16}$$

where $\bar{n}_f = \langle a_f^+ a_f\rangle$, $\quad \bar{\xi} = \langle b_f^+ b_f\rangle$.

The homogenous part of (16) is the equation for the energy of bound states of two excitons. The two-particle Green's function has the form

$$\langle\langle a_{f_1}^+ a_{f_2}^+;\, a_{g_1} a_{g_2}\rangle\rangle_E = \sum_{\alpha\beta} C_{\alpha\beta}\, U_\alpha(f_1 f_2)\, U_\beta^*(g_1 g_2) \tag{17}$$

under the condition

$$\sum_{f_1 f_2} U_\alpha (f_1 f_2)\, U^*_\beta(f_1 f_2) = \delta_{\alpha\beta}\,. \tag{18}$$

To solve the energy equation a variational method is used. The variational function $U_\alpha$ which is chosen as the Fourier-component of the biexciton function is given by

$$U_\alpha(f_1, f_2) = \frac{\delta_{k\,r}(f_1 + f_2 + k)}{\sqrt{V}} \left(\frac{2\pi}{\alpha}\right)^{3/4} e^{-\frac{|f_1 - f_2|^2}{16\alpha}} \tag{19}$$

($k$ quasimomentum of the biexciton, $\alpha$ variational parameter).

For simplicity we put $k = 0$. The bound-state energy which is obtained with the function (19) is given by

$$\begin{aligned}
E_\alpha &= 2\Delta + 2\nu(0)\frac{N}{V} + \sum_{f_1} e^{-\frac{f_1^2}{2\alpha}} \Bigg\{ 2T_{f_1} - \frac{1}{N_a} \times \\
&\times \sum_{p} \left[ \frac{\theta^2(f_1 - p)}{T_p - T_{f_1} + B_{p-f_1}} + \frac{\theta^2(f_1 + p)}{T_p - T_{f_1} + B_{p+f_1}} \right] + \frac{1}{N_a} \times \\
&\times \sum_p \Bigg( \Bigg[ \frac{\nu(f_1 - p)}{V_0} + \frac{\nu(f_1 + p)}{V_0} + \frac{2\theta^2(f_1 - p)\, B_{f_1 - p}}{(T_{f_1} - T_p)^2 - B^2_{f_1 - p}} + \frac{2\theta^2(f_1 + p)\, B_{f_1 + p}}{(T_{f_1} - T_p)^2 - B^2_{f_1 + p}} \Bigg] \times \\
&\times \bar{n}_p - \left[ \frac{2\xi_{p-f_1}(T_p - T_{f_1})}{(T_p - T_{f_1})^2 - B^2_{p-f_1}} + \frac{2\bar{\xi}_{p+f_1}(T_p - T_{f_1})}{(T_p - T_{f_1})^2 - B^2_{p+f_1}} \right] \Bigg) \Bigg\} \frac{\hbar^3}{V} \left(\frac{2\pi}{\alpha}\right)^{3/2} + \\
&+ \frac{\hbar^3}{N_a V} \left(\frac{2\pi}{\alpha}\right)^{3/2} \sum_{p\, f_1} e^{-\frac{f_1^2 + p^2}{4\alpha}} \times \\
&\times \left[ \frac{\nu(p - f_1)}{2V_0} + \frac{\nu(p + f_1)}{2V_0} + \frac{\theta^2(p - f_1)\, B_{p - f_1}}{(T_p - T_{f_1})^2 - B^2_{p - f_1}} + \frac{\theta^2(p + f_1)\, B_{p + f_1}}{(T_p - T_{f_1})^2 - B^2_{p + f_1}} \right] \times \\
&\times (1 + \bar{n}_{f_1} + \bar{n}_{-f_1})\,.
\end{aligned} \tag{20}$$

## 4. Two Limiting Cases

To elucidate the role of the direct exciton-exciton interaction and exciton-phonon interaction in the formation of biexcitons two limiting cases are examined:

1. Only direct exciton-exciton interaction is taken into account.
2. Only exciton-exciton interaction through acoustic or optical phonons is taken into account.

Numerical estimates are given for $Cu_2O$ and a molecular crystal.

$$\begin{aligned}
1.\ E_\alpha &= 2\Delta + 2\nu(0)\frac{N}{V} + \frac{1}{V}\sum_{f_1} e^{-\frac{f_1^2}{2\alpha}} \Bigg\{ 2T_{f_1} + \frac{1}{V}\sum_p [\nu(f_1 - p) + \\
&+ \nu(f_1 + p)]\, \bar{n}_p \Bigg\} \hbar^3 \left(\frac{2\pi}{\alpha}\right)^{3/2} + \frac{\hbar^3}{2V^2}\left(\frac{2\pi}{\alpha}\right)^{3/2} \sum_{p\, f_1} e^{-\frac{f_1^2 + p^2}{4\alpha}} \times \\
&\times [\nu(p - f_1) + \nu(p + f_1)]\,(1 + \bar{n}_{f_1} + \bar{n}_{-f_1})\,.
\end{aligned} \tag{21}$$

In the estimate for $Cu_2O$ the exciton-exciton interaction curve No. 5 will be used in which the dielectric constant of the medium ($\varepsilon$) is introduced. This curve is approximated by the Morse potential

$$v(r) = \frac{D_0}{\varepsilon} \{e^{2\delta(1 - r/R_0)} - 2\, e^{\delta(1 - r/R_0)}\} \tag{22}$$

($R_0$ equilibrium distance $= 15$ Å; $D_0$ biexciton binding energy $= 0.0106$ eV; $\delta = 2$; $\varepsilon = 9$).

The Fourier-component of the interaction energy is

$$\nu(p) = 16\,\pi\, R_0^3 \frac{D_0}{\varepsilon} \delta\, e^\delta \left\{\frac{e^\delta}{(4\,\delta^2 + p^2 R^2)^2} - \frac{1}{(\delta^2 + p^2 R^2)^2}\right\}, \tag{23}$$

where $R = R_0/\hbar$.

A Maxwell distribution is accepted for excitons: $n_p = A\, e^{-\frac{p^2}{2 m_{ex} k T}}$ and $A$ is determined by the condition $\sum_p \bar{n}_p = N$. $E_\alpha$ was estimated using the following values of the parameters: $m_{ex} = 1.16\, m_0$, $n_{ex} = 10^{15}$ cm$^{-3}$, $r_{ex} = 12$ Å, $T = 4$ °K. $E_\alpha - 2\,\Delta$ has a minimum at $\alpha = 10^{-45} \frac{\text{g}^2\,\text{cm}^2}{\text{s}^2}$ and is equal to $10^{-6}$ eV. In this case the exciton-exciton interaction does not lead to the formation of biexcitons.

In the estimate for a molecular crystal curve No. 6 is used in which the square of the refractive index $n_0^2 = 3$ is introduced. For the following values of parameters: $r_{ex} = 5$ Å, $m_{ex} = 100\, m_0$, $V = 27 \times 10^{-24}$ cm$^3$, $m_e = \frac{m_{ex}}{3}$, $m_h = \frac{2\, m_{ex}}{3}$, the direct exciton-exciton interaction leads to the formation of biexcitons. The binding energy is found to be 0.0008 eV.

2. The terms connected with the direct exciton-exciton interaction in (20) are neglected.

Following [12] the $\theta(p)$ for the optical and acoustic branch, respectively, are given by

$$\left.\begin{aligned} \theta(p) &= \left[2\,\pi \left(\frac{1}{n_0^2} - \frac{1}{\varepsilon}\right) \hbar\, \omega_0 \frac{e^2}{V_0} \frac{\hbar^2}{p^2}\right]^{1/2} \{-\Phi_h(p) + \Phi_e(p)\}, \\ \theta(p) &= \left(\frac{2\,p}{9\,M\,u}\right)^{1/2} \{-C_h \Phi_h(p) + C_e \Phi_e(p)\}, \qquad \Phi_i(p) = \frac{1}{\left[1 + \frac{\mu^2}{m_i^2} p^2 \frac{r_{ex}^2}{4\,\hbar^2}\right]^2}. \end{aligned}\right\} \tag{24}$$

Here $\omega_0$ means the limiting frequency of optical phonons, $V_0$ the volume of the unit cell, $\mu$ the reduced mass of the exciton, $m_i$ the electron or hole effective mass, $r_{ex}$ the exciton radius, $M$ the ionic mass, $C_e$ and $C_h$ the deformation potential constants of electron and hole respectively, and $u$ the acoustic velocity in the crystal).

$E_\alpha$ is tabulated for values of parameters which correspond to $Cu_2O$:

$$m_e = 0.36\, m_0, \quad m_h = 0.8\, m_0, \quad r_{ex} = 12\ \text{Å}, \quad V_0 = (1.84)^3 \times 10^{-24}\ \text{cm}^3,$$

$$\varepsilon = 9, \quad n_0^2 = 4, \quad u = 5 \times 10^5\ \frac{\text{cm}}{\text{s}}; \quad \omega_0 = 5 \times 10^{13}\ \text{s},$$

$$T = 4\ °\text{K}; \quad C_e = 15\ \text{eV}; \quad C_h = 10\ \text{eV}.$$

In the case of acoustic phonons $E_\alpha - 2\,\Delta$ has a minimum at $\alpha = 10^{-46} \frac{\text{g}^2\,\text{cm}^2}{\text{s}^2}$ and is equal to $-0.000185$ eV. For two separated excitons with quasi-momenta $f = 0$, one finds $2\,(E_{ex} - \Delta) = -0.000186$ eV. Therefore the exciton-exciton interaction through acoustic phonons does not lead to the formation of biexcitons. An analogous calculation for optical phonons gives $E_\alpha - 2\Delta = -0.01140$ eV at $\alpha = 10^{-45}\,\frac{\mathrm{g^2\,cm^2}}{\mathrm{s^2}}$ and $2(E_{\mathrm{ex}} - \Delta) = -0.01142$ eV. The exciton–exciton interaction through optical phonons does not lead to the creation of biexcitons either. This calculation shows that in $Cu_2O$ biexcitons cannot exist. As follows from (20) the condition for biexciton formation approximately may be written in the form

$$\sum_{\boldsymbol{f}} e^{-\frac{f^2}{2\alpha}} \left\{ T_{\boldsymbol{f}} - \frac{1}{N_a} \sum_{\boldsymbol{p}} \frac{\theta^2(\boldsymbol{p}-\boldsymbol{f})}{T_{\boldsymbol{p}} - T_{\boldsymbol{f}} + B_{\boldsymbol{p}-\boldsymbol{f}}} + \frac{1}{N_a} \sum_{\boldsymbol{p}} \frac{\theta^2(\boldsymbol{p}-\boldsymbol{f})\, B_{\boldsymbol{p}-\boldsymbol{f}}}{(T_{\boldsymbol{p}} - T_{\boldsymbol{f}})^2 - B^2_{\boldsymbol{p}-\boldsymbol{f}}}\, e^{\frac{f^2 - p^2}{4\alpha}} \right\} < -\frac{1}{N_a} \sum_{\boldsymbol{p}} \frac{\theta^2(\boldsymbol{p})}{T_{\boldsymbol{p}} + B_{\boldsymbol{p}}}. \tag{25}$$

The condition (25) implies that the third term which describes exciton–exciton interaction through phonons is negative. This requirement is satisfied when $(T_{\boldsymbol{p}} - T_{\boldsymbol{f}})^2$ is smaller than $B^2_{\boldsymbol{p}-\boldsymbol{f}}$ and may be neglected. Therefore a formation of biexcitons may be expected in crystals where the exciton mass is large for example, in molecular crystals [13].

Let us make such a calculation of $E_\alpha$ with the parameter values which were used above in the estimates for a molecular crystal, and $C_{\mathrm{e}} = C_{\mathrm{h}} = 10$ eV. In this case the exciton–exciton interaction through acoustic phonons leads to the formation of biexcitons. The biexciton binding energy is equal to 0.049 eV.

Now we will give an expression for the cross section for elastic scattering between two excitons in Born's approximation. In this calculation the exciton–exciton interaction potential curve from Section 2 is approximated by a Morse potential. The initial and final states of two excitons, in centre-of-mass variables, are chosen in the form

$$\left.\begin{aligned} \Psi_{\boldsymbol{k}} &= \sqrt{\frac{\mu}{2\hbar k}}\,\{e^{i\boldsymbol{k}\boldsymbol{R}} + e^{-i\boldsymbol{k}\boldsymbol{R}}\}\,, \\ \Psi_{\boldsymbol{k}'} &= \frac{1}{\sqrt{2}}\left(\frac{1}{2\pi}\right)^{3/2} \{e^{i\boldsymbol{k}'\boldsymbol{R}} + e^{-i\boldsymbol{k}'\boldsymbol{R}}\}\,; \qquad \boldsymbol{k} = \frac{1}{2}(\boldsymbol{k}_1 - \boldsymbol{k}_2) \end{aligned}\right\} \tag{26}$$

($\boldsymbol{k}_1$, $\boldsymbol{k}_2$ quasi-momenta of excitons; $R$ coordinate of relative motion; $\mu$ reduced mass of two similar excitons).

In this case the cross section is given by

$$\begin{aligned} \sigma = \frac{256\pi\beta^2 D_0^2 \mu^2}{\hbar^4 \varepsilon^2} \Bigg\{ & e^{4\beta R_0} \left[ \frac{k^4 + 3k^2\beta^2 + 3\beta^4}{384\beta^6 (k^2 + \beta^2)^3} + \frac{1}{64\beta^2 (2\beta^2 + k^2)^2 (\beta^2 + k^2)} - \right. \\ & \left. - \frac{1}{8k^2 (4\beta^2 + 2k^2)^3} \ln \frac{\beta^4}{(\beta^2 + k^2)^2} \right] + e^{2\beta R_0} \left[ \frac{2}{3}\, \frac{(16k^4 + 12k^2\beta^2 + 3\beta^4)}{\beta^6 (4k^2 + \beta^2)^3} + \right. \\ & \left. + \frac{1}{\beta^2 (\beta^2 + 2k^2)^2 (\beta^2 + 4k^2)} + \frac{1}{8k^2 (\beta^2 + 2k^2)^3} \ln \frac{(\beta^2 + 4k^2)^2}{\beta^4} \right] - \\ & - 2 e^{3\beta R_0} \left[ - \frac{5\beta^2 + 4k^2}{24\beta^6 (\beta^2 + k^2)(\beta^2 + 4k^2)} + \frac{4}{9\beta^6 (4k^2 + \beta^2)} + \frac{1}{27\beta^6 k^2} \times \right. \\ & \times \ln \frac{k^2 + \beta^2}{4k^2 + \beta^2} - \frac{3}{8\beta^2 (5\beta^2 + 4k^2)(\beta^2 + 4k^2)(\beta^2 + k^2)} + \\ & \left. + \frac{4}{\beta^2 (\beta^2 + 4k^2)(5\beta^2 + 4k^2)^2} - \frac{1}{k^2 (5\beta^2 + 4k^2)^3} \ln \frac{\beta^4}{(\beta^2 + k^2)(\beta^2 + 4k^2)} \right] \Bigg\}. \end{aligned} \tag{27}$$

In the special case $k = 0$ we have

$$\sigma = \frac{256\,\pi\,D_0^2\,\mu^2}{\hbar^4\,\beta^6\,\varepsilon^2}\left\{\frac{e^{4\beta R_0}}{64} + 4\,e^{2\beta R_0} - \frac{1}{2}\,e^{3\beta R_0}\right\}. \tag{28}$$

For a numerical calculation curve No. 6 is used. The polarization of the medium is taken into account by introducing the dielectric constant $\varepsilon$ into the exciton-exciton interaction potential.

Substituting into (28) $\varepsilon = 4.2$, $\mu = 10^{-27}$ g, $\beta R_0 = 2.13$, $R_0 = 9$ Å, $D_0 = 0.021$ eV we obtain $\sigma = 1.8 \times 10^{-14}$ cm$^2$.

For very large $k \gg \beta$

$$\sigma \approx \frac{256\,\pi\,D_0^2\,\mu^2}{\hbar^4\,\beta^4\,k^2\,\varepsilon^2}\left\{\frac{e^{4\beta R_0}}{384} + \frac{e^{2\beta R_0}}{6} - 0.04\,e^{3\beta R_0}\right\} \tag{29}$$

which, for the chosen numerical values of parameters, leads to $\sigma \approx 3.37/k^2$.

I am indebted to S. A. Moskalenko for proposing the treatment of this problem and for helpful discussions.

## References

[1] M. A. LAMPERT, Phys. Rev. Letters **1**, 450 (1958).
[2] S. A. MOSKALENKO, Optika i Spektroskopiya **5**, 147 (1958).
[3] E. HYLLERAAS and A. ORE, Phys. Rev. **71**, 493 (1947).
[4] D. B. CHESSNUT, J. chem. Phys. **41**, 472 (1964).
[5] M. TRLIFAJ, Czech. J. Phys. **13**, 631 (1963).
[6] R. C. CASSELLA, J. appl. Phys. **34**, 1703 (1963).
[7] F. VARSANYI and G. DIEKE, Phys. Rev. Letters **7**, 442 (1961).
[8] H. HELLMAN, Acta physicochim. URSS **1**, 938 (1935).
[9] S. A. MOSKALENKO and K. B. TOLPYGO, Zh. eksper. teor. Fiz. **36**, 149 (1959).
[10] S. A. MOSKALENKO, P. I. KHADZI, A. I. BOBRYSHEVA, and A. V. LELYAKOV, Fiz. tverd. Tela **5**, 1444 (1963).
[11] H. W. WYLD and B. D. FRIED, Ann. Phys. (USA) **23**, 374 (1963).
[12] Y. TOYOZAWA, Progr. theor. Phys. **20**, 53 (1958).
[13] S. YOMOSA, J. Phys. Soc. Japan **18**, 10 (1963).

*(Received March 10, 1966)*

See Short Note by D. Rühlicke

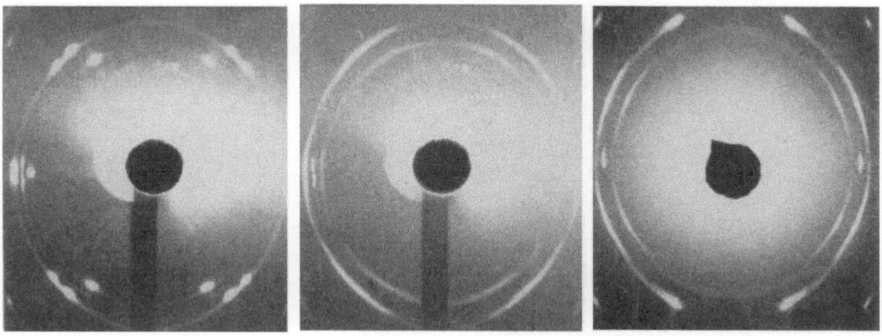

a b c

Fig. 1. Ausbildung der Drehkristallreflexe für verschiedene Verformungstemperaturen (Bruchstelle) a) 125 °C, b) 150 °C, c) 200 °C (Reflexfolge von innen nach außen $\{10\bar{1}0\}$, (0002), $\{10\bar{1}\bar{1}\}$)

Fig 3.. Gleitlinien für makroskopische Sekundärgleitung im System $(\bar{1}\bar{1}01)$; 1/3 $[\bar{1}\bar{1}20]$ (×125)

# *Short Notes*

phys. stat. sol. 16, K1 (1966)

Institut für Metallkunde und Materialprüfung
der Bergakademie Freiberg

## Zum Gleitmechanismus plastisch verformter Magnesiumeinkristalle oberhalb Raumtemperatur

Von
D. RÜHLICKE

Zur Beurteilung der Bedeutung sekundärer Gleitprozesse bei der plastischen Verformung von h.k.p. Metallen wurde als Ergänzung zu früheren Untersuchungen (1) der makroskopische Gleitmechanismus von Mg-Kristallen oberhalb der Raumtemperatur studiert.

Geeignet präparierte und für Einfachgleitung orientierte einkristalline Proben ($40^{\circ} < \chi_0 < 50^{\circ}$) wurden im Temperaturbereich von 100 bis 300 °C durch Zug verformt und die Orientierungsänderung der Stabachse nach schrittweiser Dehnung mittels Drehkristallverfahren bestimmt. Für $T < 150$ °C ergab sich in Analogie zu den Versuchen bei Raumtemperatur (1) alleinige Betätigung des Basisgleitsystems (0001); 1/3 $\langle 11\bar{2}0\rangle$. Definiert man als Beginn makroskopischer Sekundärgleitung die erste, nicht im Sinne der Basisgleitung stattfindende Verschiebung der Röntgenreflexe, so ergibt sich eine Einsatztemperatur von 150 °C. Fig. 1 zeigt Ausschnitte aus Drehkristalldiagrammen für 3 verschiedene Verformungstemperaturen. Auf der für 150 °C gültigen Aufnahme (b) ist die beginnende Aufspaltung der $\{10\bar{1}0\}$- und $\{10\bar{1}1\}$-Interferenzen deutlich sichtbar, was auf Orientierungsänderungen des Gitters infolge Betätigung sekundärer Gleitsysteme zurückzuführen ist. Die Verhältnisse sind bei höheren Temperaturen entsprechend deutlicher ausgeprägt (c). Für $T > 150$ °C lassen sich die verschiedenen Gleitmechanismen auf Grund unterschiedlicher Gitterdrehungen einwandfrei trennen, wobei für Sekundärgleitung eine äquivalente, aber um 60° versetzte Gleitrichtung mit dem Burgersvektor 1/3 $\langle 11\bar{2}0\rangle$ betätigt wird. Aus der für $T < 200$ °C röntgenographisch nachge-

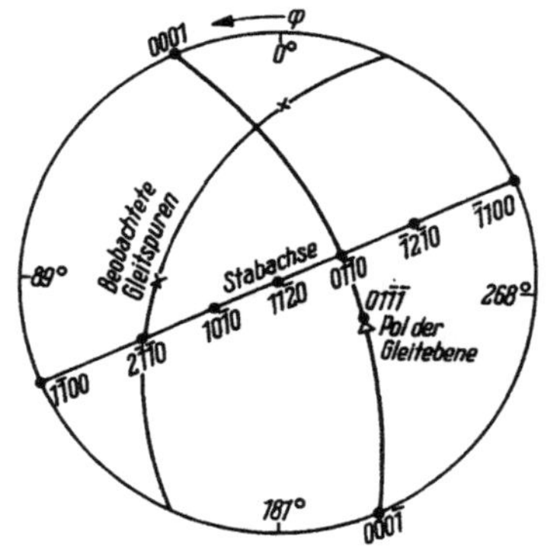


Fig. 2. Stereographische Auswertung der sekundären Gleitebene

wiesenen Sekundärgleitung resultieren jedoch nur geringe Abgleitungsbeträge. Diese Ergebnisse sind mit den von Yoshinaga und Horiuchi (2) entwickelten Vorstellungen verträglich, wonach für T < 250 °C die Beweglichkeit der Schraubenversetzungen infolge Versetzungsreaktionen blockiert wird, während im Temperaturbereich von 250 bis 350 °C die thermisch aktivierte Bewegung von Sprünge enthaltenden Schraubenversetzungen für die Vergrößerung des Verformungsvermögens verantwortlich sein soll.

Zur Indizierung der sekundären Gleitebene wurde ein quadratischer Mg-Kristall (4x4 $mm^2$; Stabachse ∥ <11$\bar{2}$0>) bei 275 °C durch Zug verformt, die entstandenen Gleitlinien lichtmikroskopisch vermessen und zusammen mit der Kristallorientierung stereographisch projiziert (Fig. 2). Fig. 3 zeigt ein Beispiel für Sekundärgleitlinien, deren Aussehen für nicht dichtestgepackte Ebenen des h.k.p. Gitters charakteristisch ist. Die ermittelte aktive Gleitebene des Typs {10$\bar{1}$1}, entspricht dem pyramidalen Gleitsystem (1$\bar{1}$01); 1/3 [$\bar{1}\bar{1}$20], für das Švábová (3) die maximale $\tau$-Komponente angegeben hat.

Das Verhältnis der Schubspannungskomponenten $\tau_{(1\bar{1}01)}/\tau_{(0001)}$ im Pyramiden- und Basisgleitsystem als Funktion der Kristallorientierung ist durch die Beziehung

$$\tau_{(1\bar{1}01)}/\tau_{(0001)} = N(\chi) = K\,(\sin\alpha\cot\chi + \cos\alpha) \qquad (1)$$

gegeben, wobei $\alpha$ = arctan c/a, $\chi$ der Winkel zwischen (0001) und der zugbeanspruchten Kristallachse sowie K eine nur von c/a abhängige Konstante ist. Gleichung (1) läßt sich unter Berücksichtigung der für die verwendeten Ausgangsorientierungen zulässigen Vereinfachung $\chi \sim \lambda$ aus den in (3) entwickelten Orientierungsfunktionen ableiten. Die errechnete Abhängigkeit $\tau_{(1\bar{1}01)}/\tau_{(0001)} = N(\chi)$ ist für den interessierenden Orientie-

ig. 4. Verteilung der Schub-
pannungen im Pyramiden- und
asisgleitsystem als Funktion
er Kristallorientierung

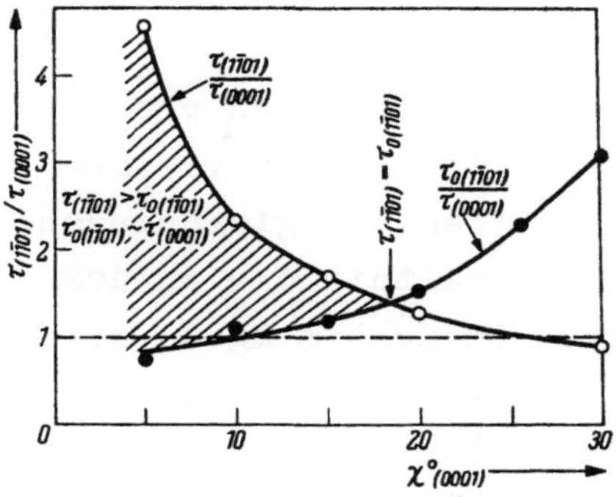


ungsbereich in Fig. 4 dargestellt. Zum Vergleich wurde außerdem as für 200 °C experimentell bestimmte Spannungsverhältnis $\tau_{o(1\bar{1}01)}/\tau_{(0001)}$ eingetragen ($\tau_{o(1\bar{1}01)}$ ist die kritische Schub-pannung für Sekundärgleitung nach (2); $\tau_{(0001)}$ ist der jewei-ige Fließspannungswert für Basisgleitung nach (4)). Im schraf-ierten Gebiet sollte makroskopische Sekundärgleitung auftreten, a hier die im Gleitsystem (1Ī01); 1/3 [ĪĪ20] wirkende Schub-pannung größer als die kritische Schubspannung des gleichen ystems ist ($\tau_{(1\bar{1}01)} > \tau_{o(1\bar{1}01)}$) und außerdem die zur Aufrecht-rhaltung der konkurrierenden Basisgleitung notwendigen Fließ-pannungswerte die Größe von $\tau_{o(1\bar{1}01)}$ erreichen ($\tau_{(0001)} \sim \tau_{o(1\bar{1}01)}$). Im Experiment wird makroskopische Sekundärgleitung ber erst für $\chi \sim 5^{o}$ beobachtet ($\tau_{(1\bar{1}01)} \gg \tau_{(0001)}$). Als rklärung erscheint das anormale Verfestigungsverhalten bei yramidaler Gleitung (2), bedingt durch die charakteristische usbildung dieser Gleitebenen sowie die latente Verfestigung es Sekundärsystems, sinnvoll.

Gleitebenen des Typs {10Ī1} sind bei der plastischen Ver-ormung von Mg und Mg-Legierungen bei und oberhalb Raumtempe-atur wiederholt beobachtet worden (1, 2, 5). Diese Pyramiden-benen 1. Art und 1. Ordnung sollten die wichtigsten Sekundär-leitebenen bei höheren Temperaturen sein, was bei geometri-cher Betrachtungsweise verständlich ist. Speziell zur makrosko-ischen Nebengleitung von Mg-Kristallen mit mittlerer Ausgangs-rientierung liegen nur von Schmid (4) vergleichbare Resultate or. Die kritische Temperatur betrug 250 °C. Yoshinaga und oriuchi (5) haben für den Beginn der Sekundärgleitung von

MgLi-Kristallen einen Temperaturbereich von 100 bis 200 °C angegeben.

Durch Verformen von MgCd-Kristallen mit 2 at% Cd wurden im wesentlichen die gleichen Ergebnisse erhalten, was sich auf das für diesen Legierungsgehalt nur geringfügig geänderte Achsenverhältnis c/a zurückführen läßt. Entsprechende Untersuchungen im System Mg-Cd sind in Vorbereitung.

Der Autor dankt den Herren Prof. Dr. F. Günther, Dr. M. Boček und Dr. G. Hötzsch für fördernde Diskussionen sowie Herrn Dipl.-Met. P. Birkholz für die sorgfältige Durchführung der röntgenographischen Arbeiten.

## Literatur

(1) D. RÜHLICKE, phys. stat. sol. 8, 93 (1965).

(2) H. YOSHINAGA und R. HORIUCHI, Trans. Japan Inst. Metals 5, 14 (1964).

(3) M. ŠVÁBOVÁ, phys. stat. sol. 3, 1819 (1963).

(4) E. SCHMID, Z. Elektrochem. 37, 447 (1931).

(5) H. YOSHINAGA und R. HORIUCHI, Trans. Japan Inst. Metals 4, 134 (1963).

(Received May 9, 1966)

phys. stat. sol. 16, K5 (1966)

University of Delaware, Physics Department Newark, Delaware

# The Influence of Oxygen in the Ultrahigh Vacuum Range on Electrical Properties of CdS[1)]

By

K.W. BÖER and R. SCHUBERT

The influence of oxygen on the electrical properties of CdS single crystals has been investigated by a number of authors (1 to 4) in a pressure range between one atmosphere and $10^{-5}$ Torr, and it has been observed that oxygen reduces the photoconductance, especially if the oxygen is ionized (2). These effects are slow at room temperature, with time constants of the order of minutes for adsorption and of hours for desorption, and can be explained as chemisorption at the surface. Adsorption and desorption depend on the electron concentration at the surface and therefore can be influenced by optical excitation and consequent shifts in the quasi-Fermi level as has been shown by Mark (3, 4). The change in the electrical conductance is explained by changes in space charge layers.

In this note, measurements of adsorption and desorption of oxygen in the range between $2\times10^{-12}$ and $5\times10^{-6}$ Torr and corresponding influences on photoconductance of undoped CdS single crystal platelets shall be reported.

Adsorption of oxygen at outgassed (vacuum heat treated) CdS, results in a slow almost irreversible decrease of the photoconductance in agreement with (3, 4).

However, with an electron source present, e.g., as it is when simultaneously operating a mass spectrometer, (located directly in front of a CdS crystal) a new influence of oxygen on the conductance can be observed. The photoconductance <u>increases</u> after crystal exposure to oxygen (see Fig. 1), with

---

1) Work supported in part by the National Aeronautics and Space Administration and by the U.S. Army Research Office in Durham.

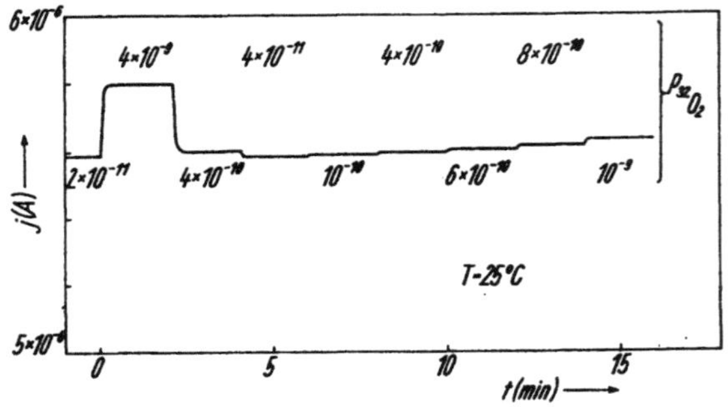


Fig. 1. Increase and decay of photocurrent ( ~10 ft-c white light) after stepwise increase (and decrease of the oxygen partial pressure as observed from the mass spectroscope (room temperature). Crystal dimensions $3\times9\times0.1$ mm$^3$ (gap electrode, applied voltage 1 V)

Table 1

Partial pressures (iongetterpump on)

| $p_{total}$ | $\approx 3 \times10^{-9}$ | $\approx 2 \times10^{-8}$ |
|---|---|---|
| $p_2$ | $8.5\times10^{-10}$ | $4.2\times10^{-9}$ |
| $p_4$ | 4.0 | 3.5 |
| $p_{12}$ | 0.3 | |
| $p_{15}$ | 0.1 | |
| $p_{16}$ | 13.0 | 1.6 |
| $p_{17}$ | 0.1 | 1.4 |
| $p_{18}$ | 0.5 | |
| $p_{19}$ | 0.1 | |
| $p_{28}$ | 3.3 | 4.5 |
| $p_{32}$ | 0.1 | 6.2 |
| $p_{40}$ | 0.2 | |
| $p_{44}$ | 0.5 | |

Mass spectrum at a total pressure of $3\times10^{-9}$, and $2\times10^{-8}$ Torr after the pressure was raised by leaking-in ultra pure oxygen (99.996 %)

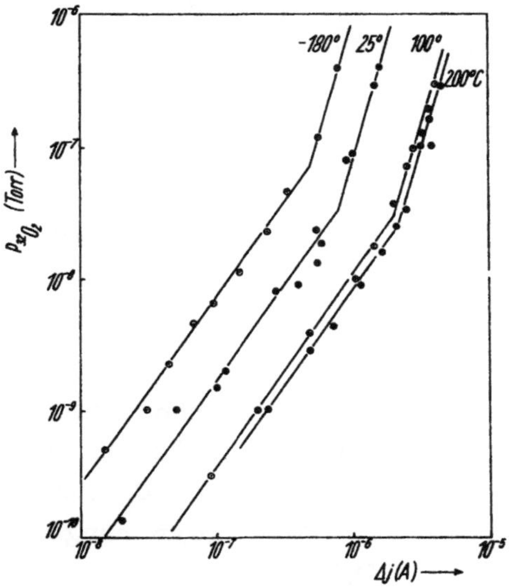


ig. 2. Change of photocurrent t 10 ft-c for different oxygen artial pressures at four temeratures

time constant[2)] for adsorpion and desorption of ≤ 0.2 s ven at −180 °C, i.e. as fast s changes in partial pressure f oxygen can be detected by he mass spectrometer.

In Fig. 2, the change of hotoconductance of the CdS rystal is plotted versus the hange of oxygen pressure for ifferent temperatures. One observes changes of the photourrent well above the noise level with changes in the partial ressure of oxygen down to the low $10^{-12}$ Torr range. This ffect is about six orders of magnitude more sensitive than as been previously reported for changes in the conductance f CdS as a function of oxygen pressure.

In Table 1 a mass analysis of the remaining gas in the ystem at a pressure of $3x10^{-9}$ Torr is given. After backfilling o $2x10^{-8}$ Torr with ultra pure oxygen, the partial pressure f other gases such as $H_2$, He, and CO are also increased to he same order of magnitude as oxygen, indicating enhanced esorption from the walls and the ion pump via exchange with xygen.

In order to distinguish the reported effect from the nfluence of gases other than oxygen, backfilling with ultraure $H_2$ and $CO_2$, which are known to influence the photoconuctance at higher pressures was carried out. The results are hown in Fig. 3 (solid curves).

) Time lag between indicated $O_2$ pressure changes in the mass spectrograph and current changes of the CdS crystal. The total change in current as shown in Fig. 1 is considerably slower because of slow changes in pressure.

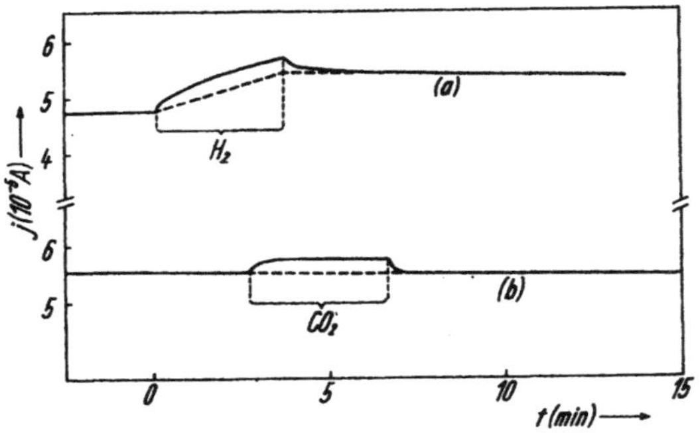


Fig. 3. Change of photocurrent at 10 ft-c after changing the $H_2$ partial pressure from $10^{-9}$ to $10^{-6}$ Torr
(a) and the $CO_2$ partial pressure from the $10^{-10}$ to the $10^{-8}$ Torr range
(b) dashed curves corrected for zero oxygen partial pressure using Fig. 2

Table 2

| $p_{total}$ | $\approx 1x10^{-6}$ | $\approx 2\ x10^{-8}$ |
|---|---|---|
| $p_2$ | $1000x10^{-9}$ | $0.8x10^{-9}$ |
| $p_4$ | 9 | 0.2 |
| $p_{12}$ | | 2.4 |
| $p_{16}$ | 1 | 2.6 |
| $p_{17}$ | 5 | |
| $p_{18}$ | | 0.4 |
| $p_{28}$ | 20 | 10.0 |
| $p_{32}$ | 3 | 1.6 |
| $p_{44}$ | | 3.4 |
| $p_{16} + p_{32}$ | $4x10^{-9}$ | $4.2x10^{-9}$ |

Mass spectrum at a total pressure of $10^{-6}$ Torr after pressure was raised by leaking-in ultra pure hydrogen (99.999 %) and at a total pressure of $2x10^{-8}$ Torr after leaking-in ultra pure carbon dioxide (99.995 %)

A mass spectroscopic analysis of the gas in the system or both backfillings is given in Table 2. In both cases the artial pressure of oxygen has risen to $4\times10^{-9}$ Torr.

After correcting the results of Fig. 3 by the current hanges for oxygen as given in Fig. 2, it is seen that there s no additional influence of $CO_2$ left (dashed curve in Fig. 3). he rapidly rising and decaying part super-imposed on the $H_2$ ffect also disappears after correcting for zero oxygen partial ressure. However, the slow increase in photoconductance with $_2$ backfilling is due to hydrogen in agreement with (2).

At the present time, the mechanism leading to the observed ery sensitive specific oxygen influence of the photoconductance s not fully transparent. The fact that an acceleration voltage or electrons in the mass spectrometer of at least 11 V is eeded for the described effect indicates the possibility of threshold energy of this order of magnitude for electrons. owever, the complex geometrical arrangement of the present ystem prevents an unambiguous analysis, which therefore should e postponed until further measurements with a system now nder construction are made.

From the present observations, one might conclude that olecular oxygen becomes charged: $O_2 + e \rightarrow O_2^-$ and migrates to he CdS surface where discharging and immediate desorption akes place. It thereby temporarily increases the current hrough the crystal, and raises the Fermi-level, causing fast esorption in agreement with (3).

## References

1) W. MUSCHEID, Ann. Phys. (Germany) 13, 322 (1953).
2) H. BERGER, K.W. BÖER, and E.-H. WEBER, Z. Phys. 158, 501 (1960).
3) P. MARK, J. Phys. Chem. Solids 25, 911 (1964); 26, 959 (1965); 26, 1767 (1965).
4) P. MARK, RCA Rev. 26, 461 (1965).

(Received May 23, 1966)

phys. stat. sol. 16, K11 (1966)

Institute of Physics, Czechoslovak Academy of Sciences, Prague

# Temperature Dependence of the FMR Linewidth of Iron Single-Crystal Platelets

By

B. HEINRICH and Z. FRAIT

The relaxation mechanism in the process of ferromagnetic resonance (FMR) in metals is not yet well understood. We can discuss the suitability of various relaxation theories only by comparing the theoretical resonance linewidths with values obtained experimentally. Usually the dependence of the linewidth on some external parameter is measured, e.g., the frequency dependence (1, 2) or the temperature dependence (3, 4).

We have studied the temperature dependence of the FMR linewidth of single-crystal platelets of pure iron, and have evaluated the temperature dependence of the relaxation time. The samples were grown by Kotrbová and Hauptman by a vapour-solvent method (5, 6), the surfaces of the platelets were of mirror-like quality and of the (100) type. The length of their sides was of the order of tenths of a mm, the thickness was in the range 0.8 to 1.2 μm. The measurements of microwave absorption were performed in the 3 cm wavelength region (f = 9.25 GHz), the static magnetic field was parallel to the platelet plane and to the [100] direction.

The temperature dependence of the linewidth obtained from experiment is plotted in Fig. 1. These results are in good agreement with Rodbell's measurements on iron whiskers (3). The linewidth values are determined from the inflection points of the derivative of the microwave absorption curve. (The absorption is proportional to the root of the real part of the complex experimental permeability $\mu_R$, or to the real part of the surface impedance Z of the sample (7).) The effect of surface pinning in the case of pure iron is negligible (6, 8), therefore we suppose that the resonance lines are broadened by an exchange-conductivity mechanism (7) in addition to a microphysical relaxation mechanism. We can see

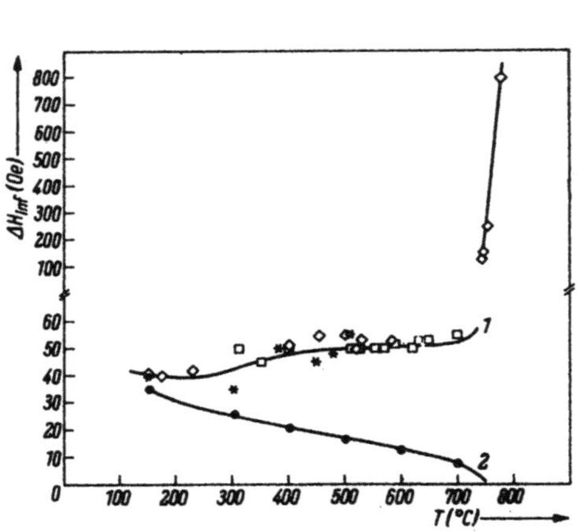


Fig. 1. Temperature dependence of the absorption linewidth (measured between the inflection points of the derivative curve) of iron single-crystal platelets (curve 1). Curve 2 shows computed width values due to exchange-conductivity broadening only

□ 1 PLT (0.8 μm);
* 2 PLT (1.0 μm);
◇ 3 PLT (1.2 μm)

from Fig. 1 that at low temperatures (150 to 200 °C) the exchange-conductivity broadening is a principal mechanism. Its influence decreases with increasing temperature, and the relaxation mechanism dominates at high temperatures (T near the Curie point).

We shall suppose, in agreement with other authors (1, 8), that the relaxation damping caused by a microphysical mechanism can be described by the Landau-Lifshits term:

$$\frac{1}{\gamma}\frac{d\mathfrak{M}}{dt} = \mathfrak{M} \times \mathfrak{H} + \left(\frac{2A}{M^2}\right)\mathfrak{M} \times \nabla^2 \mathfrak{M} - \left(\frac{\lambda}{\gamma M^2}\right)\mathfrak{M} \times (\mathfrak{M} \times \mathfrak{H}) . \qquad (1)$$

(This agrees with conclusions drawn from the general quantum-thermodynamic method (9).)
The temperature dependence of $\lambda$ (see Fig. 2) was obtained from experiment using the convenient relations given by Rodbell (8) We have used the temperature dependence of the saturation magnetization given by Bozorth (10), the temperature dependence of the electrical conductivity (11), the value of the effectiv exchange constant $A = 2.0 \times 10^{-6}$ erg/cm (15), and its temperature dependence in the form (12)

$$A_T = A_{20^\circ C}\left(\frac{M(T)}{M_{20^\circ C}}\right)^2 , \qquad (2)$$

where M(T) is the saturation magnetization at temperature T. The validity of Rodbell's linewidth formulas is limited in our

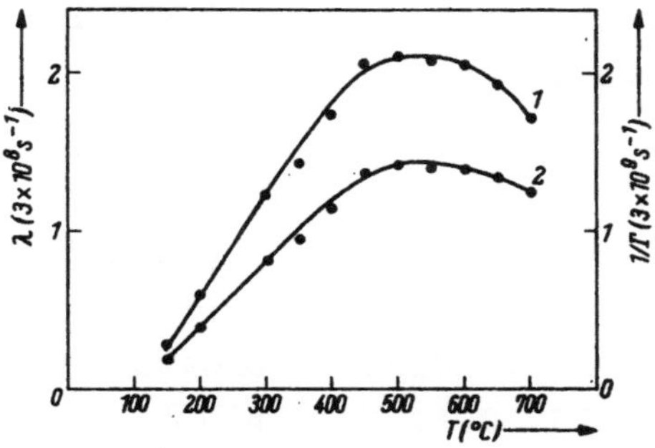


ig. 2. Temperature dependence of the Landau-Lifshits damping onstant (1) and the recipro- al value of relaxation time /Γ (2)

ase by the relation $H/4\pi M(T) \ll 1$ (H is the static internal ield at resonance) to temperatures up to 700 °C. The following ormula relates λ to the microphysical relaxation time Γ (13)

$$\frac{1}{\Gamma} = \lambda(\frac{\omega}{\gamma}) \frac{1}{M(T)} \left[1 + \left(\frac{4\pi M(T)\gamma}{\omega}\right)^2 \frac{1}{4}\right]^{1/2}, \qquad (2)$$

here: γ spectroscopic splitting ratio, ω microwave frequency. is equivalent to the Bloch-Bloembergen perpendicular rela- ation time $T_2$; this can be seen by comparing the permeability ensor components computed from the macroscopic theory (e.g. 1)) and from the quantum-mechanical theory (9).

A small part of the linewidth may be caused also by a ertain amount of demagnetizing field inhomogeneity, as the eometry of the sample differs from an ideal ellipsoid. Such roadening will decrease with temperature. Because the value f the relaxation factor at low temperature (150 °C) in our ase agrees well with room-temperature measurements performed n bulk samples (1) which are homogeneously magnetized, we eel that the influence of demagnetizing inhomogeneity on the inewidth of thin iron platelets can be neglected. The results how that the relaxation time Γ decreases with temperature up to 500 °C) and grows in the region where the saturation agnetization is strongly temperature-dependent. The sudden ncrease of the linewidth near the Curie point region T > 750 °C) is explained by the ferromagnetic-paramagnetic ransition: in the paramagnetic state dipole-dipole broadening s significant (14).

The authors thank J. Kaczer for helpful comments on the easurements, Z. Hauptman and M. Kotrbová for the preparation of the iron platelets, and R. Gemperle for performing the thick-ness measurements.

## References

(1) Z. FRAIT and H. MCFADEN, Phys. Rev. 139, 1173 (1965).
(2) S.M. BHAGAT, L.L. HIRST, and J.R. ANDERSON, Technical Reports (University of Maryland) No. 507, 472 (1965).
(3) D.S. RODBELL, Growth and Perfection of Crystals, Ed. C.A. NEUGEBAUER, J.B. NEWKIRK, and D.A. VERMILYEA, Wiley, New York 1958 (p. 247).
(4) A.S. RODBELL, Phys. Rev. Letters 13, 471 (1964).
(5) Z. HAUPTMAN and M. KOTRBOVÁ, Czech. J. Phys. B15, 64 (1965).
(6) Z. FRAIT, D. FRAITOVÁ, M. KOTRBOVÁ, and Z. HAUPTMAN, Czech. J. Phys., in press.
(7) W.S. AMENT, and G.T. RADO, Phys. Rev. 97, 1558 (1955).
(8) A.S. RODBELL, Physics 1, 279 (1965).
(9) S.V. TYABLIKOV, Metody kvantovoi teorii magnetizma, Izd. Nauka, Moscow 1965 (§ 38).
(10) R.M. BOZORTH, Ferromagnetism, D. van Nostrand Co. 1951.
(11) Gmelins Handbuch der anorganischen Chemie, Verlag Chemie GmbH Weinheim und Berlin.
(12) ZUBAREV, Uspekhi Fizicheskikh Nauk 71, 1 (1960).
(13) BERK, J. appl. Phys. 28, No. 2 (1957).
(14) J.H. VAN VLECK, Phys. Rev. 74, 1168 (1948).
(15) Z. FRAIT and M. ONDRIS, Izv. Akad. Nauk SSSR, ser. fiz. 28 No. 3 (1964).

(Received May 9, 1966)

phys. stat. sol. 16, K15 (1966)

Institute of Physics, Academy of Sciences of the Azerbaidzhan SSR, Baku

# Infrared Absorption of $A^{III}B^{VI}$ Single Crystals

By

G.A. AKHUNDOV and T.G. KERIMOVA

The optical properties of $A^{III}B^{VI}$ layer semiconductors n the infrared spectral region have not been completely tudied so far. In the present report transmission spectra f GaS, GaSe, InSe, and TlSe single crystals in the range .75 to 25 µm are given for the first time.

GaS, GaSe, and InSe single crystals were grown by slightly reezing, in the presence of an invariable temperature gradient cross the ingot, and TlSe single crystals were grown by verical zone melting.

GaS, GaSe, and InSe crystals have a hexagonal structure, hereas TlSe crystals have a tetragonal structure.

Single-crystal specimens of GaS, GaSe, and InSe with lane-parallel mirror surfaces were prepared by cleaving from large ingot perpendicular to the c-axis, and TlSe specimens arallel to this axis. Transmission spectra were obtained at 00 °K by means of an infrared spectrometer. The dimensions f the investigated samples varied in the ranges : 12 to 22 mm n length; 5 to 10 mm in width; 0.08 to 0.6 mm in thickness. he samples were illuminated perpendicular to the corresponding leavage plane.

Fig. 1 shows transmission spectra of GaS and GaSe. As an be seen, GaS and GaSe have wide transmission-ranges .75 to 14 µm and 0.75 to 17 µm, respectively, and absorption

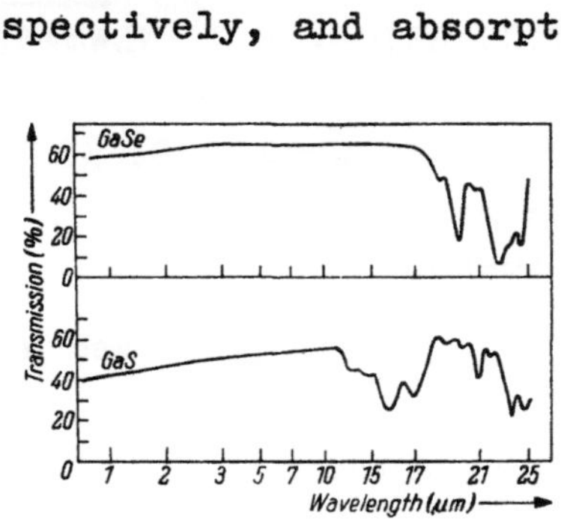


ig. 1. Transmission spectra of aS and GaSe single crystals at 00 °K

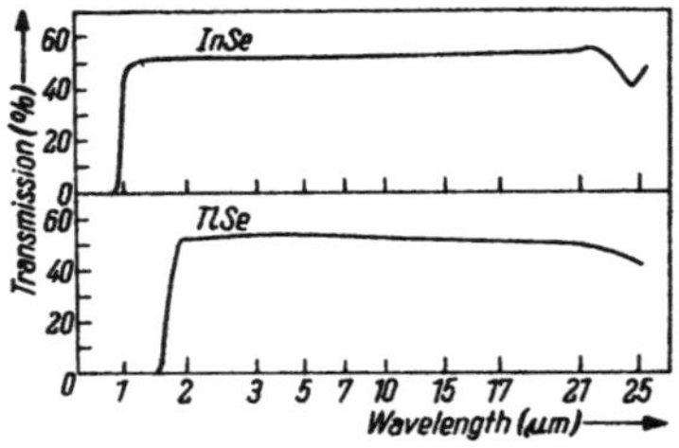


Fig. 2. Transmission spectra of InSe and TlSe single crystals at 300 °K

bands in the range 14 to 25 μm. It was found that the intensity of these bands does not vary from sample to sample even if impurities are introduced, and this fact confirms that these absorption bands are probably due to lattice vibrations. This assumption seems to be confirmed also by the fact that these crystals possess a considerable part of ionic binding.

Transmission spectra of InSe and TlSe are given in Fig. 2. It is seen that these single crystals, similarly as GaS and GaSe, have a wide range of transparency. In contrast to GaS and GaSe, the fundamental absorption edge in the case of InSe and TlSe lies in the infrared spectral region. The width of the forbidden gap determined from the fundamental absorption edge is found to be 1.2 and 0.8 eV for InSe and TlSe, respectively. These values are in good agreement with reported data (1, 2).

The investigation of the influence of impurities on the absorption and reflection of layer semiconductors will be continued.

The authors would like to thank Dr. G.B. Abdullaev for many helpful discussions.

## References

(1) J.C. BRICE, P.C. NEWMAN, and H.C. WRIGHT, Brit. J. appl. Phys. 9, 110 (1958).

(2) G.A. AKHUNDOV, G.B. ABDULLAEV et al., Izv. Akad. Nauk Azerbaidzhan SSR, ser. fiz., tekh. i matem. nauk 3, 107 (1964).

(Received May 31, 1966)

phys. stat. sol. 16, K17 (1966)

Electrical Engineering Department,
Massachusetts Institute of Technology

## Antiferromagnetism in $Mn_3B_2O_6$, $Co_3B_2O_6$, and $Ni_3B_2O_6$

By

R.E. NEWNHAM, R.P. SANTORO, P.F. SEAL, and G.R. STALLINGS

The orthoborates of manganese, cobalt, and nickel are iso-tructural with the mineral kotoite, $Mg_3B_2O_6$. Crystallographic easurements on the Mg, Co, and Ni salts have been reported y Sadanaga (1), Berger (2), and Götz (3). The kotoite struc-ure consists of a hexagonal close-packed oxygen arrangement, ith boron in triangular coordination and divalent metal ions n octahedral sites. Lattice parameters are given in Table 1; he space group is Pnmn, orthorhombic, with Z = 2.

Polycrystalline specimens of the transition-metal kotoites ere synthesized by solid state reaction. Stoichiometric mix-ures of vitreous $B_2O_3$ and the metal oxide were fired in air at 00 °C. Reaction products were identified from X-ray diffracto-eter patterns. Neutron-diffraction samples were prepared from sotope $B^{11}$, thus avoiding the large capture cross-section of $^{10}$.

Table 1

agnetic and crystallographic data for the transition-metal kotoites

| | $Mn_3B_2O_6$ | $Co_3B_2O_6$ | $Ni_3B_2O_6$ |
|---|---|---|---|
| color | tan | violet | green |
| a (Å) | $5.67_5$ | $5.46_2$ | $5.39_5$ |
| b (Å) | $8.69_5$ | $8.43_6$ | $8.39_1$ |
| c (Å) | $4.64_8$ | $4.52_9$ | $4.49_0$ |
| $T_N$ (°K) | 35 | 30 | 49 |
| $p_{eff}$ ($\mu_B$) | $6.1_8$ | $5.2_9$ | $3.0_7$ |
| $\theta$ (°K) | 185 | 63 | 5 |

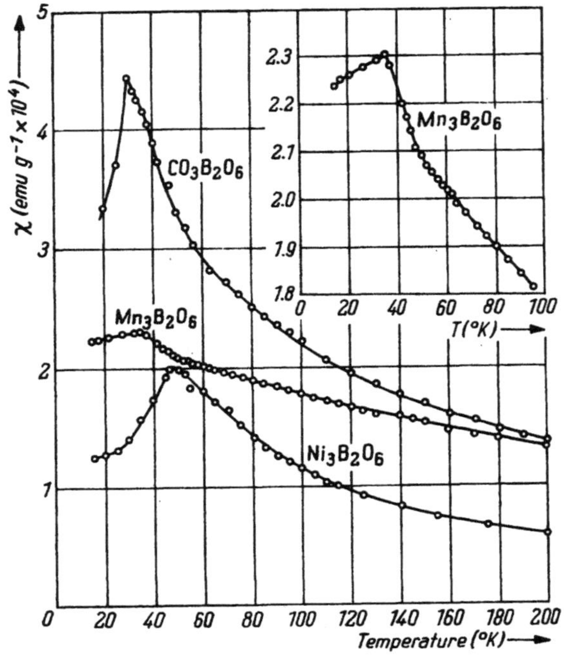


Fig. 1. Magnetic susceptibility of polycrystallin nickel, cobalt, and manganese orthoborates

Magnetic susceptibility measurements for $Mn_3B_2O_6$, $Co_3B_2O_6$, and $Ni_3B_2O_6$ were made with a vibrating sample magnetometer. From 300 °K down to liquid nitrogen temperature, the data follow a Curie-Weiss law. $\chi = C/(T + \theta)$. Values of the effective atomic moment ($p_{eff}$) and the paramagnetic Curie temperature ($\theta$) are given in Table 1. Paramagnetic-antiferromagnetic transitions occur at lower temperatures, as shown in Fig. 1. $Ni_3B_2O_6$ and $Co_3B_2O_6$ exhibit normal susceptibility behavior with $\chi(0)/\chi(T_N) \approx 2/3$, typical of a polycrystalline antiferromagnetic. The Néel points probably occur near the maxima at 30 °K ($Co_3B_2O_6$) and 49 °K ($Ni_3B_2O_6$). The low temperature susceptibility of manganese orthoborate is more unusual. It shows a small break near 55 °K, rises to a peak at 35 °K, and decreases only slightly at lower temperatures. Similar susceptibility curves have been observed for $Mn_2SiO_4$ and $Fe_2SiO_4$, where two antiferromagnetic transitions occur (4).

Neutron diffraction patterns of the three compounds were taken at room temperature and 6 °K, above and below the transition temperatures. Magnetic reflections indicative of long-range order occurred in the low-temperature powder patterns. An analysis of the $Co_3B_2O_6$ neutron data has been reported previously (5). $Ni_3B_2O_6$ gives a similar pattern, which can also be indexed on a $\underline{a}$ x 2$\underline{b}$ x 2$\underline{c}$ magnetic unit cell, four times th size of the chemical cell. Only magnetic reflections with ndices k and l odd were observed. Powder pattern intensity ata is insufficient for a complete determination of the comlex magnetic structures. The magnetic reflections of $Mn_3B_2O_6$ ould not be indexed on the chemical unit cell or any simple uperlattice.

## References

1) R. SADANAGA, X-rays 5, 2 (1948).
2) S.V. BERGER, Acta Chemica Scand. 3, 660 (1949).
3) W. GÖTZ, Naturwissenschaften 50, 567 (1963).
4) R.P. SANTORO, R.E. NEWNHAM, and S. NOMURA, J. Phys. Chem. Solids 27, 655 (1966).
5) R.E. NEWNHAM, M.J. REDMAN, and R.P. SANTORO, Z. Krist. 121, 418 (1965).

(Received May 31, 1966)

phys. stat. sol. 16, K21 (1966)

epartment of Solid State Physics, Charles University, Prague

# Interband Faraday Effect in CdTe Single Crystals

By

M. ZVÁRA, F. ŽALOUDEK, and V. PROSSER

In recent years much attention has been paid to the mag- eto-optical properties of semiconductors. The Faraday effect n particular, as a dispersive magneto-optical phenomenon, is ery sensitive to the band structure of semiconductors in a agnetic field.

This paper gives experimental results of Faraday rotation easurements on p-type CdTe near the absorption edge at 100 and 00 °K. Some of the results have been reported at the Selenium- ellurium Symposium in New York (1).

The single beam method has been used for detecting Faraday otation. The schematic diagram of the experimental arrangement s shown in Fig. 1. Light from the source A (100 W tungsten amp) after passing through the grating monochromator B CF4NI Optica Milano, linear dispersion 16 Å/mm in the investi- ated region) and polarizer C, has been chopped by a mechani- al chopper D at a frequency of 250 Hz. The set of mirrors $_1 ... E_6$ enables measurements of both the Faraday and Voigt ffects, when the sample F is in the magnetic field G–G'. The ight then enters the analyser H whose plane of polarization as a fixed angle of 45° with that of the polarizer. The small hanges caused by the rotation of the plane of polarization re detected by a photomultiplier I (Zeiss M 12 F 35, cooled y solid $CO_2$), or with a Mullard PbS detector. The signal is ecorded by a lock-in amplifier and potentiometric recorder.

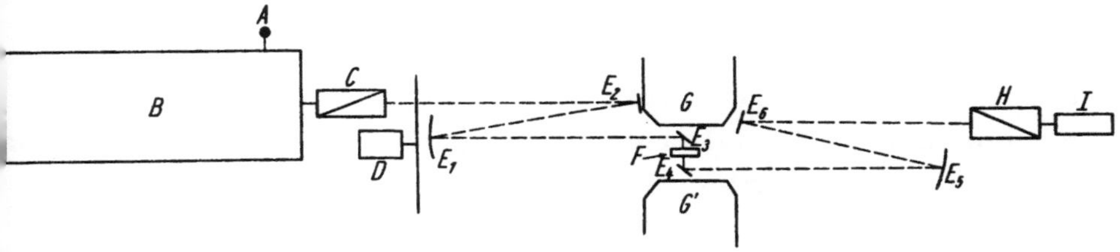


ig. 1. Schematic diagram of the experimental arrangement

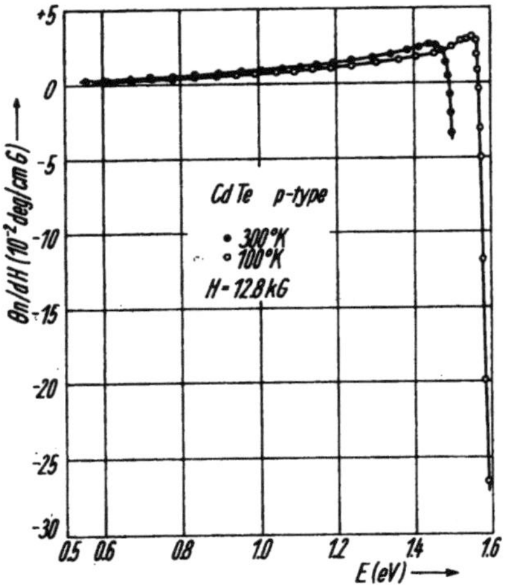


Fig. 2. Energy dependence of the specific Faraday rotation multiplied by the refractive index, θn/Hd, in the absorption edge region at 300 and 100 °K in p-type CdTe

From these intensity differences the angle of rotation can be evaluated (2).

Faraday rotation has been measured in high resistivity ($\varrho$ = $10^8$ Ωcm) p-type CdTe single crystals grown by a molten zone method. Plates with a thickness of about 10 μm for the absorption region and about 1 mm for the transmitting region were prepared by grinding and optical polishing. They were mounted strain-free in the common type of cryostat including mirrors $E_3$ and $E_4$. The measured energy dependence of Faraday rotation in the absorption edge region at 300 and 100 °K is shown in Fig. 2. The positive angles of rotation at 300 and 100 °K are shown in Fig. 3a and 3b together with the measurements of Marple (3) and Ebina et al. (4).

The spectral dependence of the Faraday rotation relative sharply drops from positive to negative values after passing zero at E = 1.491 or 1.569 eV. A similar effect has been observed in Ge (5), GaAs, GaSb (6) and some other $A^{III}B^{V}$ compounds. Of the $A^{II}B^{VI}$ semiconductors, CdSe so far shows a similar dependence (7). In other compounds (CdS, ZnS, ZnSe, ZnTe, ZnO (see (4, 8, 9)) steadily rising positive values of rotation have been observed when approaching the absorption edge. The measurements of Marple (3) and Ebina et al. (4) on CdTe crystals do not show any indication of a sign reversal.

In agreement with Ebina et al. (4) the experimental values in the region $E < E_g$ cannot be fitted with either theoretical expressions for direct transitions (10, 11). The main reason for this discrepancy could be possibly found in the complicated

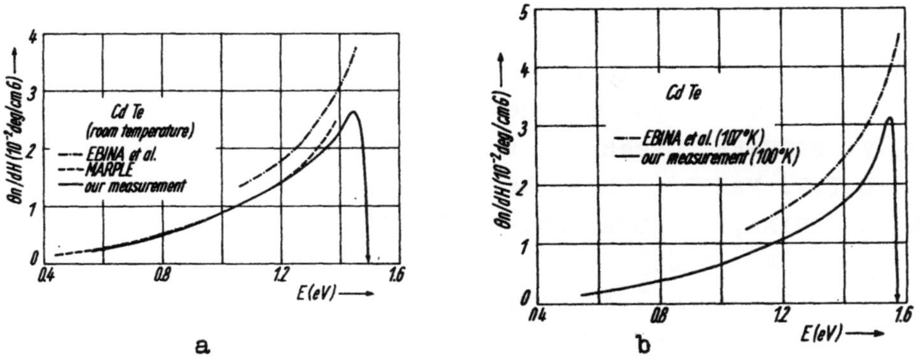


a b

ig. 3. Comparison of our results of Faraday rotation with results of Marple (3) and Ebina et al. (4) at 300 (a) and 100 °K (b). For greater distinctness the Faraday rotation is plotted only in the region of positive values

and structure of CdTe which has a relatively large spin-orbital plitting of the valence band. As a result of the dispersive haracter of the Faraday effect, both intraband optical tranitions (transitions $V_1 \rightarrow V_3$ in the valence band of p-type rystals) and interband transitions between the first Landau evels will occur in the spectral region under examination.

The shape of the Faraday rotation curves near the aborption edge indicates a strong spin-orbital interaction accoring to the following consideration. In the sequence of compounds of cadmium with group VIb elements, CdS, CdSe, CdTe, nly CdS shows no sign reversal near $E_g$, in mixed CdS-CdSe rystals, however, the tendency to change sign has been oberved when increasing the Se:S ratio (7). The sign reversal s then quite obvious in the case of CdTe. In the same sequence he spin-orbital interaction increases, giving the following atio for $S = \Delta_0/E_g$ ($\Delta_0$ spin-orbital splitting energy)
$S_{CdS} : S_{CdSe} : S_{CdTe} = 0.026 : 0.23 : 0.50$ (see e.g. (12)).

For comparison with theory in detail, more precise data n band structure parameters of CdTe are essential, as well s further measurements on both n- and p-type materials at nergies near the absorption edge. Measurements of this type are in progress.

References

(1) V. PROSSER, Č. KOŇÁK, J. DILLINGER, and M. ZVÁRA, Selenium Tellurium Symposium, New York 1965.
(2) V. PROSSER, Czech. J. Phys. B15, 128 (1965).
(3) D.T.F. MARPLE, Phys. Rev. 129, 2468 (1963).
(4) A. EBINA, T. KODA, and S. SHIONOYA, J. Phys. Chem. Solids 26, 1947 (1965).
(5) H. PILLER and V.A. PATTON, Phys. Rev. 129, 1169 (1963).
(6) H. PILLER, Proc. Internat. Conf. Phys. Semicond., Paris 1964 (p. 297).
(7) M. BALKANSKI, E. AMZALLAG, and D. LANGER, phys. stat. sol. 11, K59 (1965).
(8) M. BALKANSKI, E. AMZALLAG, and D. LANGER, J. Phys. Chem. Solids 27, 299 (1966).
(9) E. JAHNE, phys. stat. sol. 10, K35 (1965).
(10) L.M. ROTH, Phys. Rev. 133, A 542 (1964).
(11) I. BOSWARVA, R.E. HOWARD, and A.D. LIDIARD, Proc. Roy. Soc A269, 125 (1962).
(12) M. CARDONA, J. Phys. Chem. Solids 24, 1543 (1963).

(Received June 2, 1966)

phys. stat. sol. 16, K25 (1966)

aterials Research Laboratory, Martin Company, Orlando, Florida
a) and Department of Physics, Battersea College of Technology,
London (b)

## Crystallographic Formulae for Hexagonal Lattices - Simplifications

By
H.M. OTTE (a) and A.G. CROCKER (b)

In an earlier paper (1) a Table was given of crystallo-
raphic formulae for hexagonal lattices. This Table is repro-
uced here in revised form. The symbols have the same meanings
s before (1), namely:

represents the Miller or Miller-Bravais indices of a direction.
represents the Miller or Miller-Bravais indices of a plane.
is the identity distance along a direction d.
is the interplanar spacing of planes with normal $n^*$.
is the angle between two directions, $d_1$ and $d_2$.
is the angle between two plane normals $n_1^*$ and $n_2^*$.
is the angle between a direction d and a plane normal n.

The principal modifications that have been made occur in
ne column for the 4-axis, 4 place formulae which are subject
o the restrictions $u + v + t = h + k + i = 0$. Use of these
elations permit the formulae to be simplified. Misprints and
ranscription errors have also been corrected. One correction
hould also be made in the text: the first bracket of equation
(8) for $\cos \rho$ should read $(a_{ij}\, u^i\, v^j)$.

We would like to take this opportunity of bringing to the
ttention of the readers Frank's paper (2), of which we were
naware at the time we checked our proofs, and which appeared
n print about the same time as ours (1). Frank's approach is
lightly different from that adopted by us and no attempt is
ade in his paper to summarize the formulae in tabular form.

### References

1) H.M. OTTE and A.G. CROCKER, phys. stat. sol. 9, 441 (1965).
2) F.C. FRANK, Acta cryst. 18, 862 (1965).

(Received June 2, 1966)

Table
Crystallographic Formulae for Hexagonal Lattices

| | Ortho-hexagonal | 3-axis hexagonal |
|---|---|---|
| D | [pqr] | UVW |
| P | (efg) | (HKL) |
| d | $a\left[p^2 + 3q^2 + \gamma^2 r^2\right]^{1/2}$ | $a\left[U^2 - UV + V^2 + \gamma^2 W^2\right]^{1/2}$ |
| n | $a(e^2 + \frac{1}{3}f^2 + \gamma^{-2}g^2)^{-1/2}$ | $a(\frac{4}{3}\left[H^2 + HK + K^2\right] + \gamma^{-2}L^2)^{-1/2}$ |
| cos φ | $d_1^{-1}d_2^{-1}a^2(p_1p_2 + 3q_1q_2 + \gamma^2 r_1r_2)$ | $d_1^{-1}d_2^{-1}a^2\left[U_1U_2 - \frac{1}{2}(U_1V_2 + U_2V_1) + V_1V_2 + \gamma^2 W_1W_2\right]$ |
| cos ψ | $n_1n_2a^{-2}(e_1e_2 + \frac{1}{3}f_1f_2 + \gamma^{-2}g_1g_2)$ | $\frac{4}{3}n_1n_2a^{-2}\left[H_1H_2 + \frac{1}{2}(H_1K_2 + H_2K_1) + K_1K_2 + \frac{3}{4}\gamma^{-2}L_1L_2\right]$ |
| cos η | $nd^{-1}(pe + qf + rg)$ | $nd^{-1}(UH + VK + WL)$ |

| | 4-axis hexagonal | |
|---|---|---|
| | 4-place (Miller-Bravais) indices | 3-place (Miller) indices |
| D | [uvtw] | [uv·w] |
| P | (hkil) | (hk·l) |
| d | $a\left[\frac{3}{2}(u^2 + v^2 + t^2) + \gamma^2 w^2\right]^{1/2}$ | $a\left[3(u^2 + uv + v^2) + \gamma^2 w^2\right]^{1/2}$ |
| n | $a(\frac{2}{3}\left[h^2 + k^2 + i^2\right] + \gamma^{-2}l^2)^{-1/2}$ | $a(\frac{4}{3}\left[h^2 + hk + k^2\right] + \gamma^{-2}l^2)^{-1/2}$ |
| cos φ | $d_1^{-1}d_2^{-1}a^2\left[\frac{3}{2}(u_1u_2 + v_1v_2 + t_1t_2) + \gamma^2 w_1w_2\right]$ | $3d_1^{-1}d_2^{-1}a^2\left[u_1u_2 + \frac{1}{2}(u_1v_2 + u_2v_1) + v_1v_2 + \frac{1}{3}\gamma^2 w_1w_2\right]$ |
| cos ψ | $n_1n_2a^{-2}(\frac{2}{3}\left[h_1h_2 + k_1k_2 + i_1i_2\right] + \gamma^{-2}l_1l_2$ | $\frac{4}{3}n_1n_2a^{-2}\left[h_1h_2 + \frac{1}{2}(h_1k_2 + h_2k_1) + k_1k_2 + \frac{3}{4}\gamma^{-2}l_1l_2\right]$ |
| cos η | $nd^{-1}\left[hu + kv + it + lw\right]$ | $nd^{-1}\left[(2u + v)h + (u + 2v)k + lw\right]$ |

a and $c = \gamma a$ are the lattice parameters.

phys. stat. sol. 16, K27 (1966)

Central Electricity Generating Board,
Berkeley Nuclear Laboratories, Berkeley

# Grain Boundary Impurity Diffusion

By
G.B. GIBBS

Theoretical analyses of grain boundary self diffusion have een presented by Fisher (1), Whipple (2) and Suzuoka (3). heir implications have been thoroughly discussed by Le Claire 4). Experimental techniques for measuring grain boundary difusion coefficients are reviewed in a book by Leymonie and acombe (5). They all lead to values for a quantity A which he above analyses indicate to be the product $D'\delta$, where D' s boundary diffusivity and $\delta$ is boundary width. The purpose f the present note is to point out that for impurity diffuion $A = \alpha D'\delta$, where $\alpha$ is the solute segregation factor.

Writing c' and c as concentrations of tracer atoms inside nd outside a boundary respectively, the diffusion equations n these two regions are

$$\frac{\partial c'}{\partial t} = D' \frac{\partial^2 c'}{\partial y^2} + \frac{2D}{\delta} \frac{\partial c}{\partial x} , \qquad (1)$$

$$\frac{\partial c}{\partial t} = D \left\{ \frac{\partial^2 c}{\partial x^2} + \frac{\partial^2 c}{\partial y^2} \right\} , \qquad (2)$$

here D is the volume diffusion coefficient. The concentrations and c' are related at the boundary interface by (6)

$$c' = (c \exp \Delta F/RT)/(1 + c \exp \Delta F/RT) = \alpha c, \qquad (3)$$

here $\Delta F$ is the solute-grain boundary binding energy. Equations 1) and (3) combine to give

$$\partial c/\partial t = D' \left( \frac{\partial^2 c}{\partial y^2}\right) + (2D/\alpha\delta) \left(\frac{\partial c}{\partial x}\right) \qquad (4)$$

s a limiting condition on the volume diffusion equation (2) t the boundary.

ince the self-diffusion analyses suppose $\alpha$ = unity, their reults may be applied generally to impurity diffusion if $\delta$ is replaced throughout by $\alpha\delta$. In particular,

$$A = \alpha D' \delta . \tag{5}$$

The importance of solute segregation at grain boundaries has been emphasised in a different connection by Mortlock (7). However it is usually overlooked in discussion of the relative A-values for different impurities, although variations in $\alpha$ may be comparable with changes in D'. For a weakly-segregating impurity $\alpha \approx \exp \Delta F/RT$. Writing $\Delta F = \Delta U - T \Delta S$,

$$A = D_0' \delta \ \exp - \Delta S/R \ \exp - (P_{g_i b_i} - \Delta U)/RT. \tag{6}$$

Reported variations in the overall activation energy of A between different impurities in a given solvent are often only $\sim 1$ to 3 kcal $mol^{-1}$ which could easily be accounted for entire by $\Delta U$.

A semi-empirical equation relating $\Delta F$ in a wide-angle grain boundary with solute size factor and elastic moduli of solvent and impurity has been given by McLean (6). Typically for magnesium in aluminium, $\Delta F$ is $\approx 1.2$ kcal $mol^{-1}$ and $\alpha$ is $\sim 4.5$ at 400 $^{\circ}K$. Segregation effects are particularly importan for interstitial impurities where $\Delta F$ may be $\approx 15$ kcal $mol^{-1}$. As yet, no theoretical model is available for the additional possibility that impurity atoms alter the boundary width.

For a small-angle grain boundary which may be described as a dislocation array, $\Delta F$ may be obtained from the formula of Bilby (8) as indicated by Queisser et al. (9).

This note is published by permission of the Central Electricity Generating Board.

References

(1) J.C. FISHER, J. appl. Phys. 22, 74 (1951).

(2) R.T.P. WHIPPLE, Phil. Mag. 45, 1225 (1954).

(3) T. SUZUOKA, Trans. Japan Inst. Metals 2, 25 (1961).

(4) A.D. LE CLAIRE, Brit. J. appl. Phys. 14, 351 (1963).

(5) C. LEYMONIE and P. LACOMBE, Radioactive Tracers in Physical Metallurgy, Chapman and Hall, London 1963 (p. 81).

(6) D. MCLEAN, Grain Boundaries in Metals, Clarendon Press, Oxford 1957 (p. 118).

(7) A.J. MORTLOCK, Acta metall. 8, 132 (1960).
(8) B.A. BILBY, Proc. Phys. Soc. A63, 191 (1950).
(9) H.J. QUEISSER, K. HUBNER, and W. SHOCKLEY, Phys. Rev. 123, 1245 (1961).

(Received June 6, 1966)

phys. stat. sol. 16, K31 (1966)

Institute of Physics, Academy of Sciences of the Azerbaidzhan SSR, Baku

## Determination of the Oxygen Content in Selenium

By

G.B. ABDULLAEV, S.I. MEKHTIEVA, G.M. ALIEV, D.SH. ABDINOV, and T.G. KERIMOVA

From investigations of the effect of Mn, Cd, Sb, Tl, and a admixtures on the electrical and thermal properties of $B_5$ rade high purity selenium (99.99999 %) before and after de-xidation (in solid and liquid states) it is concluded that xygen in selenium creates acceptor levels, and that the above-entioned admixtures act as deoxidizing ones by forming oxygen omplexes of the types MnO, CdO, $Sb_2O_3$ etc. (1, 2).

It is the aim of the present paper to check the validity f this assumption by measuring optical absorption of selenium n the infrared spectral region and determine the amount of xygen in selenium.

Measurements were carried out at room temperature on $B_5$ rade amorphous selenium.

Amorphous specimens were produced by abrupt cooling of melt at liquid nitrogen temperature. In this way plane-paral-el plates were obtained with a thickness of 0.75 to 1.0 mm.

Transmission spectra were taken by the use of a double-eam infrared spectrophotometer in the wavelength range 0.75 o 25 μm.

Transmission spectra for the above specimens are shown n the figure presented here, where curves 1, 2, 3, and 4 refer o ordinary (non-deoxidized) selenium, deoxidized selenium, nd selenium with Cd and Mn admixtures, respectively.

It is shown that in ordinary selenium (curve 1) absorption ands are observed at 11.03, 13.60, 17.65, and 20.45 μm with nergies of 0.11, 0.009, 0.07, and 0.06 eV, respectively.

After deoxidation (curve 2) or introduction of 0.15 at% d and Mn (curves 3 and 4, respectively) the absorption bands t 11.03, 13.60 and 17.65 μm disappear, but the intense band t 20.45 μm remains. The absorption spectrum of amorphous se-

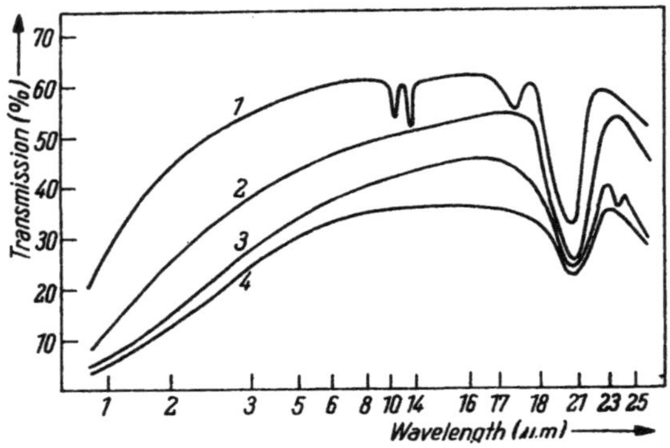


Fig. 1. Spectral transmission of amorphous selenium
1 - ordinary Se
2 - deoxidized Se
3 - Se + 0.15 at% Cd
4 - Se + 0.15 at% Mn

lenium has been considered in many works (3 to 10), but the results differ considerably from each other.

In one of these works (3), in addition to some weak absorption bands, two strong bands have been observed at 13.5 and 20.5 μm, these are believed to be fundamental bands determined by the Se-Se bond.

In another work (4) practically no single band has been found.

In (5) it is reported that the measurements of (4) are correct and refer to pure selenium, while the absorption band at 20.5 μm found in (3) is attributed to the presence of admixtures in selenium or in another modification of it.

One of the above works (6) covers measurements of absorption spectra for the purest selenium within the range 2 to 25 μm; absorption bands have been found at 13.4 and 20.4 μm. Bands at 13.5 and 20.5 μm have also been observed in (10). In this work it is indicated that the absorption bands at 13.5 and 17.9 μm are due to the presence of $SeO_2$ in selenium.

The results of the present work show that the band at 20.45 μm is a proper absorption band of amorphous selenium determined by the Se-Se bond.

Bands at 11.03, 13.60, and 17.65 μm are due to the presence of an oxygen admixture in selenium which disappears after deoxidation.

We have not observed absorption bands up to 10 μm which

re obtained in other works (3, 7), as our material with a con-
ent of metallic and non-metallic admixtures of less than
$)^{-5}$ weight% was purer than that used by the authors of the
bove works. This shows that these absorption bands obtained
the above authors (3, 7) are connected with admixtures.

As the admixtures of Cd and Mn act as deoxidizing ones
forming oxygen complexes in selenium, the absorption bands
t 11.03, 13.60, and 17.65 μm disappear in these cases too.

The absorption band at ≈23 μm observed in the case of
morphous selenium with an admixture of Cd is apparently due
the variation of the cadmium-oxygen bond.

It can be supposed that also an absorption band corres-
onding to the MnO complex should be observed, but this lies
resumably beyond the range of our measurements.

By taking into account that oxygen in selenium exists pre-
ominantly as $SeO_2$ (2, 10), the number of $SeO_2$ particles can
e calculated from the maximum of the absorption coefficient
$\alpha_{max}$) as in (11):

$$N_o = \frac{9n}{(n^2 + 2)^2} \frac{3\mu c}{\pi e^2} \alpha_{max} H, \qquad (1)$$

here n the refractive index of amorphous selenium, $\mu$ the
educed mass of the oscillator ($SeO_2$), e the electronic charge,
the velocity of light, $\alpha_{max}$ the value of the absorption co-
fficient at 11.03, 13.60, and 17.65 μm, and H the half-width
in eV) of these absorption bands.

Taking into account that $n = 2.5$ (12), $\mu = 22.77\ m_p$,
$= 4.8 \times 10^{-10}$ CGSU, $c = 3 \times 10^{10}$ cm/s, $\alpha_{max} = 9.87\ cm^{-1}$, and
$= 8.82 \times 10^{-3}$, $3.45 \times 10^{-3}$, and $8.66 \times 10^{-4}$ eV for the absorption
ands at 11.03, 13.60, and 17.65 μm, respectively, we obtain
rom the height of the three absorption bands), for the total
umber of $SeO_2$ particles in selenium a value of $\approx 8.23 \times 10^{18}\ cm^{-3}$,
nd for the oxygen concentration $\approx 5.49 \times 10^{18}\ cm^{-3}$.

## References

1) G.B. ABDULLAEV, S.I. MEKHTIEVA, D.SH. ABDINOV, and G.M. ALIEV, phys. stat. sol. 2, 891 (1965).

(2) G.B. ABDULLAEV, D.SH. ABDINOV, and G.M. ALIEV, Dokl. Akad Nauk Azerbaidzhan SSR 21, 18 (1965).
(3) H.A. GEBBIE and C.G. CANNON, I.O.S.A. 42, 277 (1952).
(4) R. FRERICHS, I.O.S.A. 43, 1153 (1953).
(5) I. LECOMTE, Spectroscopie dans l'infrarouge, Flügge, Handb. d. Phys. 26, 643 (1958).
(6) F. ECKART and H. BERG, Naturwissenschaften 45, 335 (1958).
(7) H. GOBRECHT and A. TAUSEND, Z. Phys. 173, 54 (1963).
(8) A. VASKO, Silikaty 4, 169 (1960).
(9) I. SRB and A. VASKO, Czech. J. Phys. 11, 664 (1961).
(10) A. VASKO, phys. stat. sol. 8, K41 (1965).
(11) W. KEISER, P.H. KECH, and C.F. LANGE, Phys. Rev. 101, 126 (1956).
(12) R.S. CALDWELL and H. FAN, Phys. Rev. 114, 664 (1959).

(Received June 7, 1966)

phys. stat. sol. 16, K35 (1966)

Institute of Metal Physics, Academy of Sciences of the USSR, Moscow

# Influence of Spatial Inhomogeneities on the Nuclear Polarization due to Direct Current in Semiconductors

By

V.P. KALASHNIKOV

Clark and Feher (1) have found experimentally that the steady-state nuclear magnetization induced by the flow of hot carriers in a semiconductor depends not only on the even powers of the external electric field, as the theory of the effect for a homogeneous semiconductor predicts (2, 3), but also contains odd powers of E, which are important at low field strength and may result in a change of sign of the nuclear magnetization with the reversal of the electric field.

It is possible to explain this phenomenon by the influence of spatial inhomogeneities in the distribution of conduction electrons.

The effective Hamiltonian of carriers in crossed electric and magnetic fields (two-band model of a semiconductor (4); notations as in (2))

$$\mathcal{H} = \frac{1}{2m}\left(\mathfrak{p} - \frac{e}{c}\mathfrak{A}\right)^2 + e\varphi(\mathfrak{r}) - \mu\vec{\sigma}\,\mathfrak{H}(\mathfrak{r}) + C\left[\mathfrak{E}(\mathfrak{r}), \mathfrak{p} - \frac{e}{c}\mathfrak{A}\right]\vec{\sigma} + \mathcal{H}_{int}$$

$$\mathfrak{H}(\mathfrak{r}) = \mathrm{rot}\,\mathfrak{A}\,; \qquad \mathfrak{E}(\mathfrak{r}) = -\vec{\nabla}\varphi(\mathfrak{r}) \tag{1}$$

$$C = \frac{c\mu}{\varepsilon_g}\,\frac{g-2}{g}\,\frac{\Delta_{so} + 2\varepsilon_g}{\Delta_{so} + \varepsilon_g}$$

leads to the following coupled set of kinetic equations for the density $n(\mathfrak{p},\mathfrak{r})$ and magnetization $M_\alpha(\mathfrak{p},\mathfrak{r})$ of carriers

$$\frac{\partial n}{\partial t}(\mathfrak{p},\mathfrak{r}) + \frac{p_\alpha}{m}\frac{\partial}{\partial r_\alpha} n(\mathfrak{p},\mathfrak{r}) + e\left\{E_\alpha(\mathfrak{r}) + \frac{1}{mc}\left[\mathfrak{p},\mathfrak{H}(\mathfrak{r})\right]_\alpha\right\}\frac{\partial}{\partial p_\alpha} n(\mathfrak{p},\mathfrak{r}) +$$

$$+ \frac{1}{\mu}\left\{\mu\frac{\partial H_\beta(\mathfrak{r})}{\partial r_\alpha}\frac{\partial}{\partial p_\alpha} + C\left[\mathfrak{p}, \frac{\partial\mathfrak{E}(\mathfrak{r})}{\partial r_\alpha}\right]_\beta\frac{\partial}{\partial p_\alpha} + \frac{eC}{c}\left[\mathfrak{E}(\mathfrak{r}), \left[\mathfrak{H}(\mathfrak{r}), \frac{\partial}{\partial\mathfrak{p}}\right]\right]_\beta + \right. \tag{2}$$

$$\left. + C\left[\mathfrak{E}(\mathfrak{r}), \frac{\partial}{\partial\mathfrak{r}}\right]_\beta\right\} M_\beta(\mathfrak{p},\mathfrak{r}) = \left\{\frac{\partial n(\mathfrak{p},\mathfrak{r})}{\partial t}\right\}_{coll}\,;$$

$$\frac{\partial M_\alpha(p,r)}{\partial t} + \frac{p_\beta}{m}\frac{\partial M_\alpha(p,r)}{\partial r_\beta} + e\left\{E_\beta(r) + \frac{1}{mc}\left[p, \mathcal{H}(r)\right]_\beta\right\}\frac{\partial M_\alpha(p,r)}{\partial p_\beta} +$$

$$+ \frac{2\mu}{\hbar}\left[\mathcal{H}(r) + \frac{C}{\mu}\left[p, \mathcal{E}(r)\right], \mathfrak{M}(p,r)\right]_\alpha + \mu\left\{\frac{\partial H_\alpha(r)}{\partial r_\beta}\frac{\partial}{\partial p_\beta} + C\left[p, \frac{\partial \mathcal{E}(r)}{\partial r_\beta}\right]_\alpha \frac{\partial}{\partial p_\beta} +\right. \tag{3}$$

$$\left. + \frac{eC}{c}\left[\mathcal{E}(r), \left[\mathcal{H}(r), \frac{\partial}{\partial p}\right]\right]_\alpha + C\left[\mathcal{E}(r), \frac{\partial}{\partial r}\right]_\alpha\right\} n(p,r) = \left\{\frac{\partial M_\alpha(p,r)}{\partial t}\right\}_{Coll};$$

$$n(p,r) = \frac{1}{2}\,Sp_\sigma\,\hat{f}(p,r)\,; \qquad M_\alpha(p,r) = \frac{M}{2}\,Sp_\sigma\,\sigma^\alpha \hat{f}(p,r)$$

$\hat{f}(p,r)$ being the quantum-mechanical distribution function.

If spin-orbital coupling is neglected, equations (2) and (3) coincide with the results of Silin (5). Using the approximation adopted in (2), we may neglect the spin-dependent terms in (2). Assuming that $\hat{f}(p,r) = \hat{f}_0(p - m v(r), r)$ and $\hat{f}_0(p,r) = \hat{f}_0(-p,r)$ we obtain the rate equation for the total spin magnetization of the carriers

$$\frac{\partial M_\alpha(r)}{\partial t} + v(r)\nabla_r M_\alpha(r) + \frac{2\mu}{\hbar}\left[\mathcal{H}(r), \mathfrak{M}(r)\right]_\alpha - \mu C\,\mathrm{rot}_\alpha\left\{n(r)\,\mathcal{E}(r)\right\}$$

$$+ \frac{2Cm}{\hbar}\left[\left[v(r), \mathcal{E}(r)\right]\mathfrak{M}(r)\right]_\alpha = \left\{\frac{\partial M_\alpha(r)}{\partial t}\right\}_{Coll}; \tag{4}$$

$$n(r) = \int\frac{dp}{(2\pi\hbar)^3}\,n(p,r)\,; \qquad M_\alpha(r) = \int\frac{dp}{(2\pi\hbar)^3}\,M_\alpha(p,r)$$

Taking into account the relation

$$M_\perp \approx M_z\frac{mV_\perp^2}{\varepsilon_g}\,\frac{g-2}{g}\,\frac{\Delta_{so} + 2\varepsilon_g}{\Delta_{so} + \varepsilon_g}\,, \qquad V_\perp \approx \frac{cE}{H}$$

we find that the last term on the left side of (4) makes a contribution to the balance equation of longitudinal magnetization $M_z$ having the small factor $\approx(mV_\perp^2/\varepsilon_g)^2$. So we shall further omit this term. For the stationary case the saturation parameter S which determines the steady-state field-enhanced nuclear magnetization

$$\frac{I(\mathcal{E})}{I(0)} = \frac{T_0}{T}\left\{1 - \frac{\Delta_s}{\Delta_n}\left(S - \frac{mV^2}{3\bar{\varepsilon}}\right)\right\}\frac{\omega_{sn}}{\omega_{sn} + \omega_{nl}} + \frac{\omega_{nl}}{\omega_{sn} + \omega_{nl}} \tag{5}$$

ecomes

$$S = S_{coll} - S' - S'' =$$

$$= S_{coll} - \frac{\tau_s}{\chi H}\left\{ \mathcal{W}(\varkappa)\, \nabla_{\varkappa} M(\varkappa) + C\mu\left[\mathcal{E}(\varkappa), \vec{\nabla} n(\varkappa)\right]_z \right\}. \qquad (6)$$

It is obvious that the quantities S' and S" which appear ue to the spatial inhomogeneities of the system, are odd unctions of the external electric field E.

Since the solution (6) of (2) and (3) is valid provided $(\bar{\varepsilon}\tau_s^2 / mL^2)^{1/2} < 1$, L being the mean spatial extension of the inhomogeneities, we obtain

$$S' \approx \frac{V\tau_s}{L} < 1 , \qquad S'' \approx \frac{V_\perp \tau_s}{L} \frac{\bar{\varepsilon}}{\varepsilon_g} \frac{g-2}{g} \frac{\Delta_{so} + 2\varepsilon_g}{\Delta_{so} + \varepsilon_g} < 1 \qquad (7)$$

But in (5) S is multiplied by the large factor $\Delta_s/\Delta_n$, so the contribution of (S' + S") in $\frac{I(E)}{I(0)}$ is of the order of ten even if these corrections to the saturation parameter are small (for $V \approx 10^4$, $\tau_s \approx 10^{-7}$, $L \approx 1$ CGSU, $S' + S'' \approx 10^{-3}$). This is in agreement with the observed magnitude (1) of the odd part of nuclear magnetization.

## References


(1) W. CLARK and G. FEHER, Phys. Rev. Letters 10, 134 (1963).
(2) V.P. KALASHNIKOV, phys. stat. sol., in press.
(3) M. WEGER, Phys. Rev. 132, 581 (1963).
(4) Y. YAFET, Solid State Physics 14, 1 (1963).
(5) V.P. SILIN and A.A. RUKHADZE, Elektromagnitnye svoistva plazmy i plasmapodobnykh sred, Moscow 1961.


(Received June 8, 1966)

phys. stat. sol. 16, K39 (1966)

nstitut für Angewandte Mathematik und Mechanik der Deutschen
kademie der Wissenschaften zu Berlin und Institut für Magne-
ische Werkstoffe der Deutschen Akademie der Wissenschaften zu
Berlin, Jena

# erechnung der Energie, Magnetisierungsverteilung und Ausdehnung einer Kreuzblochlinie

Von
H. BÄURICH

Bei der unvollständigen Ummagnetisierung dünner magneti-
cher Schichten bilden sich unterhalb einer gewissen Schicht-
icke $D_u$ Néelwände, oberhalb einer Schichtdicke $D_o$ Blochwände
us. Im Übergangsgebiet bilden sich kompliziertere Wandstruk-
uren, die aus Teilen beider Typen bestehen. Zum Teil sind
ies Néelwände verschiedener Drehrichtung, die durch Blochli-
ien getrennt sind (1). Dabei entstehen zwei Typen von Bloch-
inien, die Kreisblochlinie und die Kreuzblochlinie. Die Aus-
ehnung und Magnetisierungsverteilung der Kreisblochlinie wurde
m Jahre 1965 von Feldtkeller und Thomas (2) berechnet, während
ies für die Kreuzblochlinie bis jetzt noch nicht getan wurde.

Ziel dieser Arbeit soll es sein, die Energie, Magnetisie-
ungsverteilung und Ausdehnung einer Kreuzblochlinie zu be-
echnen. Dabei wurden folgende Annahmen gemacht:

1. Die Komponenten der Magnetisierungsvektoren in der Schichtebene bilden Hyperbeln der Gleichung
$$x^2 - y^2 = C. \qquad (1)$$
2. Der Einfluß der Anisotropie in der Kreuzblochlinie wurde vernachlässigt.
3. Die Magnetisierungsrichtung soll nicht von z (Achse senkrecht zur Schichtebene) abhängen.
4. Die Schichtdicke D soll sehr klein sein gegenüber der Ausdehnung der Schicht.

Mit diesen Voraussetzungen ergibt sich für die Energie
er Kreuzblochlinie
$$E = E_A + E_F + E_R + E_{FR}. \qquad (2)$$

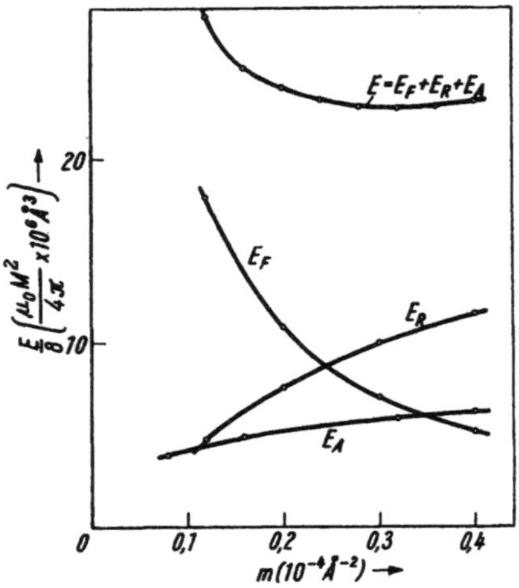


Fig. 1. Abhängigkeit der Austauschenergie $E_A$, der Raumladungsenergie $E_R$ und der Oberflächenladungsenergie $E_F$ sowie der Gesamtenergie E einer Kreuzblochlinie von m, wenn die Magnetisierungskomponente in z-Richtung (senkrecht zur Schichtebene) mit

$$M_z = Me^{-m(x^2+y^2)}$$

angenommen wird

Dabei bedeutet : $E_A$ die Austauschenergie, $E_F$ die Energie der magnetischen Oberflächenladung, $E_R$ die Energie der magnetische Raumladung und $E_{FR}$ die Wechselwirkungsenergie zwischen magnetischer Oberflächenladung und Raumladung.

Setzen wir

$$M_z = M \sin\vartheta, \tag{3}$$

($\vartheta$ Winkel zwischen Magnetisierung M und Schichtebene), so ergibt sich für

1. die Austauschenergiedichte $\varepsilon_A$

$$\varepsilon_A = A\left[\left(\frac{\partial\vartheta}{\partial x}\right)^2 + \left(\frac{\partial\vartheta}{\partial y}\right)^2 + \frac{1}{x^2+y^2}\cos^2\vartheta\right], \tag{4}$$

(A Austauschkonstante)

2. die Oberflächenladungsdichte

$$\sigma_F = \pm\mu_0 M \sin\vartheta, \tag{5}$$

3. die Raumladungsdichte

$$\varrho = -\mu_0 \operatorname{div} \mathfrak{M} = -\frac{2\mu_0 M x y \cos\vartheta}{\left(\sqrt{x^2+y^2}\right)^3} - \frac{\mu_0 M\left(y\frac{\partial\vartheta}{\partial x} - x\frac{\partial\vartheta}{\partial y}\right)\sin\vartheta}{\sqrt{x^2+y^2}}. \tag{6}$$

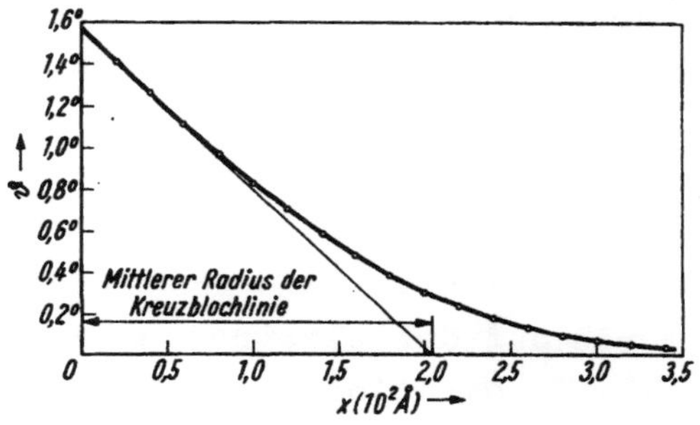


Fig. 2. Abhängigkeit des Winkels $\vartheta$ zwischen Magnetisierung und Schichtebene einer Kreuzblochlinie von der Entfernung vom Mittelpunkt (x-Richtung). Schichtdicke D = 200 Å (Ni-Fe-Schicht; 80% Ni, 20% Fe)

Mit diesen Größen erhalten wir für die in (2) enthaltenen Energien

$$E_A = D \int\int \varepsilon_A \, dx \, dy, \tag{7}$$

$$E_F = \frac{1}{8\pi\mu_0} \iint \frac{\sigma_F(x',y')\,\sigma_F(x,y)}{|\mathfrak{r}' - \mathfrak{r}|} \, dF' dF =$$

$$= \frac{\mu_0 M^2}{4\pi} \iiiint \left[ \frac{1}{\sqrt{(x'-x)^2 + (y'-y)^2}} - \frac{1}{\sqrt{(x'-x)^2 + (y'-y)^2 + D^2}} \right] \times$$

$$\times \sin\vartheta(x'y') \sin\vartheta(xy) \, dx' dy' dx \, dy, \tag{8}$$

$$E_R = \frac{1}{8\pi\mu_0} \iint \frac{\varrho(x',y')\,\varrho(x,y)}{|\mathfrak{r}' - \mathfrak{r}|} \, dV' dV \tag{9}$$

$$E_{FR} = \int V_R \, \sigma_F \, dF = 0, \tag{10}$$

da $V_R$ senkrecht zur Schichtebene in beiden Richtungen symmetrisch ist ($V_R$ magnetisches Potential der Raumladung).

Zur Berechnung der Ausdehnung der Kreuzblochlinie muß das Minimum der Energie $E = E_A + E_F + E_R$

bei Variation der Funktion $\vartheta(x,y)$ bestimmt werden. Dieses Variationsproblem wurde in erster Näherung nach der Ritzschen Methode mit dem Ansatz

$$\sin\vartheta = e^{-m(x^2+y^2)} \tag{11}$$

(m Variationsparameter) gelöst.

Die Rechnung wurde auf der elektronischen Rechenmaschine ZRA1 durchgeführt. Die Integration wurde vom Mittelpunkt der Kreuzblochlinie bis zu den Seiten eines Quadrates der Seitenlänge 2a = 800 Å ausgeführt. Am Rand ist der Winkel $\vartheta < 2{,}3^o$ (für m = $0{,}2\times10^{-4}$ Å$^{-2}$). Fig. 1 zeigt die Abhängigkeit der Energien $E_A$, $E_F$, $E_R$ und $E = E_A + E_F + E_R$ von m. Die Energien beziehen sich auf das Volumen $4a^2D$. Bei der Rechnung wurden die Werte a = 400 Å, D = 200 Å, $\sqrt{2A/\mu_o M^2} = 50$ Å (Ni-Fe-Schicht; 80% Ni, 20% Fe) benutzt.

Für $m = m_o = 0{,}30\times10^{-4}$ Å$^{-2}$ hat die Gesamtenergie ein Minimum.

Fig. 2 zeigt die Abhängigkeit des Winkels $\vartheta$ von x bei y = 0. Führen wir für die Ausdehnung der Kreuzblochlinie eine ähnliche Definition ein, wie sie Feldtkeller und Thomas (2) bei der Kreisblochlinie verwendet haben (Fig. 2), so ergibt sich mit (11) nach Fig. 2

$$x_o = \frac{\pi}{2\sqrt{2m}}$$

und mit $m = m_o = 0{,}30\times10^{-4}$ Å$^{-2}$

$$x_o \approx 200 \text{ Å}.$$

Für die Energie ergibt sich

$$E_{min} \approx 180 \frac{\mu_o M^2}{4\pi} \times 10^6 \text{ Å}^3.$$

Vergleicht man die Ergebnisse mit denen, die Feldtkeller und Thomas für die Kreisblochlinie erhalten haben ($r_o \approx 80$ Å, $E_{min} \approx 70\mu_o M^2/4\pi \times 10^6$ Å$^3$), so ergibt sich für die Kreuzblochlinie etwa die 2,5-fache Ausdehnung und die 2,6-fache Energie der Kreisblochlinie.

Zum Schluß möchte ich Herrn Dr. Andrä für die wertvolle Anleitung bei der Bearbeitung dieses Themas danken.

## Literatur

(1) E. FELDTKELLER und E. FUCHS, Z. angew. Phys. 18, 1 (1964).
(2) E. FELDTKELLER und E. THOMAS, Phys. kondens. Materie 4, 8 (1965).

(Received June 8, 1966)

phys. stat. sol. 16, K45 (1966)

II. Physikalisches Institut der Universität München

# Vollständige Lösung des magnetoelastischen Problems von Bereichstrukturen mit zyklischer Magnetisierung

Von

H. SPREEN und CH. SCHWINK

Untersuchungen der Bereichstruktur von Nickeleinkristallen führten uns auf folgendes magnetoelastische Problem:

Nach Beobachtungen mit der Bittertechnik besteht die ferromagnetische Bereichstruktur in stabförmigen ⟨100⟩-Nickeleinkristallen aus "zusammengesetzten Platten", die senkrecht zur Stabachse angeordnet sind (1, 2). Im einfachsten Fall ist nach unserer Meinung eine zusammengesetzte Platte aus vier Elementarbereichen aufgebaut, die den Querschnitt in vier gleiche Sektoren unterteilen. Jeder der Sektorbereiche ist in kleinen Feldern nahezu in einer leichten Richtung magnetisiert (Fig. 1). Solch eine Bereichstruktur steht unter magnetoelastischen Spannungen: Man erkennt anhand der Verschiebungsvektoren $\vec{u} = (u,v,w)$ (Fig. 1), daß im Falle frei entspannter Bereiche an allen Bereichsgrenzen zwischen den Sektorbereichen eine Lücke mit einem Winkel $\gamma = -\lambda_{111}$ klafft und daß außerdem die den Sektorbereichen gemeinsame Achse in jedem Bereich in eine andere Richtung verkippt ist. Es ergibt sich damit die Frage, wie groß die magnetoelastische Energiedichte dieser Bereichanordnung ist und ob sie nicht so groß ist, daß sie die Struktur von Fig. 1 sehr unwahrscheinlich macht. Wir beantworten in dieser Mitteilung nur einen Teil der Frage, der aber das wesentlich Neue des gestellten gegenüber bekannten magnetoelastischen Problemen zeigt: Wir behandeln nur jene Spannungen der Plattenstruktur, die jeden Sektorbereich auf den vollen Winkel spreizen. Die Linearität der Elastizitätstheorie im Hookeschen Bereich ermöglicht die Abspaltung dieser Teilfrage. Bei einem ⟨100⟩-Eisenkristall entsprechender Bereichstruktur träte überhaupt nur diese Teilfrage auf.

In der Bereichstruktur des ⟨100⟩-Nickelkristalls gibt es, abgesehen von möglichen Störungen im Plattenzentrum, nur Blochwände zweiter Art (3). Die in Achsenrichtung aufeinander-

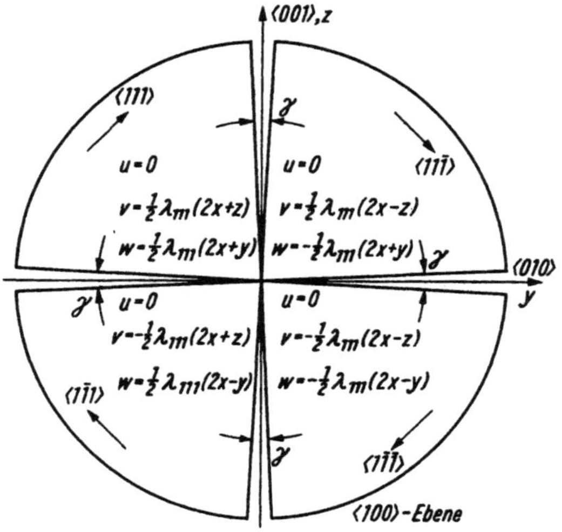


Fig. 1. Bereichstruktur mit zyklischer Magnetisierung, gezeichnet sind die Projektionen der Magnetisierungsvektoren {111}. Die vier Sektorbereiche sind frei entspannt, in jedem sind die rechtwinkligen Komponenten des Verschiebungsvektors eingetragen

folgenden Platten werden durch $109^{\circ}$(100)-Wände und die Sektorbereiche innerhalb einer Platte durch $71^{\circ}$(110)-Wände getrennt. Die Blochwände zweiter Art fügen die angrenzenden Bereiche spannungsfrei aneinander und stellen damit - im Gegensatz zu den Blochwänden erster Art - keine Quellen von Spannungsfeldern dar. Die auftretenden Spannungen sind in der Gesamtanordnung der Bereiche begründet.

Wir setzen im folgenden voraus, daß das Medium elastisch isotrop ist und durch den Elastizitätsmodul E und die Poissonsche Zahl $\mu$ charakterisiert wird. Wir vereinfachen die mathematische Behandlung des Problems dadurch, daß wir die vier frei entspannten Sektorbereiche durch Drehung um die Kristallachse zusammenschieben und irgendwo in einem Schnitt senkrecht zur Platte eine einzige Lücke mit einem Winkel $4\gamma = -4\lambda_{111}$ klaffen lassen (Fig. 2).

Wir rechnen in Zylinderkoordinaten: Der Verschiebungsvektor $\mathfrak{u}$, der den magnetostriktiven Effekt rückgängig macht, hat die Komponenten u (axial), p (radial) und q (azimutal). Die Komponenten des Spannungstensors sind $\sigma_{xx}$, $\sigma_{xr}$, ... usw. Der Verschiebungsvektor $\mathfrak{u}$ muß der Differentialgleichung

$$\Delta\mathfrak{u} + \frac{1}{1-2\mu} \operatorname{grad} \operatorname{div} \mathfrak{u} = 0$$

genügen. Die Randbedingungen der Platte sind (vgl. Fig. 2):

Fig. 2. Das aus Fig. 1 hervorgehende Problem in Zylinderkoordinaten

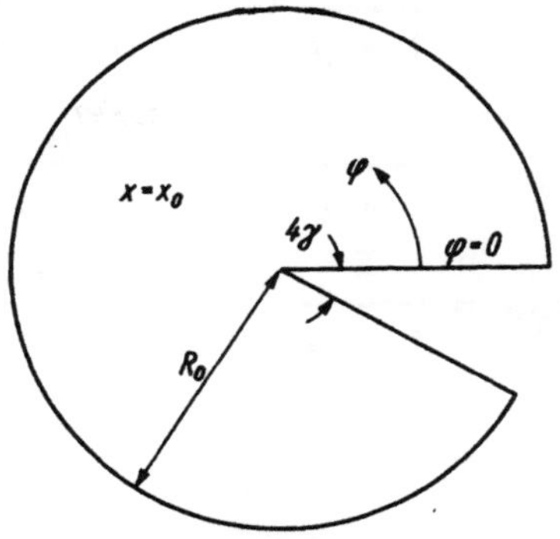


$$x = x_0 : u = Ax;\ \sigma_{xr} = \sigma_{x\varphi} = 0;\ \int_0^{2\pi}\int_0^{R_0} \sigma_{xx} r\, dr d\varphi = 0$$

$$x = R_0 : \sigma_{rr} = \sigma_{xr} = \sigma_{r\varphi} = 0$$

$$\varphi = 0 : q = 0;\ \sigma_{x\varphi} = \sigma_{r\varphi} = 0$$

$$\varphi = 2\pi : q = -4\lambda_{111} r.$$

Sie besagen, daß

1. an der gesamten Oberfläche der eingeschnittenen Platte keine Schubkräfte auftreten,
2. an der freien Oberfläche des Kristalls ($r = R_0$) keine Normalkraft wirkt,
3. der Einschnitt in die Platte geschlossen wird,
4. die $109^\circ$(100)-Wände der Platte senkrecht zur Kristallachse bleiben und
5. die Kraft auf die Platte in Richtung der Kristallachse verschwindet.

Es erscheint physikalisch sinnvoll, daß die Lösung des Differentialgleichungssystems rotationssymmetrisch ist:

$$u = 0;\ p = \frac{\lambda_{111}}{\pi} r \left(\frac{1 - 2\mu}{1 - \mu} \ln \frac{r}{R_0} - 1\right);\ q = \frac{2\lambda_{111}}{\pi} r\varphi .$$

Dabei ist

$$0 \leq r \leq R_0 .$$

Mit $\operatorname{div} \check{u} = p_r + \frac{1}{r} p + \frac{1}{r} q_\varphi + u_x$ kommt

$$\sigma_{xx} = \frac{E}{1+\mu}\left(u_x + \frac{\mu}{1-2\mu}\,\mathrm{div}\,\check{u}\right) = -\frac{\lambda_{111}E}{\pi(1-\mu^2)}\left(2\mu \ln\frac{r}{R_o} + \mu\right),$$

$$\sigma_{rr} = \frac{E}{1+\mu}\left(p_r + \frac{\mu}{1-2\mu}\,\mathrm{div}\,\check{u}\right) = -\frac{\lambda_{111}E}{\pi(1-\mu^2)}\ln\frac{r}{R_o} ,$$

$$\sigma_{\varphi\varphi} = \frac{E}{1+\mu}\left(\frac{1}{r}p + \frac{1}{r}q_\varphi + \frac{\mu}{1-2\mu}\,\mathrm{div}\,\check{u}\right) = -\frac{\lambda_{111}E}{\pi(1-\mu^2)}\left(\ln\frac{r}{R_o}+1\right),$$

$$\sigma_{xr} = \sigma_{x\varphi} = \sigma_{r\varphi} = 0.$$

Über die Energiedichte

$$f = \frac{1+\mu}{2E}\left[\frac{1}{1+\mu}(\sigma_{xx}+\sigma_{rr}+\sigma_{\varphi\varphi})^2 + 2(\sigma_{xr}^2+\sigma_{x\varphi}^2+\sigma_{r\varphi}^2-\sigma_{xx}\sigma_{rr}-\sigma_{rr}\sigma_{\varphi\varphi}-\sigma_{\varphi\varphi}\sigma_{xx}\right.$$

erhält man die über den Kristall gemittelte Energiedichte $\bar{f}$:

$$\bar{f} = \frac{1}{\pi R_o^2}\int_0^{R_o}\int_0^{2\pi} f\,r\,dr\,d\varphi = \frac{E\lambda_{111}^2}{2\pi^2(1-\mu^2)} .$$

Mit den für Nickel bei Zimmertemperatur geltenden ungefähren Werten $E = 20\,000$ kp/mm$^2$, $\lambda_{111} = -2{,}4\times10^{-5}$, $\mu = 0{,}3$ berechnet man $\bar{f} = 64$ erg/cm$^3$. Das ist eine für die Bereichstrukturen von Nickel geringe Energiedichte. Zum Vergleich beträgt die mittlere Energiedichte der magnetostriktionsfreien Bereichstruktur im ⟨110⟩-Kristall etwa 50 erg/cm$^3$ (2).

Der geringe Wert für $\bar{f}$ rührt daher, daß $\sigma_{xx}$ und $\sigma_{\varphi\varphi}$ als Funktionen von r zwischen 0 und $R_o$ das Vorzeichen wechseln. Es liegt also keine homogene Verzerrung vor. Das ist das charakteristische Merkmal des betrachteten Spannungszustands.

Wenn die Lücke mit einem Winkel $4\gamma = -4\lambda_{111}$ im Widerspruch zu den Randbedingungen durch eine homogene Verzerrung der einzelnen Sektorbereiche geschlossen würde, ergäbe sich die Energiedichte

$$\bar{f}' = \frac{E\lambda_{111}^2}{4(1+\mu)} = 220 \text{ erg/cm}^3.$$

Die vorstehende Rechnung läßt sich analog auch für Eisen durchführen. Unsere vollständige Lösung läßt erkennen, daß nur im Zentrum der Platten große elastische Spannungen herrschen, während der größere Volumenanteil der Platten nur geringfügig verzerrt ist.

## Literatur

(1) CH. SCHWINK, H. SPREEN, E. BUNGENSTAB, G. DEDIÉ und O. GRÜTER, IEEE Trans. Magn., erscheint demnächst; CH. SCHWINK und H. SPREEN, phys. stat. sol. 10, 57 (1965).

(2) H. SPREEN, phys. stat. sol., erscheint demnächst.

(3) G. RIEDER, Abh. Braunschw. Wiss. Ges. 11, 20 (1959).


(Received June 8, 1966)

phys. stat. sol. 16, K51 (1966)

Physics Department, Allahabad University, Allahabad

# On the Specific Heat Anomaly in Superconducting Lead[1)]

By

J.S. RAJPUT and A.K. GUPTA

A $T^3$ contribution of electronic specific heat in superconducting lead has been observed by Keesom and Van der Hoeven (1). The BCS theory predicts a much more rapid exponential decay of the electronic contribution. A possible explanation to this specific heat anomaly may be thought in terms of a temperature dependent electronic effective mass as a result of electron-phonon interactions in superconductors with strong electron-phonon coupling. However, we wish to point out that howsoever extreme variations of the electronic effective mass with temperature can not lead to a satisfactory explanation of the $T^3$ contribution of electronic specific heat in superconducting lead.

We assume the BCS type one particle excitation spectrum $E_k = (\epsilon_k^2 + \Delta_k^2)^{1/2}$ with $\Delta k(0) = 2.1\ kT_c$ for Pb. The excitations obey the Fermi statistics, the density of states function is $N(E_k)$. We calculate the temperature dependent total energy of excitations in the superconducting state

$$U_s(T) = \int_{\Delta k}^{\infty} E_k f(E_k) N(E_k)\, dE_k, \tag{1}$$

where $f(E_k)$ is the Fermi function

$$f(E_k) = (1 + e^{\beta E_k})^{-1}, \quad \beta = 1/kT.$$

The normal state energy is

$$U_n(T) = \int_0^{\infty} \epsilon_k N(\epsilon_k) f(\epsilon_k)\, d\epsilon_k,$$

which is obtained as

$$\begin{aligned} U_n(T) &= \pi^2 N(0)/3 \int_0^{\infty} \frac{\epsilon k}{1 + e^{\beta \epsilon k}}\, d\epsilon_k \\ &= \pi^2 N(0) k^2 T^2/3 = \frac{1}{2}\gamma T^2, \end{aligned} \tag{2}$$

---

1) Sponsored, in part, by C.S.I.R. India.

where N(0) is the density of states at the Fermi surface. From normal state energy, we obtain a normal state specific heat,

$$C_n(T) = dU_n(T)/dT = \gamma T,$$

$\gamma$ is a conventional electronic specific heat parameter.

We now calculate $U_s(T)$ assuming a temperature dependent electronic effective mass. Introducing $m^*/m = a(T)$, the ratio of effective mass to free electron mass, and $\Delta_k(T)/\Delta_k(0) = b(T)$, $U_s(T)$ is obtained as an integral

$$\frac{U_s(T)}{U_N(T_c)} = \frac{(2.1)^2}{a^2} \int_0^\infty \frac{(y^2 + a^2b^2)^{1/2}}{1 + e^{[(2.1)/at]/[(y^2 + a^2b^2)^{1/2}]}}\, dy, \qquad (3)$$

t being the reduced temperature.

Now we investigate it in the limits of small a and very large a.

Case 1): When $a \ll 1$, the ratio is

$$\frac{U_s(T)}{U_n(T_c)} = \frac{(2.1)^2}{a^2} \int_0^\infty \frac{y}{e^{2.1y/at}}\, dy = t^2 \qquad (4)$$

which is independent of the effective mass, and proportional to the square of t, while experimentally $U_s(T)/U_n(T_c)$ should vary as $t^4$.

Case 2): When $a \gg 1$, the integral is transformed into another form by slight manipulation; substituting

$$\frac{ab}{(y^2 + a^2b^2)^{1/2}} = z$$

we get,

$$\frac{U_s(T)}{U_N(T_c)} = \frac{4.41}{a^2}\, a^2b^2 \int_0^1 \frac{e^{-2.1b/zt}}{z^2\,(z^2 - 1)^{1/2}}\, dz \qquad (5)$$

which is equal to $Ab^2e^{-Bb/t}$, where A and B are constants independent of a and b; as in case 1), the result is independent of the effective mass ratio a and no $T^4$ dependence can be obtained consequently.

## References

(1) P.H. KEESON and VAN DER HOEVEN Jr., Phys. Letters (Netherlands) 3, 360 (1963).

(Received June 8, 1966)

phys. stat. sol. 16, K55 (1966)

Institute of Physics, Polish Academy of Sciences, Warsaw (a)
and Institute of Physics, Warsaw University, Warsaw (b)

# The Effect of Higher Bands on the Effective Masses in InSb and InAs

By

J. KOŁODZIEJCZAK (a) and S. ŻUKOTYŃSKI (b)

The effect of higher bands on the band parameters in III-V compounds was analysed in 1963 by Cardona (1). In our earlier paper (2) we have developed his approach and calculated the effective masses using the newest results for the band parameters of the fourth group semiconductors (3) and making account of the linear terms in the energy of heavy holes. Unfortunately after Cardona we made a mistake in the definition of the band parameter M.

The proper definition of this parameter is (4)

$$M = - \frac{Q^2}{\varepsilon(\Gamma_{15c}) - \varepsilon(\Gamma_{15v})}$$

All of the analytical results presented in paper (2) are correct, but the numerical values for holes are by some 20% too high. At the end of this note we present the tables of the corrected numerical values of the effective masses. In general, these new values agree quite well with available experimental data. In this note we would like to compare the experimentally obtained electron effective masses for high electron concentration in InSb and InAs with our calculations. Fig. 1 shows

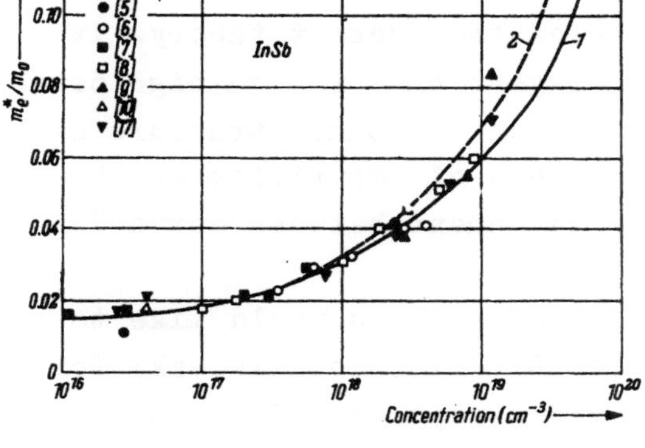


Fig. 1. Dependence of electron effective mass in InSb on concentration. Theoretical curves:
1 —— effect of higher bands not included
2 --- effect of higher bands included

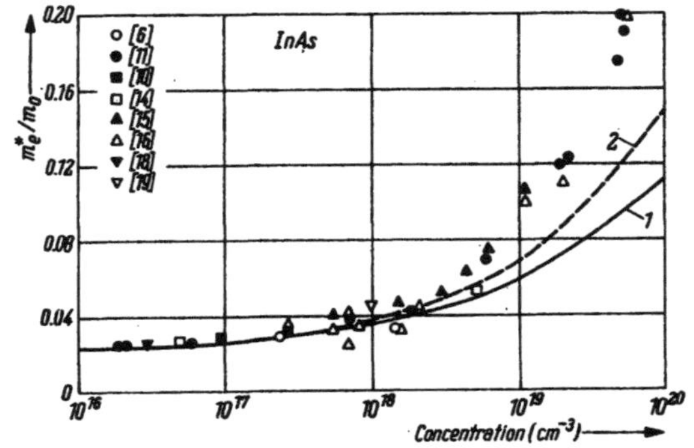


Fig. 2. Dependence of electron effective mass in InSb on concentration
Theoretical curves:
1 —— effect of higher bands not included
2 - - - effect of higher bands included

the experimental points for InSb obtained by a number of authors (5 to 10, 17). The solid line presents the effective mass calculated without taking account of higher bands. The dotted line is based upon our present results. It can be seen that a quite good agreement is obtained.

It was shown by Korenblit et al. (11) that in InAs the experimental values of electron effective mass cannot by explained by the theory including only conduction-valence band interaction. Basing on Kane's theory (12) and our earlier paper (2) we have calculated the effect of higher band interactions on the electron effective mass. The results as well as the experimental points are shown in Fig. 2. It is seen that for very high concentrations the effect of higher bands does not fully explain the strong increase of the effective mass. It seems reasonable to suggest that until the experimental data concerning the energy gaps and/or the effective mass of electrons in InAs are not revised, the hypothesis of Ginter (13) that the effect of higher band interactions completely explains the effective mass is too optimistic within the limits of the present theory. It seems however quite reasonable to suggest that for so high concentrations the effect of impurities on the band structure cannot be neglected. Possibly the influence of impurities on the band structure could improve the agreement between curve 2 and the experimental points on Fig. 2.

The authors would like to express their gratitude to Prof. Dr. I.M. Tsidilkovski for indicating the mistake in our previous paper and for his interest in our calculations.

Table 1

| | $m_1$ | $m_t$ | $m_d$ | $m_{100}$ | $m_{111}$ | | $m_{110}$ | |
|---|---|---|---|---|---|---|---|---|
| GaP | 1.324 | 0.1478 | 0.7733 | 0.2314 | 0.3222 | 0.1478 | 0.1761 | 0.4423 |
| AlSb | 1.012 | 0.1817 | 0.8116 | 0.2700 | 0.3492 | 0.1817 | 0.2132 | 0.4288 |
| GaAs | 0.9717 | 0.1763 | 0.7847 | 0.2615 | 0.3378 | 0.1763 | 0.2067 | 0.4139 |
| InP | 0.8691 | 0.1721 | 0.7439 | 0.2522 | 0.3211 | 0.1721 | 0.2010 | 0.3867 |
| GaSb | 0.7340 | 0.1535 | 0.6517 | 0.2233 | 0.2817 | 0.1535 | 0.1789 | 0.3357 |
| InAs | 0.7152 | 0.1634 | 0.6736 | 0.2345 | 0.2916 | 0.1634 | 0.1896 | 0.3419 |
| InSb | 0.5989 | 0.1402 | 0.5732 | 0.2004 | 0.2482 | 0.1402 | 0.1625 | 0.2898 |

Effective masses of heavy holes in the ellipsoidal range (in units of free electron mass)

Table 2

| | $m_d$ | $m_{100}$ | $m_{111}$ | $m_{110}$ | $m_e$ (0) | $m_{lh}$ (0) | $\alpha$ | $\gamma$ |
|---|---|---|---|---|---|---|---|---|
| GaP | 0.6851 | 0.3974 | 0.6851 | 0.5801 | 0.1231 | 0.1037 | −3.573 | 2.366 |
| AlSb | 0.6707 | 0.4455 | 0.6707 | 0.5955 | 0.0904 | 0.0883 | −2.999 | 2.011 |
| GaAs | 0.6471 | 0.4311 | 0.6471 | 0.5751 | 0,0756 | 0.0794 | −3.094 | 2.002 |
| InP | 0.6019 | 0.4119 | 0.6019 | 0.5397 | 0.0606 | 0.0687 | −3.194 | 1.919 |
| GaSb | 0.5205 | 0.3624 | 0.5205 | 0.4693 | 0.0409 | 0.0445 | −3.597 | 1.864 |
| InAs | 0.5261 | 0.3767 | 0.5261 | 0.4786 | 0.0233 | 0.0271 | −3.409 | 1.769 |
| InSb | 0.4448 | 0.3209 | 0.4448 | 0.4057 | 0.0135 | 0.0143 | −3.984 | 1.743 |

Effective masses of heavy holes in the warped range. Electron and light hole masses at $k = 0$ (in units of free electron mass) and the band parameters $\alpha$ and $\gamma$

## References

(1) M. CARDONA, J. Phys. Chem. Solids 24, 1543 (1963).

(2) J. KOŁODZIEJCZAK, S. ŻUKOTYŃSKI, and H. STRAMSKA, phys. stat. sol. 14, 471 (1966).

(3) W. FAWCETT, Proc. Phys. Soc. 85, 931 (1965).

(4) M. CARDONA, J. Phys. Chem. Solids 26, 1351 (1965).

(5) V.V. GALAVANOV, O.V. EMELYANENKO, and F.P. KESAMANLY, Fiz. tverd. Tela 5, 616 (1963).

(6) W.G. SPITZER and H.Y. FAN, Phys. Rev. 106, 882 (1957).

(7) S.D. SMITH, T.S. MOSS, and K.W. TAYLOR, J. Phys. Chem. Solids 11, 131 (1959).

(8) P. BYSZEWSKI, J. KOŁODZIEJCZAK, and S. ŻUKOTYŃSKI, phys. stat. sol. 3, 1880 (1963).

(9) L. SNIADOWER, J. RAUŁUSZKIEWICZ, and R.R. GAŁĄZKA, phys. stat. sol. 6, 549 (1964).

(10) E.D. PALIK, S. TEITLER, and R.F. WALLIS, J. appl. Phys. 32, 2132 (1961)

(11) L.L. KORENBLIT, D.V. MASHOVETS, and S.S. SHALYT, Fiz. tverd. Tela 6, 559 (1964).

(12) E.O. KANE, J. Phys. Chem. Solids 1, 249 (1957).

(13) J. GINTER, phys. stat. sol. 14, K131 (1966).

(14) M. CARDONA, Phys. Rev. 121, 752 (1961).

(15) D. GEIST, Z. Phys. 157, 335 (1959).

(16) S.A. SHULMAN and YU.I. UKHANOV, Fiz. tverd. Tela 7, 952 (1965).

(17) F.P. KESEMANLY, YU.V. MALTSEV, D.N. NASLEDOV, YU.I. UKHANOV, and A.S. FILIPCHENKO, Fiz. tverd. Tela 8, 1176 (1966).

(18) S.S. SHALYT, Fiz. tverd. Tela 3, 2887 (1961).

(19) I.G. AUSTIN, J. Electronics and Control 8, 167 (1960).

(Received June 10, 1966)

phys. stat. sol. 16, K59 (1966)

.V. Physikalisches Institut der Humboldt-Universität zu Berlin
und Physikalisch-Technisches Institut der Deutschen Akademie
der Wissenschaften zu Berlin

# A Simple Asymptotic Formula for the Motion of Simultaneous Electric Field and Conductivity Domains in Semiconductors

By

W.E. WILHELM

In a homogeneous semiconductor characteristic layer-like inhomogeneities of electric field and conductivity occur simultaneously, if, as was shown in a previous paper (1), the conductivity decreases stronger than linearly with increasing field strength. These field inhomogeneities may be attached to the electrodes or travel from cathode to anode (n-type material), giving rise to current oscillations described in the literature for a variety of semiconductors, ranging in frequency from about $10^{-4}$ Hz (CdS (2, 3)) to $10^{9}$ Hz (GaAs (4, 5)).

The above mentioned instability criterion in the form $dj/dE < 0$ (negative differential conductivity), controlling the formation of regions of high and low field and conductivity in the semiconductor has been simply deduced by Ridley (6) from electrodynamics using Poisson equation and continuity equation. On the other hand a simple general expression for the velocity of these inhomogeneities has not been deduced so far from the same basis assumptions. However this is hardly to be expected, because in general the time constants of the models leading to a negative differential conductivity should enter the velocity formula.

Nevertheless in this note it will be shown that it is possible to obtain a simple asymptotic formula for the motion in terms of electrodynamics (i.e. independent of a special model), assuming that the conductivity instantaneously responds to changes of the field. The travelling field and conductivity inhomogeneity is assumed to have the special shape of a stationary moving domain, in the rigorous sense: piecewise constant field and conductivity, constant width of the high field

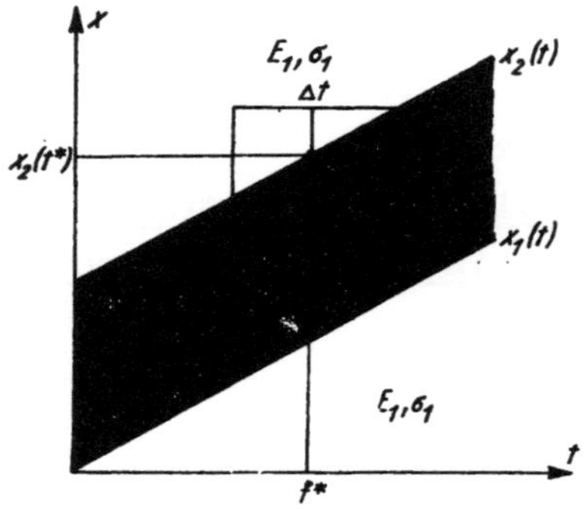


Fig. 1

region (constant d.c. voltage). The simplification of piecewise constancy has been introduced by Ridley (6) and Stöckmann (7) in discussing the stationary case.

We presume the existence of such a domain and ask for its travelling velocity. Since space charges are absent, a consideration of the Poisson equation is irrelevant. The surface charge density $\eta$ due to the discontinuities of E may be calculated from $\varepsilon\varepsilon_0(E_2 - E_1) = \pm\eta$ with $\eta \lessgtr 0$ at $x_1(t)$ and $x_2(t)$, respectively (cf. Fig. 1). As is also obvious from this figure the diffusion current as well as the Maxwellian displacement current are significant at the discontinuities. Therefore the current equation in monopolar semiconductors has to be used in the following form

$$J = \sigma E - U_T \frac{\partial\sigma}{\partial x} + \varepsilon\varepsilon_0 \frac{\partial E}{\partial t} \qquad (1)$$

$$\left[ E = (-E,0,0), \quad J = (-J,0,0), \quad U_T \equiv kT/e \right].$$

The total current J is a constant parameter in the present case instead of $j = \sigma E - U_T \frac{d\sigma}{dx}$ in the stationary theory.

Integrating (1) at an arbitrary moment $t^*$ along the interval $\Delta x$ containing the discontinuity, e.g. at $x_2(t)$, we find

$$J\Delta x = (\sigma_1 E_1 + \sigma_2 E_2)\frac{\Delta x}{2} - U_T \int_{x_2(t^*)-\frac{\Delta x}{2}}^{x_2(t^*)+\frac{\Delta x}{2}} \frac{\partial\sigma}{\partial x}\, dx + \varepsilon\varepsilon_0 v \int_{t^*-\frac{\Delta t}{2}}^{t^*+\frac{\Delta t}{2}} \frac{\partial E}{\partial t}\, dt .$$

The integration is carried out by means of Dirac's $\delta$-function. In the limit $\Delta x \longrightarrow 0$ it follows

$$0 = -U_T(\sigma_1 - \sigma_2) + \varepsilon\varepsilon_0 v(E_2 - E_1).$$

Thus, the velocity of the domain is found to be

$$v = \frac{U_T}{\varepsilon\varepsilon_0} \frac{\sigma_1 - \sigma_2}{E_2 - E_1} . \qquad (2)$$

This expression may asily be re-written using he above definition of a tationary moving domain

$$J = \sigma_1 E_1 = \sigma_2 E_2, \qquad (3)$$

o that equation (2) reuces to the equivalent orms

$$v = \frac{U_T/E_2}{\tau_{R1}} \qquad (4a)$$

r

$$v = \frac{U_T/E_1}{\tau_{R2}} \qquad (4b)$$

ith $\tau_R = \varepsilon\varepsilon_o/\sigma$ . Hence, hese relations for the elocity of a stationary oving domain, whose sign corresponds to the motion towards he anode (n-type material) connect field strength and dielecric relaxation time inside and outside the domain.[1)]

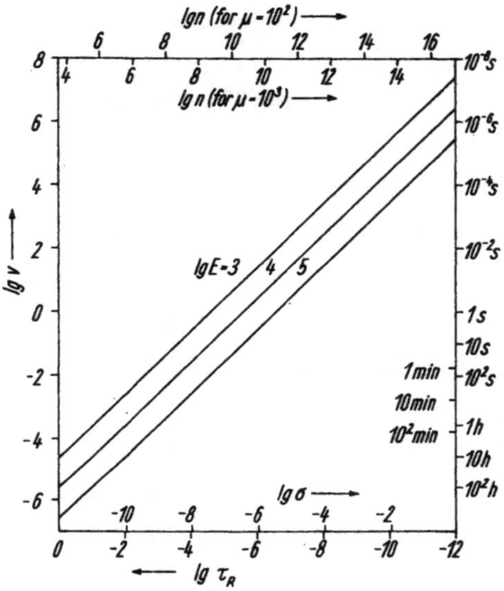


Fig. 2

Obviously, in sufficiently long specimens current oscilations are connected with domains moving periodically, whose requency is given by their reciprocal transit time $\nu = v/L$ L electrode distance). According to the large range of conuctivity in different semiconducting materials a wide range f frequency should be expected. In Fig. 2 the domain velocity (cm/s) and the transit time (for L = 1 cm) are plotted as unctions of the dielectric relaxation time for the fieldtrengths $10^3$, $10^4$, and $10^5$ V/cm, at room temperature. For $\varepsilon = 10$ and $\mu = 10^2$ and $10^3$ $cm^2/Vs$ the conductivity in $\Omega^{-1}cm^{-1}$ nd the electron concentration n in $cm^{-3}$ are also indicated. omparing with real substances, for instance CdS and GaAs,

1) It can be shown (8) that (1) splits into two quasilinear differential equations of the first order, if the assumption is made that $J = \sigma(x,t)\cdot E(x,t)$ holds throughout. Their solutions correspond to simultaneous nonlinear simple waves $E(x,t)$ and $\sigma(x,t)$ moving with an analogous velocity $v = (U_T/\varepsilon\varepsilon_o)(\sigma/E)$.

frequencies are obtained partly close to those observed experimentally (2 to 5).

A more refined tratment of the problems discussed here is in progress.

The author wishes to thank Dr. E. Gutsche for his interest and for the discussion of the manuscript.

References

(1) K.W. BÖER and W.E. WILHELM, phys. stat. sol. 3, 1704 (1963).
(2) K.W. BÖER, H.-J. HÄNSCH, and U. KÜMMEL, Z. Phys. 155, 170 (1959).
(3) K.W. BÖER and W.E. WILHELM, phys. stat. sol. 4, 237 (1964).
(4) J.B. GUNN, Proc. Internat. Conf. Phys. Semicond., Plasma Effects in Solids, Paris 1964 (p. 199).
(5) G.F. DAY, Bull. Amer. Phys. Soc. 10, 383 (1965).
(6) B.K. RIDLEY, Proc. Phys. Soc. 82, 954 (1963).
(7) H. KIESS and F. STÖCKMANN, phys. stat. sol. 4, 117 (1964).
(8) W.E. WILHELM, to be published.

(Received June 10, 1966)

phys. stat. sol. 16, K63 (1966)

Metals and Ceramics Division, Oak Ridge National Laboratory,
Oak Ridge, Tennessee

# Environmental Effects on the Diffusion of $Ta^{182}$ in B.C.C. Titanium[1)]

By
J. ASKILL

The unusual features of self and solute diffusion in the etals titanium, zirconium, and uranium are now well established. ey are a) non-linear Arrhenius plots and b) high diffusion tes for both self and solute diffusion with low values of $D_o$ d Q.

It was suggested by Kidson (1) that these phenomena can explained by two mechanisms of diffusion - single vacancy high temperatures and an enhancement of the vacancy concen- ation by oxygen-vacancy complexes at low temperatures. The rconium self diffusion data that Kidson used to demonstrate is possibility were obtained from experiments carried out normal laboratory vacua of between $10^{-5}$ and $10^{-6}$ Torr. Ex- riments in high vacua have been carried out using high purity dide titanium and $Ta^{182}$ as the tracer. Various combinations pre-anneal and diffusion anneal conditions were used : ) Argon anneal with vacuum pre-anneal, b) Vacuum anneal with cuum pre-anneal, c) Vacuum anneal with high vacuum pre-anneal, d d) high vacuum anneal with high vacuum pre-anneal. The gon anneals were carried out under a pressure of about two mospheres of high purity argon (batch analysis 1 ppm $O_2$, ppm $N_2$ + CO). The vacuum anneals and pre-anneals were carried t in a Ta resistance furnace in vacua of between $10^{-5}$ and $)^{-6}$ Torr, and the high vacuum anneals were carried out in cua of between $5 \times 10^{-8}$ and $8 \times 10^{-9}$ Torr in a special glass gh vacuum furnace with the sample heated by an R.F. induction ater. The temperatures were measured with a calibrated opti- l pyrometer. Standard serial sectioning, using a lathe, was ed for the analysis of the samples. The results are given in

---

) Research sponsored by the U.S. Atomic Energy Commission under contract with Union Carbide Corporation.

Table 1

| Temperature ($^{o}$C) | Diffusion coefficient ($cm^2/s$) | Pre-anneal | Diffusion anneal |
|---|---|---|---|
| 914 | $1.86 \times 10^{-10}$ | V | A |
| 951 | $3.43 \times 10^{-10}$ | V | V |
| 990 | $5.69 \times 10^{-10}$ | H V | H V |
| 1023 | $6.06 \times 10^{-10}$ | V | V |
| 1102 | $1.32 \times 10^{-9}$ | V | A |
| 1210 | $3.43 \times 10^{-9}$ | H V | H V |
| 1230 | $3.16 \times 10^{-9}$ | V | A |
| 1330 | $7.47 \times 10^{-9}$ | V | A |
| 1348 | $9.64 \times 10^{-9}$ | V | A |
| 1387 | $1.13 \times 10^{-8}$ | V | A |
| 1447 | $1.72 \times 10^{-8}$ | H V | V |
| 1517 | $3.21 \times 10^{-8}$ | V | A |
| 1560 | $4.38 \times 10^{-8}$ | V | A |
| 1596 | $6.03 \times 10^{-8}$ | V | A |

V Vacuum, A Argon, H V High Vacuum

Table 1 and are shown in Fig. 1.

Some other experiments on diffusion in titanium in a poor vacuum ($10^{-2}$ to $10^{-3}$ Torr) have also shown no change in the diffusion rates even though the hardness of the material was considerably increased by the anneal (2). Thus the vacuum environments for the diffusion anneal from pressures of $10^{-2}$ to $10^{-9}$ Torr, and argon environments up to two atmospheres pressure appear to have no effect on the tracer diffusion rates in titanium.

According to Kidson's theory, in the low temperature region where the concentration of extrinsic vacancies (those supposedly introduced with the oxygen) is much greater than the thermal equilibrium value, the measured activation energy is just the energy of motion of the vacancies. Assuming typical values for the factors in the $D_o$ term gives a required concentration of extrinsic vacancies of about 0.1 at%. Chemical

Fig. 1. Environmental effects on diffusion of $Ta^{182}$ in b.c.c. Titanium

● Argon anneal, vacuum pre-anneal

○ Vacuum anneal, vacuum pre-anneal

□ Vacuum anneal, high vacuum pre-anneal

△ High vacuum anneal, high vacuum pre-anneal

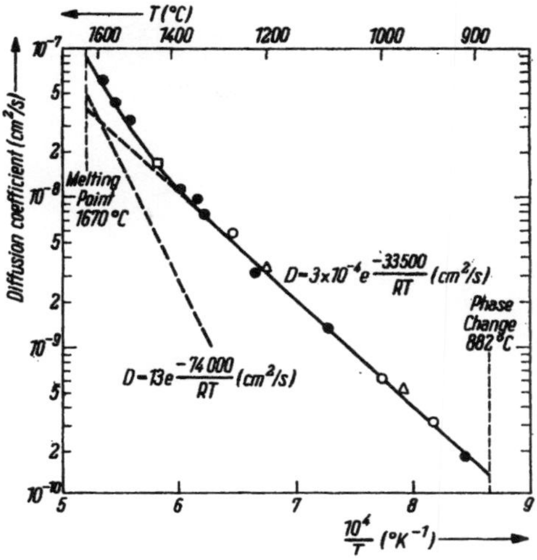


analysis of the samples after diffusion anneal gave about $2x10^{-3}$ at% oxygen. These results indicate that the high diffusion rates in $\beta$-titanium are not due to oxygen impurities but are a result of some intrinsic property of titanium itself, possibly its low "bond strength" as reflected by thermal expansion coefficients and elastic moduli (3).

The author is indebted to Mr. J. Ogle of this laboratory for help in the high vacuum experiments.

## References

(1) G.V. KIDSON, Canad. J. Phys. 41, 1563 (1963).

(2) D. GRAHAM, private communication.

(3) J. ASKILL and G.B. GIBBS, phys. stat. sol. 11, 557 (1965).

(Received June 11, 1966)

# *Pre-printed Titles and Abstracts*

Papers to be published in "physica status solidi"
Vol. 16, No. 2

Review Article

J.R. DALE, Mullard Research Laboratories, Redhill, Surrey, lloyed Semiconductor Heterojunctions

Original Papers

M. TOMÁŠEK, Institute of Physical Chemistry, Czechoslovak cademy of Sciences, Prague, Some Remarks about Electronic orrelation in Crystal Surfaces: The Ideal (100) Surface of a emi-Infinite Diamond-Like Crystal The simple qualitative ethod for electronic correlation analysis in crystal surfaces, ecently suggested by the author (1) and by Davison and Amos 2), is extensively discussed. As an example, the effect of nterelectronic repulsion on the behaviour of electrons locaized near the ideal (100) surface of a diamond-like crystal s studied. Splitting of surface state energy levels of elecrons with ↑ and ↓ spins is found, and possible physical conseuences of this splitting are mentioned. Surface states lying n "forbidden" bands differ from volume states in their sensiivity against interelectronic repulsion. Surface states with nergies lying in volume state energy bands seem to have the ame sensitivity as volume states.

1) M. TOMÁŠEK, Surface Science, to be published.
2) S.G. DAVISON and A.T. AMOS, J. Chem. Phys. 43, 2223 (1965).

Z. ŠROUBEK, Institute of Radio Engineering and Electronics, zechoslovak Academy of Sciences, Prague, On the Theory of the uperexchange Interaction in Ionic Crystals: $KNiF_3$ Within he framework of the Heitler-London theory, studies are made n the superexchange in ionic crystals with particular refeence to $KNiF_3$. Using the results of the generalized quantitaive theory of covalency, the transfer of an electron from cation to cation and the transfer of two electrons from the anion to the cations can analysed in detail without any preliminary presumption; numerical calculations are carried out. The antiferromagnetic superexchange constant given by this calculation agrees very well with the experimentally obtained data.

H. LEMKE, Physikalisch-Technisches Institut der Deutschen Akademie der Wissenschaften zu Berlin, Abteilung Kristallelektronik, Zur Zeitabhängigkeit raumladungsbegrenzter Injektionsströme in Halbleitern Es wird die Zeitabhängigkeit von Raumladungsströmen unter Berücksichtigung eines Haftterms untersucht. Für die Füllung und Entleerung der Haftstellen werden Näherungslösungen angegeben, aus denen die Einfangzeit $\tau_c$ und Emissionszeit $\tau_e$ bestimmt werden können. Bei hinreichend kleiner Spannung ändert sich der Injektionsstrom exponentiell mit der Zeitkonstante $\tau_c$. Nach Abschalten der Injektionsspannung werden die freien Ladungsträger von dem durch haftende Raumladung verursachten Feld aus dem Kristall herausgeschwemmt, wodurch die thermische Ladungsträgerkonzentration $p_o$ auf den Minimalwert $p = (\tau_R/\tau_e)p_o$ ($\tau_R$ dielektrische Relaxationszeit) absinken kann. Bei starker Füllung des Haftterms gegenüber dem thermischen Gleichgewicht ändert sich dabei die haftende Raumladungskonzentration exponentiell mit der Zeitkonstante $\tau_e$. Bei schwacher zusätzlicher Haftstellenfüllung ist die zeitliche Änderung der haftenden Raumladung konstant und hängt von der im thermischen Gleichgewicht vorhandenen Besetzung ab.

H. LEMKE, Physikalisch-Technisches Institut der Deutschen Akademie der Wissenschaften zu Berlin, Abteilung Kristallelektronik, Zeitabhängigkeit raumladungsbegrenzter Injektionsströme in Fe-dotiertem p-Silizium Es wird über Impulsmessungen raumladungsbegrenzter Ströme bei $p^+$-p-$p^+$-Kristallen mit eindiffundierten Anreicherungskontakten berichtet. Aus der Zeitabhängigkeit des Injektionsstroms kann die Einfangwahrscheinlichkeit $r_c(H - H_o^+)$ bestimmt werden. Bei Vorliegen einer gegenüber der thermischen Löcherkonzentration großen haftenden Raumladungskonzentration erfolgt die Haftstellenentleerung ei stark abgesenkter Löcherkonzentration, wodurch der Re-rapping-Einfluß vernachlässigbar ist. Für die zeitliche Ände-ung der haftenden Raumladungskonzentration ist das Verhältnis on injizierter getrappter Trägerdichte $\Delta H^{+}(o)$ zu der im ther-ischen Gleichgewicht vorhandenen Zentrenbesetzung $H_{o}^{+}$ maßgebend. ei hoher zusätzlicher Besetzungskonzentration erfolgt die ntleerung nach einer Exponentialfunktion, wodurch eine direkte estimmung der Emissionswahrscheinlichkeit $r_{e}$ ermöglicht wird. ei kleiner injizierter Trägerdichte ist die zeitliche Ände-ung der haftenden Raumladung konstant und proportional zu $_{e}H_{o}^{+}$. Aus den Emissionswahrscheinlichkeiten erhält man für den irkungsquerschnitt (Löcher) des Eisenzentrums $g\Theta_{p}$ = 0,7x10$^{-15}$ cm$^{2}$ (g statistischer Gewichtsfaktor).

S. CERESARA, T. FEDERIGHI, Istituto Sperimentale dei Me-alli Leggeri, Novara, and F. PIERAGOSTINI, SORIN, Saluggia, eterminatiоn of Diffusion Coefficients in Metals by a Resisto-etric Method - Application to the Diffusion of Zn in Al

method for determining the diffusion coefficient in dilute olid solutions by electrical resistivity measurements is escribed. The analytical approach of the diffusion problem, elated to the selected experimental conditions, includes the umerical calculation of the theoretical curves suitable for irect comparison with resistometric data. Results obtained or the diffusion parameters of Zn in pure Al indicate the igh precision and efficiency of the method.

N.K. ARCHIPOVA, S.M. KLOTSMAN, A.N. TIMOFEEV, and .CH. TRACHTENBERG, Institute of Metal Physics, Academy of ciences of the USSR, Sverdlovsk, Effect of a DC Field on the attice Diffusion of Silver-110 in Copper and Gold A study is ade of the effect of a direct current, with density of about 00 to 150 A/mm$^{2}$ on the diffusion of silver-110 in gold 99.99%) and copper ("high purity" grade) at temperatures bove 800 °C. Diffusion of silver-110 is measured by the re-idual activity method. The direction of the electromigration nd the magnitude of the activated ion effective charge is clear vidence for the electron drag. The effective charge decreases inearly with temperature increase. The temperature coefficient of the effective charge of silver in gold and copper is higher than the temperature coefficient of the matrix conductivity.

S.K. SIDOROV and A.V. DOROSHENKO, Institute of Metal Physics, Academy of Sciences of the USSR, Sverdlovsk, On the Magnetic Structure of Some Alloys of Transition Metals A study is made of binary alloys, in which atoms of one of the components, and unlike atoms, interact ferromagnetically whereas the interaction between atoms of the other component is antiferromagnetic. Under these conditions a magnetic structure arises which involves a noncollinear alignment of the atomic magnetic moments. Expressions are given for calculating the average magnetic moment per atom and its dependence upon the concentration and the state of ordering. The results are compared with experimental data for Ni-Mn, Ni-Fe, and Pd-Fe alloys.

M. ATTARDO, Brookhaven National Laboratory, Upton, New York, and J.M. GALLIGAN, Columbia University, New York, A Field Ion Microscope Study of Neutron Irradiated Tungsten A study of the damage introduced into tungsten by fast neutrons, using field ion microscopy, is described. This consisted of examining the type of damage as reflected in the vacancy content of the material and the removal of the vacancy content after various annealing treatments. Evidence is presented which shows that vacancies are removed from tungsten in stage IV. The observations also indicate the presence of depleted zones in the irradiated tungsten.

D.A. BLACKBURN, Centre d'Études Nucléaires de Saclay, Gif-s-Yvette, Essonne, Concentration de défauts au voisinage des interfaces dans un métal soumis à des gradients de potentiel électrique ou de température Si un flux de lacunes dans un métal est causé par l'application d'un champ, la concentration des lacunes autour d'une barrière transversale peut être différente de la concentration dans le volume du métal. Les effets prévus sont variables avec la condition de l'interface: si le contact est mauvais la concentration sera partout celle du volume, si il est bon on s'attend à certaines variations. Avec l'hypothèse d'un bon contact la dimension de la perturbation doit décroître exponentiellement avec la distance

.e la barrière avec une longueur caractéristique $\lambda$, le parcours ioyen des lacunes. L'écart entre les concentrations à la sur-'ace de la barrière et dans le volume du métal, par rapport à .'équilibre, doit être $\pm(ZEe/kT)\lambda$ pour un gradient de potentiel ;lectrique E, et $\pm\alpha/kT^2(E_f + E_t)\lambda$ pour un gradient de tempéra-;ure $\alpha$. $E_f$ et $E_t$ sont les énergies de formation et de transport .es lacunes, et Ze est la charge effective portée par une la-;une. Le sens est tel qu'on a un effet positif du côté de la ›arrière vers lequel les lacunes arrivent.

A.G. SAMOILOVICH, M.V. NITSOVICH, and V.M. NITSOVICH, ;hernovtsy State University, On the Theory of Anisotropic Ther-ioelectric Power in Semiconductors Possible mechanisms for ;he appearance of an anisotropic thermoelectric power in semi-›onductors are examined using simple models. It is shown that ın anisotropic thermoelectric power may arise both as a result ›f combined phonon and impurity scattering, and in the presence ›f two types of carriers if at least one type of carrier has ın anisotropic effective mass. The effect of different para-ıeters on the magnitude of the anisotropy in the thermoelectric ›ower is examined and it is shown that for reasonable values ›f these parameters the anisotropy may be significant.

K.H.J. BUSCHOW and J.F. FAST, Philips Research Laborato-·ies, N.V. Philips' Gloeilampenfabrieken, Eindhoven, Magnetic ınd Structural Characteristics of Some Equiatomic Rare-Earth ;ermanides Magnetic and structural data are given for eleven ·are earth germanium intermetallic compounds represented by ;he formula RGe. The FeB type of structure is observed for ł = La and Ce, while the CrB type is found for R = Pr, Nd, Sm, ;d, Tb, Dy, Ho, Er, and Y. The compounds LaGe and YGe are ?auli paramagnets. The compounds PrGe and NdGe are found to ›ecome ferromagnetically ordered upon cooling, whereas the ›ompounds CeGe, SmGe, GdGe, TbGe, DyGe, and ErGe order anti-'erromagnetically. A field-induced transition from the anti-:erromagnetic state to a ferrimagnetic state is observed for :he compound HoGe at 4.2 $^{\circ}K$ when studied in moderate fields; vhen pulsed field up to 80 kOe is used a second transition point is observed. The susceptibility at high temperatures of most of the compounds follows the Curie-Weiss law with effective moments close to those of the trivalent free rare earth ions. In the case of SmGe, a contribution to the susceptibility from higher multiplets is observed.

H.A. WRIEDT and S. ARAJS, Edgar C. Bain Laboratory for Fundamental Research, United States Steel Corporation, Research Center, Monroeville, Ferromagnetic Curie Temperatures of Some Iron-Zinc Solid Solutions The preparation of homogeneous iron-zinc alloys by zinc-vapor impregnation of iron is described. Ferromagnetic Curie temperatures of the b.c.c. solid solutions containing 11.2, 14.4, and 20.1 at% zinc were determined to be 981.3, 957.0 and 925.0 $^{o}$K, respectively. These measurements are in good agreement with the results obtained by Fallot (1) but not with those of other investigators. Zinc lowers the Curie temperature of iron linearly by about 5.8 deg/at% zinc. This lowering is inconsistent with the simple molecular field predictions.

(1) M. FALLOT, Ann. Phys. (Germany) 7, 420 (1937).

M.J. STOWELL and T.J. LAW, Tube Investments Research Laboratories, Hinxton Hall, Cambridge, Interrupted Vapour Deposition of Epitaxial Gold Films Grown inside an Electron Microscope During the growth of epitaxial gold films inside an electron microscope, the gold vapour stream was interrupted in order to study changes in the deposit structure which occur in the absence of incident vapour. It is shown that rapid shape changes of the deposit islands characteristic of "liquid-like" coalescence continue when deposition is stopped and that further coalescences do occur. The results of these experiments give added confirmation to the thesis of Pashley et al. (1) that the "liquid-like" shape changes of coalescing deposit islands are brought about by surface self diffusion motivated by capillarity forces.

(1) D.W. PASHLEY, M.J. STOWELL, M.H. JACOBS, and T.H. LAW, Phil. Mag. 10, 127 (1964).

J. TREUSCH and R. SANDROCK, Institut für Theoretische Physik (II) der Universität Marburg, Energy Band Structures of

Selenium and Tellurium (Kohn-Rostoker Method) The band structures of hexagonal Se and Te are calculated using a group theoretical treatment based on the Kohn-Rostoker method (1). With certain assumptions the method yields continuous wave functions with continuous derivatives. The energy gaps agree qualitatively with experiment. The polarization dependence of optical transitions can be explained by the present band models. Hole effective masses are estimated, and it is found that $m_{\perp}$ is about four times as large as $m_{\parallel}$ in both substances. The variation of the Se energy gap with pressure is investigated and found to fit experiments.

(1) W. KOHN and N. ROSTOKER, Phys. Rev. 94, 1111 (1954).

A.A. GLAZER, A.P. POTAPOV, R.I. TAGIROV, and YA.S. SHUR, Institute of Metal Physics, Academy of Sciences of the USSR, Sverdlovsk, On Particularities of Magnetic Properties of Thin Films with Exchange Anisotropy Investigations are made of the magnetic properties of multilayer magnetic films consisting of Fe-Ni-Mn and permalloy which have exchange anisotropy. It is found that a unidirectional anisotropy and an additional uniaxial anisotropy appear simultaneously. At the same time rotational hysteresis losses which do not vanish even in fields as high as 10 kOe, and a large magnetic after-effect are observed. The domain structure observations show that thin film remagnetization starts by means of a bidirectional rotation, and after the appearance of domains of opposite polarity the process of wall movement proceeds. The easy direction of the unidirectional anisotropy cannot be rotated at room temperature even in fields of 180 kOe.

E. STEINBEISS, Institut für Magnetische Werkstoffe der Deutschen Akademie der Wissenschaften zu Berlin, Jena, Untersuchungen über Ummagnetisierungsprozesse in Rechteckferriten The effect of an uniaxial anisotropy energy on the field dependence of the switching time is investigated. Several models for magnetization reversal are discussed.

V.A. ZUEV and V.G. LITOVCHENKO, Institute of Semiconductors, Academy of Sciences of the Ukrainian SSR, Kiev, Theory of the Photovoltaic Effect in Semiconductors The detailed theory of "surface photovoltage" is developed for the case of small signals in Si and Ge. In addition to the usual factors considered (such as the spreading of space charge during illumination etc.) other important effects such as the trapping of photocarriers on the surface and the photoconductivity in the space charge region are taken into consideration. $\Delta$n involves the surface trapping parameters in addition to the electron-hole pair generation rate. A convenient set of approximate equations are chosen, from the usual relations, to describe the dependence of the photo e.m.f. on temperature, surface potential and the parameters of the local centres. It is shown that under appropriate simplifying assumptions the expressions obtained reduce to those formerly described.

A.S. ANTONOV, Joint Institute for Nuclear Research, Dubna, On the Mechanism of Lithium Ion Drift in the Electrical Field of a p-n Junction in Silicon The mechanism of lithium ion drift in the electric field of a p-n junction is studied. In particular, the lithium concentration in the drift region and the change in the boundaries of the drift region with time are found. The distribution of the lithium concentration is calculated theoretically.

CHR. LEHMANN and P. SIGMUND, Institut für Reaktorwerkstoff Kernforschungsanlage Jülich, On the Mechanism of Sputtering It is shown that spot patterns in single crystal sputtering do not necessarily arise as a consequence of focussed collision sequences. Spots can be obtained if it is assumed that only the surface of a solid has a regular structure. The spot width is determined by the energy distribution in the collision cascade and the potential barrier at the surface. The contribution of focusons to sputtering is negligible at low bombarding energies because of the small penetration depth of the ions. Recent experimental results indicate that the average path length of focusons contributing to sputtering is extremely small even at high bombarding energies, so that the proposed sputtering mechanism should be dominant over the whole energy range. This result appears to be consistent with collision cascade theory if forward scattering is taken into account.

M.M. SHUKLA and B. DAYAL, Department of Physics, Banaras Hindu University, Varanasi, Lattice Vibrations of Aluminium on the Basis of Krebs' Model The vibration spectrum of aluminium is studied using Krebs' model (1). The phonon dispersion curves for the three symmetry directions [100], [110], and [111] are presented and are found to be in very good agreement with neutron scattering measurements. The (Θ - T) curves calculated from Krebs' model agree very well with experiment using a screening parameter deduced from the Bohm-Pines theory.

(1) K. KREBS, Phys. Letters (Netherlands) 10, 12 (1964).

M. BANCIE-GRILLOT et P. BOURTAYRE, Centre National de la Recherche, Scientifique Faculté des Sciences, Paris, Émission photoluminescente visible du gadolinium inclus dans le réseau cristallin du sulfure de zinc Nous précisons ici certains caractères de l'émission de raies visibles du sulfure de zinc activé par du gadolinium, que nous avons récemment découverte (1). Elle est constituée de 46 raies fines, s'étendant entre 4.664 et 7.057 Å; six d'entre elles, formant un système secondaire, disparaissent par adjonction de minimes traces de cobalt et sont atténuées dans les cristaux mixtes de zinc et de cadmium. Cette émission ne peut être attribuée à l'ion trivalent $Gd^{3+}$, seule forme considérée chimiquement stable de ce lanthanide. Il est en effet connu que le premier état excité de cet ion est, énergétiquement, à près de 4 eV audessus de son état fondamental correspondant à la configuration électronique $4f^7$. Dans le réseau cristallin assez fortement covalent du sulfure de zinc, le gadolinium capterait solidement un électron, son état stable non excité étant probablement $4f^7 5d^1$.

(1) M. BANCIE-GRILLOT et P. BOURTAYRE, Communication au Colloque de Spectroscopie du Solide, Montpellier, Novembre 1965, J. Phys. Radium, sous press.

P. GUYON et J. LAJZEROWICZ, Laboratoire de Physique générale, Faculté des Sciences, Grenoble, Mesures du temps de retournement ferroélectrique de $KH_2PO_4$ Des mesures du temps de retournement en régime impulsionnel ont été faites et mettent en évidence, à faibles champs, un retournement en deux temps. D'autre part, de nouvelles mesures du champ coercitif sont également présentées.

D.J. BARBER, University of Essex, Colchester, Electron Microscopy of Irradiation-Induced Defect Clusters in Magnesium Fluoride The passage of electrons through thin crystals of magnesium fluoride produces point defects. These defects are mobile at temperatures normally attained under the electron beam, and they can give rise both to dislocation loops and to precipitate platelets. The aggregation of these radiation-induced defects can be controlled by using a cooled specimen stage in the electron microscope. This allows the growth and characteristics of the defect clusters to be studied. Dislocation loops lie on {011} planes and they sometimes show stacking fault contrast; their Burgers vectors are $\frac{1}{2}\langle 01\bar{1}\rangle$ and $\langle 01\bar{1}\rangle$. Prolonged irradiation can produce precipitates on and within the loops. Higher beam intensities produce precipitates directly. The nature of the loops and precipitates is discussed in relation to the production of point defects by ionization mechanisms and irradiation-stimulated migration.

V.V. RATNAM, Physics Department, Indian Institute of Technology, Kharagpur, On the Electrical and Magnetic Propertie of Coloured and Uncoloured Calcium Fluoride Crystals The electrical and magnetic properties of uncoloured calcium fluoride crystals and of specimens of these crystals coloured by X-rays and electrolysis, are reported. The dc conductivity and dielectric loss measurements of uncoloured and X-ray coloured $CaF_2$ crystals show that in a freshly prepared crystal 1, the number of free negative ion vacancies is very small and 2, the anti-Frenkel defects which are initially present are in the associated form. These points explain why, for irradiation at room temperature, the $\alpha$-band (the F-band of this system) is not observed in optical absorption. The nature of the traps is also discussed in the context of the observed changes in the dc conductivity accompanying the thermal bleaching of the colour centres in an X-irradiated crystal. The temperature variation of the magnetic susceptibility indicates that X-irradiation produces both paramagnetic and diamagnetic colour entres. From the results of measurements on electrolytically oloured crystals, the oscillator strength of the F-band is stimated as about $10^{-2}$; this also explains why the formation f the $\alpha$-band is not detected for room temperature X-irradi-tion.

V.V. RATNAM, Physics Department, Indian Institute of Tech-ology, Kharagpur, On the Colour Centres and X-Ray Luminescence f Calcium Fluoride Crystals A study is made of differences n the colouration spectrum of calcium fluoride crystals ob-ained by X-ray irradiation at different temperatures and by lectrolysis. The changes occurring in the X-ray luminescence pectrum due to the use of different irradiation temperatures re also considered. It is shown that the $\alpha$-and $\beta$-bands, usu-lly only observed in additive and electrolytic colouration, an also be produced by X-ray irradiation if the irradiation emperature is between 180 and 250 °C. The formation of the -band (the F-band of this system) is difficult to detect at ower irradiation temperatures because of 1, the small number f free negative ion vacancies at these temperatures and 2, he low oscillator strength associated with this band. The hanges observed in the X-ray luminescence spectrum for different rradiation temperatures are explained and the associated lectronic processes are discussed briefly.

F. BIRAGNET, J. DEVENYI, G. CLERC, O. MASSENET, R. MONT-ORY, and A. YELON, Laboratoire d'Electrostatique et de Physi-ue du Métal, Grenoble, Interactions between Domain Walls in oupled Films A study is made of the domain wall structure f coupled uniaxial ferromagnetic films separated by a thin on-magnetic layer. It is found that the walls in the two ilms are not always superimposed. After demagnetization in he hard direction they are separated but tend to cross fre-uently. After demagnetization in the easy direction, they may e separated and parallel or they may be superimposed. An ex-lanation of this behavior in terms of wall-quasi-wall inter-ctions is considered.

W. BRODKORB und W. HAUBENREISSER, Institut für Magnetische erkstoffe der Deutschen Akademie der Wissenschaften zu Berlin,

Jena, Zur Theorie des Ferromagnetismus magnetisch anisotroper Schichten mit kubischer Kristallstruktur (II) A calculation is made of the ferromagnetic and the paramagnetic Curie temperature of a Fe-(100)-film as a function on the number of atomic layers. The theoretical results, using a molecular field model, and a Green's function method, are compared with experimental data.

M. MUSA, Institute of Physics of the R.S.R. Academy, Bucharest, Some Results Concerning the $Co^{++}$ Site Symmetry in Alkali Halide Crystals Spectral studies were made of the symmetry of the $Co^{++}$ site in LiCl and KBr host lattices. It was found that in LiCl and KBr the symmetry around the $Co^{++}$ ion is octahedral and tetrahedral, respectively. To support the above statement the spectra were compared with those of $CoCl_2$, $Cs_3CoCl_5$, and $Cs_2CoBr_4$. It was found that the spectrum of RbCl-Co(Zn)$Cl_2$ shows a more pronounced structure than that of RbCl-$Co^{++}$ single-crystals grown from the melt.

J. VAN LANDUYT, Solid State Physics Department, S.C.K., Mol, Determination of the Displacement Vector at the Anti-Phase Boundaries in Rutile by the Contrast Experiments in the Electron Microscope Contrast experiments are used to determine by transmission electron microscopy and electron diffraction the displacement vector associated with the anti-phase boundaries in rutile. The criteria derived elsewhere (1) for the contrast at $\pi$-boundaries are applied, and the conditions for extinction of the contrast are also used to determine the vector. A vector of the type $\frac{1}{6}\langle 211\rangle$ is found, which is in disagreement with the $\frac{1}{2}\langle 101\rangle$ vector previously proposed (2). The method used here is general and could be used to find the displacement vector at a stacking fault or an anti-phase boundary in other materials.

(1) J. VAN LANDUYT, R. GEVERS, and S. AMELINCKX, phys. stat. sol. 7, 519 (1964).
(2) J. VAN LANDUYT, R. GEVERS, and S. AMELINCKX, phys. stat. sol. 7, 307 (1964);
K.A.G. ASHBEE, R.E. SMALLMAN, and G.K. WILLIAMSON, Proc. Roy. Soc. A276, 542 (1963).

CHR. NANEV, Institut für Physikalische Chemie der Bulgarischen Akademie der Wissenschaften, Sofia, Elektronenmikroskopische Untersuchung der Wachstumsmorphologie von Zink- und Kadmiumeinkristallen

The platinum-carbon replica technique is used to study the surface morphology of zinc and cadmium single crystals grown from their vapours. It is shown that the surface relief depends on the growth conditions. Special attention is given to the temperature dependence of the equilibrium form of the crystals. In this connection an explanation is given of the good agreement between the experimental data and the results obtained previously when the equilibrium form was determined by direct microscopic observations of the initial growth forms (1).

(1) I.N. STRANSKI, Z. phys. Chem. B38, 451 (1938).

M.I. MUKHERJEE and H.N. BOSE, Department of Physics and Meteorology, Indian Institute of Technology, Kharagpur, Nonradiative Destruction of F-Centres in X-Irradiated KCl Doped with Anionic Impurities

Thermoluminescent behaviour and room temperature stability of F-centres produced by X-rays in KCl doped with some anionic impurities are studied. A low percentage of $KNO_3$, KOH, and $K_2CO_3$ enhances, but high a percentage inhibits the formation of F-centres. In lightly doped samples the F-centres in the dark and at room temperature decay with time in two stages by some nonradiative process. Thermoluminescence intensities of these samples are found to be weak. In the case of $KNO_3$ doping, infrared absorptions at 7.1 μm and 7.7 μm due to $(NO_3)^-$ and $(NO_2)$ ions decrease upon irradiation by X-rays suggesting decomposition of the complexes. It is suggested that diffusion of some decomposition products with trapped holes is responsible for the decay of F-centres with time at room temperature. The activation energies for the diffusion processes are of the order of 0.98 and 1.05 eV in all these samples. With 1 mole% $KNO_3$, almost no F-centres are detectable. It is estimated that in this case the average distance between one nitrate or nitrite ion and an F-centre is of the order of 12 atomic units.

R. GRANGER, Laboratoire de Magnétisme et de Physique du Solide, C.N.R.S., Bellevue, Distribution électronique anisotrope dans les semiconducteurs purs au voisinage de l'avalanche An analysis is made of the electron distribution function in homogeneous pure semiconductors in high electric fields E, and this is used to interpret experimental results on intrinsic impact ionization across an energy gap $\varepsilon_i$. When the electron concentration is lower than about $10^{15}$ $cm^{-3}$, inter-electronic collisions are not dominant and the introduction of an electron temperature is not valid. In these circumstances scattering by acoustic or optical phonons must be considered in detail. In the latter case, a part of the distribution (for energy $\varepsilon > \varepsilon_0(E)$), consists of electrons with momenta along the direction of the electric field; these "ballistic" electrons are mainly responsible for the impact ionization, since they are readily accelerated to energies above $\varepsilon_i$. The steady-state electron distribution can be sub-divided into three energy regions: $\varepsilon < \varepsilon_0$, $\varepsilon_0 < \varepsilon < \varepsilon_i$, $\varepsilon > \varepsilon_i$. The transition rates between these regions are calculated. The populations of the three regions are then obtained as functions of the electric field and the current-voltage characteristics are thus deduced. The results are applied to several semiconductors and the breakdown fields are calculated for various temperatures. Although the anisotropy of the distribution is relatively weak, it is by no means negligible, due to a difference of five orders of magnitude between the relaxation time and the lifetime of the carriers. A comparison is made between the theory and experiments on InSb.

J.A. CUNDALL and A.P. KING, Mullard Research Laboratories, Redhill, Surrey, Comparison of Magnetic Measurements and Theoretical Predictions for Nickel-Iron Films over a Wide Composition Range A comparison is made of the theoretical predictions of several recent models of polycrystalline magnetic films with the experimental results obtained for nickel-iron films in a wide composition range. The models considered are the micro-magnetic and the random anisotropy models. Magnetic measurements were made of the remanence distribution and the easy direction domain wall coercive force. The results suggest that no single theory adequately describes the properties of the films over the whole composition range although all have a limited success for different parts of the range.

YA.A. IOSILEVSKII, Institute of Solid State Physics, Academy of Sciences of the USSR, Moscow, Mössbauer Effect in Some Polyatomic Crystals Expressions are obtained for the probability of the Mössbauer effect and the thermal shift in the energy of a $\gamma$-photon for binary crystalline compounds $AB_{r-1}$. These compounds include crystals of the type NaCl or CsCl, crystals of completely ordered alloys ($AuCu_3$, $AlFe_3$ etc.), and monatomic crystals with isolated impurity atoms. It is assumed that the force constants depend only on the differences of the position vectors of the atoms. The treatment includes the two cases in which nucleus A or nucleus B is resonant. The case in which atom A (mass $M_o$) and atom B (mass M) differ considerably in mass is analysed in greater detail. For $M_o/M \ll 1$ or $M_o/M \gg 1$ the fraction $e^{-Z_o}$ of recoilles emissions (or absorptions) of $\gamma$-rays by atom A and the thermal shift $\delta E_o$ are independent of $M_o/M$ for given interatomic forces. Values of these quantities are hence determined by the appropriate limiting expressions ($M_o$ = const, $M \to \infty$ or $M \to 0$). Thus the finite Mössbauer effect for an atom surrounded by light atoms is independent of the mass of the latter. In addition there exists a limit to the gain in the probability of the Mössbauer effect and in the thermal shift for an atom which can be obtained by surrounding it with increasingly heavier atoms. It is also shown that the probability of the effect and the thermal shift for the atoms A, for the given interactions, are essentially independent of r. The quasiclassical expansions of $Z_j$ and $E_j$, for the j-th atom in a unit cell, are obtained for more complicated compounds of the type $A^{(1)}_{r_1} A^{(2)}_{r_2} \ldots A^{(n)}_{r_n}$ ($r_1 + r_2 + \ldots + r_n = r$) by the method of moments.

R.T. DELVES, Chelsea College of Science and Technology, London, The Theory of the Stability of the Solid-Liquid Interface under Constitutional Supercooling (I) This paper discusses the stability of the solid-liquid interface when a crystal is grown from a melt which contains a segregating component. The transient motion of the interface in response to a thermodynamic fluctuation in concentration is calculated using the time dependent diffusion equations. An expansion is made in powers of the amplitude of the fluctuation, the first order terms are found exactly, and some important higher order terms are included. The interface is either stable and returns to its original flat shape or else distorts exponentially with time into the cellular shape. The stability condition is exactly the original constitutional supercooling criterion of Tiller et al. (1). Physical and mathematical arguments show that the interface is either stable or unstable and that no steady state slightly perturbed interface can exist. This paper is restricted to the special case of infinite thermal diffusivity, and does not include effects on an atomic scale.

(1) W.A. TILLER, I.W. RUTTER, K.A. JACKSON, and B. CHALMERS, Acta metall. 1, 428 (1953).

V.A. BURAVIKHIN, V.G. KAZAKOV, V.S. KHRISTOSENKO und P.I. KRUKOVER, Irkutsker Pädagogisches Institut, Irkutsk,
Der Einfluß elastischer einachsiger Spannungen auf die Ummagnetisierungsprozesse dünner Permalloyschichten Es wurde der Einfluß äußerer einachsiger Spannungen auf die kritischen Felder untersucht, die die Prozesse der quasistatischen Ummagnetisierung von Permalloyschichten charakterisieren: die Koerzitivkraft $H_c$, das Anisotropiefeld $H_k$, das Keimbildungsfeld $H_z$, das Startfeld eines Keimes $H_s$, das minimale Feld der Wandverschiebung $H_o$ und das Feld der Zerstörung der Bereichsstruktur $H_p$. Es wurde festgestellt, daß die Veränderungen der Felder $H_z$ und $H_s$ unter der Wirkung einer äußeren Belastung in starkem Maße vom Randprofil der Schichten abhängen. Die Felder $H_o$ und $H_p$ nehmen bei Deformation für Schichten dünner als $\sim$1000 Å zu; für Schichten dicker als $\sim$1000 Å ändern sie sich praktisch nicht. Die Koerzitivkraft läßt sich (in Abhängigkeit vom Randprofil der Schicht und von der Größe der Deformation) als mittleres Wandverschiebungsfeld wie auch als Keimbildungsfeld bestimmen.

V.A. BURAVIKHIN und V.I. POPOV, Irkutsker Pädagogisches Institut, Irkutsk, Über die Struktur und die Polarität von Bereichsgrenzen in ferromagnetischen Schichten

L.V. KIRENSKII, T.P. ISOTOVA, N.M. SALANSKII und N.S. CHISTYAKOV, Physikalisches Institut der Sibirischen Abteilung der Akademie der Wissenschaften der UdSSR, Krasnoyarsk, Magnetische Mehrfachschichten

R. DIETZ and W. HELLENTHAL, Physikalisches Institut der Universität Münster, Recovery Processes in Nickel Thin Films between 20 and 300 °C Evaporated nickel thin films about 500 Å thick are prepared by condensation at a rate of 5 Å/s onto glass substrates. Recovery processes are then studied by measuring the electrical resistance whilst the annealing temperature is increased linearly with time. This method gives the activation energy E and kinetic g of the recovery processes in individual samples. An investigation of the dependence of E on the pressure p during evaporation, controlled by emission of nitrogen gas, shows a reduction in E of about 1/4 as p is increased to around $10^{-4}$ Torr. By extrapolation of E to p = 0 a value of E = (1.05 ± 0.07) eV is obtained. The annealing kinetic appears to follow g = 2. These values apply to temperatures between 60 and 110 °C. In the temperature range from about 230 to 270 °C another recovery process is observed with an activation energy E of approximately 1.4 eV.

S.G. CHRISTOV, Institute of Physical Chemistry of the Bulgarian Academy of Sciences, Sofia, On the General Theory of Electron Emission from Metals A theory of the emission of electrons from metals is developed on the basis of a general expression for the total transmission probability of an incident electron on the potential barrier at the metal surface. This integral expression is expressed as a sum of three conveniently defined integrals. In this way formulas are obtained for the current density for all temperature and field conditions of practical interest. The maximum error in the calculation of the total current by this method is about 10%. This theory includes as limiting cases thermionic and field emission as well as some earlier results for the emission over a narrow intermediate range of conditions. The limits of applicability of the theory in its general form are outside those of practical importance and include temperatures from 0 to 3500 °K and fields from 0 to $10^8$ V/cm.

R. GERBER, Institut für Festkörperphysik der Tschechoslowakischen Akademie der Wissenschaften, Prag, L. MICHALOWSKY, K. MOTZKE und E. STEINBEISS, Institut für Magnetische Werkstoffe der Deutschen Akademie der Wissenschaften zu Berlin, Jena, Über die induzierte Anisotropieenergie manganhaltiger Ferrite Es wurde die Konstante $K_u$ der induzierten einachsigen Anisotropieenergie von Manganferriten ($Mn_{1+x}Fe_{2-x}O_{4+\gamma}$) bei 88 °K in Abhängigkeit von x und $\gamma$ gemessen. Dabei ergab sich, daß $K_u$ im Bereich $0<x\leq 0,8$ linear von x abhängt und mit abnehmendem Sauerstoffgehalt der Proben zunimmt. Außerdem zeigte sich, daß Zusätze von Ni, Zn und Mg eine Abnahme von $K_u$ bewirken, die der Konzentration dieser Ionen proportional und von der Ionensorte nahezu unabhängig ist. Eine einheitliche Deutung der Meßergebnisse konnte bisher nicht gefunden werden.

P. THOMA and G. VAUBEL, Physikdepartment der Technischen Hochschule München, Electrical Properties of Anthracene Single Crystals at Very Low Temperatures

V. GEROLD and H. HABERKORN, Max-Planck-Institut für Metallforschung, Institut für Metallkunde, Stuttgart, On the Critical Resolved Shear Stress of Solid Solutions Containing Coherent Precipitates A quantitative estimate of the increase $\Delta\tau_0$ of the critical resolved shear stress is given for the case of solid solutions containing spherical and coherent precipitates. The interaction between edge dislocations and the coherency strain fields is believed to be the predominant curin mechanism. The calculations are compared with measurements on Al-Zn and Cu-Co alloys. For small particles ($R/b < 20$) the dislocations mainly cross the precipitates, whereas for larger particles the Orowan mechanism becomes predominant.

K.-P. CHIK, Institut für Physik am Max-Planck-Institut für Metallforschung, Stuttgart, und Institut für theoretische und angewandte Physik der Technischen Hochschule Stuttgart, Electron Diffraction Contrast of Small Stacking Fault Tetrahedr

ith Applications to Quenched Gold The characteristic diffrac-
ion contrast of small stacking fault tetrahedra obtained by
naging with a {220} reflection can be used to distinguish small
nresolved stacking fault tetrahedra from dislocation loops.
ne application of these results to quenched gold shows that
acancy clusters exist already in their early stage as stacking
ault tetrahedra.

K.W. BÖER, Physics Department, University of Delaware,
ewark, Delaware, J.W. FEITKNECHT, and D.G. KANNENBERG, Allen-
radley Co., Laboratory for Applied Electronic Research,
ilwaukee, Wisconsin, Properties of Recrystallized Evaporated
dS Layers Recrystallization of evaporated CdS layers in an
tmosphere containing $N_2$, HCl, $O_2$, Cu, and CdS at 630 °C are
nvestigated using electron microscopic and electron probe
nalysis, and show an oxygen enhanced single crystal growth
ith preferably (001) parallel to the substrate surface. Photo-
onducting layers of unusual high frequency response for modu-
ated light excitation (35% value at about 100 kHz), low elec-
ronic noise (above 60 Hz due to density fluctuation) and high
hotosensitivity at 170 °K, with a maximum gain factor of $10^8$
nd a "useful" light/dark current ratio (after 1 s decay) of
$0^7$ at 100 ft-c are described. These layers have a trap density
f $\leq 3 \times 10^{11}$ $eV^{-1} cm^{-3}$ between 0.3 and 0.4 eV and show with
00 ft-c bias illumination at room temperature electron lifetime
30 μs) determined relaxation.

H. VOGT und H. HAPP, Physikalisches Institut der Universität
reiburg i. Br., Das Ultrarotspektrum von $NaNO_2$ zwischen 30
nd 150 μm in der ferro- und paraelektrischen Phase Die
rbeit befaßt sich mit dem Ultrarotspektrum des Ferroelektri-
ums $NaNO_2$, dessen Phasenübergang bei 163 °C ein Order-Disorder
bergang ist. Das Reflexionsspektrum wird im Bereich der äußeren
R-aktiven Gitterschwingungen (30 bis 150 μm) gemessen, und
war für Polarisationsrichtungen des elektrischen Feldes parallel
u den drei kristallographischen Achsen. In der ferroelektrischen
hase treten fünf Reflexionsbanden auf. Eine von diesen ent-
pricht auf Grund der Gittersymmetrie einer reinen Translations-
chwingung der Ionen, während die anderen zu Kombinationen von

Translationsbewegungen und Torsionen der $NO_2^-$-Ionen gehören. Bei Erhöhung der Kristalltemperatur von 30 bis 200 °C wird eine starke Reduktion aller Reflexionsbanden beobachtet.

L. GOUSKOV, G. LECOY, C. LLINARES et M. SAVELLI, Centre d'Etudes d'Electronique des Solides, associé au C.N.R.S., Faculté des Sciences de Montepellier, Études de la durée de vie et du bruit de génération-recombinaison controles par les dislocations dans le germanium Une étude du bruit de génération-recombinaison et de la durée de vie mesurée par décroissance de la photoconductivité est effectuée sur une série d'échantillons des types p et n de germanium dopé en impuretés peu profondes. Les dislocations créés à la naissance du cristal, sont représentées par un niveau accepteur situé environ à 0.15 eV au dessous du bas de la bande de conduction, jouant le rôle de centre recombinant. La variance des porteurs et les durées de vie mesurées sont, dans le domaine 250 a 350 °K dues aux processus du type Shockley-Read de transitions bande à bande, par l'intermédiaire de ces niveaux de dislocations.

## Original Papers published in "Fizika Tverdogo Tela" Vol. 8, No. 5

A.S. ANTONOV, Moscow, Impurity and Electric Field Distri-
ution in the Drift Region of Silicon p-i-n Detectors The
mpurity concentration and the distribution of the electric
ield in the drift region of silicon p-i-n detectors are deter-
ined from measurements of the junction capacity as a function
f the inverse voltage. A general expression for the electric
ield strength is obtained from the experimental results.

P.P. KUZMENKO and P.A. SUPRUNENKO, Kiev, Estimation of
he Phase Transformation Heat and of the Vacancy Formation
nergy in Simple Metals In this calculation it is assumed
hat the major contribution to the thermal effect of the tran-
ition arises from a change in the internal energy at T = 0 °K.
he internal energy is calculated using the potential of the
etal in the free electron approximation and the virial theorems.
t is shown that for univalent metals the melting heat is pro-
ortional to the Fermi energy and to the relative volume change.
he calculated values agree with experiment. For multivalent
etals it is necessary to take into account the change in the
nternal energy due to the redistribution of the valence elec-
rons over the energy bands. The redistribution values are
stimated for allotropic transitions in Sn and Tl, and for the
elting of Al.

S.V. AIRAPETYANTS, M.N. VINOGRADOVA, I.N. DUBROVSKAYA,
.V. KOLOMOETS, and I.M. RUDNIK, Leningrad, Valence Band Struc-
ure of Heavily Doped PbTe The energy gap between the two
alence bands and the heavy-hole effective mass in PbTe are
etermined from electrical measurements in p-type samples with
arrier concentrations (estimated from the Hall effect) of
$x10^{18}$ to $1.4x10^{20}$ $cm^{-3}$. The change of energy gap $\Delta E$ with tem-
erature T, obtained from thermo e.m.f. measurements, has the
orm: $\Delta E = \Delta E_o - \alpha T$, where $\alpha = 2x10^{-4}$ $eVdeg^{-1}$ and $E_o = 0.18$ eV.

N.G. BASOV, O.V. BOGDANKEVICH, P.G. ELISEEV, and
.M. LAVRUSHIN, Moscow, $GaP_xAs_{1-x}$ Laser with Fast Electron

Excitation Coherent Radiation at 7040 Å is observed in a $GaP_xAs_{1-x}$ crystal subjected to fast electron excitation at liquid nitrogen temperature. The emission spectra, the directional diagram, and the dependence of the radiation intensity on the current density of the electron beam are given.

I.M. DYKMAN and P.M. TOMCHUK, Kiev, Distribution Function and Mobility of Electrons in Polar Semiconductors with Non-Parabolic Bands A theoretical study is made of the interaction of conduction electrons with polar lattice vibrations of constant energy $\hbar\omega_0 = k\Theta$ ($\omega_0$ frequency, $\Theta$ optical phonon "temperature") in a donor-type polar semiconductor with a non-parabolic band. It is assumed that the lattice temperature $T_0 \gg \Theta$. It is shown that the normal breakdown condition does not occur when the effective mass increases with increasing energy $\varepsilon$ and the dependence of the energy $\varepsilon$ on the momentum p is given by $\varepsilon \sim p^\nu$, with $\nu < \frac{4}{3}$. When the band has the form $p = \sqrt{2m\varepsilon(1 + \frac{\varepsilon}{\varepsilon_g})}$ as in InSb, the electron temperature increases monotonically with field (m electron effective mass at the bottom of the band, $\varepsilon_g$ energy gap). The instability of the solution for fields F higher than the critical field $F^*$ is eliminated and breakdown due to this instability is not possible. The conductivity and the mobility obtained for this type of band are different from those obtained for a parabolic band. With increasing field the mobility decreases. This is related to the increase of the average effective mass of the electrons with increasing electron temperature.

V.S. LVOV, Leningrad, Galvanomagnetic and Thermomagnetic Effects in Deformed n-Type Germanium The effect of a deformation on the galvanomagnetic and thermomagnetic effects in n-Ge is studied theoretically for the case of non-quantizing magnetic fields, taking into account only the redistribution of the carriers over the extrema with conservation of the total number of carriers. The calculations are made for uniaxial deformation with the stress either in the direction of the current or the temperature gradient. For the three orientations of the magnetic field ([111], [110], [100]) the results are given in a form which facilitates comparison with experiment.

V.S. LVOV and T.V. SMIRNOVA, Leningrad, Tensogalvanomag-
etic Effects in n-Type Germanium Measurements are made of he dependence of the tenso effects on the magnetic field in -Ge with electron concentration $\approx 10^{13}$ $cm^{-3}$ at 77 °K. The esults are compared with the theory developed in (1). The ffect of intervalley scattering on the tensogalvanomagnetic ffects is analyzed. It is shown that for isotropic scattering he effect of intervalley scattering on these effects is in- ependent of the directions of either the magnetic field or he current. The anisotropy factor $K_\tau = \tau_\parallel / \tau_\perp$ with respect to he relaxation times $\tau_\parallel$ and $\tau_\perp$ is shown to be energy-dependent. urther a study is made of the Hall effect, which is an even unction of the magnetic field and is related to the non-sphe- ical character of the energy bands.

1) V.S. LVOV, Fiz. tverd. Tela 8, 1375 (1966).

YA.A. ROZNERITSA and A.G. CHEBAN, Kishinev, Theory of hermo-Field Ionization of F-Centres A theoretical study is ade of the multiphonon thermal ionization of the excited tates of an F-centre under the action of an external electric ield. An expression is derived for the probability of multi- honon thermo-field ionization. The dependence of the quantum ield of the internal photoeffect on the electric field strength s studied for KCl crystals and found to be in good agreement ith experimental results.

G.M. ZVEREV and A.I. SMIRNOV, Study of the Spin-Lattice elaxation of $Er^{3+}$ in $CdF_2$ and $CaF_2$ Single Crystals in the requency Range 3 to 70 GHz Measurements are made of the re- axation time $\tau_1$ of $Er^{3+}$ in $CdF_2$ and $CaF_2$ single crystals in he frequency range 3 to 70 GHz using an impulse saturation ethod. In weak magnetic fields $\tau_1$ first increases linearly ith increasing frequency, passes through a maximum and then ecreases sharply. In the case of $CdF_2$ the experimental re- ults for T = 4.2 °K (where the relaxation is due to direct rocesses), can be described approximately by the expression $\tau_1^{-1} = \omega = 5x10^4\ \nu^{-1} + 0.86\ \nu^2$ ($\nu$ frequency in GHz, $\omega$ in $s^{-1}$).

A.A. BERDYSHEV, Sverdlovsk, Ferromagnetic Semiconductors with Exchange Coupling via the Conduction Electrons The possibility of the existence of a new "photoferromagnetic" effect is discussed for a certain class of antiferromagnetic and paramagnetic semiconductors. A study is made of the magnetic properties of a semiconducting ferromagnetic with exchange coupling via the conduction electrons, using the molecular field method. It is shown that in such a ferromagnetic the ferromagnetic range is limited by an upper and a lower Curie point whereas the paramagnetic susceptibility obeys a different relation from that applicable in common ferromagnetics. The predicted effects may be observed in alloyed oxides, sulphides, selenides, tellurides, antimonides, and bismuthites of the transition metals and also in mixed (via the anions and cations) compounds of the transition metals with the above mentioned elements.

N.V. MILESHKINA, I.L. SOKOLSKAYA, and L.B. KIS, Leningrad, Emission Properties of Germanium at Different Faces of a Tungsten Single Crystal Investigations are made of the emission current from separate regions of a Ge-covered W-surface by varying the surface density of germanium and the temperature in a field emission microscope with a mobile point emitter. Peculiarities connected with the appearance of semiconductor properties can be observed in some regions of the layer due to the discontinuous character of the covering.

L.I. DATSENKO, O.N. EFIMOV, and A.M. ELISTRATOV, Kiev and Leningrad, Study of Defect Interaction by Anomalous X-Ray Transmission The interaction of vacancies with impurity atoms, vacancies, and dislocations, and the defect interaction occurring in the decay of a solid solution of Cu in Ge, are studied by analyzing the integral characteristics of anomalous X-ray transmission. The intensity is measured by a two-crystal spectrometer in the Bragg-Laue position for (220) reflections of Ge and $CuK_\alpha$ radiation. A quantitative discussion is given of the possible mechanisms for the interactions considered.

E.G. BROVMAN and YU. KAGAN, Phonon Spectrum of White Tin The lattice vibrations of white tin are studied theoretically n the basis of the Born-Karman model. Force constants are ised which form the total dynamic matrix of the vibrations or the first three coordination spheres. The interaction vith the next three coordination spheres is approximated by wo force constants. The dispersion curves for symmetric rientations, the frequency distribution function of the phoon spectrum, the heat capacity, and the probability of the össbauer effect are obtained and are compared with experiiental results. On the basis of the theoretical results it is ossible to explain the anomaly of the heat capacity observed t low temperatures and the anomalous sign of the anisotropy f the Mössbauer effect for temperatures between 77 and 400 $^{o}$K. t is found that as T $\to$ 0 the anisotropy practically vanishes; change of sign may occur.

V.I. MAKUKHA and B.M. TSAREV, Moscow, lectron Emission of Alkali Earth Metals on Tungsten, Iridium, nd Rhodium The behaviour of alkali earth metal films (Ba, r, Ca) and of their mixed layers on W, Ir, and Rh single rystals is investigated using an electron projector. The preerential adsorption regions and the work functions for Ba, r, Ca on W, Ba and Ca on Ir, and Ba on Rh are determined.

A.S. BORSHCHEVSKII, YA.A. OKSMAN, and V.N. SMIRNOV, igh-Frequency Electroluminescence in GaAs and GaSe The interal high-frequency electroluminescence of polycrystalline amples of GaAs and GaSe is investigated. The light intensity eaches a value 100 $\mu$Wcm$^{-2}$ at 77 $^{o}$K in polycrystalline aAs. The frequency and amplitude dependences of the brightness f the high-frequency electroluminescence in GaAs and GaSe re similar. The ac photoconductivity is measured in GaSe. The esults indicate hopping photoconductivity and this is diffiult to understand in view of the assumed impact ionization echanism of the high-frequency electroluminescence. It is uggested that multi-stage ionization is possible in semiconuctors with a high density of point defects.

S.A. POLTINNIKOV and E.I. TURKEVICH, Leningrad, Some Magetic Properties of Nickel-Cadmium Ferrites Measurements are made of the temperature dependence of the permeability and of the magnetic spectra of nickel-cadmium ferrites with the composition $Ni_{1-\delta}Cd_{\delta}Fe_2O_4$ with eight values of $\delta$ between 0.1 and 0.8. The results are compared with data pertaining to nickel-zinc ferrites.

N.A. GORBATYI and S. KHASHIMOVA, Tashkent, Emission and Adsorption Characteristics of the System W-La The behaviour of lanthanum on a single crystal of tungsten is studied with a field emission microscope. It is found that the work function of the system W-La passes through a minimum as the degree of covering increases. The average work function for optimal covering is $(2.2 \pm 0.1)$ eV, and the average heat of vaporization is $(5.1 \pm 0.2)$ eV. The system is found with optimal covering to be temperature-stable in the range 1400 to 1800 $^{\circ}$K. In this case La reduces the work functions of the (112)-, (111)- and (116)-faces of W.

V.G. BARYAKHTAR and V.A. SANINA, Scattering of Spin Waves by Paramagnetic Impurities in Antiferromagnetics A calculation is made of the relaxation time of spin waves in antiferromagnetics due to the scattering of the spin waves by impurities. Antiferromagnetics with "easy axis"-anisotropy and with "easy plane"-anisotropy are considered.

I.N. DUBROVSKAYA and YU.I. RAVICH, Leningrad, Investigation of the Non-Parabolicity of the Conduction Band of PbTe by Thermo E.M.F. Measurements in Strong Magnetic Fields The thermo e.m.f. in a strong magnetic field is measured in n-PbTe with electron concentrations from $2 \times 10^{18}$ to $1 \times 10^{20}$ $cm^{-3}$ at about 80 $^{\circ}$K. The density of states is determined as a function of energy, and the Fermi level as a function of concentration. The measured curves are discussed on the basis of two simple models which describe the non-parabolicity of the conduction band. The effective energy gap and the density of states effective mass at the bottom of the conduction band are determined.

N.N. GERASIMENKO, YU.V. LOBURETS, G.F. POLYAKOV, and L.S. SMIRNOV, Novosibirsk, Study of Recombination Radiation in CdS due to Fast Electron Excitation The spectra and kinetics of recombination radiation are studied in undoped CdS single crystals which are excited by a pulsed beam of electrons with energies 200 to 400 keV. The spectrum depends on the irradiation dose when the electron energy exceeds the threshold for defect formation. The kinetics of the green (5000 to 5600 Å) and red (6000 to 10000 Å) bands are found to be quite different.

O.V. KONSTANTINOV and A.L. EFROS, Leningrad, High-Injection in a Non-Degenerate p-n Junction A theoretical study is made of high injection in a non-degenerate p-n junction. The solution obtained here is more exact than the drift approximation of Herring (1). A relation between the injected concentration and the current is found. The current-voltage characteristic is calculated.

(1) C. HERRING, Bell System techn. J. 28, 401 (1949).

V.P. DENKS, Tartu, Effect of Electric Field on Photoluminescence in Ionic Crystals An experimental method is suggested for separating non-carrier (exciton) and carrier-generating (electron-hole pair) electron excitation in ionic crystals. The method is applied to four alkali iodides doped with thallium.

E.F. GROSS, S.A. PERMOGOROV, and B.S. RAZBIRIN, Leningrad, Motion of Free Excitons and the Exciton-Phonon Interaction The possibility of realizing the effect of the kinetic energy of freely moving excitons in the spectrum of exciton luminescence is considered. It is found that the differences between zero-phonon, one-phonon, and two-phonon exciton annihilation processes are related to the different formulation of the momentum conservation law for these processes. These considerations are used in the analysis of the exciton luminescence spectrum of CdS single crystals in the temperature range 4 to 77 $^{o}$K. It is shown that in this case the excitons can be regarded as free quasi-particles, having a Maxwellian distribution with respect to the kinetic energy, which interact with phonons and obey the momentum conservation law.

A.P. OVCHINNIKOV and B.M. TSAREV, Moscow, Adsorption of Cesium at Tungsten Single Crystal Faces A probe method of making field emission measurements from single crystal faces is described. The method is used to investigate adsorption and emission properties of cesium films on various faces of tungste single crystals.

A.I. MITSEK, Sverdlovsk, Theory of Spin Waves in Ferromagnetics The values of $\langle S^z_k S^z_{-k} \rangle$ for small $k$ are calculated for the Heisenberg model by the two-time Green's function method ($S^z_k$; $S^z_{-k}$ are projections of the spin density on the quantization axis; $k$ is the wave vector). It is shown that there must exist in ferromagnetics a branch of "longitudinal" spin waves with a linear dispersion law for small $k$. The temperature dependence of the magnetic anisotropy constants is discussed on the basis of the results obtained.

F.F. VITMAN, M.V. MASTEROVA, and V.P. PUKH, Leningrad, Effect of Temperature on the Strength of Etched Quartz Glass in its High-Strength State Quartz glass is observed in its high-strength state (at about 300 $kgmm^{-2}$) when etched in hydrofluoric acid. The temperature dependence of the strength of quartz is investigated. It is shown that the strength of quartz is temperature independent in the interval 50 to 400 $^{\circ}C$ when the effect of irreversible thermal and mechanical damage and the influence of moisture on the strengthening are neglected.

YU.M. CHASHCHINOV and V.A. MOKIEVSKII, Leningrad, Twinning and Morphology of Epitaxial GaAs Layers on Ge The direction of the orientated growth, the limitation of the surface, and the twinning of epitaxial GaAs layers in the closed heterogeneous system $GaAs-GaJ_x-As_y$ are studied as functions of the orientation of the germanium support. The epitaxial layers obtained at semi-spherical supports are investigated by means of a goniometer, a photogoniometer, and chemical etching. It is found that the favourable growth direction of GaAs from the gas phase is one of the polar directions; the sign of the polarity of such orientations agrees with the sign of the direction [111]. The limit of the epitaxial layers depends on the orientation of the support and on the degree of supersaturation. The probability of formation of twin crystals is determined by the orientation of the support and increases in the succession: (111)<(100)<(110).

E.S. MASHKOVA and V.A. MOLCHANOV, Moscow, Scattering of Ions by Crystals The energy spectra of the scattering of inert gas ions with energies 25 to 30 keV at the (100)- and (114)-faces of a Cu crystal are investigated. Under certain conditions double scattering is observed. The experimental results can be described qualitatively by a simple scattering model.

V.S. VAVILOV, G.P. GOLUBEV, E.A. KORONOVA, E.L. NOLLE, and V.F. SERGIENKO, Moscow, Recombination Radiation in Diamond due to Fast Electron Excitation The radiative recombination spectrum of diamond near the absorption edge and in the visible region is studied for the case of excitation by 150 keV electrons. The spectrum of IIa-type diamond (nitrogen concentration $< 10^{18}$ cm$^{-3}$) in the fundamental absorption range consists of three bands with maxima at 235, 242.3, and 250 nm. A broad band from 320 to 580 nm is observed in the visible region of the spectrum. From the form and position of the ultraviolet bands and their dependence on excitation level and temperature it is concluded that this emission is due to exciton annihilation with simultaneous phonon emission.

E.P. SINYAVSKII, Kishinev, Estimation of Multi-Phonon Capture Cross Sections The multi-phonon capture cross section of holes at negatively charged centres in a $Cu_2O$ crystal is calculated without using the Condon approximation. Taking into account the resonance of the adiabatic potentials it is found that the capture cross section is 200 times greater than in the Condon approximation. The temperature dependence of the present cross section for thermal capture differs (in the medium temperature range) from the temperature dependence arising from the Condon approximation.

B.M. LEBED, L.YA. MUKHA, V.I. MOSEL, and A.G. TITOVA, Ferromagnetic Resonance in Garnet Single Crystals of $Bi_{0.5}Ca_{2.5}Fe_{3.75}V_{1.25}O_{12}$ The temperature dependence of the linewidth $\Delta H$ of ferromagnetic resonance is measured in a polished spherical single crystal of $Bi_{0.5}Ca_{2.5}Fe_{3.75}V_{1.25}O_{12}$ in the temperature interval 4.2 to 530 $^{\circ}$K for five fixed frequencies between 495 and 9250 MHz. The minimal linewidth $\Delta H = 0.5$ Oe occurs at 4.2 °K and 9250 MHz for the [111]-direction. For the [100]-direction and the same temperature $\Delta H = 1.3$ Oe. On decreasing the frequency the anisotropy of $\Delta H$ decreases, and the minimal linewidth occurs at higher temperatures.

N.P. SAZHIN, M.G. MILVIDSKII, V.B. OSVENSKII, and O.G. STOLYAROV, Moscow, Effect of Doping on the Plastic Deformation of GaAs Single Crystals The effect of donor and acceptor impurities on the yield point of GaAs single crystals is investigated experimentally. It is shown that donors (Te) strengthen the crystals but acceptors (Zn) have only a small effect on the mechanical properties. The effect of doping on the temperature and velocity dependence of the upper yield point of GaAs is also studied. The results are compared with data for elemental semiconductors and are discussed on the basis of dislocation theory.

E.E. GODIK, Moscow, Effect of Electric Field on the Capture and Scattering of Holes in Boron-Doped Silicon Measurements are made of the conductivity and Hall constant in boron-doped silicon for electric fields from 0.5 to 300 $Vcm^{-1}$ and different temperatures (4.2 to 18 °K) under conditions of optical hole generation. A nonlinear dependence of the current density on the electric field strength is observed. This is assumed to be related to the decrease of the capture coefficient and the hole mobility in the electric field. The electric field strength and temperature dependence of the hole capture coefficient is investigated. It is shown that these dependences can be modified by changing the concentration of the compensating impurities. The results are discussed on the basis of the theory of cascade phonon capture (1, 2).

(1) M. LAX, Phys. Rev. 119, 1502 (1960).
(2) A.L. MCWORTHER and D.R. HAMANN, Phys. Rev. 134, A250 (1964)

D.G. SANNIKOV, Moscow, Theory of Dielectric Dispersion in Polar Solids An expression is obtained for the dispersion of the dielectric constant in polar solids. The dispersion is caus by the vibration of rigid dipoles in the equilibrium position under the influence of an external electric field. The calculations are performed by the kinetic equation method assuming small displacements of the dipoles from elastic equilibrium positions.

A.A. GRINBERG and N.I. KRAMER, Leningrad, Photon-Photon Scattering in Semiconductors and Insulators Photon-photon scattering in insulating crystals is studied theoretically. The phenomenon is caused by the polarizability of the crystal due to virtual transitions of electrons from the valence to the conduction band. The calculations are made in a two-band approximation. The scattering cross section of photons with energies $\hbar\omega \ll E_g$ is proportional to $\omega^6$ ($E_g$ band gap). For $\hbar\omega_1 + \hbar\omega_2 \leq E_g$ the cross section is of the order of $10^{-28}$ $cm^2$.

YU.N. NOVIKOV and F.I. POLOVIKOV, Kachinsk, Electric Charges in Polymethylmetacrylate under Compression The polarization charges which are generated in polymethylmetacrylate by different compression processes are investigated. As the temperature varies from 25 to 180 °C the state of polarization is affected by the magnitude of the strain, by the degree of deformation, and by the deformation rate. The magnitude of the charge depends on the rate and degree of deformation. A stable charge is formed for deformations greater than 75% when during the orientation of the molecules a rotation of the polar groups takes place.

A.I. GUBANOV and N.A. GUSHCHINA, Leningrad, Treatment of the Disordered Chain by the LCAO-Method The energy spectrum of the one-dimensional disordered chain is calculated using the LCAO-method. The behaviour of the moduli and phases of the coefficients involved in the development of the wave function of the system is studied in order to predetermine the dependence of the wave function on the next order in every point of the chain.

M.I. DAUNOV, Makhach-Kala, Thermomagnetic Nernst-Ettinghausen Effect as an Indicator of the Scattering Mechanism It is shown that in the intrinsic conductivity region the longitudinal Nernst-Ettinghausen effect can be used to indicate the dominant scattering mechanism. This method is applied to various substances.

M.I. DAUNOV, Makhach-Kala, Effect of Mixed Conductivity on the Temperature Dependence of the Kinetic Coefficients in Semiconductors The effect of mixed conductivity on the temperature dependence of the kinetic coefficients is considered for semiconductors with a ratio of electron to hole mobility greater than unity. In p-type semiconductors the mixed conductivity is found to determine the extrema of the kinetic coefficients and the signs of the dimensionless fields of the longitudinal and transverse Nernst-Ettinghausen effects. In n-type semiconductors the effect of mixed conductivity is smaller, and only in certain cases is it possible to determine the extremum of the dimensionless field of the longitudinal Nernst-Ettinghausen effect.

N.Z. MIRYASOV and V.A. SEMKINA, Moscow, Effect of Low $Co^{2+}$ Concentrations on the Ferromagnetic Resonance of Ni-Zn-Cr Ferrites after Thermomagnetic Treatment The temperature dependence of the resonance absorption linewidth 2 $\Delta H$, the resonance field $H_r$, and the anisotropy field $H_a$ in polycrystalline ferrites of $Co_x(NiZn)_{0.8-x}Cr_{0.2}Fe_2O_4$ (with x = 0, 0.02, 0.04, 0.06, 0.08) is investigated between 77 and 520 °K after thermomagnetic treatment.

I.E. BOLOTOV and E.A. MURAVEV, Sverdlovsk, Initial Stage of Crystallization and Growth Mechanism of Selenium Spherolites The crystallization of selenium spherolites in thin amorphous layers (about 100 Å thick) is investigated by electron microscopy and microdiffraction. In the initial stage of the single crystal growth a regular lattice bending is observed which is accompanied by a decrease of the lateral face area with maximum surface tension. As a result a spherolite is formed with a concentric arrangement of the c-axis in the hexagonal cell (axis of spiral "molecules" of selenium). The lattice bending can be explained by the different values of surface tension on different faces. This phenomenon is also observed in fibrous crystals. An analogy is drawn between the crystallization of spherolites in selenium and in polymers.

D.I. KHOMSKII, Moscow, Limits of the Spectrum of Disordered ystems Using a general criterion concerning the properties f matrices (Hadamard's criterion) it is possible in some cases o find the limits of the spectrum of disordered systems. For ne- and three-dimensional disordered alloys the limit is alculated using the tight-binding approximation. The distri- ution condition of the optical and acoustic branch of the ibration spectrum is considered for an amorphous medium. The ne-dimensional disordered Kronig-Penney model (system of -function wells with variable depth and distance) is also tudied. In the first case a simple proof is obtained for the heorem of Luttinger (1) concerning the gap in the energy pectrum. In the second case a condition for the existence f an energy gap and its limits are found.

1) J.M. LUTTINGER, Philips Res. Rep. 6., 303 (1951).

E.P. SVIRINA, O.A. MALIKOVA, and M.A. MALIKOVA, Moscow, all- and Nernst-Ettinghausen Effect in Ferrites with an Excess f Iron Ions The temperature dependence of the spontaneous nd classical Hall- and Nernst-Ettinghausen effect is investi- ated for monocrystalline ferrites of Mn, Ni, and Co containing n excess of iron ions. The ratio of the spontaneous and clas- ical Hall coefficients depends on the square of the spontaneous agnetization as the temperature is varied. An empirical rela- ion between the Hall- and Nernst-Ettinghausen coefficients is roposed. It is suggested that the temperature variation of he spontaneous Hall coefficient in ferrites depends only on he temperature dependence of the carrier concentration and arrier scattering at spin inhomogeneities.

V.V. MAKAROV and N.N. PETROV, Leningrad, Effect of Ion ombardment on the Cathodoluminescence of SiC The effect of on bombardment on the cathodoluminescence spectra of SiC (6H) ingle crystals is studied. It is found that $K^+$ ion bombardment nd annealing at 1000 °C causes a sharp increase in the intensity nd a significant change of the spectral distribution of the athodoluminescence. A system of narrow intense lines and bands s observed in the range 2.64 to 2.45 eV (half-width 0.002 to .01 eV). Narrow lines also appear between 2.86 and 2.69 eV, and between 2.52 and 2.2 eV, but not all of these lines result from bombardment. In the latter energy range the cathodoluminescence spectra of the opposite side of the crystals show some differences.

L.A. ANGELOVA, V.S. VAVILOV, and A.E. YUNOVICH, Moscow, Radiative Recombination in GaP for Electric Current- and Fast-Electron-Beam-Excitation The dependence of the radiative recombination spectra of GaP crystals on the excitation level and impurity concentration is investigated for carrier excitation both by injection through a p-n junction and by fast electron bombardment.

## Short Notes published in "Fizika Tverdogo Tela"
## Vol. 8, No. 5

YA.O. DOVGII and B.F. BILENKII, Lvov
Study of the Fundamental Absorption Edge of α-HgS Single Crystals

YU.K. VISHCHAKAS, G.B. YUSHKA, A.D. PETRAVICHUS, and A.YU. MATULENIS, Vilnyus
Kinetics of the Longitudinal Space-Charge-Limited Photocurrent in Amorphous Selenium

V.S. LVOV and T.V. SMIRNOVA, Leningrad
Thermomagnetic Effects in Deformed n-Ge

K.A. VERKHOVSKAYA and V.M. FRIDKIN, Moscow
Anomalous Temperature Shift of the Fundamental Absorption Edge of $BaTiO_3$ Single Crystals in the Phase Transition Region

Z.I. URITSKII, G.V. SHUSTER, and A.I. KOZINSKAYA, Sverdlovsk
Theory of Optical Absorption by Carriers in a Quantizing Magnetic Field

A.A. PTASHCHENKO, V.V. SERDYUK, and I.A. KUZMENKO, Odessa
Infrared Quenching of Extrinsic Photoconductivity in CdS

A.D. VOLCHEK, K.S. GARGER, and M.I. KORSUNSKII, Alma-Ata
Illumination Characteristics of the Anomalous Photoconductivity of Amorphous Selenium under Constant Illumination

V.I. KRYUK, R.I. MINTS, and V.S. KORTOV, Sverdlovsk
Exoemission from Polished Ge- and Si-Surfaces

S.A. POLTINNIKOV, Leningrad
Initial Permeability of Vanadium Substituted Yttrium-Iron Garnets

K.A. MINAEVA, B.A. STRUKOV, and V.A. KOPTSIK, Moscow
Ultrasonic Relaxation Absorption near the Phase Transition Temperature in $PbMg_{0.5}W_{0.5}O_3$

A.S. GERSHUN, L.A. SYSOEV, and B.L. TIMAN, Kharkov
Some Peculiarities of the Current-Voltage Characteristics of Thin CdS Single Crystals with Non-Ohmic Contacts

## List of Short Notes

## List of Short Notes - Continued

*Contents — Continued*



**Short Notes**

(listed on the last pages of the issue)

**Pre-printed Titles and Abstracts**

of papers to be published in this or in the Soviet journal „Физика Твердого Тела" (Fizika Tverdogo Tela).

## Manuscripts and letters should be sent to

physica status solidi, 102 Berlin, Neue Schönhauser Str. 20
or Internationaler Buchversand GmbH, 1 Berlin 19, Eschenallee 21.

The object of this journal is to publish original work in the general field of solid state physics. Each issue will be published on the first day of the month. Items published will fall into the following categories:

1. Review articles on topical subjects in the field of solid state physics written by arrangement with the editors.
2. Original papers on new results of research, which have not been submitted for publication elsewhere, after being reviewed. (The latest date for acceptance of contributions is the 8th day of each month, 50 days before the delivery of the journal.)
3. Short notes on topical results of research. These will be published in the photographically reproduced part of the journal and are the sole responsibility of the author. (The latest date for acceptance of manuscripts for this part is the 12th day of each month, 16 days before the delivery of the journal.)
4. Pre-printed titles and abstracts of papers to be published in this or in the Soviet journal „Физика Твердого Тела" (Fizika Tverdogo Tela), the latter translated into English.

## Editorial Information

Manuscripts may be submitted in English, German, French, and Russian. Authors are asked to observe the following rules:

1. Manuscripts should be typewritten on one side only, with at least double spacing. For both review articles and original papers two copies should be sent. Manuscripts of original papers and short notes should not exceed 20 and 3 pages, respectively.
2. The author's name, initials and institution should be given.
3. An abstract of about ten lines, and, if possible, a translation of it into one of the above languages is required.
4. Drawings should be enclosed on separate sheets of paper; photographs should be submitted, on white paper, only for review articles and original papers.
5. Captions for the figures should be submitted on separate sheets of paper, and special instructions with regard to the size of illustrations should be given on these.
6. References should be numbered in order of appearance and should be given, on a separate sheet, in the form used in this journal.
7. Units, symbols, etc. should be used in accordance with the recommendations given by the IUPAP.

The authors will be sent two galley proofs of review articles and original papers. Proof reading of short notes and pre-printed titles and abstracts will be done the editorial staff.

For each original paper, and short note, a total of 75 reprints is available free of charge. The number of reprints free of charge per review article is 25. In addition review articles will be remunerated.

Bei Fragen zur Produktsicherheit wenden Sie sich bitte an:
If you have any questions regarding product safety,
please contact:

Walter de Gruyter GmbH
Genthiner Straße 13
10785 Berlin
productsafety@degruyterbrill.com